Electronic scanning radar antenna (photo courtesy of Thomson-CSF)

THE MICROWAVE ENGINEERING HANDBOOK

VOLUME 2

Microwave Technology Series

The *Microwave Technology Series* publishes authoritative works for professional engineers, researchers and advanced students across the entire range of microwave devices, sub-systems, systems and applications. The series aims to meet the reader's needs for relevant information useful in practical applications. Engineers involved in microwave devices and circuits, antennas, broadcasting communications, radar, infra-red and avionics will find the series an invaluable source of design and reference information.

Series editors:
Michel-Henri Carpentier
Professor in 'Grandes Écoles', France,
Fellow of the IEEE, and President of the French SEE

Bradford L. Smith
International Patents Consultant and Engineer
with the Alcatel group in Paris, France,
and a Senior Member of the IEEE and French SEE

Titles available

1. **The Microwave Engineering Handbook Volume 1**
   Microwave components
   Edited by Bradford L. Smith and Michel-Henri Carpentier

2. **The Microwave Engineering Handbook Volume 2**
   Microwave circuits, antennas and propagation
   Edited by Bradford L. Smith and Michel-Henri Carpentier

3. **The Microwave Engineering Handbook Volume 3**
   Microwave systems and applications
   Edited by Bradford L. Smith and Michel-Henri Carpentier

4. **Solid-state Microwave Generation**
   J. Anastassiades, D. Kaminsky, E. Perea and A. Poezevara

5. **Infrared Thermography**
   C. Gaussorgues
   Translated by D. Häusermann and S. Chomet

THE MICROWAVE ENGINEERING HANDBOOK

VOLUME 2

Microwave circuits, antennas and propagation

Edited by

Bradford L. Smith

International Patents Consultant and Engineer
with the Alcatel group in Paris, France, and
Senior Member of the IEEE and the French SEE

and

Michel-Henri Carpentier

Professor in the 'Grandes Écoles', France,
Fellow of the IEEE and President of the French SEE

CHAPMAN & HALL

London · Glasgow · New York · Tokyo · Melbourne · Madras

Published by Chapman & Hall, 2–6 Boundary Row, London SE1 8HN

Chapman & Hall, 2–6 Boundary Row, London SE1 8HN, UK

Blackie Academic & Professional, Wester Cleddens Road, Bishopbriggs, Glasgow G64 2NZ, UK

Van Nostrand Reinhold Inc., 115 5th Avenue, New York NY10003, USA

Chapman & Hall Japan, Thomson Publishing Japan, Hirakawacho Nemoto Building, 6F, 1–7–11 Hirakawa-cho, Chiyoda-ku, Tokyo 102, Japan

Chapman & Hall Australia, Thomas Nelson Australia, 102 Dodds Street, South Melbourne, Victoria 3205, Australia

Chapman & Hall India, R. Seshadri, 32 Second Main Road, CIT East, Madras 600 035, India

First edition 1993

Typeset in 10/12 pt Times by Pure Tech Corp., India.
Printed in Great Britain by the University Press, Cambridge.

ISBN 0 412 45670 2 0 442 31628 3 (USA)

A catalogue record for this book is available from the British Library

Library of Congress Cataloging-in-Publication data available

Contents

Contributors

Jean Anastassiades, Senior Engineer, Thomson-CSF, Meudon, France

J. P. Aubry, Senior Engineer, CEPE, Argenteuil, France

Michel-Henri Carpentier, Vice-President, Scientific and Technical Director (retired), Thomson-CSF, Paris, France, presently Professor in 'Grandes Écoles', Fellow of the IEEE and President of the French SEE

Yves Charlet, Senior Engineer, Thomson-CSF, Massy Palaiseau, France

M. Depré, Senior Engineer, Alcatel Telspace, Nanterre, France

Serge Drabowitch, Scientific Consultant, Thomson-Radant, Chatenay-Malabry, France

Gerard Forterre, Scientific Director, Tekelec, Montreuil sous Bois, France

J. P. Humbert, Commercial Director, Hewlett-Packard, Les Ulis, France

Jean-François Jouen, Senior Engineer, Thomson-CSF, Orsay, France

Didier Kaminsky, Senior Engineer, Dassault Électronique, Saint Cloud, France

Dominique Picard, Assistant Professor, Supelec, Gif sur Yvette, France

Christian Rumelhard, Professor, Centre National des Arts et Métiers, Paris, France

PART ONE
FUNCTIONS FOR MICROWAVE CIRCUITS

1

Ferrite phase shifters

Gerard Forterre

1.1 INTRODUCTION

Phase shifters are the devices most often used in phased-array radar antennas (and similar systems such as antennas used in microwave landing systems (MLS)) in order to introduce the proper phase modulation along the radiating array. Controlling this phase modulation allows rapid modification at will in either the direction of the main radiation (electronic scanning) or the shape of the radiated antenna pattern. Phase shifters are used for other purposes too, for instance to phase modulate a radio frequency continuous wave (RFCW) transmitted signal as it is achieved in some microwave radio links or in some radar systems transmitting a digitally phase modulated signal.

The first use of microwave phase shifters was probably made in the radar systems installed in the 1960s on aircraft carriers such as the US Navy Enterprise. Until that time electronic scanning was obtained by changing the transmitted frequency in a dispersive array. (Let us recall that the famous precision approach radars (PAR) or ground control approach (GCA) radar systems were formerly designed using a kind of electronic scanning obtained by mechanically changing the thickness of a waveguide, in order to change the phase difference between consecutive slots along that waveguide.) Phase shifters of the USN Enterprise radar systems were very large ferrite phase shifters, achieving electronic scanning in bearing (azimuth).

The first phase shifters using p-i-n diodes (see Vol. 1, ch. 9) as switching elements were probably used in France in 1963 in small CW radar systems transmitting pseudo-randomly digitally phase modulated signals. From that time onwards ferrite phase shifters and p-i-n diode phase shifters have been developed in parallel and more or less in competition by the various radar designers.

The advantages of ferrite phase shifters are basically the relatively low insertion loss they introduce (0.5 to 1 dB one way to be compared with 1.5 to 2.5 dB for p-i-n diode phase shifters), as well as their relative ability to withstand high power (compared to p-i-n diode phase shifters which handle high levels of power with greater difficulty). But ferrite phase shifters are large, heavy and expensive (the p-i-n diode phase shifters are small, light and less expensive), and ferrite phase shifters are basically temperature sensitive (while p-i-n diode phase shifters are naturally much less sensitive to temperature changes). The apparent equilibrum between the advantages and disadvantages of the two technologies leads to the parallel use of both of them depending on the company designing the equipment.

When the number of necessary phase shifters is small, because only one dimension electronic scanning is requested (for instance only in elevation) and moreover when the electronic scanning angle is reduced, each phase shifter has to withstand a very high level of power and ferrite phase shifters are generally used (for instance, rotary field phase shifters). When the number of phase shifters is very large (some thousands), which is often the case in large radar antennas achieving electronic scanning in elevation and in azimuth, cost considerations are essential and p-i-n diode phase shifters are generally preferred (since they have to withstand only a very small percentage of the total transmitted power).

In the future, the progressive utilization of microwave monolithic integrated circuits (MMICs) organization, where phase shifters are used on transmission before power amplifiers and on reception and hence do not handle powerful signals, will probably slowly (but only slowly) lead to the abandonment of ferrite phase shifters and hybrid p-i-n diode phase shifters in the best systems.

A ferrite phase shifter is a passive two-port device which provides variable insertion phase in a microwave signal path, without changing its physical length. Phase shift is related to variations in ferrite permeability which result from changes in the magnetic driving field. Special coils and drivers are needed to produce the exact driving current necessary to obtain the required phase shift. Progress in solid-state drivers and digital memories allows very precise control of ferrite phase shift over the temperature range and frequency band.

Many microwave structures can provide phase shift, but only a few are efficient and are used in phased array antennas. Among these structures, some include a closed magnetic polarizing circuit, which behaves as a permanent magnet after the driving current impulse disappears. The phase varies only during the driving pulse and remains 'latched' afterwards.

Before an extensive study of currently used ferrite phase shifters, a short review of their applications in radar is proposed and the main characteristics of ferrite phase shifters are described. The origin of phase shift is analysed as function of ferrite properties and propagation modes. Various possible structures are reviewed after a description of some interesting parts such as polarizers.

1.2 APPLICATIONS IN RADAR

Since 1970 practical ferrite phase shifters have been designed and used in phased array radars. In '1D' phased arrays, there is only one ferrite phase shifter for each antenna row. They have to withstand relatively high power levels; volume and weight are not very restricted.

In '2D' phased arrays, there is one ferrite phase shifter before each radiator. The maximum cross-section of each phase shifter is limited by the computed distance between radiators, which depends on the scanning angle and the frequency bandwidth of the phased array antenna. In order to minimize weight and dissipated power in the microwave element and its associated power driver, it is necessary to miniaturize both.

Radar applications require a reciprocal antenna. This means that the ferrite phase shifter must present the same phase for the transmit and the receive paths. Electronically scanned arrays for multifunction radars require very short phase switching time, at least for beam switching to another direction after any transmit/receive period. Many ferrite phase shifters meet this need.

Multi-polarization transmitted signals are sometimes also required. It is a more difficult problem to design a phase shifter which can accept either linear or circular polarization. A phase shifter element based on circular or square waveguide supports such modes, but special radiators and polarizers must be added to achieve the function.

Phase dispersions and loss variations versus driving field or frequency have an impact on side-lobe levels. Insertion phase and phase shift variations may be precisely controlled with the help of a sophisticated digital driver which takes into account frequency and temperature dispersion.

As a conclusion, requirements for phase shifters are strictly dependent on the planar phased array in which they will be used.

1.3 MICROWAVE PROPERTIES OF FERRITES

Ferrite is the name given to a very large family of ceramic ferrimagnetic materials, such as those with the 'spinel' or 'garnet' crystalline structures, made of ferric and various other metallic oxides. Rectangular loop microwave ferrites are ceramics with high relative dielectric permittivity ε_F (12 to 19) and high quality factor Q. The product of Q with frequency $F(Q \cdot F)$ may be 50 000 GHz for the best ferrites. Thus it is possible to use fully filled ferrite waveguides, which have low losses and lead to the smallest possible phase shifter cross-sections.

The permeability of ferrite polarized by a magnetic field $\boldsymbol{H}_o$ is not scalar but described by the Polder tensor (μ). If the $+z$-axis is aligned with magnetic polarizing field $\boldsymbol{H}_o$, the relation between microwave magnetic flux density vector $\boldsymbol{b}$ and magnetic field vector $\boldsymbol{h}$ is:

$$\boldsymbol{b} = (\mu)\,\boldsymbol{h} \qquad \text{with} \quad (\mu) = \mu_0 \cdot \begin{bmatrix} \mu & -jK & 0 \\ jK & \mu & 0 \\ 0 & 0 & \mu_z \end{bmatrix}$$

where μ, μ_0 and K depend mainly on ferrite saturation magnetization $\boldsymbol{M}_s$, internal polarizing field $\boldsymbol{H}_o$, local magnetization $\boldsymbol{M}(\boldsymbol{H}_o)$, shape and frequency, but also on many other ceramic and microwave parameters (Green and Sandy, 1974). Approximate formulae are given in Appendix 1.A.

At any frequency, μ and K, under a magnetic field $\boldsymbol{H}_R$ related to that frequency (Appendix 1.A), present a resonance called the gyromagnetic resonance. Around this resonance, there is a frequency/magnetic field domain where magnetic losses are higher than outside, the domain is known as the spinwave manifold, where spinwaves, eigenmodes of magnetic media, may exist. Its maximum width is:

$$\Delta F_{SWM} = 0.7 \cdot g \cdot 4\pi M_s$$

in MHz, where g is the Lande factor. Ferrite materials have a natural gyromagnetic resonance due to their internal anisotropy field $\boldsymbol{H}_{anis}$. Without any polarizing field, high losses are observed in a frequency domain extending up to F_M. If the saturation magnetization is given in gauss and noted as $4\pi M_s$, F_M (in MHz) is approximately:

$$F_M = 1.4\, g(4\pi M_s).$$

Under a polarizing field, outside the spinwave manifold, microwave magnetic losses vary mainly as a function of the ratio F_M/F (see Appendix 1.A).

Uniform eigenmodes of ferrite material have circularly polarized magnetic fields around the magnetic polarizing field $\boldsymbol{H}_o$. For such modes, the ferrite is birefringent and the permeability becomes scalar but is different for each direction. They are:

$$\mu_+ = \mu - K \qquad \text{and} \qquad \mu_- = \mu + K$$

Figure 1.1(a) shows their variation versus frequency at a given internal polarizing field $\boldsymbol{H}_o$, and Fig. 1.1(b) shows, at a given frequency, the variation versus internal magnetic flux $\boldsymbol{B} = \mu_0 [\boldsymbol{H}_o + \boldsymbol{M}(\boldsymbol{H}_o)]$.

Two magnetic polarization cases give very different properties and devices: 'low field' or 'below resonance' devices, where $\boldsymbol{H}_o$ is lower than the magnetic field $\boldsymbol{H}_R$ creating the gyromagnetic resonance at the operating frequency; and 'high field' or 'above resonance' devices, in the other case. Since an efficient ferrite phase shifter needs to be driven by a weak magnetic field, it needs to be a low field device where the internal driving field is quite small. Therefore the ferrite material is never saturated.

1.4 MAGNETIC PROPERTIES OF FERRITES

Microwave ferrites are polycrystalline ceramics, the magnetization $\boldsymbol{M}$ of which varies with the applied internal field $\boldsymbol{H}_o$. Their $\boldsymbol{M}(\boldsymbol{H}_o)$ laws present hysteresis curves as 'square' as that of permanent magnets, but with very much smaller applied fields. This means that the remanent magnetization $\boldsymbol{M}_r$ is not

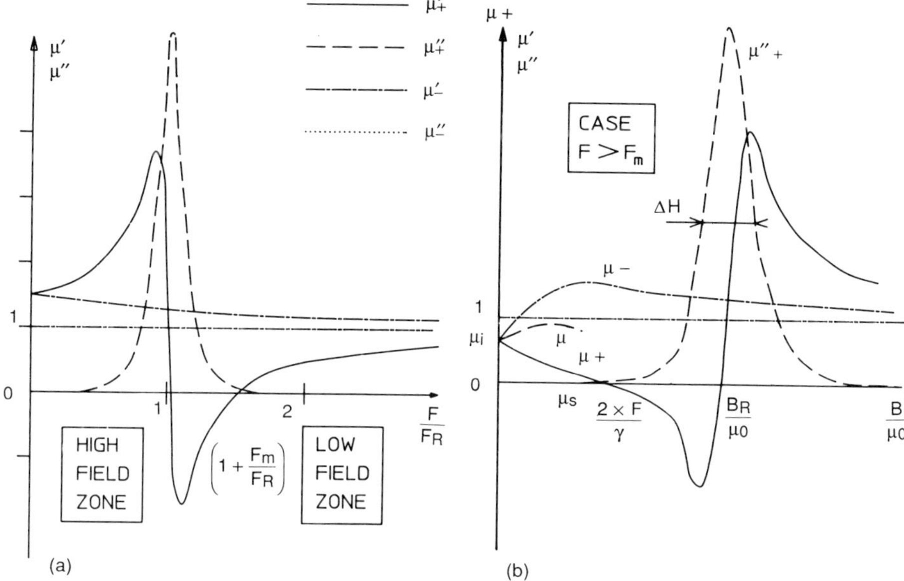

Fig. 1.1 Relative complex permeabilities of ferrite for circularly polarized microwave fields; (a) versus frequency, (b) versus internal flux density.

very much smaller than the saturation magnetization $\boldsymbol{M}_s$ (70% to 80%). The squareness of the hysteresis loop (ratio of remanent magnetization over maximum magnetization for a giving driving field) may be controlled by various metal ion substitutions in the ferrite lattice. Reducing ferrite magnetostriction is also a way to improve the squareness ratio and to stabilize it with regard to temperature. Ferrite grinding and firing processes must give a very reproducible hysteresis loop in order to obtain a yield compatible with a mass production of phase shifters.

This property can be best used in a closed magnetic path, sometimes created by a special toroid shape for the microwave ferrite part. Thus the microwave ferrite part can act as a permanent magnet, whose magnetization may be set at any value between $-\boldsymbol{M}_r$ and $+\boldsymbol{M}_r$. This is the secret to achieving any phase value in a latching phase shifter, and therefore no energy is required to hold the device at a given state.

The intrinsic switching time of a ferrite is related to magnetic domain wall motion in the polycrystalline ferrite ceramic, which is the predominant mechanism by which magnetization reversal takes place in a ferrite. Domain wall motion has a limited speed, as if the magnetic medium was viscous. Delay caused by inertia of domain walls is about 10 to 20 ns.

To achieve full magnetization reversal in a toroid, it is necessary to apply a driving field $\boldsymbol{H}_s$ slightly higher than the ferrite applied field $\boldsymbol{H}_c$. If the driving field $\boldsymbol{H}$ is higher than $2\boldsymbol{H}_c$, an approximate formula gives the switching time τ between $-\boldsymbol{M}_r$ and $+\boldsymbol{M}_r$ states:

$$\tau = Sw/(H - H_s)$$

where Sw is the switching coefficient of the ferrite material. Typical values of Sw, for rectangular loop ferrites, ranges between 0.5 to 1.5 Oe · μs. As $\boldsymbol{H}_c$ values range between 0.3 to 1.5 Oe, it is usually very easy to achieve an intrinsic switching time of less than one microsecond. Switching time to an intermediate magnetization value is shorter than τ.

All magnetic parameters vary with temperature. Most ferrites exhibit a reduction of their magnetic parameters when temperature increases, up to their Curie point where their ferrimagnetism disappears. Some ferrites present a type of compensation for their saturation magnetization curve $\boldsymbol{M}_s(T)$ near room temperature, whose effect is associated with a reduced $\boldsymbol{M}_s$. Temperature effects must be compensated for if one wants to obtain phase shift stability in the operating temperature range. As most parameter relations to temperature are non-linear, ferrite phase shifters have found applications in phased arrays only since the arrival of low cost digital electronics.

1.5 PHASE SHIFT AND PROPAGATION MODES

The main characteristics are determined from a combination of the properties of the magnetically polarized ferrite and the type of microwave magnetic field polarization in the propagating line.

If microwave magnetic fields in the guiding structure are circularly polarized around the $\boldsymbol{H}_0$ axis, and if the circular polarization reverses with energy propagation direction, the microwave propagation coefficient is different for the two possible propagation directions.

It may be noticed that such magnetic field patterns can be observed with $-\boldsymbol{H}_0$ parallel to propagation direction, i.e. longitudinal polarization; and $+\boldsymbol{H}_0$ orthogonal to propagation direction, i.e. transverse polarization. For circular or square waveguides, this can be achieved using I/O polarisers. With $\boldsymbol{H}_0$ along the propagation axis, the non-reciprocal effect is as described in Fig. 1.2.

Maximum phase variation is produced in fully filled waveguides supporting circularly polarized propagation modes, such as circular waveguides. In this case, the ratio of direct β_+ and reverse β_- propagation coefficients has the highest possible value:

$$\beta_+/\beta_- = (\mu_+/\mu_-)^{\frac{1}{2}}.$$

Other waveguide types present only some areas where magnetic field polarization is circular around $\boldsymbol{H}_0$, as in rectangular waveguide. Their efficiency as phase shifter structures are reduced but they offer other advantages such as high power capability.

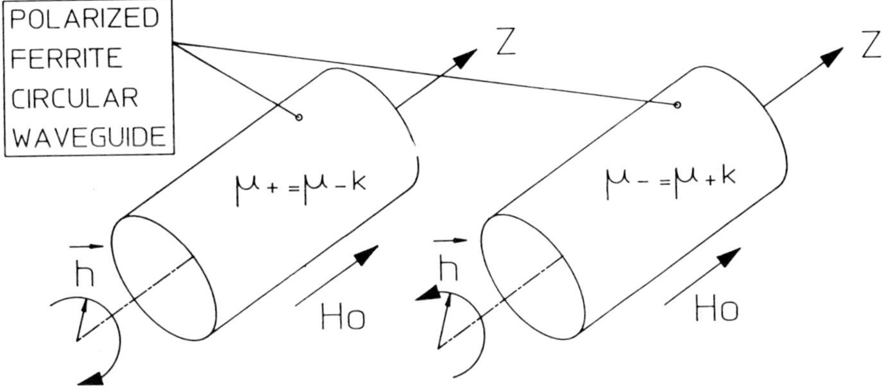

Fig. 1.2 Scalar permeabilities of circularly polarized propagation modes.

For a rectangular waveguide and magnetic polarizing field orthogonal to the propagation direction, such an effect is quite a natural property of the first propagating TE_{10} mode. Even in rectangular waveguide loaded with multiferrite or dielectric slabs parallel to the smallest walls of the waveguide, there are two areas where microwave magnetic field is circularly polarized (Gardiol, 1970).

Non-reciprocal effects are described in Fig. 1.3 for the case of two thin ferrite slabs located in the two circular polarization planes. Circular

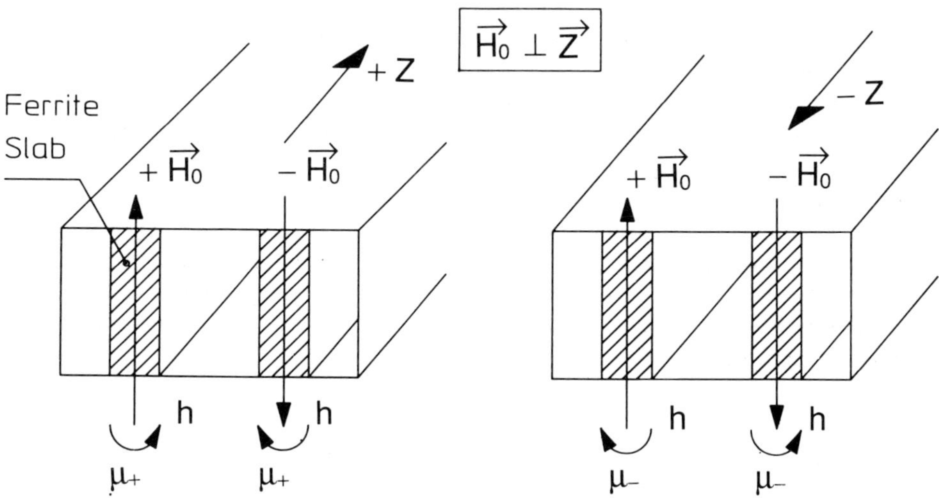

Fig. 1.3 Twin ferrite slabs in rectangular waveguide; (a) direction of propagation: $+Z$, (b) direction of propagation: $-Z$.

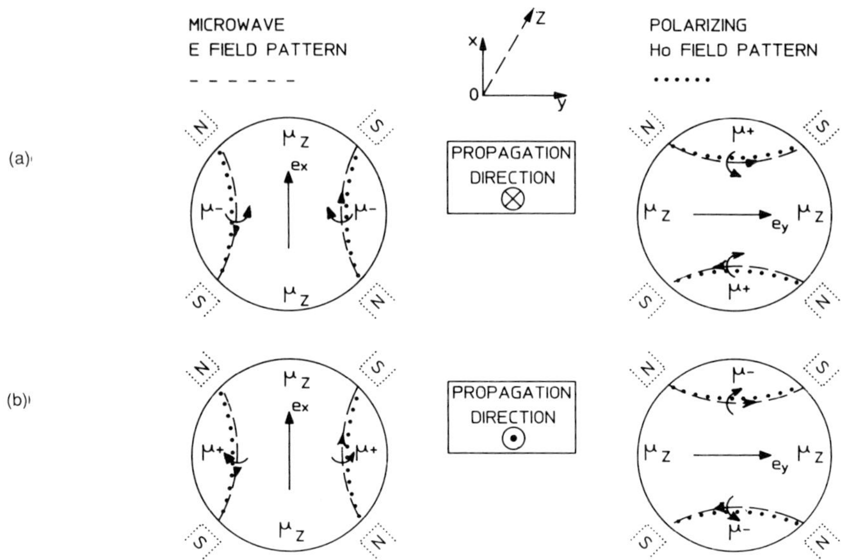

Fig. 1.4 Areas with circular microwave magnetic polarization in ferrite filled circular waveguide magnetized by a quadruple magnet system (a) propagation in + Z direction, (b) propagation in − Z direction.

waveguides may also be efficient with the TE_{11} mode and tranverse magnetic polarizing fields. The most efficient outline is as the diagram of Fig. 1.4. A quadrupole magnet system produces tranverse magnetic polarization. The circular waveguide may be completely filled with ferrite. For such a configuration, there are two orthogonal linear microwave polarizations which can be considered as 'eigenmodes' of the structure: linear polarization is maintained; and average permeabilities are scalar and different for the two orthogonal polarizations. They permute with propagation reversal. Their directions are the two bisectors of pole piece axes. As for TE_{10} mode in rectangular waveguide, microwave magnetic field polarization reverses if propagation direction does.

Due to the number of possible cases, each structure must be analysed separately for defining phase shift possibilities. Depending on the type of microwave field polarization, a ferrite phase shifter section may be reciprocal or non-reciprocal versus the propagation direction. Fast switching non-reciprocal phase shifters may nevertheless be used for reciprocal antennas. They must be switched at twice the pulse repetition frequency (PRF): for the transmit path, one direction is fixed for $\boldsymbol{H}_0$ and $\boldsymbol{M}(\boldsymbol{H}_0)$, and just after transmission, reverse $\boldsymbol{H}_0$ direction is selected. This gives the same phase for the receive path and the transmit path.

To summarize, there are four classes of ferrite phase shifters: reciprocal or non-reciprocal, latching or non-latching. Overall switching time of a ferrite phase shifter depends on the class of the ferrite phase shifter, the driver flux energy and the coil inductance.

1.6 FERRITE MATERIAL SELECTION

Ferrite material selection is always a compromise, due to the great number of incompatible goals. The main requirements are:

1. small cross-section and weight: the highest possible $4\pi \boldsymbol{M}_s$ and $\boldsymbol{M}_r$;
2. temperature stability: high Curie point and low $\Delta 4\pi \boldsymbol{M}_s/\Delta T$;
3. low insertion loss: low $4\pi \boldsymbol{M}_s$ and $\Delta \boldsymbol{H}_{eff}$;
4. bandwidth: high $4\pi \boldsymbol{M}_s$ and $\boldsymbol{M}_r$;
5. high level of transmitted power: low $4\pi \boldsymbol{M}_s$ and high $\Delta \boldsymbol{H}_k$;
6. switching time and energy: low $\boldsymbol{H}_c$ and $\boldsymbol{M}_r$;
7. low cost ceramic part: composition, process and yield.

As a selection guide, Fig. 1.5 indicates variations versus saturation magnetization over frequency, of magnetic losses, dielectric losses ($Q \cdot F$ is quite constant), conductive losses in metal walls, and non-reciprocal efficiency, the ratio of free space propagation constants at $\boldsymbol{M}_r$. Peak power capability is limited by the excitation of parametric spinwaves when the microwave magnetic field is high enough. Above a power threshold, losses increase sharply and the ferrite acts as a limiter. The critical magnetic field is given by the equation:

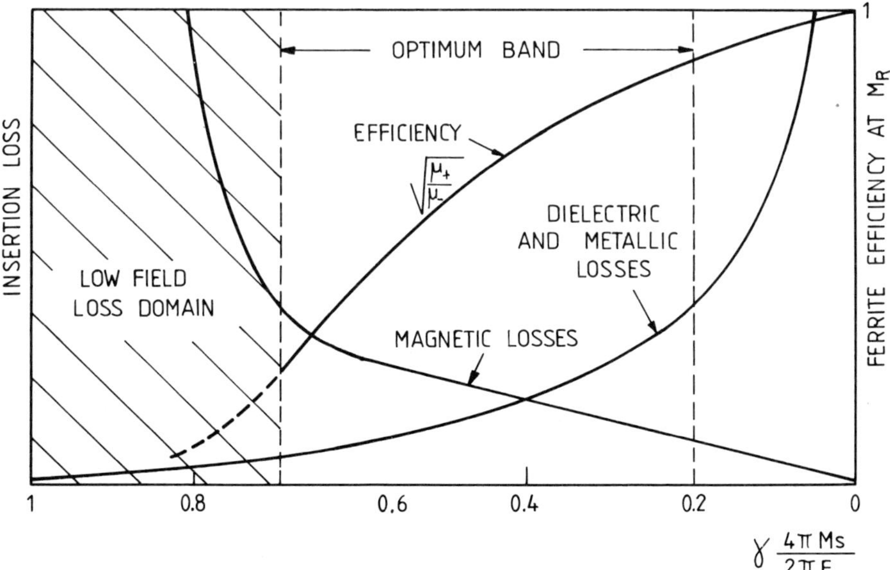

Fig. 1.5 Guide to selection of ferrite material.

$$h_{crit} = k \cdot \Delta H_k \cdot F/F_m$$

where ΔH_k is the spinwave linewidth of the ferrite and k is a function of polarizing field such that $k \geqslant 1$.

Increasing the peak power capability is very simple: one can reduce the ratio F/F_m, increase the spinwave linewidth, or, reduce microwave magnetic field intensity in the ferrite loaded guide, which is the same as increasing transverse dimensions of the waveguide. Therefore, the choice of ferrite also depends on acceptable cross-section, length and mass for the phase shifter.

1.7 FLUX DRIVER AND PHASE PRECISION

A driving magnetic field may be created by a coil directly wound onto the toroid. Most often this coil has a single loop. Adjusting the driving field may set the ferrite magnetization to any value $\boldsymbol{M}'$ between $+\boldsymbol{M}_s$ and $-\boldsymbol{M}_s$ as in Fig. 1.6(a). The remanent magnetization value $\boldsymbol{M}'_r$ observed when this driving field vanishes, is quite close to the setting magnetization for field amplitudes lower than $\boldsymbol{H}_{OM}$.

To overcome the hysteresis effect, a set of two driving pulses is sent to the ferrite assembly, as in Fig. 1.6(b). Before any phase is set, a reset pulse is applied with, for example, negative polarity. This pulse drives the magnetization to $-\boldsymbol{M}_r$, a value which acts as a reference to define the energy for the

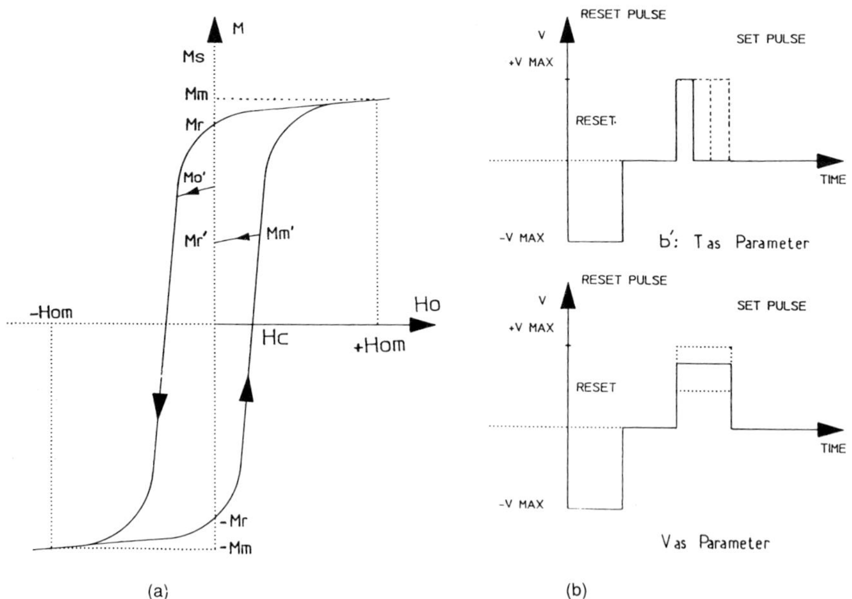

Fig. 1.6 Magnetization driving process; (a) magnetization loop $\boldsymbol{M}(\boldsymbol{H}_o)$, (b) RESET-SET driver pulses.

positive set pulse, which drives the magnetization to M'_m. After the set pulse, the magnetization decreases slightly back to the remanent value M'_r.

Actually, it is more realistic to discuss the energy needed to overcome the hysteresis effect than to define the magnetic field $\boldsymbol{H}_o$ needed to obtain a magnetization value. Switching energy may be applied to the driving coil as a voltage pulse of given amplitude V and duration τ. This energy allows the magnetic flux Φ through the ferrite toroid section to vary

$$\Delta\Phi = \int_0^\tau V \, dt.$$

Therefore, magnetization control is the same as magnetic flux control. Due to the limited velocity of magnetic wall displacement through the magnetic crystallites, in the case where the driving voltage is constant during the switching time, the magnetization proceeds at a uniform rate until it reaches the maximum value. The relation between flux Φ and phase shift appears quite linear and is not very dependent on the shape of the hysteresis curve.

This is the reason for the choice of the flux driving principle to adjust phase shift and this process is the most powerful for precise phase control. Flux may vary with applied voltage or with pulse duration; they must remain higher than values which give respectively high enough magnetic field or pulse width higher than the ferrite intrinsic switching time. This energy is roughly proportional to the ferrite volume V_F, to $\boldsymbol{B}_r$, and to the hysteresis loop width, and increases with $1/\tau$. Approximate calculation can easily be made with this simplified hypothesis: pulse length limited to τ; constant voltage and driving current; equivalent circular base of the cylindrical toroid.

The result for the driving energy W_τ (in Joules) between states $\boldsymbol{B}_{r-}$ and $\boldsymbol{B}_{r+}$ is:

$$W_\tau = 10^{-7} \cdot V_F \cdot B_r \cdot (S_W/\tau + H_S)/8.$$

As a conclusion, the power driver must have at least the following three main functions.

1. To maintain Φ constant during the set operation, for a given phase state and for all temperatures, for all specified external voltage variations, through all life and environmental conditions. A secondary coil, identical to the driving one, may help to do that.
2. To compensate for deviations of phase shift from the linear law versus Φ and for insertion phase variations.
3. In the case of a non-reciprocal phase shifter, to deliver the same pulse sequence with direct/inverted polarities, following the transmit/receive selected path.

Phase shift versus temperature and frequency is mainly related to variations of ε, μ and K, the two last parameters also depending on $\boldsymbol{M}_r$ variations. As magnetization fall-back varies with temperature, this effect is to be evaluated. Secondarily, it depends on guiding structure sensitivity to the same parameters and on mechanical tolerances. When precise phase values are requested over a wide temperature range or frequency band, phase shifter temperature and frequency information must be sent to the driver. In that case, insertion phase dispersion needs to be measured and compensated.

A practical way of achieving such tasks is to use a digital processor based electronic control of the driving pulse voltages. Present constructions use eight or sixteen bits for phase quantization and correction tables loaded in PROM (programmable read only memory). That technology leads to 'three to ten bits' phase shifters, with overall precision at least of the same order as the last bit value. That method is cheaper than that of trying to adjust each phaser, as PROM loading may be achieved very quickly by an automatic microwave measurement bench, directly coupled to the digital electronic circuitry.

1.8 SHORT DESCRIPTION OF VARIOUS POSSIBLE STRUCTURES

Every guiding line having a ferrite insert presents phase variations under an external driving field. In fact, the ferrite loaded guide is only one part of a phase shifter. A complete ferrite phase shifter must be made before having an overview of its real interest for phased arrays and knowing if it could be cheap enough in large quantities. A phase shifter is made of:

1. matched coupling structures to radiating elements and feeders;
2. polarization analysers, if necessary;
3. reciprocal or non-reciprocal polarization transformers, if necessary;
4. ferrite phase shift section;
5. spurious mode suppressors, if necessary;
6. driving and control coils;
7. external yoke, if necessary;
8. cooling structure;
9. electronic driver, often directly associated with each phase shifter.

Extensive studies and prototypes have been made on many structures which seemed to be interesting. Possible guiding structures are:

1. rectangular waveguide;
2. square waveguide;
3. circular waveguide;
4. coaxial or strip line;
5. microstrip lines;
6. dielectric waveguide;
7. helicoidal waveguide.

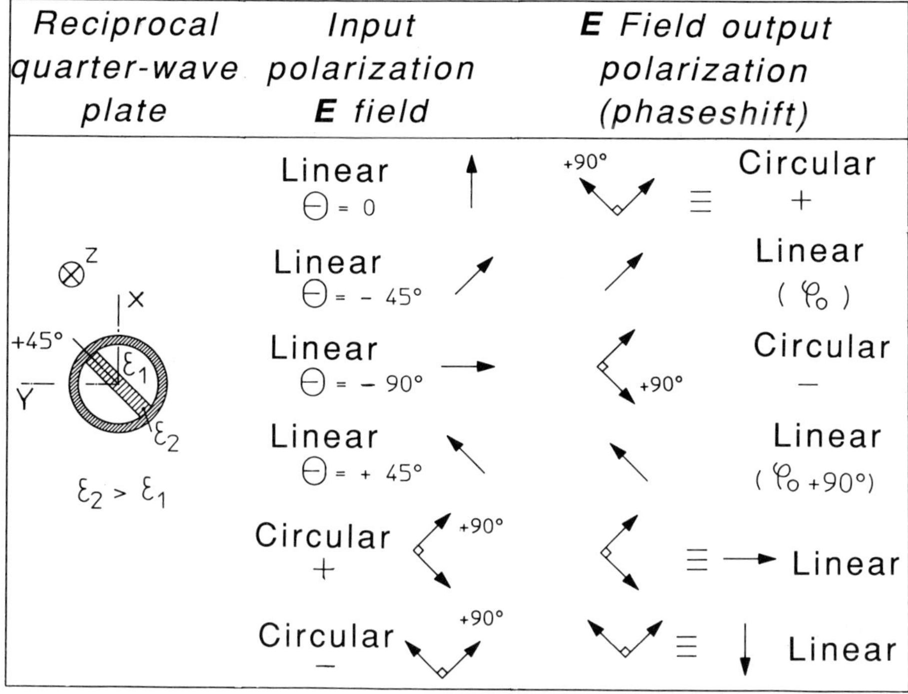

Fig. 1.7 Response of a reciprocal quarter-wave plate polarization transducer.

Magnetic polarizing fields can be parallel or orthogonal to the guide axis. A toroid form for microwave ferrite is compatible with partially filled rectangular waveguides, microstrip lines and dielectric lines. Reciprocal phase shifters are easily built with symmetric ferrite loading and longitudinal magnetic polarization for TEM lines or rectangular waveguides. But other known structures appear inadequate or less efficient than the three structures further described.

Circular waveguide and square waveguide have advantages, as they support both circular and linear polarized modes. It is possible to design reciprocal or non-reciprocal polarization transducers for transforming linearly polarized input signals into circularly polarized waves or for the inverse function. Different types of circular waveguide devices act as a polarizer.

1. Having an axial dielectric slab, a quarter-wave long polarization transducer is reciprocal, with responses as on Fig. 1.7, where + circular polarization refers to anti-clockwise polarization.
2. With the help of a quadrupole polarizing magnet system, a 'quarter-wave plate' ferrite circular waveguide is non-reciprocal with responses as on Fig. 1.8.

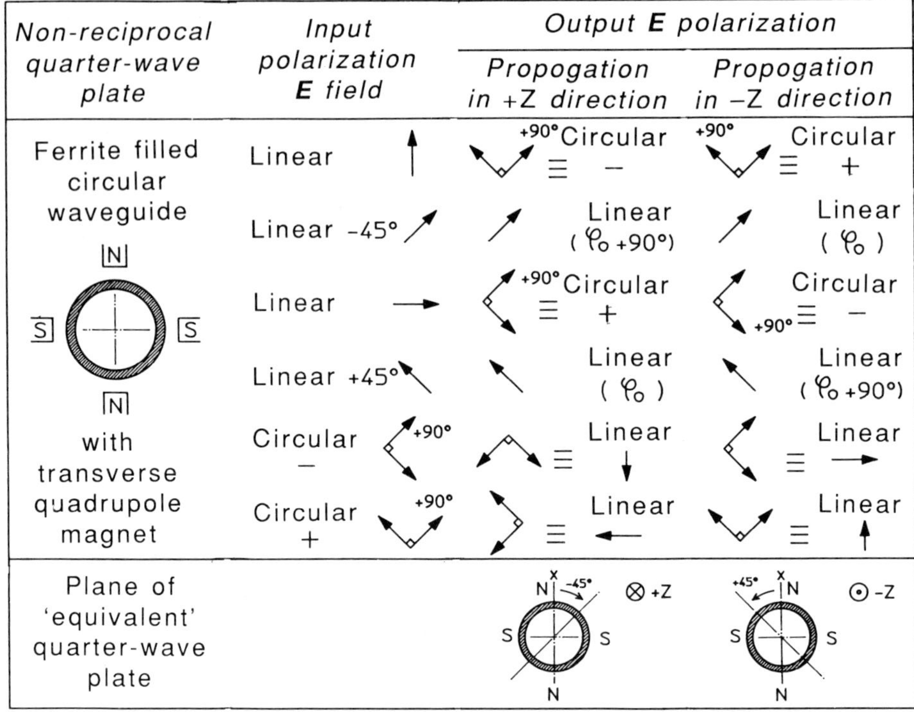

Fig. 1.8 Response of a non-reciprocal quarter-wave plate polarization transducer.

3. With a quadrupole polarizing magnet system, a 'half-wave plate' ferrite circular waveguide is reciprocal for circularly polarized modes, such as the dielectric mode. Both permute the polarization sign.

Possible structures for a complete ferrite phase shifter appear very numerous but only a few are really used. Three are described further.

1.9 NON-RECIPROCAL RECTANGULAR WAVEGUIDE PHASE SHIFTER

This high performance phase shifter is made following the design of twin slabs loading a rectangular waveguide, the idealized model being as described above. From this model, two different designs are used, having a closed magnetic path. The first one proposed has a long square loop ferrite toroid with a dielectric insert in its axial hole as shown in Fig. 1.9(a). The second consists of two identical long toroids separated by a dielectric slab as shown in Fig. 1.9(b).

Both designs have similar phase shift efficiency, even if ferrite loading seems different. In the two toroids design, the magnetic return path of each

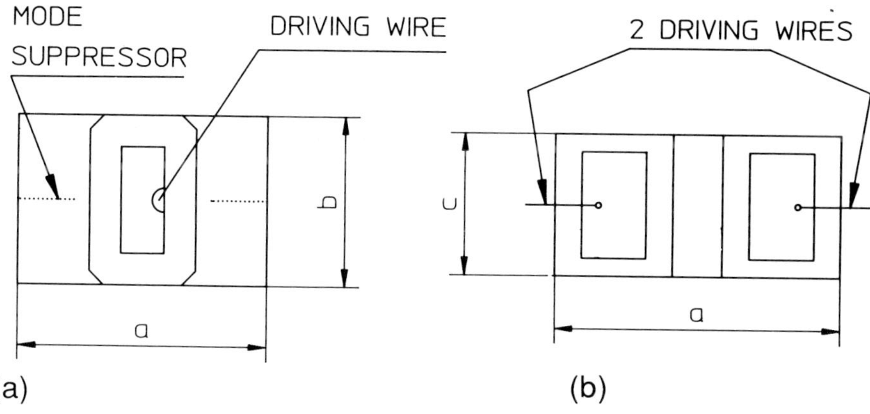

Fig. 1.9 Non-reciprocal rectangular waveguide ferrite phase shifters; (a) centred toroid, (b) dual toroids.

toroid is located along a small wall of the waveguide. At this location microwave magnetic fields are quite linearly polarized, and do not modify the non-reciprocal effect due to the ferrite parts laying along the centred dielectric slab. If, in both cases, ferrite and dielectric slabs have the same dimensions, phase propagation constants and phase shifts are quite close, but with better efficiency for the two toroids solution.

The driving coil is made of only one wire located in the hole of each toroid. When a current flows in that wire, it generates a circumferential magnetic field. This field produces a magnetic flux in the ferrite proportional to the driving field. The ratio between flux and driving field is the low frequency scalar permeability of the ferrite.

As these structures are non-reciprocal versus propagation path, the following process is defined to achieve the same phase for both directions. A driving pulse RESET/SET sequence is applied before transmit pulse, the RESET pulse polarity having one sign. Just after the transmit pulse, another driving pulse sequence is sent for reversing magnetization, i.e. the same driving pulse sequence but with opposite signs of polarities. Both are compatible with 2D phased array requirements.

The propagation modes are very similar to dielectric LSE_{10} modes (Gardiol, 1970), with perturbations due to the latching wire. Exact calculation of the propagation constant ($\alpha\pm+j\beta\pm$), including losses, is not possible, due to rather complex cross-sections of these designs. Through computer analysis, an approximate approach based on twin slab geometry gives acceptable results for $\beta+$, $\beta-$ and losses (Gardiol, 1973). Empirical adjustments improve results. Mode matching analysis without losses leads to more precise values for $\beta+$ and $\beta-$ (Hauth, 1986). The feasibility of finite elements computation has been demonstrated (Forterre, Giesbers, Laroche, 1987).

1.9.1 Comparison between one and two toroids approach

One toroid phase shifter

1. Small bandwith design due to phase variation with frequency and spurious modes;
2. high peak power capability;
3. spurious higher modes sensitivity due to the drive wire located in an area with the highest electrical field.

Two toroid phase shifter

1. Uniform ferrite and dielectric loading at centre of the structure which reduces the losses;
2. wide bandwidth design, up to an octave;
3. external metal walls may be deposited over this compact element;
4. highest allowable permittivity for dielectric slab; the consequences are:
   (a) reduced dimensions (cross-section and length) and mass;
   (b) reduced switching time due to the smallest ferrite volume;
   (c) reduced insertion losses due to high Q dielectric material;
   (d) improved insertion phase and phase shift stability;
   (e) better cooling conditions for ferrite toroids;
   (f) highest mean power capability;
5. easier to machine parts with good precision;
6. reduced sensitivity to spurious modes, case of medium power designs;
7. less phase sensitivity versus temperature and power;
8. less phase dispersion in mass production leading to reduced cost;
9. more difficult to match to coupling structure and radiators: matching needs two or three transformer sections for large bandwidth.

Due to these numerous advantages, the design with two toroids remains alone even if older designs are still produced.

1.9.2 Phase sensitivity

Phase shift is directly proportional to $\boldsymbol{M}$ and K/μ. Frequency sensitivity of the phase shift may be adjusted by a proper choice of the dielectric constant of the central slab and the thicknesses of the central slab and the ferrite toroids. Phase stability can be very good up to over an octave bandwidth, but with problems arising from high order modes. The choice is mostly based on different criteria, such as losses, small cross-section or high power capability.

1.9.3 Frequency band

Applications are reported from 2 to 60 GHz and prototypes up to 90 GHz. Below 2 GHz, lack of material limits the application more than any other reason.

1.9.4 High power limits

Exact evaluation requires the knowledge of microwave magnetic field amplitudes in the structure having ferrite parts. For design with two toroids an approach to the peak power limiting threshold is possible due to the quite uniform dielectric load; it is approximately given by the formula:

$$P = 60ab(h_{\text{crit}})^2$$

(in W), where a, b are the waveguide cross-section dimensions in inches and h_{crit} is the critical microwave magnetic field intensity in oersted. Design can be done in a way that allows good cooling of the ferrite toroid. High mean power designs have been reported but not for phased array. They lead to great cross-section and weight, with small bandwith.

1.9.5 Main performances

Performance values of possible realizations are given in Table 1.1. More refined solutions or specially designed phase shifters may present better or different figures, due to the great variety of solutions.

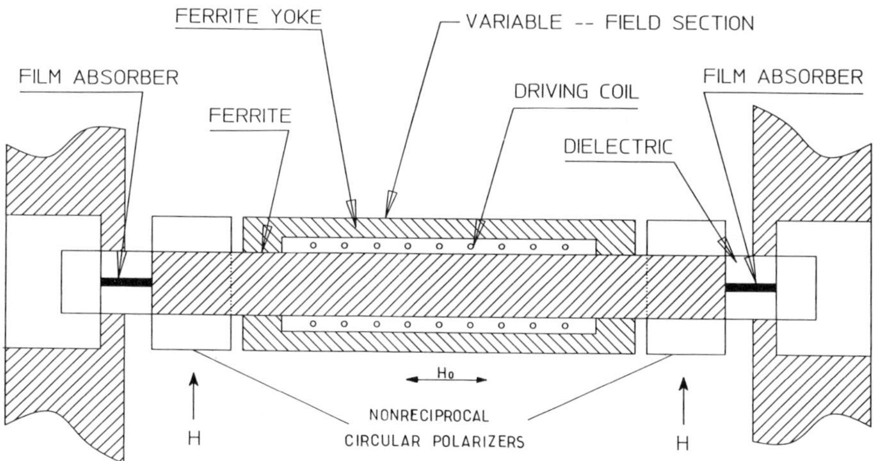

Fig. 1.10 Dual mode phase shifter configuration.

1.10 RECIPROCAL DUAL MODE PHASE SHIFTER

A dual mode phase shifter configuration is given on Fig. 1.10. The main part is a metal-plated, cylindrical ferrite piece, whose section may be circular or

Table 1.1 Dual toroid non-reciprocal ferrite phase shifter

Frequency (GHz)	3		5	10	17	35	45	60
Bandwidth (%)	6	15	6	6	6	6	5	3
Insertion loss (dB)								
Mean value $<\alpha>$	0.5	0.8	0.6	0.5	0.5	0.9	1	1.6
Peak value α_M	0.7	1	0.8	0.7	0.8	1.2	1.3	2.2
$\Delta\alpha$ *with BIT value*	0.1	0.2	0.1	0.1	0.1	0.2	0.2	0.3
$\Delta\alpha(F, T)$	0.3	0.4	0.3	0.3	0.3	0.4	0.4	0.5
Input VSWR (mini, dB)	23	20	23	23	20	20	18	18
Power peak (W)	2000	40 000	800	300	250	150	120	100
mean (W)	400	800	160	100	80	50	30	20
Temperature range (operating) (°C)	0–50	− 10–+ 60	− 20–+ 70	− 40–+ 85	− 40–+ 85	− 40–+ 85	− 40–+ 85	− 40–+ 85
BIT of phase	3–8	6	3–8	3–8	3–8	3–6	3–6	3–6
Switching time (μs) (RESET + SET)	16	20	6	3	2	1	1	1
Switching energy (μJ)	1200	3000	450	100	50	30	25	25
Toroid length (mm)	100	150	100	50	30	20	23	23
Application	2D	1D	2D	2D	2D	2D	2D	2D

rectangular. At both ends are located non-reciprocal quarter-wave polarizers with field patterns orthogonal to each other. A rectangular loop ferrite yoke closes the magnetic path which creates a longitudinal field in the central part of the ferrite cylinder. The device may be latched if the metal wall layer is thin enough.

The basic parts of this configuration operate with two circularly polarized modes, each mode responding to one propagation direction, performance which is at the origin of its name. This structure supports two orthogonal linear polarizations, but even if phase shifts are reciprocal, they are different for each polarization. In the case of only one linear polarization applications, most often there are film absorbers located between transitions to feeder and polarizer, for eliminating cross-depolarization in polarizers and the ferrite phase shift section.

Reciprocal phase shift comes from association of the two non-reciprocal polarizers which transform linear polarization into a circular one, and a non-reciprocal phase shift section where propagating modes are circularly polarized. With such an arrangement, either propagating direction has the same spatial configuration for $\boldsymbol{H}_0$ and the circular microwave magnetic field, therefore the same permeability and the same propagation coefficient. Such a phase shifter is a **reciprocal** and **latching** device. Its transverse dimensions may be very small and it can be fully compatible with 2D phased array requirements.

Main points of the design are to select the material and to choose the diameter which give acceptable performance for the smallest weight and driving energy (Boyd, 1970).

1.10.1 Phase sensitivity

Variations of $\boldsymbol{M}_s$ and $\boldsymbol{M}_r$ directly affect phase shift values. Needed phase shift must be obtained at the highest operating temperature, power effect included. Phase shift varies proportionally to waveguide wavelength, therefore it is very sensitive to frequency. It is the same with the ferrite polarizers.

1.10.2 Frequency band

Main applications are in 2D phase arrays. Applications are reported from C-band to 60 GHz, which frequency appears as a limit to that technology.

1.10.3 Power limits

These devices are mainly optimized to obtain small section, low mass and low power consumption. They are not concerned with the high power problem, that they are not able to solve.

1.10.4 Main performance characteristics

Performance values of possible realizations are given in Table 1.2.

Table 1.2 Dual mode reciprocal ferrite phase shifter

Frequency (GHz)	3	5	10	17	35	60
Bandwidth (%)	6	6	6	5	5	5
Insertion loss (dB)						
Mean value $\langle\alpha\rangle$	0.6	0.6	0.6	0.7	1.0	1.5
Peak value α_M	0.9	0.9	0.9	1	1.4	2
$\Delta\alpha$ *with BIT value*	0.2	0.2	0.2	0.2	0.25	0.3
$\Delta\alpha(F, T)$	0.3	0.3	0.3	0.4	0.5	0.6
Input VSWR (mini, dB)	20	20	20	20	16	13
Power peak (W)	1000	500	300	200	100	50
mean (W)	300	80	40	20	10	5
Temperature range (operating) (°C)	0–50	−20–+50	−40–+70	−40–+85	−40–+85	−40–+85
BIT of phase	3–5	3–5	3–5	3–5	3–5	3–4
Switching time (μs) (RESET + SET)	200	150	100	70	50	30
Switching energy (μJ)	800	400	200	100	50	20
Length (mm)	140	80	42	38	38	30
Application	2D	2D	2D	2D	2D	1D

1.11 ROTARY FIELD PHASE SHIFTER

A rotary field phase shifter configuration is given on Fig. 1.11. First is a dielectric tranformer section which is the location of a matched absorber element which can dissipate cross-polarized energy (resistive film or coupled

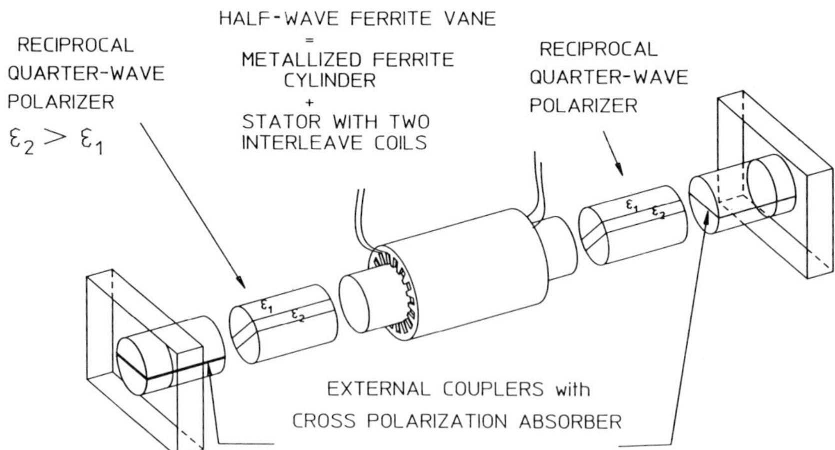

Fig. 1.11 Exploded view of reciprocal ferrite rotary vane phase shifter.

line terminated by a high power load). Next is a reciprocal dielectric quarter-wave plate which transforms input linear polarization into a circular one. The ferrite phase shift section is an equivalent half-wave plate, made of a metallized ferrite cylinder, magnetically polarized by a transverse quadrupole electromagnet, the length of which is great enough to create the half-wave differential phase shift. After this section, energy flows through symmetric parts as at the input. Although the internal waveguide is a circular one, this reciprocal device transmits only one linear polarization.

The electromagnet is very similar to the stator of a motor, but with two separate interleaved windings, each generating a four pole field pattern; the angle between them is 45°. They are designated as **sine** and **cosine** windings because of the bias field patterns generated by their respective excitations. With cylindrical coordinate definition, the radial fields in a point around the rod circumference, $\boldsymbol{B}_s$ and $\boldsymbol{B}_c$ respectively, are defined by the angle ϕ. They are:

$$\boldsymbol{B}_s = \boldsymbol{B}_{so} \sin 2\phi \qquad \text{and} \qquad \boldsymbol{B}_c = \boldsymbol{B}_{co} \cos 2\phi.$$

If we define an electrical angle θ and let the field magnitudes vary as follows:

$$\boldsymbol{B}_{so} = \boldsymbol{B}_o \sin \theta \qquad \text{and} \qquad \boldsymbol{B}_{co} = \boldsymbol{B}_o \cos \theta.$$

The resultant radial field at the periphery of the ferrite rod is:

$$\boldsymbol{B} = \boldsymbol{B}_o (\sin \theta \sin 2\phi + \cos \theta \cos 2\phi) = \boldsymbol{B}_o \cos [2(\phi - \theta/2)].$$

This indicates that, by varying sine and cosine field amplitudes with the θ law, the resulting field is a quadrupole field also, but with field pattern rotated by $\theta/2$ angle. Variations of field amplitudes with θ depends on the number of poles and on the turn distribution in the windings.

Therefore, the rotary field structure operates identically as the mechanical rotating circular waveguide phase shifter described by Fox (1947) where the dielectric half-wave plate, which reverses the circular polarization, is replaced by the ferrite half-wave equivalent plate (Sultan, 1971). That means that phase shift through the devices is equal to twice the field pattern rotation angle, i.e. the electrical angle θ used for defining the amplitudes of currents in the two windings. Practical relations for designing the 'ferrite half-plate' may be found in Boyd (1975). The magnetic materials of the stator are laminated high permeability iron sheets, preferred to ferrite materials, less efficient and very difficult to produce with the very precise dimensions needed.

1.11.1 Phase sensitivity

Two phenomena reduce the frequency effects on phase shift. Firstly, the ferrite 'half-wave' plate is polarized at relatively high and fixed magnetic field. Secondly, the I/O polarizers and absorbers eliminate the cross-depolarized waves: phase shift of remaining energy is more a consequence of the 'rotation' angle, not very influenced by the exact value of the half-wave differential phase shift. Another consequence is that the phase shift relation versus θ is linear. The device is theoretically and experimentally reciprocal.

1.11.2 Frequency band

Products are available from 2 to 18 GHz, with nominal bandwidth of 10 to 18%.

1.11.3 High power limits

Cooling of the ferrite cylinder is easier than in the dual mode structure. They can meet the power requirements of 1D radars, especially since cross-section is also less of a problem in such a radar.

1.11.4 Main performance characteristics

Performance values of possible realizations are given in Table 1.3.

Table 1.3 Rotary field reciprocal ferrite phase shifter

Frequency (GHz)	3	5	10	17
Bandwidth (%)	14	10	10	5
Insertion loss (dB)				
Mean value <α>	0.40	0.4	0.5	0.5
Peak value α_M	0.6	0.6	0.7	0.7
Δα *with BIT value*	0.1	0.1	0.1	0.1
Δα(*F*, *T*)	0.3	0.3	0.3	0.3
Input VSWR (mini, dB)	14	16	18	17
Power peak (W)	40 000	25 000	4000	2000
mean (W)	600	250	60	40
Temperature range (*operating*) (°C)	0–50	– 20–+ 50	– 40–+ 70	– 40–+ 70
BIT of phase	6–8	6–8	6–8	6–8
Switching time (μs) (RESET + SET)	300	250	200	150
Drawing power (W)	0.8	0.75	0.5	0.35
Length (mm)	200	125	80	50
Application	1D	1D	1D	1D

1.12 ANALYSIS OF ORIGINS OF ERRORS

Two types of errors have great importance to phase array performance: phase errors, which directly affect precision on target direction and increase side lobe levels; and amplitude errors, which mainly act on side lobes.

As phase arrays have a great number of phase shifters, error influences appear through their mean values and their dispersion laws rather than their peak values. Another fact is that errors do not have the same importance for all phase shifters. Following their position in the arrays, impact on global error varies, and it is possible to accept greater peak dispersion, but only for a part of phase shifters.

1D phase arrays, having only one ferrite phase shifter for each row of radiators, need more bits of quantizations, therefore errors need to be very small, at least as small as half the value of the smallest bit of phase shift, so that the last bit remains significant.

1.12.1 Various types of phase shift dispersion

Insertion phase dispersion

Insertion phase depends on waveguide dimensions, ferrite permittivity and mean permeability. All of these parameters have unavoidable production dispersions, let us say some thousandths or hundredths of each. As the insertion phase is three to five times greater than the phase shift, one degree of precision is less then one thousandth of insertion phase, which is actually 10 to 50 times lower than observed production lot dispersions.

It also depends on the driver stability, the driving process and how RESET is made. Hysteresis may lead to dispersion of reference phase. Temperature and frequency variations tend to be the same for all the phase shifters, so they most often have no importance.

Errors on each bit

Their origins are described in the previous paragraph, where it is clearly explained how it is possible to take into account magnetization variations, hysteresis cycle shape dispersions or variations and frequency effects, these in all the operating temperature range.

Non-linearity

With such a driver type, non-linearity of the relation between phase shift and the driving parameter is easily corrected.

Non-reciprocity errors

As non-reciprocal phase shifters act as reciprocal phasers because of reversal of polarization between transmit and receive time, in every case global reciprocity is very important and needs to be controlled. Every mechanical or microwave asymmetry may destroy reciprocity, particularly in polarizers.

Mismatching phase errors

Mismatch effects between phasers, feeders and radiators can give noticable phase variation if they are not stable with bit number and temperature.

Note that ferrite remanent magnetization varies with temperature in a very reproducible way. If the phase shift temperature variation law is identical for all phase shifters, correction may be the same for all phase shifters. For high mean power devices, that means an individual thermal sensor for achieving precise phase control. Ferrite remanent magnetization may vary with stresses due to magnetostriction effects. Proper selection or improvement of ferrite material is the best solution for this problem.

1.12.2 Variations of insertion losses

Insertion losses of a ferrite phase shifter vary with bit value, temperature and frequency, due to the microwave and magnetic nature of their origins. The greater the operating bandwith, the more important these variations become. They vary also from one phase shifter to the next, due to production dispersion of mechanical, microwave and magnetic parameters. Influence of polarizers is important, as their manufacture is rather critical. In a production lot, the ratio of peak values may be as high as two, especially in the case of low loss phase shifters.

1.13 FURTHER SOLUTIONS

Ferrite phase shifters are the results of very long and patient efforts for improving technology, reproducibility and reducing volume, mass and cost. Any improvement that leads to reduced cost or dimensions can transform the possibilities of known solutions which are not efficient at the moment. New solutions are under experimentation, especially with the dielectric waveguide concepts which are interesting in the millimetric range.

APPENDIX 1.A

Microwave permeability of polarized ferrites

The microwave permeability of polarized ferrite is described by the tensor (μ) proposed by Polder, which relates the microwave magnetic flux density $\boldsymbol{b}$ to the microwave magnetic field $\boldsymbol{h}$:

$$\boldsymbol{b} = (\mu) \cdot \boldsymbol{h}.$$

When an internal magnetic polarizing field $\boldsymbol{H}_0$ is applied in the $+z$-direction:

$$(\mu) = \mu_0 \cdot \begin{bmatrix} \mu & -jK & 0 \\ jK & \mu & 0 \\ 0 & 0 & \mu_Z \end{bmatrix}$$

where μ_0 is the permeability of free space.

Main parameters (in EMU CGS):

Gyromagnetic ratio:	$\gamma = -\mathrm{e} \cdot \mu_0 \cdot g/2\mathrm{m} = 2\pi \times 1.4 \times g$ (MHz/oe)
Lande factor:	$g = 2$, for garnets and lithium ferrites
saturation magnetization:	$4\pi M_s$, which vary with temperature
actual magnetization:	$4\pi M$, due to actual internal field H_o
external magnetic field:	H_E
internal magnetic field:	$H_o = H_E - H_{an} - N_Z . 4\pi M_s$ if uniform in body
demagnetizing coefficients:	N_X, N_Y, N_Z, if uniform internal field in body
equivalent anisotropy field:	H_{an}
microwave frequency:	F
free space resonance frequency:	$F_o = -\gamma \cdot H_o/2\pi$
saturated magnetization frequency:	$F_M = -\gamma \cdot 4\pi M_s/2\pi$
actual magnetization frequency:	$F_m = -\gamma \cdot 4\pi M/2\pi$
anisotropy frequency:	$F_{an} = -\gamma \cdot H_a n/2\pi$
resonance linewidth:	ΔH, in spinwave manifold (SWM)
effective linewidth:	ΔH_{eff}, out of SWM
loss factor frequency in SWM:	$F_L = -\gamma \cdot \Delta H/4\pi$
loss factor frequency out of SWM:	$F_L = -\gamma \cdot \Delta H_{eff}/4\pi$

Tensor components

Saturated material

$$\mu = 1 + \frac{F_M . (F_o + jF_L)}{(F_o + jF_L)^2 - F^2} = \mu' s - j\mu'' s$$

$$K = 1 + \frac{-F_M F}{(F_o + jF_L)^2 - F^2} = K'_S - jK''_S$$

$$\mu_+ = \mu - K$$

$$\mu_+ = 1 + \frac{F_M}{F_o + jF_L - F} = \mu'_+ - j\mu''_+$$

Unsaturated material
low field case: $F > F_M + F_{an}$

$$\mu_i' = \frac{1}{3} + \frac{2}{3}\sqrt{\left(1 - \left(\frac{F_M + F_{an}}{F}\right)^2\right)}$$

$$\mu' = \mu'_i + (\mu_S - \mu'_i)\left(\frac{M}{M_S}\right)^{1.5}$$

$$\mu_- = \mu + K$$

$$\mu'' = A\left(\frac{F_M + F_{an}}{F}\right)N$$

for a discussion of parameters A and N: see Green and Sandy (1974)

$$\mu_- = 1 + \frac{F_M}{F_o + jF_L + F} = \mu'_- + j\mu''_- \qquad K' = K'_S \frac{F_m}{F_M} \qquad K'' = K''_S$$

$$\mu_Z = 1 \qquad \mu_Z = \mu'_i \left(1 - \frac{M}{M_S}\right)^{2.5}$$

The Kittel resonance law for a body with uniform internal field and magnetization is:

$$2\pi F_R = \gamma\,(H_o + H_{an} - N_x\,4\pi M)\,(H_o + H_{an} - N_y\,4\pi M)^{0.5}.$$

BIBLIOGRAPHY

Boyd C. R., (1970), *Trans.* **MTT-18, (12)**, 1119.
Boyd C. R., (1975) *IEEE MTT-S Digest* 240.
Forterre, Giesbers, Laroche, (1987) *IEEE MTT-S Digest* 407.
Fox A. G., (1947), *PIRE Proceedings* 1489.
Gardiol F. E., (1970), *Trans.* **MTT-18, (8)**, 461.
Gardiol F. E., (1973), *Trans.* **MTT-21, (1)**, 57.
Green and Sandy, (1974), *IEEE Trans*, **MTT-22, (6)**, 641.
Sultan N. B., (1971), *Trans.* **MTT-19, (4)**, 348.

2

p-i-n diode phase shifters

Dominique Picard

Phase shifters are used to change the phase of microwave signals. There are many types of microwave phase shifter, such as manual set-ups using variable line lengths or waveguides with tuneable sections. One also finds phase shifters which are controlled by a voltage or current, such as ferrites or diode set-ups. These last phase shifters allow changing of the phase shift in a very short time, may be less than a few nanoseconds for diode set-ups.

Ferrites and diode phase shifters have different properties:

- ferrites allow higher power level because the power density is lower;
- they allow lower insertion losses, by the use of waveguide modes instead of the TEM mode for the lines of diode set-ups;
- the VSWR is lower because the leading structure is more homogenous.

But the diode phase shifters have the following advantages:

- the switch speed is much higher and the control circuits are easier to design;
- they are generally reciprocal;
- they are more reliable and the temperature drift is lower;
- they are smaller and lighter;
- they are less expensive.

A phase shifter is described by the classical characteristics of microwave set-ups:

- frequency bandwidth;
- input and output matching;
- insertion losses;
- maximum power;
- harmonics level;
- temperature drift.

Phase shifters also have more particular properties:

- phase shift variation with frequency;

- differential insertion losses (insertion loss variations with phase shift) also called phase modulation/amplitude modulation error;
- maximum and resolution phase shift;
- setting time;
- monotonicity: constant sign for the derivative of the phase shift in comparison with the control parameter.

The main application of diode phase shifters is phased array antennas.

2.1 DIFFERENT KINDS OF DIODE PHASE SHIFTERS

These phase-shifters use either p-n diodes (varactors) or p-i-n diodes as control elements. The p-n diode characteristics change progressively with the reverse voltage which is applied, while the p-i-n diodes have two different states which are only slightly dependent on the forward current or the reverse voltage. Thus the p-n diodes allow the creation of analogue controlled phase shifters for which phase shift is continuous, and the p-i-n diodes enter in the constitution of phase shifters with digital control for which phase shift can take discrete values. Also p-i-n diodes are preferred for high power and p-n diodes for lower power.

There are several kinds of microwave phase shifter:

1. switched line phase shifter;
2. loaded line phase shifter;
3. high-pass low-pass phase shifter;
4. I-Q vector modulator phase shifter;
5. coupler phase shifter;
6. circulator phase shifter.

The first four are transmission phase shifters, because the signal is transmitted across the control elements; and the last two are called reflection phase shifters, as the signal is reflected from the control elements.

2.1.1 Switched line phase shifter

Its principle is to modify the electric length between its input and its output by the switching of lines with different lengths. In fact it performs a time delay of the signal. The resulting phase shift varies linearly with frequency except if there is a correction in the switches. Each cell involves two two-way multiplexers and two different length lines (Fig. 2.1). Multiplexers are made with p-i-n diodes, and each cell involves at least four diodes (Fig. 2.2). For certain combinations of line length and diode reverse-biased capacitance, a signal part may go through the unchosen line and create high phase and amplitude error. One can avoid this by loading the lines with a resistance or a short circuit when unused (Coats, 1973, Lynes *et al*, 1974).

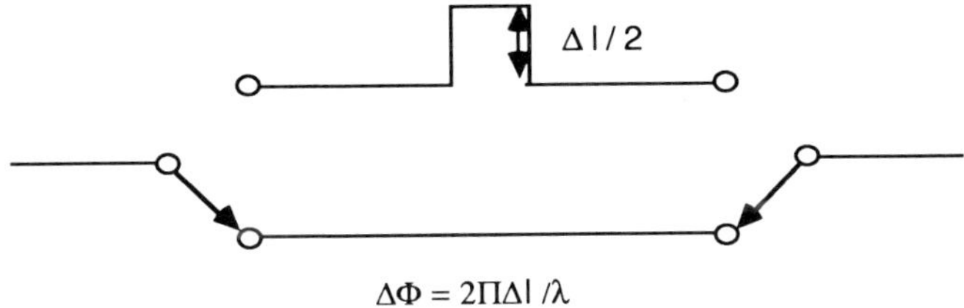

Fig. 2.1 Switched line phase shifter.

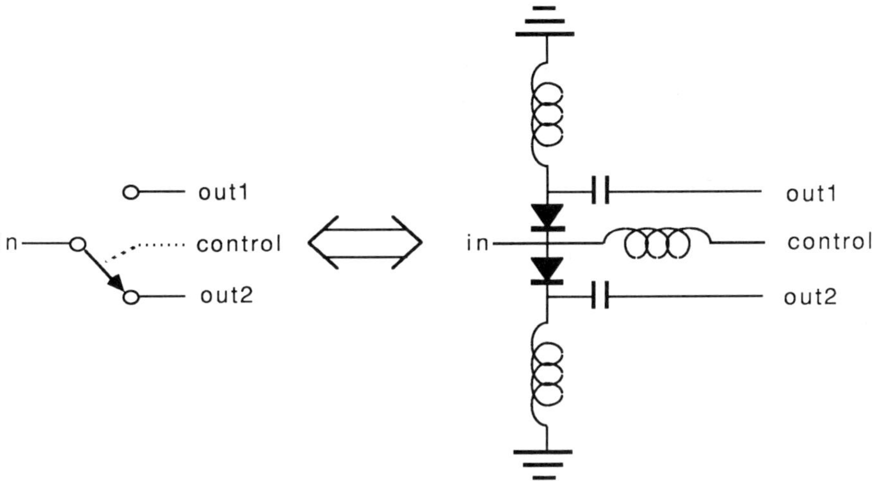

Fig. 2.2 Switch constitution.

The performance of this phase shifter depends on the switches used. Its advantages are that insertion losses due to diodes are constant; only the part due to line length differences can vary. Also, the circuit is small for low phase shift and it can work over large frequency band. The inconveniences are that four diodes are needed per stage and phase shift is proportional to frequency.

2.1.2 Loaded line phase shifter

The presence of an impedance loading a line changes the reflected and the transmitted signal. In order to modify only the transmission coefficient phase, one can use two imaginary loads (Fig. 2.3); the space between these two loads and the characteristic impedance of the line allow matching the set for two values of these loads at one frequency. The loads consist of one two-way p-i-n diode multiplexer, one inductance and one capacitor. This set-up can only perform low phase shift ($< 45°$) to assure a good matching. It is necessary to associate several cells in order to obtain the required phase shift. This com-

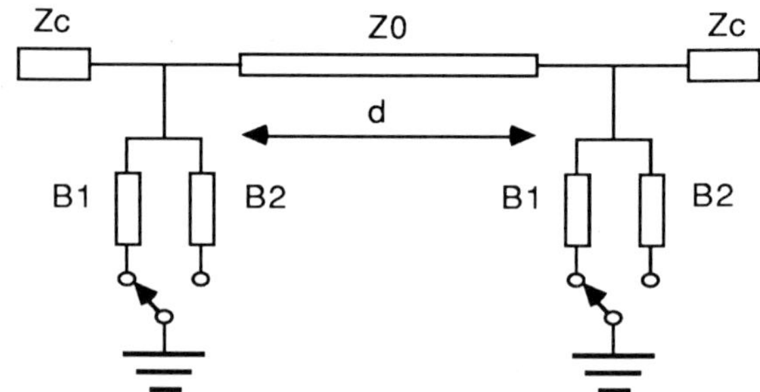

Fig. 2.3 Shunt loaded line phase shifter.

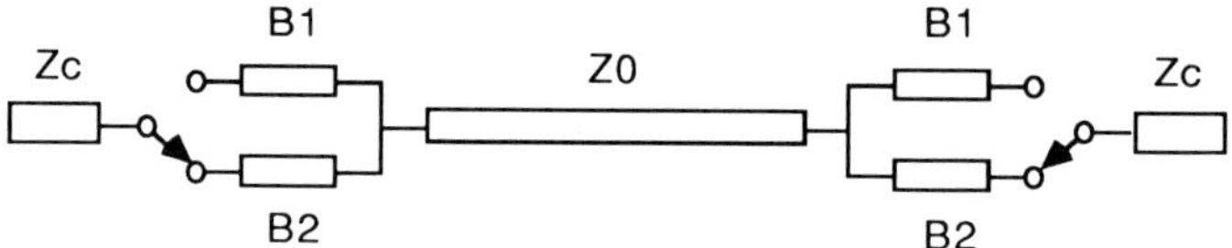

Fig. 2.4 Serial loaded line phase shifter.

plicates the manufacture of such a phase shifter but presents the advantage of a uniform distribution of applied power between each diode.

There are two different kinds of loaded line phase shifters: with parallel loads (this is the most common) and with series loads (Fig. 2.4). They usually operate over narrow frequency bandwidth on the order of 20%. This is due to the matching conditions for the largest phase shift. Over such bandwidth it is possible to have a phase shift almost independent of the frequency.

2.1.3 High-pass low-pass phase shifter

This type is not specific to microwaves. It consists of three elements associated in a π or T circuit (Fig. 2.5); and this circuit is either a low-pass or a high-pass filter depending on the state of the control diodes. A phase shift results from the difference between the insertion phases of the two filters. This set-up is more compact than the preceding one but its admissible power is lower. It is made of lumped elements, thus it can only work at the lower microwave frequencies except if it is made with monolithic microwave integrated circuit technology.

2.1.4 I-Q vector modulator phase shifter

This is shown in Fig. 2.6. The principle is to synthesize a rotating vector from two quadrature signals. One hybrid junction divides the input signal into two perpendicular components with the same amplitude. Each component is modu-

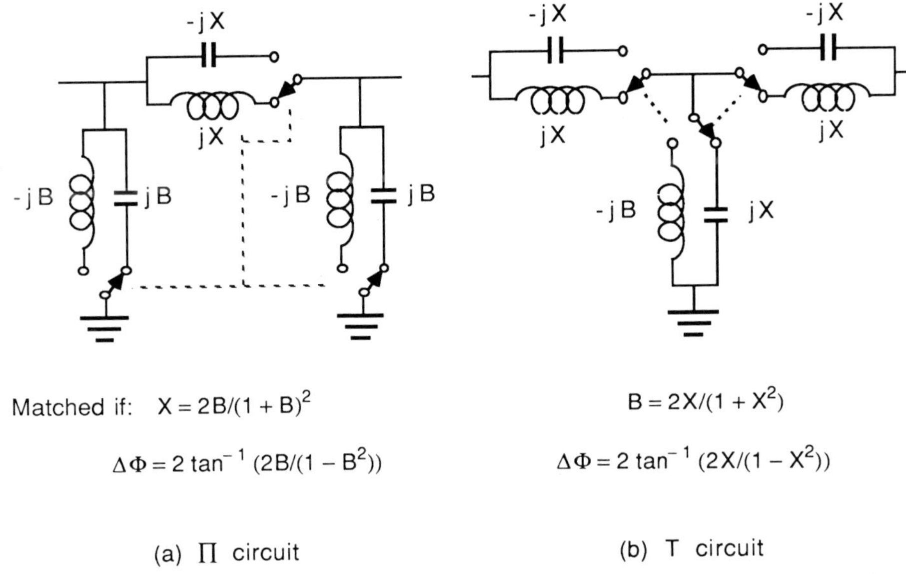

Fig. 2.5 High-pass low-pass phase shifter; (a) π circuit, (b) T circuit.

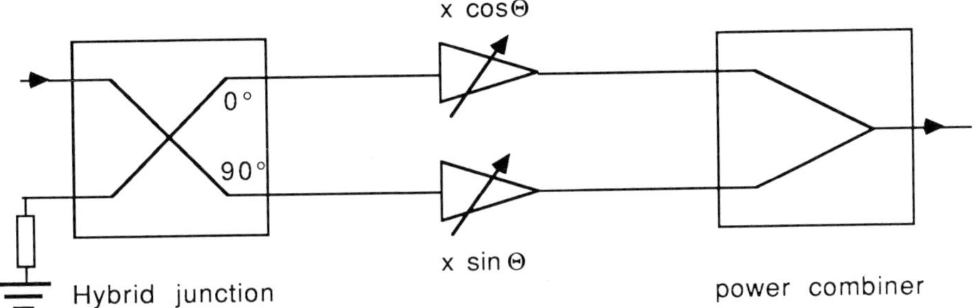

Fig. 2.6 Quadrature phase shifter.

lated in amplitude proportional to, respectively, the cosine and the sine of the desired phase shift with a biphase modulator followed by an attenuator. Then they are combined in phase and give the required phase shifted signal. This set-up can operate through a large bandwidth (3:1) but presents high insertion losses of the order of 10 dB (Tuckman, 1988). It works best at low level. The phase shift can reach 360° and is nearly frequency independent. It is complicated to make and thus expensive.

2.1.5 Coupler phase shifter

This type uses the transformation of a reflection coefficient into a transmission coefficient. Two identical controlled impedances load the ports 3 and 4 of a

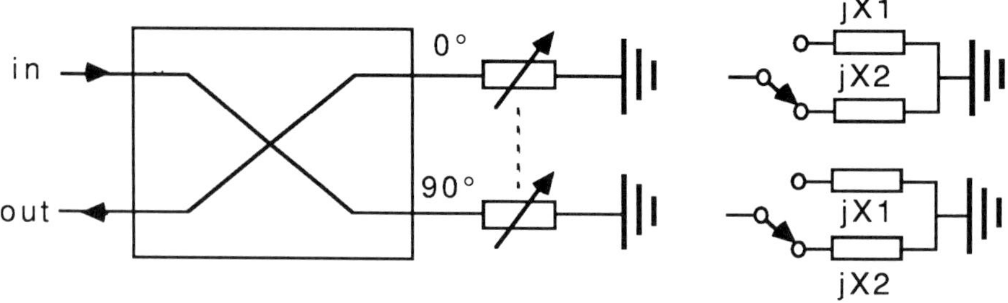

$$\Delta\Phi = 2(\tan^{-1} X_2 - \tan^{-1} X_1)$$

Fig. 2.7 Coupler phase shifter.

hybrid junction (Fig. 2.7). The input is the port 1 and the output the port 2. The input signal is divided into two perpendicular components with the same amplitude on the ports 3 and 4. These two parts are reflected on the controlled loads with the same reflection coefficient because the impedances are identical. The two reflected signals are recombined in phase on the port 2 and give the output signal in phase opposition on the port 1; and there is no reflected power on the input port. This set-up thus procures a good matching. With two imaginary impedances, the reflection coefficient amplitude is equal to unity, as is the global transmission coefficient amplitude. The phase shift depends on the value of the two loading impedances. By the use of controlled loads, the phase shift can vary from 0 to 180 °. Here it is possible to use varactors to have continuous variation of the phase shift. This set-up uses only two diodes, and each of them support half of the power. In fact the hybrid junction is not perfect, and the diodes are not identical. The consequence is that the VWSR and the insertion losses are a function of the diode polarizations and the frequency. Hybrid junctions can operate over several octaves, and one can constitute phase shifters over such bandwidths with increased insertion losses and VSWR.

2.1.6 Circulator phase shifter

The principle is very similar to that of the preceding one (Fig. 2.8) The signal comes in at port 1 and goes out at port 2. The reflection takes place on the controlled load and finally the signal comes out from port 3. Here the set-up presents the following advantage: it uses only one diode, and the imperfections are due to insertion losses and isolation faults of the circulator. But the maximum power is lower, and the weight and the cost are greater than in the case of an hybrid junction phase shifter; furthermore, it is not reciprocal.

There is another type of phase shifters which may be related to loaded line set-ups: multilayer dielectric-waveguide phase shifters (Buckman, 1977).

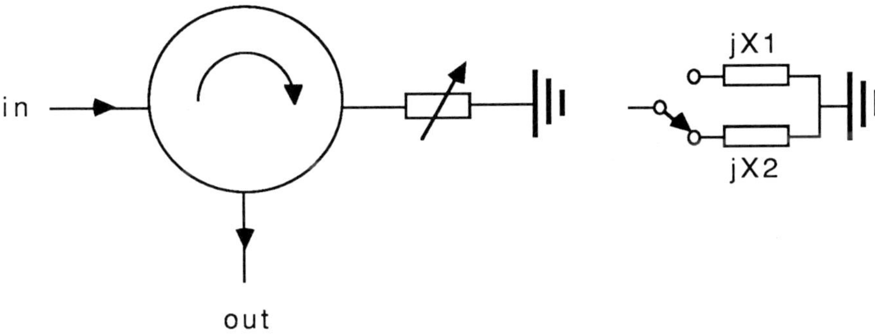

$$\Delta\Phi = 2(\tan^{-1} X_2 - \tan^{-1} X_1)$$

Fig. 2.8 Circulator phase shifter.

They come from integrated optics technology. They consist of dielectric waveguide, one dielectric layer of which is the intrinsic region of a p-i-n diode whose structure is parallel to the waveguide. The refractive index of the diode i-region depends on the polarization state of the diode. It is thus possible to change the wavelength in the waveguide, as well as its insertion phase. The limitation of such set-ups is phase shift efficiency as a function of waveguide length.

2.1.7 Switchable loads design

Several types of digital phase shifters use switchable loads: loaded line, high-pass low-pass, reflection type. These loads may be made with lumped or with line length elements (Bahl and Gupta, 1980, Atwater, 1985)

With lumped elements

Main line mounted diodes: the diode and an inductance are serially associated (Fig. 2.9(a)). When the diode is forward biased the circuit is equivalent to an inductance and a resistance mounted in series; for reverse bias the circuit is equivalent to an inductance, a capacitor and a resistance.

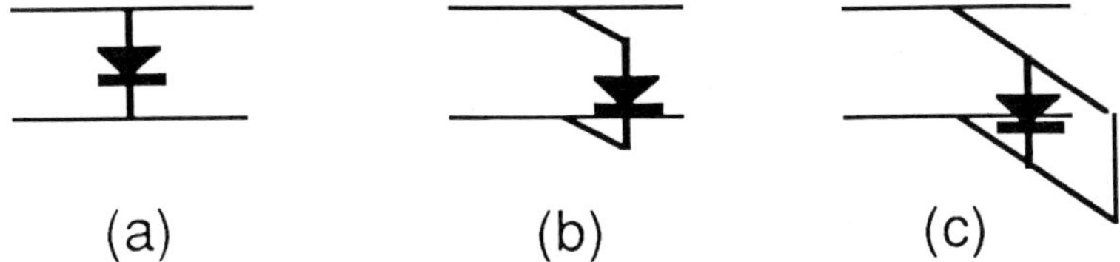

Fig. 2.9 Different diode switchable loads; (a) main line mounted, (b) stub mounted, (c) switchable stub length.

With line length elements

Stub-mounted diode or line-transformed diode impedance: the diode is set at the end of a stub (Fig. 2.9(b)). When the diode is forward biased, the equivalent circuit is a resistance at the linelength extremity, and a capacitor with a shunt resistor at the stub end.

Switchable stub length circuit: the diode is at the extremity of a first stub; a second stub terminated by a short-circuit is connected on the diode (Fig. 2.9(c)).

It is possible to transform the reflection coefficients of a two-state switchable load with a quadripole, to obtain at a given frequency the same amplitude and a given phase difference between the global reflection coefficients for the two states (Fig. 2.10). The phase shifter is then optimized at one frequency (Starski, 1977). The lumped elements circuits give wider frequency range than linelength circuits. Varactor analogue variable loads can also use such mounting systems.

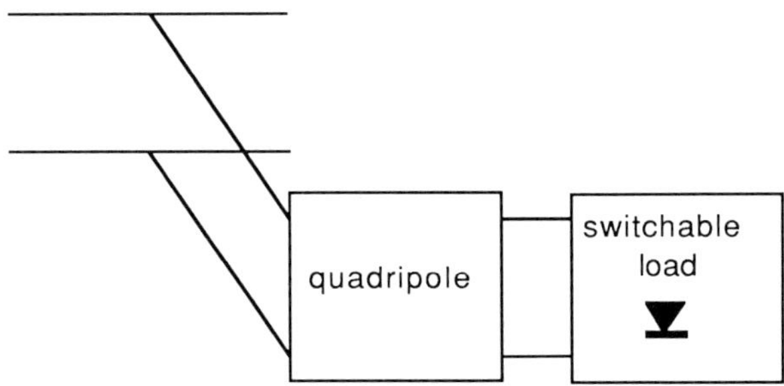

Fig. 2.10 Reflection coefficient transformation of a switchable load using a quadripole.

2.2 TECHNOLOGICAL ASPECTS

The greater part of diode phase shifters are manufactured with lines on printed circuits. There are also waveguide realizations which allow work with high power. The size, the weight and the cost of such set-ups are greater but the performance is better over narrow bandwidths. These are essentially loaded line or reflection phase shifters. One also finds phase shifters which are made with monolithic microwave integrated circuit technology. This is evidently for low power set-ups. This technology allows the manufacture of the different kinds of phase shifters except for the circulator set-ups. The diodes are often replaced by field effect transistors in MMIC realizations.

2.3 PERFORMANCE

Table 2.1 sets out the compared performance characteristics of the different types of phase shifters. The main use of these phase shifters is for phased array

Table 2.1 Compared performance characteristics of different types of phase shifters

Type	*Technology*	*Diodes type*	*Frequency bandwidth*	*Power*	*Insertion losses*	*One stage maximum phase shift*	*Control*	*Size*
Switched line	PC WG MMIC	p-i-n	Very large	Medium	Low	∞	D	Large
Loaded line	PC WG MMIC	p-i-n	Medium phase shift dependent	High	Low	45°	D, A	Large
High-pass low-pass	PC MMIC	p-i-n	Medium phase shift dependent	Medium	Low	45°	D	Small
Quadrature	PC MMIC	p-i-n	Very large	Low	High (10 dB)	360°	D	Small
Hybrid coupler	PC WG MMIC	p-i-n p-n	Large phase shift dependent	Medium	Low	180°	D, A	Medium
Circulator	PC WG	p-i-n p-n	Medium phase shift dependent	Medium	Low	180°	D, A	Medium

PC: printed circuit, WG: waveguide, MMIC: monolithic microwave integrated circuit, D: digital, A: analogue.

antennas, for which the admissible power may be important. The frequency bandwidth of these antennas is also getting larger and larger for modern radar applications.

2.3.1 Power limitations

The high power level use of phase shifters poses two problems:

1. They contain non-linear elements which may produce harmonics from the input signal. This is especially sensitive in the case of varactors for which characteristic variations are progressive, as well as for fast diodes, because the microwave signal may be sufficient to change the state (Higham, 1988).
2. Too high power level can damage the diodes either by heat, breakdown voltage or too high current density (White, 1974).

These problems are not limited to phase shifters and are discussed in the part devoted to control diodes. For high power and/or low harmonics generation, the most appropriate diodes are high biased (forward and reverse) and slow diodes. The most appropriate phase shifters are the loaded line type for printed circuit or better waveguide technology.

2.3.2 Frequency bandwidth

The wide band use of these phase shifters poses two problems: the phase shift frequency dependence and the degradation of the characteristics. In this case the possible technology is either printed circuit line or monolithic microwave integrated circuit.

For switched line set-ups, the phase shift is proportional to the frequency, and the limitations reside in the switches. For constant phase shift it is possible to use meander-lines which allow one octave for 180° phase shift and 6.5° phase error (Schiek and Köhler, 1977).

With loaded line phase shifters, the maximum bandwidth is obtained for a 90° spacing between the loads (Davis, 1974). The frequency range is about 20% for a 45° phase shift and 45% for 22.5° with a VSWR less than 1.2 and a phase shift error less than 2° (Bahland Gupta, 1980, Davis, 1974). With a VSWR less than 1.46 the bandwidth increases to 50% for 45°, 150% for 22.5° and two octaves for 11.25° (Garver, 1972).

For high-pass/low-pass phase shifters the frequency range is about an octave for a phase shift less than 45° and decreases for a larger phase shift (20% for 180°) with a VSWR less than 1.2 and phase error less than 2° (Schiek and Köhler, 1977). For larger errors, insertion losses variation of ± 2 dB and phase error of 30°, the bandwidth becomes two octaves for 180° phase shift (Ayasli *et al.*, 1984).

The phase shift of an I-Q vector modulator phase shifter is constant and the performances are good over 3:1 bandwidth with phase shift and insertion losses errors of ± 15 ° and ± 2 dB respectively (General Microwave Components and Instruments, 1986).

The frequency range of a 180 ° reflection phase shifter is limited only by the circulator or 3 dB coupler, and can work over very wide band; with a 2 ° phase error the band is one octave for 90 °, 150% for 45 °, 300% for 22.5 °, and two octaves for a 11.25 ° cell (Garver, 1972).

The bandwidth values which are given here for the three last types and for Garver (1972), are obtained with lumped elements switchable loads. It is not always possible to use such elements because of frequency or technology limitations, and in the case of linelength elements the usable bandwidth is smaller.

2.3.3 Effect of diode parameters on performance

The equivalent circuit of the p-i-n diode consists essentially of a series inductance (0.4 to 2.0 nH), a shunt capacitance (0.1 to 2.0 pF), a series resistance (0.5 to 2.0 Ω) in forward bias state, and a shunt resistance (10 kΩ) in reverse bias state. The effect of the diode parameters depends on the circuit in which it is used (Wahi and Gupta, 1976, Andricos *et al.*, 1985). But generally and due to the typical values of the different elements, the most important reactance influencing the phase shift is the shunt capacitance, while the resistances have negligible effect on this parameter and evidently increase the insertion losses.

2.4 REAL STRUCTURE OF DIODE PHASE SHIFTERS

The preceding parts show that the different kinds of phase shifters do not have the same properties and it is interesting to associate these different types to optimize the design of real set-ups. Quadrature phase shifters can give 360 ° total phase shift and good resolution with one cell. Switched line and reflection types allow a large phase shift (180 ° and 90 ° bits) with a wide frequency band while loaded line and high-pass/low-pass types have good behaviour for small phase shift (< 45 ° bits) (Burns *et al.*, 1974, Andricos *et al.*, 1985).

BIBLIOGRAPHY

Andricos, C., Bahl, J. and Griffin, E. L., (1985) C-band 6-bit GaAs monolithic phase shifter. *IEEE Trans. Microwave Theory Tech.*, **MTT-33**, 1591–6.

Atwater, H. A., (1985) Circuit design of the loaded line phase shifter. *IEEE Trans. Microwave Theory Tech.*, **MTT-33**, 626–34.

Ayasli, Y., Miller, S. W., and Hanes, L. K., (1984) Wide-band monolithic phase shifter. *IEEE Trans. Microwave Theory Tech.*, **MTT-32**, 1710–14.

Bahl, I. J. and Gupta, K. C., (1980) Design of loaded-line p-i-n diode phase shifter circuits. *IEEE Trans. Microwave Theory Tech.*, **MTT-28**, 219–24.

Buckman, A. B., (1977) Theory of an efficient electronic phase shifter employing a multilayer dielectric-waveguide structure. *IEEE Trans. Microwave Theory Tech.*, **MTT-25**, 480–3.

Burns, R. W., Holden, R. L., and Tang, R., (1974) Low cost design techniques for semiconductor phase shifters. *IEEE Trans. Microwave Theory Tech.*, **MTT-22**, 675–87.

Coats, R. P., (1973) An octave-band switched-line microstrip 3-b diode phase shifter. *IEEE Trans. Microwave Theory Tech.*, **MTT-21**, 444–9.

Davis, W. A. (1974) Design equations and bandwith of loaded-line phase shifters. *IEEE Trans. Microwave Theory Tech.*, **MTT-22**, 561–3.

Garver, R. V., (1972) Broad-band diode phase shifters. *IEEE Trans. Microwave Theory Tech.*, **MTT-20**, 314–23.

General Microwave Components and Instruments, (1986) 'Series 77–360° Phase Shifters/Frequency translators,' Catalog G25.

Higham, E., (1988) Measuring distorsion in PIN diodes. MSN & CT, 97–104.

Lynes, G. D., Johnson, G. E., Huckleberry, B. E., and Forrest, N. H., (1974) Design of a broad-band 4-bit loaded switched-line phase shifter. *IEEE Trans. Microwave Theory Tech.*, **MTT-22**, 693–7.

Schiek, B. and Köhler, J., (1977) A method for broadband matching of microstrip differential phase shifters, *IEEE Trans. Microwave Theory Tech.*, **MTT-25**.

Starski, J. P., (1977) Optimization of the matching network for a hybrid coupler phase shifter. *IEEE Trans. Microwave Theory Tech.*, **MTT-25**, 662–6.

Tuckman, M., (1988) I-Q vector modulator – the ideal control component? MSN & CT, 105–15.

Wahi, P. and Gupta, K. C., (1976) Effect of diode parameters on reflection-type phase shifters. *IEEE Trans. Microwave Theory Tech.*, **MTT-24**, 619–21.

White, J. F., (1974) Diode phase shifters for array antennas. *IEEE Trans. Microwave Theory Tech.*, **MTT-22**, 658–74.

3

Microwave mixers

Jean-François Jouen

3.1 INTRODUCTION

The primary purpose of a mixer is to translate a signal at one frequency to another where it can be amplified or processed more effectively. In order to perform the required signal treatment it is often necessary to shift signals to the frequencies where these functions can be performed best. Thus mixers have many applications in systems such as receivers, synthesizers, transmitters and so on.

A mixer is fundamentally a multiplier, Fig. 3.1 shows an ideal analogue multiplier with two sinusoids applied to it. The signal with a modulation is applied to the port commonly named the RF port. The carrier frequency ω_r is modulated by the information-bearing function $A(t)$. To the other microwave port, called the LO port, is a pure unmodulated sinusoid at frequency ω_l. Sometimes the LO is also called the pump. With some simple trigonometry the output is found to consist of modulated components at the sum and difference frequencies. The sum frequency is rejected by the IF filter, leaving only the difference. This port is called the IF port. Sometimes the sum frequency is derived instead (up-converter mixer) in this case, the RF port is used as the output port.

For microwave applications the multiplying function could be created by any non-linear devices. The drawbacks of non-linear devices are that they generate many LO harmonics and mixing products other than the desired one. The transfer function of a non-linear device can be expressed as:

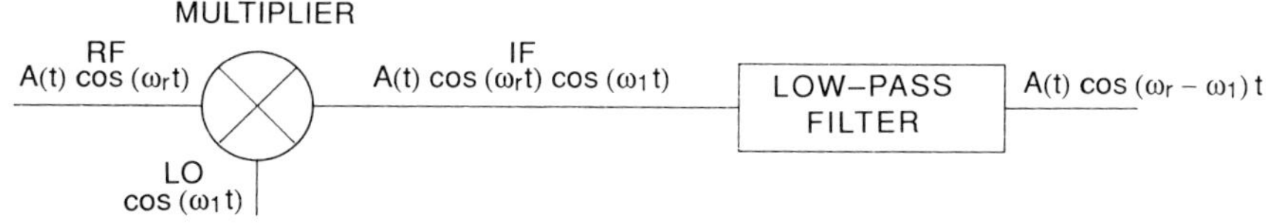

Fig. 3.1 Principles of mixing.

$$I = f(V) = a_0 + a_1 V + a_2 V^2 + \ldots + a_n V^n \tag{3.1}$$

where I and V are the device current and voltage respectively.

For a mixer, the applied voltage is a composite of an RF signal $V_v \sin \omega_p t$ and a local oscillator signal, $V_L \sin \omega_l t$ in the form $V(t) = V_R \sin \omega_r t + V_e \sin \omega_e t$. This results in an infinite series of mixing products given by:

$$I = a_0 + a_1(V_r \sin \omega_r t + V_L \sin \omega_l t) + a_2(V_r \sin \omega_r t + V_L \omega_l t)^2 +$$

$$\ldots + a_n(V_r \sin \omega_r t + V_L \sin \omega_l t)^n. \tag{3.2}$$

Analysing the first two components in equation (3.2) in the frequency domain shows that a d.c. level as well as the original signals are present in the output of the non-linear devices. The primary mixing products $(\omega_l \pm \omega_r)$ come from the second order term and are proportional to a_2 in amplitude, which explains why a non-linear device can perform a mixing function. This phenomenon also exists in amplifiers and in this case is called the intermodulation product. But these terms also generate the second harmonics of both signals. The third and higher order terms generate products of the form $n\omega_l \pm m\omega_r$ and higher order harmonics.

3.1.1 Main electrical characteristics

There are a number of basic parameters that can be used to characterize a mixer which are independent of any of the specific design or device types that will be discussed later. These are conversion loss, noise figure, intermodulation product level and dynamic range.

This conversion gain or loss of a mixer is defined as the ratio of IF output power to signal input power. It is one of most important parameters since it measures the efficiency with which a mixer converts energy at the RF frequency to another frequency, the IF. It is a function of the mixer alone and depends on the losses in the RF circuitry and the quality of the non-linear device. The noise figure of a mixer is the ratio of the signal-to-noise at the output of the mixer to the signal-to-noise ratio at the input of the mixer. The noise figure determines the low-level sensitivity of the mixer.

The level of intermodulation products determines the amount of distortion produced in the mixing process. As discussed previously, mixing products of the form $n\omega_l \pm m\omega_r$ naturally result from the non-linearity of the device. These are commonly referred to as single tone intermodulation products since they are present with only one RF signal input.

If two RF signals are present simultaneously at the mixer input, higher-order products of the form $\omega_l \pm (n\omega_{r1} \pm m\omega_{r2})$ are present at the output; these are commonly referred to as ‘two tone products’. The most important of these responses is the third order product given by:

$$IM_3 = \omega_l \pm (2\omega_{r1} \pm \omega_{r2}).$$

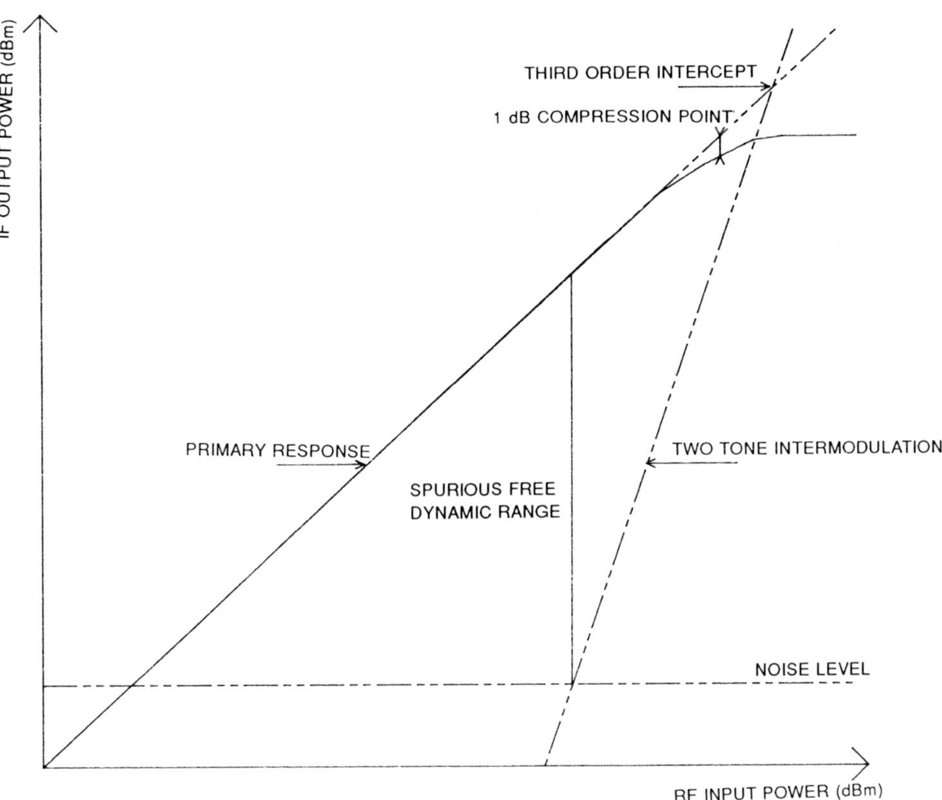

Fig. 3.2 Input output characteristic.

These products are extremely important because they will always occur with closely spaced input frequencies. To characterize the intermodulation of a mixer we refer to a parameter called 'third-order intercept point'. This is an imaginary point on an input – output characteristic that is located at the intersection of the primary response and the spurious response (Fig. 3.2). The primary response has a slope of one, while the third order product response has a slope of three. The dynamic range of a mixer is simply the power range over which the device can be used (Fig. 3.2). The low end is limited by thermal noise and the noise figure of the mixer. The sensitivity S of a mixer in dBm is given by:

$$S = -114 + N_F + \log_{10} \text{BW}$$

where N_F is the mixer noise figure, BW the bandwidth in MHz and – 114 dBm is the thermal noise in a 1 MHz bandwidth.

The maximum signal level is limited by the saturation of the mixer output; that is measured by the 1 dB compression point of the mixer. This is the input power level at which conversion gain or loss bias varied from 1 dB over the low level conversion value.

3.1.2 Devices

As we have seen before, microwave mixers require non-linear devices but only a few devices satisfy the practical requirements of mixer operation. The selected device must have a strong non- linearity low noise, adequate frequency response and repeatable electrical properties between devices. The most employed devices are diodes: p-i-n, junction diodes, point-contact diodes but used much more are the Schottky barrier diode and GaAs MESFETs. At this time, only GaAs MESFETs (single or dual-gate) can be used for microwave mixer applications. The other types of transistors such as bipolar transistors or dual gate MESFETs must be used below 1 GHz.

3.2 SCHOTTKY BARRIER DIODE MIXERS

3.2.1 Schottky barrier diodes

The rectifying junction is formed by contact between a metal and the semiconductor (Si on GaAs). The basic diode rectification theory developed by Schottky is applicable:

$$I = I_s\left[\exp\frac{qV_j}{nkT} - 1\right]$$

where I_s is the saturation current, q electron charge, T absolute temperature, k Boltzmann's constant, n ideality factor and V_j voltage across the diode junction.

The Schottky diode for the purpose of the circuit designer may be regarded as a non-linear resistance R_j given by the above equation shunted by the barrier capacitance. At low frequencies, the capacitance does not affect rectification, but at microwave frequencies its shunting action reduces the RF voltage across the barrier. Since it is impossible to tune out this capacitance at microwave frequencies with an external inductance due to the presence of R_s, it reduces the rectification efficiency and hence must be kept small.

3.2.2 Theory of diode mixers

Conversion loss

The simplest theory to explain the mixing process in a diode mixer is to consider just a single diode driven by the local oscillator into heavy forward

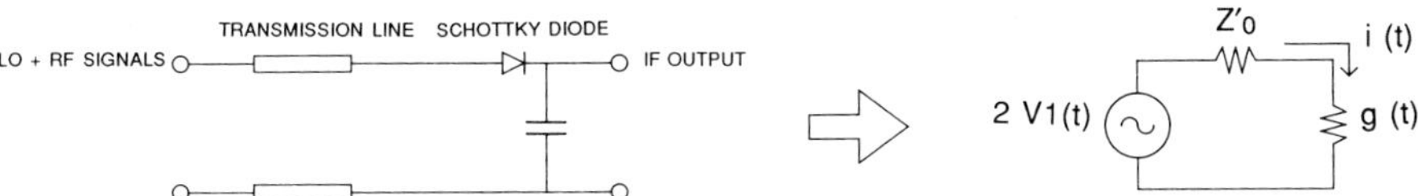

Fig. 3.3 Schottky barrier diode mixer: typical circuit and equivalent model.

conduction for nearly half a cycle and into reverse bias for the other half cycle. In this case, the diode is like a switch driven at the rhythm of the local oscillator. In other words, the reflection coefficient of the diode Γ then varies periodically as a function of time.

In this model, the only effect of the junction capacitance and other parasitics is to transform the source impedance from its actual value to some other value Z_0' in the plane of the semiconductor junction (see Fig. 3.3). Let $G(t)$ be the instantaneous junction conductance. If the applied local oscillator power is P_L then the generator voltage is:

$$2V_1(t) = 2V_L \cos \omega_1 t$$

where

$$V_L = \sqrt{2Z_0' P_L}$$

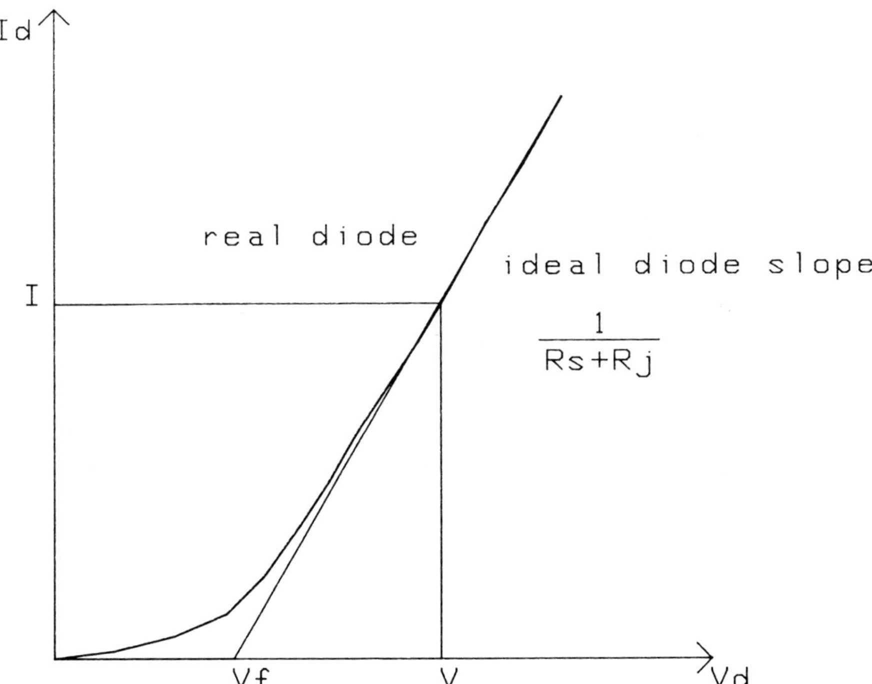

Fig. 3.4 Comparison of real diode and ideal diode I–V characteristics.

The forward diode characteristic is given by the equation:

$$i(t) = I_S \exp[(V(t) - IR_s)/0.028].$$

This equation can be approximated by a two piece linear approximation which has the diode conducting only if the voltage exceeds a forward voltage V_F (see Fig. 3.4). Therefore the low frequency conductance $G(t)$ is:

$$G(t) = \begin{cases} \dfrac{1}{R_s + R_j} & \text{if } 2V_L(t) > V_F \\ \omega^2 C_j^2 R_s & \text{otherwise.} \end{cases}$$

With this time dependent form of the diode conductance we can compute the reflection coefficient:

$$\Gamma(t) = \begin{cases} \Gamma_f & \text{if } 2V_L(t) > V_F \\ \Gamma_r & \text{otherwise.} \end{cases}$$

The waveform of the reflection coefficient Γ can be expressed as a Fourier series. When a small enough incident RF voltage $V_r \cos \omega_r t$ is applied in superposition with the LO voltage to the diode, the reflected wave $V(t)$ has the expression:

$$\Gamma V(t) = \Gamma_o V_r \cos \omega_r t + \frac{1}{2} \Gamma_1 V_r [(\cos(\omega_r - \omega_l)t + \cos(\omega_r + \omega_l)t].$$

The term in $\cos(\omega_r - \omega_l)t$ is at the IF frequency, the conversion efficiency η is the ratio of the power at this IF frequency to the incident power at ω_r.

$$\eta = \frac{4}{\pi^2} \left[1 - Z_o' \omega_l^2 C_j^2 R_s - \left(\frac{R_s + R_j}{Z_o'} \right) \right]^2 \sin\left(\frac{2\Theta}{2} \right).$$

Where Θ is the conduction angle i.e the number of electrical degrees of the LO waveform during which the diode is conducting.

The conversion loss L_c is just η expressed in dB:

$$L_c = -10 \operatorname{Log} \eta.$$

From the above expression we see that conversion loss is at minimum when $R_s = 0$. But that is physically impossible and in general low R_s means a large junction diameter and thus high C_j. So we will introduce a parameter which is essentially independent of junction diameter. The cut-off frequency, f_c, is given by:

$$f_c = \frac{1}{2\pi R_s C_j}.$$

It's more helpful for the design of a mixer to express the conversion in terms of f_c and C_j as those two parameters are given by diode manufacturers. f_c reflects the properties of materials and technology and C_j would be chosen for the frequency of the application.
Let

$$X_c = \frac{1}{\omega_c C_j}.$$

The reactance of C_j is between $Z'_o/2$ and $2Z'_o$. So the conversion efficiency is determined entirely by the ratio of LO frequency to the cut-off frequency of the diode, by the peak current which determines R_j and by the conduction angle, and that for a large range of C_j. The cut-off frequency should be as high as possible. Conduction angle and R_j are determined by LO power and forward voltage.

For high LO drive voltage levels, the conduction angle is near 180° and R_j is very small compared to Z'_o, so the conversion loss is:

$$L_c \approx 3.9\,\text{dB} + 17 + \frac{f}{f_c} + 9\frac{R_B}{Z_o}.$$

An actual single-ended mixer can be better than given by the conversion loss expression, because in our computation we haven't counted power at frequencies other than ω_r, ω_e and IF. Instead, the sum frequency and other harmonics are reflected back into the diode to be re-mixed with harmonics of the Γ wave form to produce more IF output. A complete analysis of a mixer must compute power at each frequency applied to the mixer created by itself. Such an analysis can't be simply formal so for such a purpose we must use a non-linear analysis tool like a harmonic balance program (see Volume 1, Chapter 20).

3.2.3 Noise figure

At the IF output we don't find only the wanted signal (the converted ω_r signal in the IF band) but also many sorts of random noise. The different sources of noise are as follows.

Image noise
The term image is referred to the local oscillator frequency; in other words, if the wanted signal is at frequency $f_L - f_{IF}$, a signal at $f_L + f_{IF}$ would be converted at F_{IF} with the same conversion loss. So the noise at the IF port is twice as great.

Diode thermal noise
The parasitic resistance R_s generates thermal noise.

Shot noise
Electron flow across the diode depletion layer generates shot noise. This is also directly proportional to the absolute temperature of the diode.

Excess noise

At low frequencies, the junction noise increases due to trapping of electrons. It generally decreases with the frequency and is therefore called $1/f$ noise. The effect on the noise figure of a mixer is negligible except in cases where very small IF frequencies are used (below 1 MHz for Si diodes and 30 MHz for GaAs diodes).

Other noise sources

IF noise is the contribution of the IF amplifier to the receiver noise figure. Usually mixer noise figures are measured with an IF amplifier and it's assumed that the IF amplifier has a noise figure of 1.5 dB. LO noise is the phase noise from the local oscillator which can overlap with the IF signal.

As we have seen before, noise figure is defined as the ratio of the signal-to-noise ratio at room temperature of the signal input to the mixer, to the signal at the output of the IF amplifier. For a mixer well driven by the local oscillator, thermal noise and shot noise are negligible, and so for the noise figure a mixer can be considered as an attenuator of attenuation L_c. If the mixer is followed by an IF amplifier, the overall noise figure is given by the Friis formula $N_F = L_c + N_{IF}$ where N_{IF} is the noise figure of the IF amplifier. If we need to account for the other sources of noise we can write the above formula as follows:

$$N_F = L_C + 10 \log_{10}\left(\frac{T_{eff}}{T_o}\right) + N_{IF}$$

where T_o is the room temperature and T_{eff} is the effective temperature of the diode.

3.2.4 Different mixer configurations

Single ended mixer

This is the simplest design of a diode mixer, with one diode terminating a transmission line (Fig. 3.5). All ports (RF, LO and IF) are electrically the same.

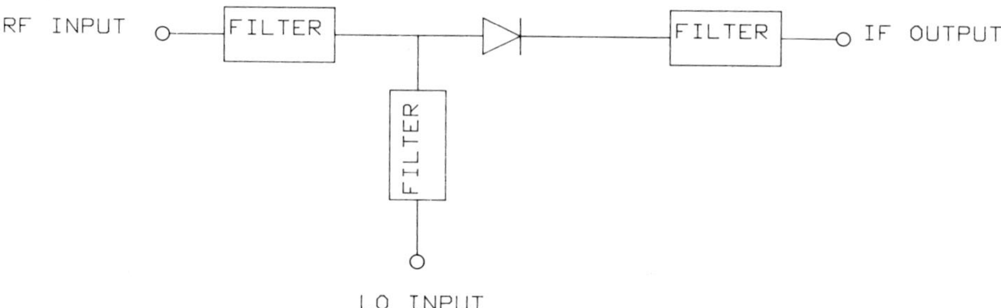

Fig. 3.5 Single ended mixer.

The interport isolations are provided by the different filters. Therefore when a high isolation must be provided between the different ports, RF, LO and IF frequencies must be clearly separated. Therefore a single ended mixer can't be a broadband mixer. From the above theory all harmonics and intermodulation products will also be present and just filters suppress them. They can operate with very low LO power (typically 0 dBm).

Single balanced mixer

Many disadvantages of a single ended mixer can be overcome by combining two circuits in a balanced configuration (Fig. 3.6). A 3 dB hybrid (either a 90° or 180°) is used to supply RF and LO power to two diodes. The advantages over single ended mixers are reduced spurious responses and convenient separation of LO and RF inputs.

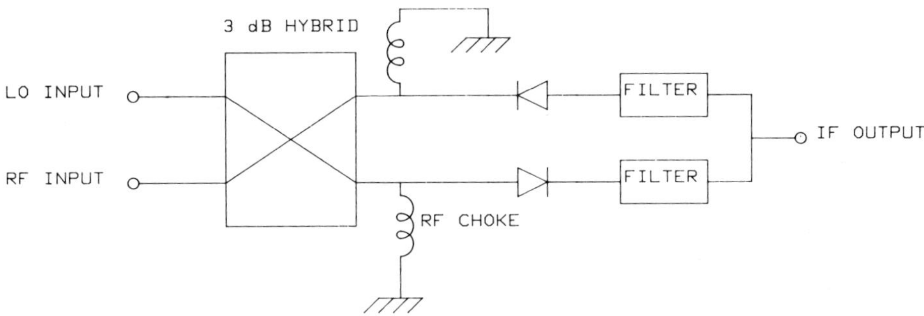

Fig. 3.6 Balanced mixer.

When a 180° hybrid is used, the even harmonics of one of the input signals are suppressed (usually the harmonics of the LO signal). The degree of spurious response suppression is controlled by the balance of hybrid and diodes. Balanced mixers designed with 90° hybrids exhibit slightly different properties. If diodes are well balanced then the VSWR of LO and RF ports is low but in this case the reflected power is transmitted to the other port and therefore RF to LO isolation is poor except if the diodes are carefully matched.

For broadband purposes we can use a 'balun' (balanced to unbalanced transitions). The two outputs of balun are equal in amplitude and at 180° in phase. Figure 3.7 shows two ways to create balanced mixers with baluns. There is no significant performance difference between these two types of mixers.

Double balanced mixers

Double balanced mixers are composed of two single balanced mixers in parallel and 180° out of phase (Fig. 3.8). Double balanced mixers generate only one quarter of the possible intermodulation products. These have odd f_r and

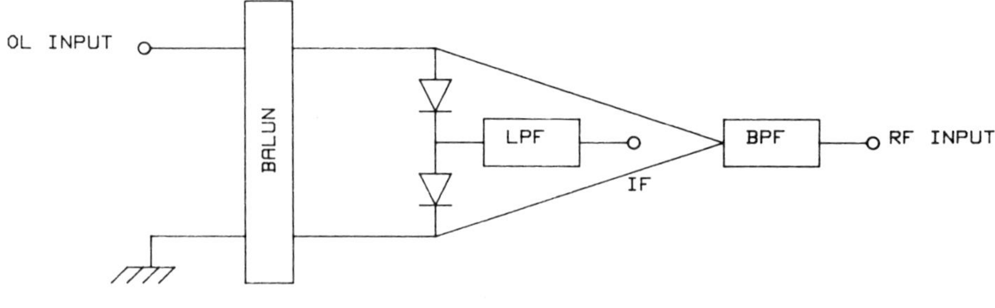

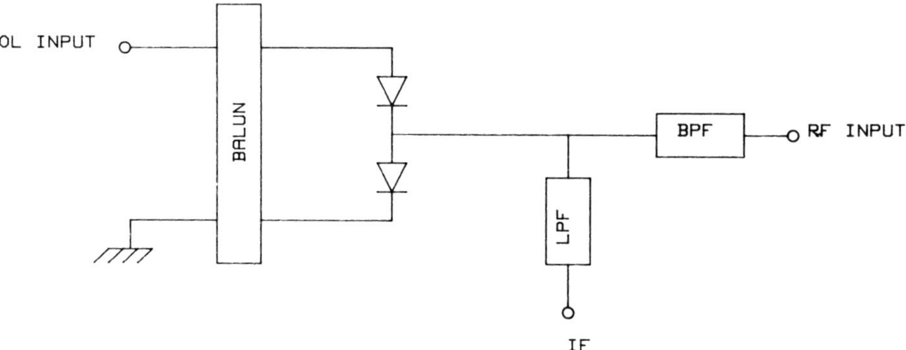

Fig. 3.7 Two versions of a single balanced mixer.

also odd f_1 harmonics. LO power for double balanced mixers is typically 3 dB higher than for single balanced mixers because twice as many diodes are used hence the 1 dB compression point is also increased. The centre top transformers shown in Fig. 3.8 are only available at low frequencies below 3 GHz. Above 3 GHz baluns are used instead of transformers. In the discussion about single balanced mixers made with baluns we have shown that there are two ways to inject RF in diodes (Fig. 3.7).

Figure 3.9 shows that combining two single balanced mixers results in either a ring or a star double balanced mixer, depending on which type of single balanced mixers are used. In the case of the star mixer, the IF port is a virtual ground for both LO and RF signals, therefore the IF bandwith can be larger than that of a ring double balanced mixer. But a ring quad mixer can be made more easily than star quad mixer.

We have seen that the IF signal is separated by a filtering action often inherent in the LO balun construction. When we want a broader bandwith IF port we must complete the balanced design of the mixer at the IF port. In this case this type of mixers are called double-double balanced mixers (Fig. 3.10). Twice the number of diodes are present in these mixers as in double balanced

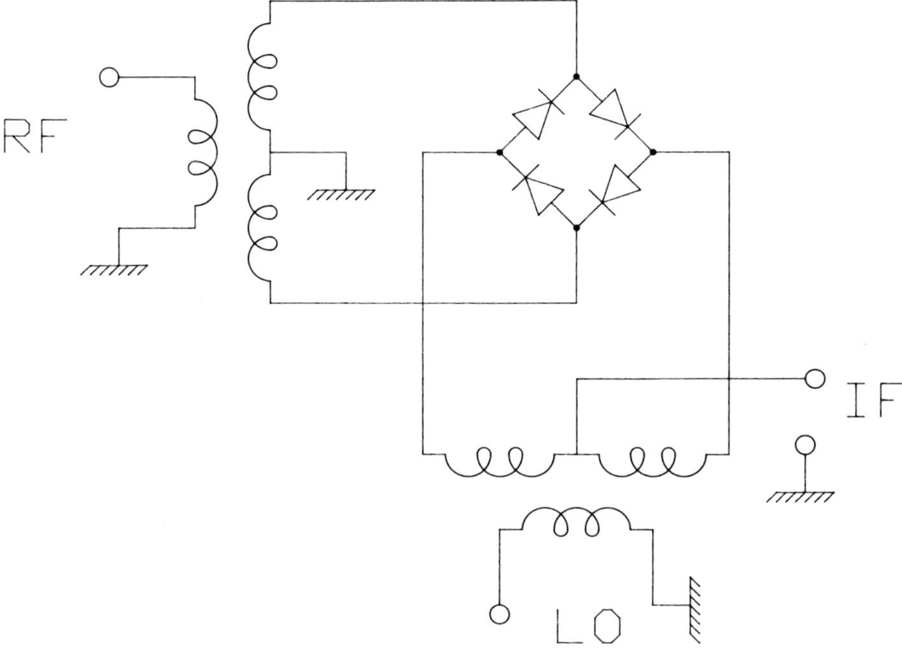

Fig. 3.8 Double balanced mixer.

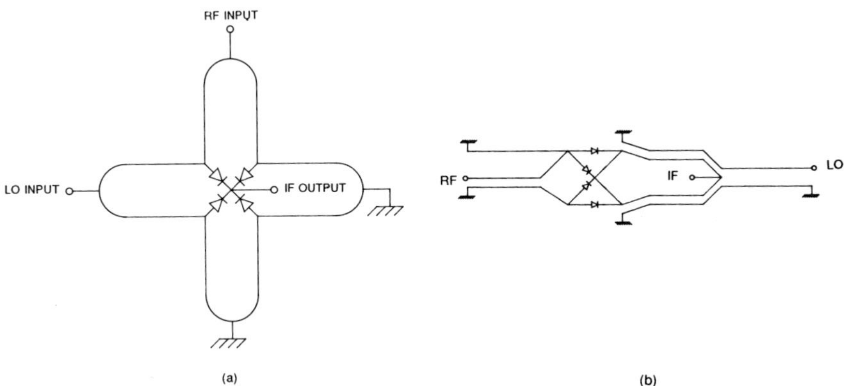

Fig. 3.9 Two versions of double balanced mixer; (a) star mixer, (b) ring quad mixer.

mixers so more LO power is required. And therefore the dynamic range is increased.

Image rejection mixers

For a given IF frequency, f_{IF}, two RF signals (one above $(f_{OL}+f_{IF})$ and one below $(f_{OL}-f_{IF})$ the LO frequency produces an IF output at f_{IF}. If one of these

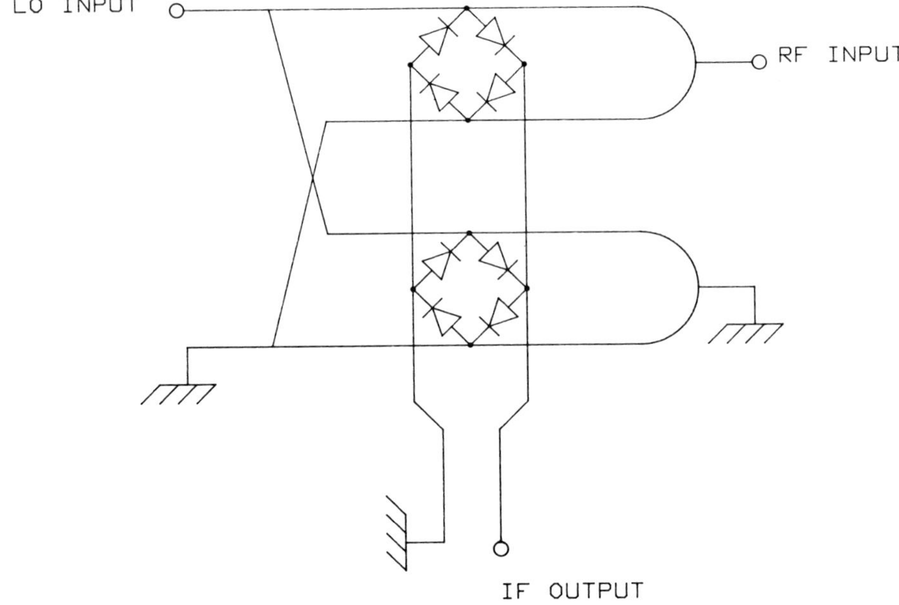

Fig. 3.10 Double double balanced mixer.

frequencies is considered to be the desired 'signal' frequency then the other is commonly termed the 'image' frequency. In many applications, it is desirable to either eliminate or distinguish the image response from the desired signal response. The simpler way to do that is to filter the unwanted signal.

That's possible only if the IF frequency is sufficiently high and the RF bandwidth narrow enough so that the signal and image frequency bandwidths do not overlap. This type of design is only suitable for narrow band applications where a high degree of image rejection (a source of noise) is required. For broadband applications, especially octave bandwidths, filtering cannot be

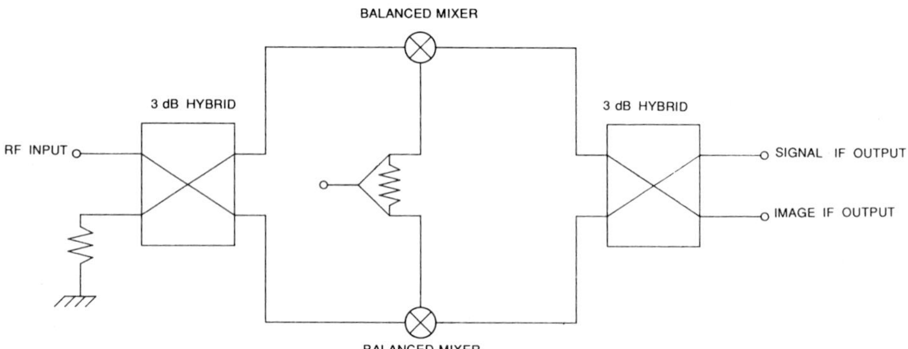

Fig. 3.11 Image rejection mixer.

used for image rejection; in this case the image frequency is rejected by phasing techniques.

Two single-balanced mixers are frequently combined to form an image-rejection mixer as shown in Fig. 3.11. The RF signal is fed to the mixers through a 90 °, 3dB hybrid while the LO signal is applied through an in-phase power splitter. The IF outputs of each mixer are then combined through a 90 ° hybrid. With this arrangement the signal frequency response appears at one output of the IF hybrid and the image at the other. Either the signal or image response can be selected by terminating the appropriate IF output port. The degree of image rejection depends on the amplitude and phase balance between the two mixers. The conversion loss of this type of mixer is higher due to the additional losses of the RF and IF hybrids. Table 3.1 gives a comparison guide.

Table 3.1 Mixer comparison guide

Mixer type	*Single-ended*	*Balanced* (90°)	*Balanced* (180°)	*Double-balanced*	*Image-reject*	*Image-recovery*
Conversion loss	Good	Good	Good	Very good	Good	Excellent
VSWR	Good	Good	Fair	Poor	Good	Good
LO, RF	Poor	Good	Fair	Poor	Good	Good
LO/RF isolation	Fair	Poor	Very good	Very good	Good	Very good
LO power required	+ 13(a)	+ 5	+ 3	+ 10	+ 7	+ 7
Spurious rejection	Poor	Fair	Fair	Good	Fair	Fair
Harmonic supression	Poor	Fair	Odd:Fair Even:Good	Very good	Even: Good Odd: Fair	Even: Good Odd: Fair
Third-order intercept	—	+ 13 dBm	+ 13 dBm	+ 18 dBm	+ 15 dBm	+ 15 dBm

3.3 MESFET MIXERS

3.3.1 Gallium arsenide MESFETs

It's possible to use GaAs MESFETs to design microwave mixers. Work has been done on both single gate MESFETs and dual gate MESFETs. Dual gate MESFETs are particularly well suited for mixer applications because they have inherently good LO/RF isolation when one gate is used to feed the LO and the other for the RF input. Figure 3.12 shows the drain I–V characteristics of a typical small signal GaAs MESFET. At very low drain-source voltages, the MESFET behaves much as a gate-controlled resistor and is suited to be operated in its linear region (Fig. 3.12a). At high drain-source voltages the current just depends, in a first approximation, on the gate-source voltage (Fig. 3.12b). These two phenomena have generated two families of MESFET

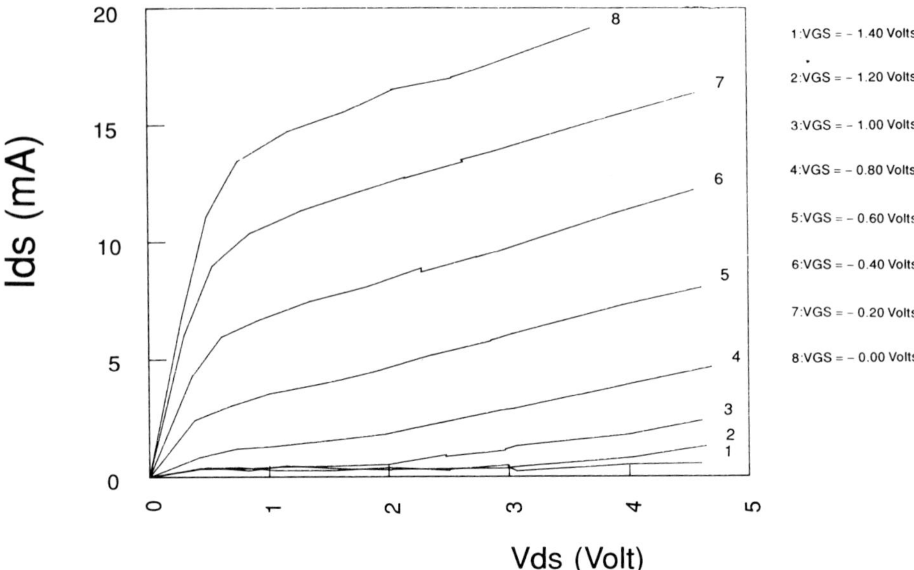

Fig. 3.12(a) Image recovery mixer.

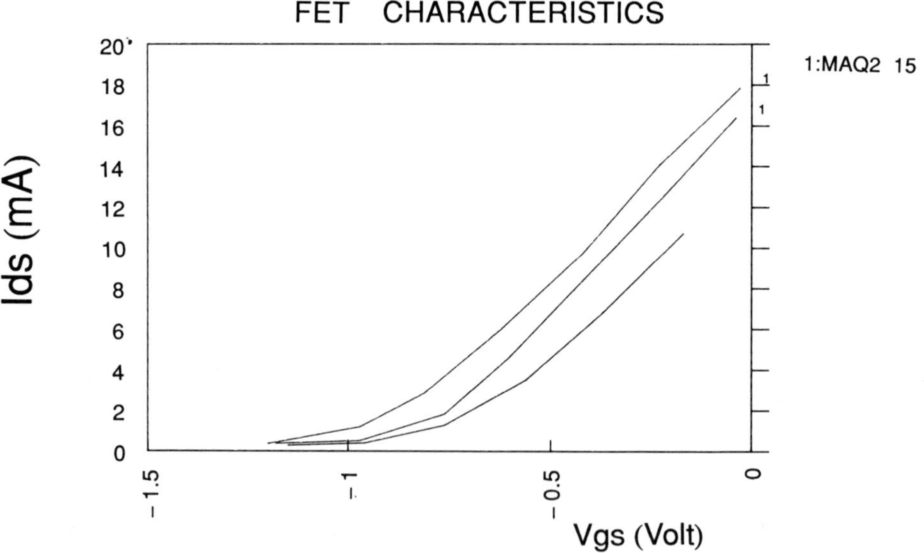

Fig. 3.12(b) Image recovery mixer.

mixers: the first is called the MESFET resistive mixers, the second is called the active MESFET mixers.

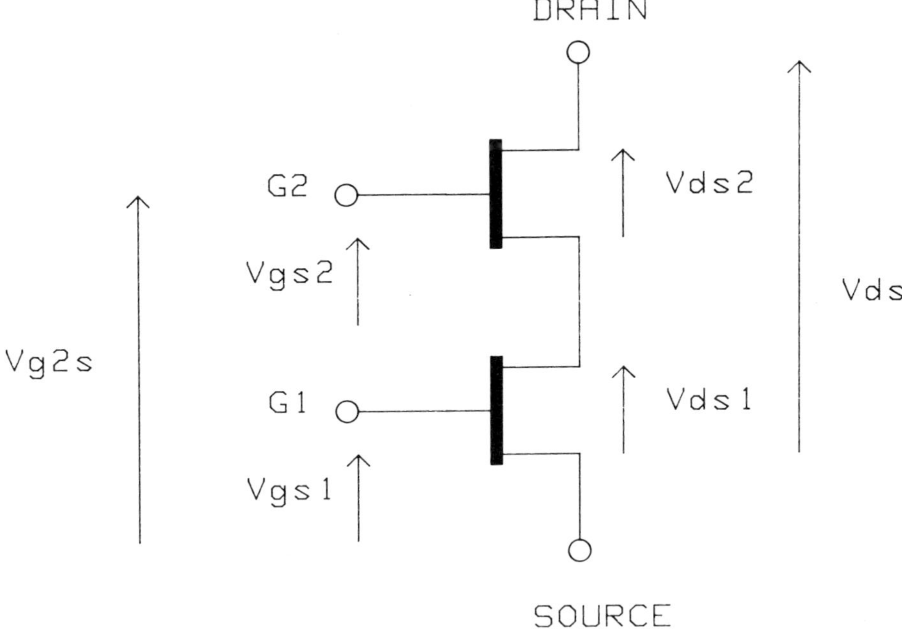

Fig. 3.13 Dual-gate MESFET model as two single-gate FET in series.

A dual-gate MESFET is similar in structure to a single-gate device, except that it includes a second gate between the first gate and the drain. The principal effect of this second gate is to control the small signal transconductance of the first gate and thereby the RF gain of the device. Dual-gate MESFETs are usually modelled as two single gate FETs in series as shown in Fig. 3.13.

3.3.2 MESFET mixer theory

MESFET resistive mixers

Figure 3.14 shows the equivalent circuit of a MESFET with no bias applied to the drain; $g(V_g)$ is the channel conductance which is weakly non-linear and is controlled only by V_g. The LO voltage and d.c. bias are applied to the gate of the MESFET. The RF signal is applied to the drain terminal and the source is grounded. The IF signal is picked by filtering the drain terminal. The channel of FET behaves as a time varying linear resistor until non-d.c. bias is applied. The FET equivalent circuit can be reduced with approximation to the equivalent circuit of a diode and it then is possible to use diode mixer theory to predict the performances of the MESFET resistive mixer. The analysis with this approximation shows that the conversion loss of a resistive mixer is of the same order as that of a diode mixer.

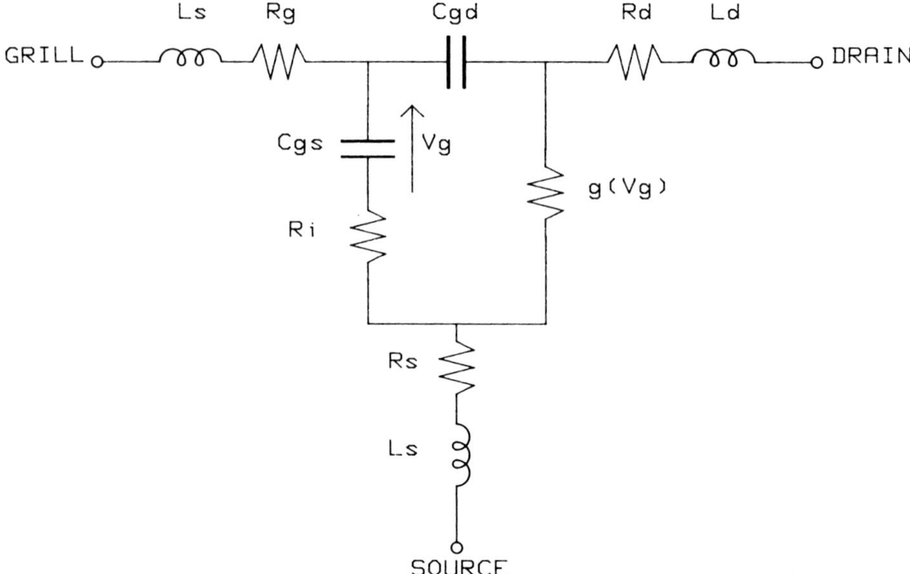

Fig. 3.14 Resistive MESFET equivalent circuit.

The advantages of MESFET resistive mixers compared with diode mixers come from the fact that the resistive MESFET acts like a linear time varying resistor. This implies that the intermodulation characteristics of such a mixer are highly improved compared with a diode mixer. Recent publications on the subject show that a double balanced MESFET resistive mixer can have a + 30 dBm input third order intercept point.

Active MESFET mixers

An exact analysis of the MESFET mixer is a very tedious work. First we must have a large signal model of the FET which includes all the bias dependent elements as shown in Fig. 3.15. Second we must use a non-linear CAD program to obtain the different parameters of the FET. But a simplified explanation of the MESFET mixer can be given when we look at the FET I–V characteristic of Fig. 3.12b. We can approximate the I_d dependence on V_{gs} as a quadratic expression. That means we can write:

$$I_d = I_{dss}\left(1 - \frac{V_{gs}}{V_p}\right)^2.$$

The tranconductance g_m of the FET is given by:

$$g_m = \frac{\delta I_d}{\delta V_{gs}} = \frac{-2I_{dss}}{V_p}\left(1 - \frac{V_{gs}}{V_p}\right).$$

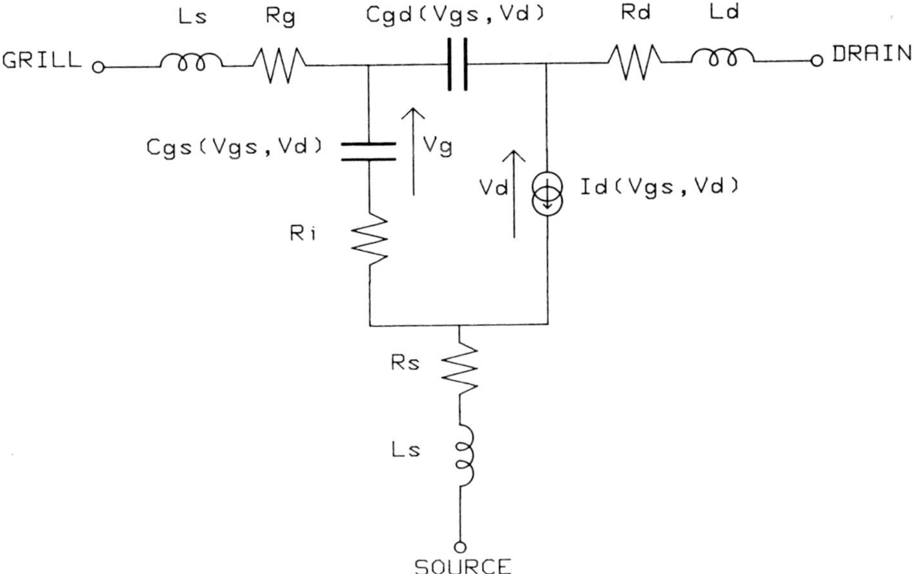

Fig. 3.15 MESFET large signal equivalent circuit.

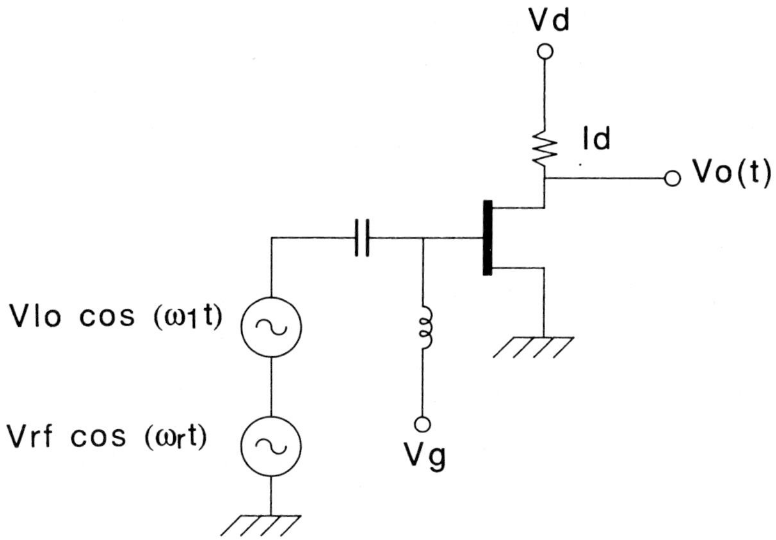

Fig. 3.16 Simplified explanation of FET mixing.

Figure 3.16 shows a simplified single-gate FET mixer where both the RF and LO power are applied accross the gate source terminal. The large signal LO modulates linearly only if the transconductance of the RF signal level is small enough. In this case we have:

$$V_{gs} = V_g + V_{LO} \cos \omega_{LO} t.$$

The time dependent transconductance $g_m(t)$ of the FET is therefore:

$$g_m(t) = \frac{-2I_{dss}}{V_p}\left(1 - \frac{V_g + V_{LO}\cos \omega_{LO} t}{V_p}\right)$$

$$= \left(\frac{-2I_{dss}}{V_p}\right)\left(1 - \frac{V_g}{V_p}\right) + \left(\frac{2Id_{ss}}{V_p^2}\right) V_{LO} \cos \omega_{LO} t$$

$$= g_{m0} + g_{m1}(t).$$

When a small RF signal is applied simultaneously a drain component appears that is given by:

$$i_d(t) = g_{m0} V_{RF} \cos \omega_{RF} t + \frac{I_{dss} V_{LO} V_{RF}}{V_p^2} \cos(\omega_{LO} - \omega_{RF})t$$

$$+ \frac{I_{dss} V_{LO} V_{RF}}{V_p^2} \cos(\omega_{LO} + \omega_{RF})t.$$

The second term in $i_d(t)$ is at the IF frequency. The third term is at the sum frequency. This analysis shows the availability of frequency conversion through a MESFET, with just the non-linearity of the transconductance, although both the gate-source capacitance (C_{gs}) and drain source resistance (R_{ds}) also are non-linear functions of bias and therefore will vary as the large signal LO power varies. The experimental results have shown that with MESFET we can obtain conversion gain but the noise figure of the MESFET active mixer is comparable with that of a diode mixer.

3.3.3 MESFET mixer configurations

The circuits we have shown with diode mixers have their counterparts in MESFET mixers. Fig. 3.17 shows a single-balanced mixer, and Fig. 3.18 shows a double balanced mixer.

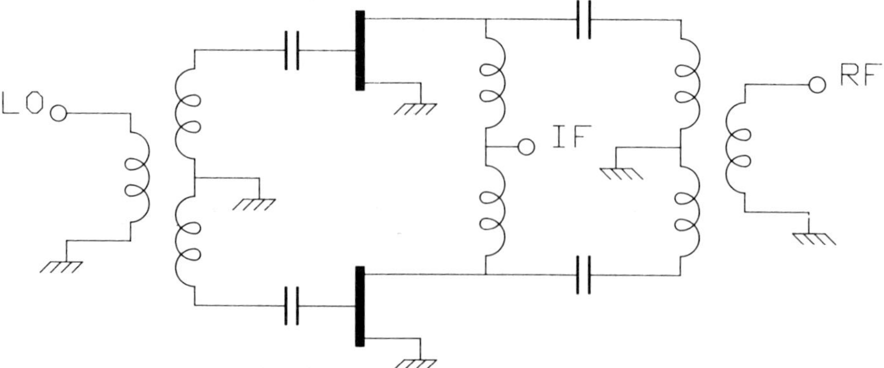

Fig. 3.17 Single balanced MESFET mixer.

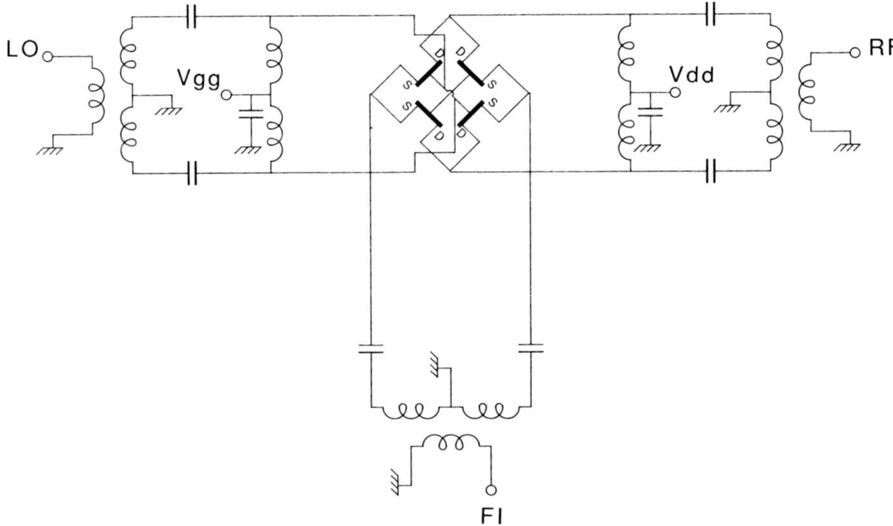

Fig. 3.18 Double balanced MESFET mixer.

3.4 MMIC MIXERS

3.4.1 Diode monolithic mixers

In the case of diode MMIC mixers we can consider two types of applications: millimetre waves, and small bandwith.

At millimetre-wave frequencies, all the circuitry required to create a mixer and also transitions between the mixer and waveguide can be implemented on a single chip of small size. For small band mixers, one finds the translation in MMIC technology of the classical balanced or double balanced circuits. The problem of increasing the bandwidth is in making baluns compatible with MMIC technology. One way to overcome this, is to make lumped element transformers or design active baluns around FET.

3.4.2 FET monolithic mixers

A lot of work has been done around FET MMIC mixers because FET is the most adequate component for MMIC technology. We can divide these mixers into two categories: reduced band and ultrawide band.

For the reduced band, MMICs are often made with double gate FETs (for the inherent gate-to-gate isolation). These mixers have conversion gain, opposite to the diode mixers who have a conversion loss. Despite the noise figure of this mixer which is comparable with diode mixers, there are many potential

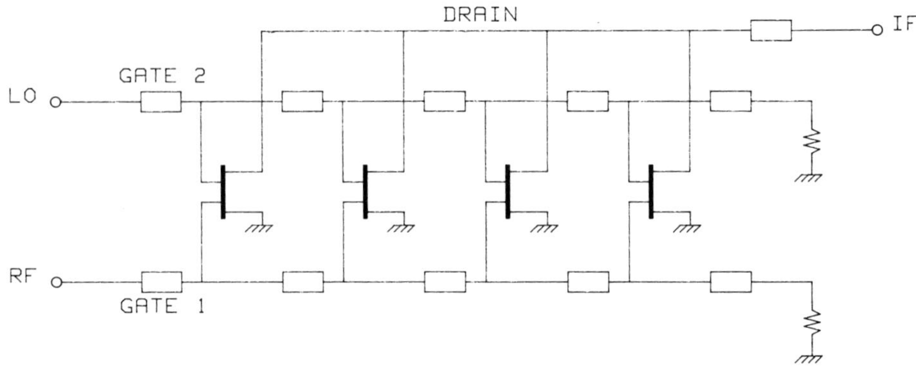

Fig. 3.19 Four stage distributed mixer.

applications for large-volume commercial systems. This is because in some cases, the local oscillator and/or the IF preamplifier can be designed on the same MMIC as the mixer. This can give a low cost front-end receiver.

For ultrawide band MMICs the problem is again the one of balun creation. The solutions can be the same as for diode MMIC or to use completely new designs that are allowed by the MMIC possibilities. One of these is the distributed mixer. A distributed mixer is formed by inserting FETs into an artificial transmission line structure that uses series inductive transmission lines to connect the shunt device capacitances C_{gs} or C_{ds} (cf. Fig. 3.19), as for a distributed amplifier. The IF appears on the drain line like any other FET mixer, however the noise figure of this distributed mixer is poor.

BIBLIOGRAPHY

Maas, S. A. (1986) *Microwave Mixers*, Artech House, Dedham, MA.

Maas, S. A. (1988) A low distortion GaAs MESFET resistive mixers *Microwave Journal*, 213.

Maas, S. A. (1988) *Non linear Microwave Circuits*, Artech House, Dedham, MA.

Saleh, A. A. M. (1971) *Theory of Resistive Mixer,* MIT Press, Cambridge, MA.

Sze, S. M. (1981) *Physics of Semiconductor Devices* 2nd edn, John Wiley and Sons,

Weiner, S. *et al.* (1988) 2 to 8 GHz Double Balanced MESFET Mixer with +30 dBm input 3rd Order Intercept *IEEE, MTT-S Microwave Symposium Digest* 1097.

4

Microwave oscillators

Didier Kaminsky

4.1 INTRODUCTION

Microwave solid-state oscillators convert dc energy to microwave signals. They are used as generators in all communication systems, radars, electronic counter-measures (ECM), etc. Though they did not exist three decades ago, they are replacing more and more low power tubes such as small klystrons, and their main advantages are low voltage supply, reliability, small weight, etc.

A solid-state microwave oscillator is composed of three parts.

A resonant structure which stores the energy and fixes the frequency. A high Q-factor is desirable.

A negative resistance system controlled by the power. For a given frequency, it is not always easy to obtain sufficient negative resistance because it is reduced as frequency and power are increased.

A coupling system able to deliver power to the load. This system must not disturb the resonance.

As oscillators are fundamentally non-linear systems, they are very difficult to analyse. Three methods can be employed to design oscillators:

'Cut and try'

This is the simplest method and is only justified to show the feasibility of a system because results cannot be reproducible as they are a function of the components, the technology, etc.

Small signal design

This is the most used method. From the measured electrical impedances or scattering matrix, oscillation conditions are calculated and matching networks are synthesized. The main disadvantage of this method is that output power, harmonic and spurious levels, FM noise, etc., cannot be calculated. Only a rough value of the oscillating frequency and range of oscillation can be given.

Non-linear analysis
This method uses measurements made at high levels in the active components. Once non-linearities are determined, calculations can take place. Up to now, this method has been very difficult to employ and is not widely used.

In this chapter, only small signal analysis will be given. In the next sections, after a presentation of the general conditions of oscillations, including FM noise definitions, characteristics of the active microwave solid-state components will be discussed. Then the different fixed frequency oscillators and electronically tuneable oscillators will be analysed and compared.

4.2 GENERAL CONDITIONS OF OSCILLATION

4.2.1 Scattering parameters analysis

Generalized oscillation conditions

An oscillator can be considered as a combination of an active multiport and a passive multiport (the embedding network) as shown in Fig. 4.1.
For the active device:

$$[b] = [S] \cdot [a] \tag{4.1}$$

and for the embedding network:

$$[b'] = [S'] \cdot [a'] \tag{4.2}$$

when the active device and the embedding network (including the loads) are connected such that port i is connected to i', then:

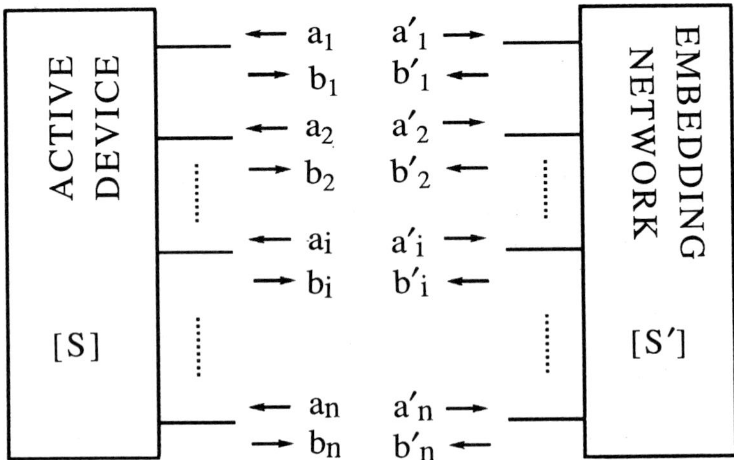

Fig. 4.1 S-parameters of active and passive multiports of an oscillator.

$$[b] = [a'] \tag{4.3}$$

$$[a] = [b'] \tag{4.4}$$

by combining (4.1) to (4.4), we obtain:

$$[a'] = [S]\,[S']\,[a']$$

or

$$([S]\,[S'] - [1]) \cdot [a'] = 0$$

where [1] is the unit matrix. Now, since $[a'] \neq 0$, $[M] = [S]\,[S'] - [1]$ is a singular matrix. Thus,

$$\det [M] = 0 \tag{4.5}$$

represents the generalized oscillation conditions.

It can be noted that in the case of a multiport defined by its impedance or admittance matrix we have for the active device

$$[V] = [Z]\,[I] \tag{4.6}$$

and for the embedding network

$$[V'] = [Z']\,[I'] \tag{4.7}$$

by connecting the port i to port i', we have:

$$[I] = -[I'] \tag{4.8}$$

$$[V] = [V'] \tag{4.9}$$

by combining (4.6) and (4.9), we obtain:

$$([Z] + [Z']) \times [I] = 0$$

since $[I] \neq 0$, the matrix $[Z] + [Z']$ is singular or

$$\det ([Z] + [Z']) = 0. \tag{4.10}$$

A similar relationship can be obtained with admittance matrix

$$\det ([Y] + [Y']) = 0 \tag{4.11}$$

(4.10) or (4.11) represents the generalized oscillation conditions.

Application to negative resistance oscillators

A negative resistance oscillator can be represented by the circuit of Fig. 4.2. By applying equation (4.10), we obtain the oscillating conditions:

$$Z_L + Z_d = 0 \tag{4.12}$$

or, by using reflection coefficients:

$$\gamma_L \times \gamma_D - 1 = 0. \tag{4.13}$$

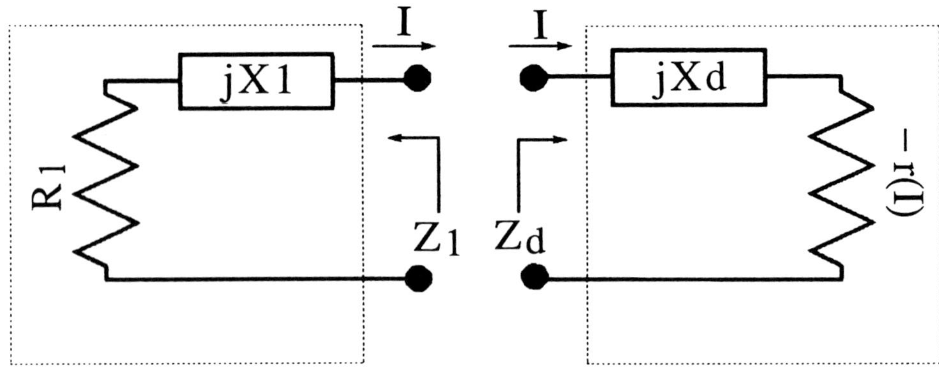

Fig. 4.2 Negative resistance oscillator topology.

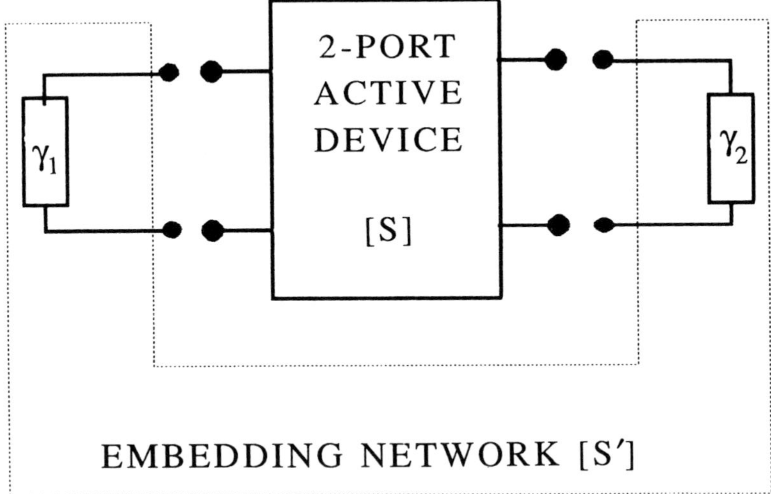

Fig. 4.3 Two-port device loaded by two impedances.

Application to a two-port device loaded by two impedances

The circuit is represented in Fig. 4.3. In this case, the oscillation conditions are given by the two well-known conditions:

$$\frac{1}{\gamma_1} = S_{11} + \frac{S_{12}S_{21}\gamma_2}{1 - S_{22}\gamma_2} \tag{4.14}$$

and

$$\frac{1}{\gamma_2} = S_{22} + \frac{S_{12}S_{21}\gamma_1}{1 - S_{11}\gamma_1} \tag{4.15}$$

4.2.2 Stability conditions

The stability conditions of an oscillator will be calculated using the negative resistance approach. The equivalent circuit of a free running oscillator is represented in Fig. 4.4, where Z_D is the non-linear impedance of the device and is a function of the RF current, and Z_L is the load impedance.

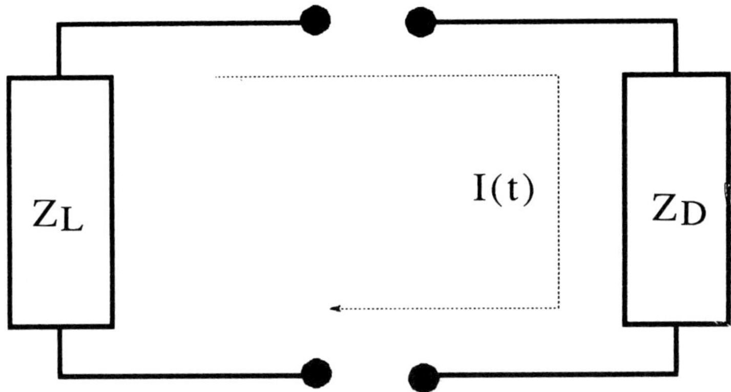

Fig. 4.4 Equivalent circuit of a free running oscillator.

The RF current $I(t)$ can be expressed by:

$$I(t) = A(t)\,e^{(pt)}$$

where p is the complex frequency $p = \alpha + j\omega$.
The oscillation condition is:

$$Z_T(A, p) = Z_L(p) + Z_D(A, p) = 0. \tag{4.16}$$

As a first approximation, we can suppose that only the real part of Z_D is a function of the RF current and is independent of the frequency.

In this case, after some computations, the two main conclusions concerning stable oscillations are that

$$\frac{\partial X_T}{\partial \omega} > 0$$

and curves $Z_T(A)$ for $\omega = \omega_o$ and $Z_T(\omega)$ for $A = A_o$ must be perpendicular.

4.2.3 Pulling

We shall see here the frequency drift due to a small change of the load which is very often $Z_o = 50\ \Omega$ in microwaves. If Z_D is the non-linear negative resistance, the oscillation conditions are

$$Z_T(A_o, \omega_o) = Z_D + Z_L = 0$$

where $Z_L \neq Z_o$.

It can be shown that, if S = VSWR of the load and Q_{ext} is the quality factor

$$Q_{ext} = \frac{\text{accumulated energy} \cdot 2\pi}{\text{power lost per cycle in the load}}$$

or

$$Q_{ext} = \frac{\omega_o}{2Z_o} \frac{\partial X_T}{\partial \omega}$$

then the frequency drift of the oscillator as the phase of the load is changed from 0 to 360° is:

$$\Delta\omega = \frac{\omega_o}{2Q_{ext}} \left(S - \frac{1}{S} \right). \tag{4.17}$$

This result is important because it shows that to reduce the pulling there are two solutions. First, to increase Q_{ext}, and second, to reduce the variation of the load seen by the oscillator either by using good loads or by isolating the oscillator from the load with isolator, amplifier, attenuators, etc.

A measurement set-up is presented in Fig. 4.5. As the phase of the load is changed by moving the sliding short, the frequency drift is recorded and the Q_{ext} can then be calculated.

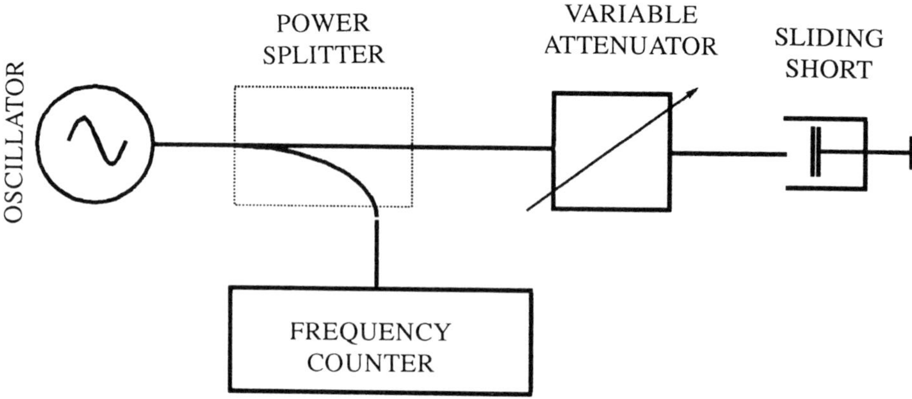

Fig. 4.5 Test set-up for frequency pulling measurements.

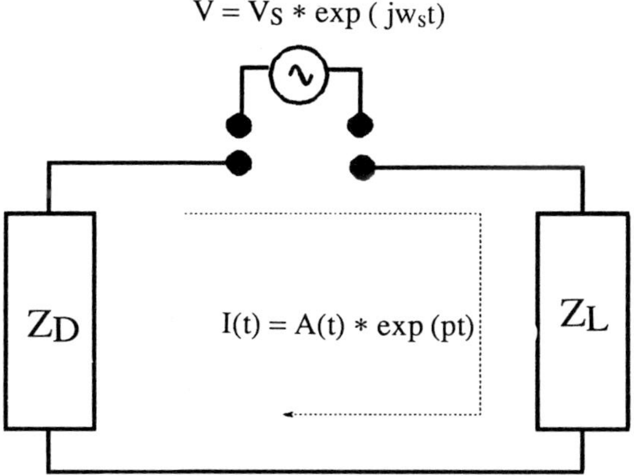

Fig. 4.6 Model used for the analysis of the synchronization of an oscillator.

4.2.4 Synchronization of an oscillator

By injecting a signal (ω) in the output of an oscillator, it is possible to force the oscillator to the same frequency. To analyse the system, we shall represent the injected signal by a microwave generator in series with two loads Z_D and Z_L (Fig. 4.6).

If

$$I(t) = (A_o + \delta A)\exp\{j[(\omega_o + \delta\omega)t + \varphi]\}$$

the conditions of oscillation are:

$$\frac{\partial Z_T}{\partial \omega}\delta\omega + \frac{\partial Z_T}{\partial A}\delta A = \frac{V_s}{A_o}\exp\{j[(\omega_s - \omega_o - \partial\omega)t - \varphi]\}. \tag{4.18}$$

When there is locking $\omega_s = \omega_o + \partial\omega$.

The maximum drift can be calculated and is given by:

$$\Delta\omega = \frac{2V_s}{A_o}\cdot\frac{\sqrt{1+\alpha^2}}{\dfrac{\partial X_T}{\partial\omega} - \alpha\dfrac{\partial R_T}{\partial\omega}} \tag{4.19}$$

where

$$\alpha = \frac{\partial X_T}{\partial A}\bigg/\frac{\partial R_T}{\partial A}$$

in the particular case where $\partial X_T/\partial A = 0$, the locking bandwidth is then:

$$\Delta\omega = \frac{V_s}{A_o}\frac{\omega_o}{Z_o Q_{ext}} \tag{4.20}$$

where

$$Q_{ext} = \frac{\omega_o}{2Z_o} \frac{\partial X_T}{\partial \omega}$$

as $P_{synch} = (1/8)\, V_s^2/R_L$ is the power of the synchronizing signal and $P_{oscill} = (1/2)\, R_L A_o^2$, the locking bandwidth can be put in the form:

$$\Delta\omega = \frac{2\omega_o}{Q_{ext}} \sqrt{\frac{P_{synch}}{P_{oscill}}}\,.$$

It is important to note that the locking bandwidth increases with the ratio of the power of the synchronizing signal to the output power of the oscillator.

4.2.5 FM noise of oscillators

The spectrum of an oscillator

The output of an oscillator jitters both in amplitude and in phase. This jitter is mainly the result of the noise produced in the active device of the oscillator.

Edson (1960) first analysed the FM noise performance of a feedback oscillator using a simple model and assuming white noise (thermal noise) to be associated with the active device. Leeson (1966) extended the analysis to include the effects of 1/f (flicker phase) and the effect of noise floor. Finally, Kurokawa (1968) analysed in detail the noise characteristics of a negative resistance oscillator.

Oscillators never generate a pure signal but produce a continuous spectrum of infinitely close frequency modulated components distributed around the fundamental frequency (the carrier). What we have called the spectrum is then a representation of the output power of the oscillator in the frequency domain as represented on Fig. 4.7, i.e., the spectral distribution of energy around the carrier f_o.

This output spectrum, due to noise, is composed of two parts.

1. Systematic noise resulting from the d.c. power or from a modulation in the bias circuits. This systematic noise can be reduced by carefully designing low frequency circuits and by choosing low noise d.c. generators.
2. Random noise due to random effects in the microwave components. This random noise depends on the amplification or oscillation mechanism, technology and materials of the active component and on the design of the circuit.

These low frequency effects, as they act directly on the microwave signal, are converted into high frequency noise around the carrier by non-linear effects.

As the number of active components and circuits is limited, oscillator characteristics are a compromise between noise and performance such as output

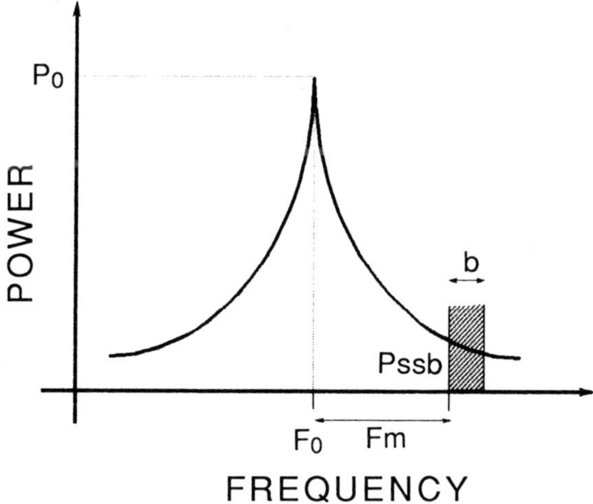

Fig. 4.7 Spectral distribution of energy around the carrier.

power, efficiency or cost. The output signal of an oscillator is then modulated either in amplitude or in frequency (or phase) (or both).

A real signal is described by:

$$V(t) = [A_o + a(t)] \cos [2\pi f_o t + \varphi(t)]$$

where $a(t)$ and $\varphi(t)$ are respectively amplitude and phase fluctuations. These fluctuations can be described in terms of long-term or short-term stability.

Long-term stability refers to slow changes in the signal due to modification of components in the circuit. It is usually expressed in a percentage or in ppm

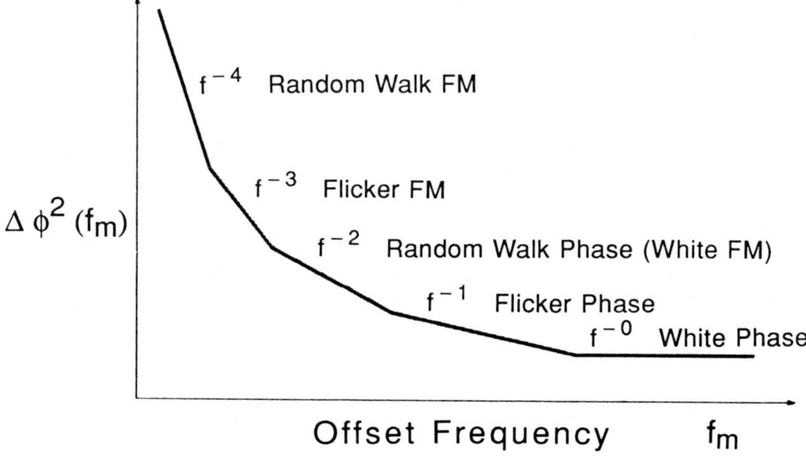

Fig. 4.8 Spectral density distribution versus offset frequency.

(parts per million) of the measured quantity (either frequency or power) for a given period of time (hours to months or even years). Short-term stability refers to changes which cannot be described as drift but are either random and/or periodic fluctuations that are shorter than some seconds.

An oscillator being a non-linear system, an amplitude variation can be converted into frequency variation. The amplitude spectrum then contributes to the phase (or frequency) spectrum of the output signal.

In the frequency domain, as represented in Fig. 4.8, the spectral density distribution of phase fluctuation as a function of the offset frequency can be described in terms of 'random walk', 'flicker' and 'white noise' corresponding to the different slopes.

Quantification of phase noise

Due to the random nature of the instabilities, the phase deviation is represented by a spectral density distribution expressed in energy within a given bandwidth b (cf. Fig. 4.7). The short-term instabilities are measured as a low-level phase modulation of the carrier.

The main different quantities that can be measured are:

$L(f_m)$, single side band (SSB) noise to carrier;
$S_\phi(f_m)$, spectral density of phase fluctuations;
$S_{\Delta f}(f_m)$, spectral density of frequency fluctuations;
$S_y(f_m)$, spectral density of fractional frequency fluctuations.

If the phase modulated signal is demodulated by using a phase detector, the output signal V is a function of the phase fluctuations of the incoming signal ($V_s = K_\phi \Delta\phi_{in}$ where K_ϕ is expressed in volts/radians).

On a spectrum analyser, the measured signal V_{rms} is proportional to $\Delta\phi_{rms}$. Then:

$$S_\phi(f_m) = \frac{\phi^2_{rms}(f_m)}{b} = \frac{V^2_{rms}(f_m)}{K^2_\phi b} = \frac{S_{vrms}(f_m)}{K^2_\phi} \tag{4.21}$$

(in rad^2/Hz) where $S_{vrms}(f_m)$ is the power spectral density of voltage fluctuations at the output of the phase detector.

$L(f_m)$, the single sideband noise to carrier is easily related to the RF power spectrum observed on a spectrum analyser. It is defined as the ratio of the power in one phase modulation sideband on a per hertz basis, to the total signal power. $L(f_m)$ is generally expressed logarithmically, in dB relative to carrier per hertz of bandwidth (dBc/Hz). $L(f)$ can be derived from $S_\phi(f_m)$ using phase modulation theory.

If we suppose a sinusoidal variation of the phase, the signal can be written:

$$V = A_o(\cos \omega_o t + \Delta\phi \sin \omega_m t)$$

where $\Delta\phi$ is the maximum deviation of the phase.

The instantaneous frequency f is:

$$f = f_o + \frac{1}{2\pi}\frac{d\varphi}{dt} = f_o + \Delta\phi f_m \cos \omega_m t.$$

The maximum deviation of the frequency is:

$$\Delta f = \Delta\phi f_m \tag{4.22}$$

or $\Delta\phi = \dfrac{\Delta f}{f_m}$ is the modulation index.

By using Bessel functions the spectrum of V can be written:

$$V(t) = A_o \sum_{n=-\infty}^{+\infty} J_n(\Delta\phi) \cos\,[\omega_o + n\omega_m]t. \tag{4.23}$$

If we suppose that $\Delta\phi \ll 1$ rad, then:

$$J_0(\Delta\phi) = 1 \qquad J_1(\Delta\phi) = \Delta\phi/2 \qquad J_{-1}(\Delta\phi) = -\Delta\phi/2$$

$$J_n(\Delta\phi) = 0 \quad \text{for } |n| \geqslant 2.$$

Equation (4.23) reduces to:

$$V(t) = A_o \cos \omega_o t + \frac{A_o\Delta\phi}{2}\cos(\omega_o + \omega_m)t - \frac{A_o\Delta\phi}{2}\cos(\omega_o - \omega_m)t \tag{4.24}$$

or

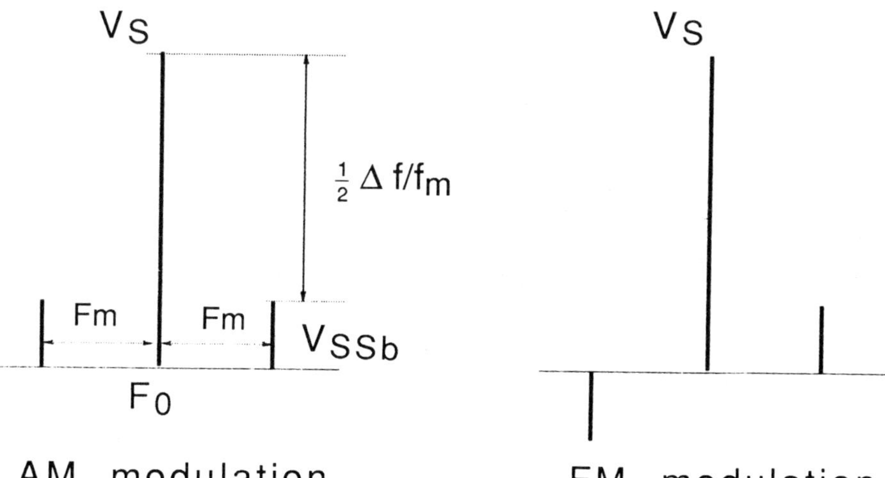

Fig. 4.9 Comparison between amplitude and phase modulation (side bands).

$$V(t) = A_o \cos(\omega_o t) - A_o \Delta\phi \sin\omega_o t \sin\omega_m t$$

$$V(t) = A_o \cos\omega_o t + A_o \Delta\phi \cos\left(\omega_o t + \frac{\pi}{2}\right) \sin\omega_m t. \quad (4.25)$$

From equations (4.24) and (4.25), it can be seen that phase modulation is in quadrature with the carrier and that the two side-band signals are in opposite phase (fig. 4.9).

The ratio of the single side band signal to the carrier is (equation 4.24).

$$\frac{V_{ssb}}{V_s} = \frac{\Delta\phi}{2}$$

or, in terms of power:

$$\frac{P_{ssb}(f_m)}{P_S} = \frac{V_{ssb}^2(f_m)}{V_S^2} = \frac{1}{4}(\Delta\phi)^2 = \frac{(\Delta\phi_{RMS})^2}{2}$$

or

$$L(f_m) \approx 20 \log \frac{\Delta\phi_{RMS}}{\sqrt{2}}.$$

Having measured the single side band power in the bandwidth b, imposed for example by the detector, it is often necessary to express it in another bandwidth (1 Hz for example).

As the power is proportional to b, then $\Delta\phi$ is proportional to $\sqrt{b}$ so that

$$\frac{\Delta\phi_1}{\sqrt{b_1}} = \frac{\Delta\phi_2}{\sqrt{b_2}}.$$

Then:

$$L_2(f_m) \approx 20 \log \frac{\Delta\phi_{rms}}{\sqrt{2}} = L_1(f_m) + 10 \log\left(\frac{b_2}{b_1}\right).$$

If, for example, the noise of an oscillator is 60 dBc in 1 kHz bandwidth, expressed in 1 Hz bandwidth the noise will be 90 dBc/Hz at the same distance from the carrier on a per hertz basis:

$$L(f_m) = \left(\frac{P_{ssb}(f_m)}{P_S}\right)_b = \frac{(\Delta\phi_{rms})^2}{2b} = \frac{S_\phi}{2} \qquad \left(\frac{\text{rad}^2}{\text{Hz}}\right). \quad (4.26)$$

The spectral density of frequency fluctuations is defined as:

$$S_{\Delta f}(f_m) = \frac{\Delta f_{rms}^2(f_m)}{b}.$$

By using equation (4.26):

$$S_{\Delta f}(f_m) = f_m^2 \frac{\Delta\phi_{rms}^2(f_m)}{b} = f_m^2 S_\phi(f_m) \qquad (\text{Hz}^2/\text{Hz}). \quad (4.27)$$

The spectral density of fractional frequency fluctuations allows direct comparison between sources of different carrier frequencies.
By defining

$$y = \frac{\Delta f(f_m)}{f_0}$$

and using equation (4.22) then

$$\begin{aligned} S_y(f_m) &= \frac{y_{rms}^2}{b} = \frac{f_m^2 \Delta\phi_{rms}^2(f_m)}{f_0^2\, b} \\ S_y(f_m) &= \frac{f_m^2}{f_0^2} S_\phi(f_m) \qquad (1/\mathrm{Hz}). \end{aligned} \tag{4.28}$$

The different measurement systems are presented in Hewlett-Packard (Seminar Handbook).

4.3 ACTIVE MICROWAVE SOLID-STATE COMPONENTS

Formerly, microwave generation was associated with vacuum tubes such as klystrons, magnetrons, etc. The characteristics of these devices were: thermionic emission, high voltages, reduced life and for some of them static magnetic field. Solid-state microwave components such as diodes or transistors have not led to the disappearance of tubes, but have opened new fields where size and consumption were the main characteristics even though output power is not very high (less than 1 to 10 W).
The usual components used are:

- Gunn diodes
- IMPATT diodes
- bipolar junction transistors (BJTs)
- field effect transistors (FETs).

The first two elements are described in chapter 10, and the last two in chapters 13 and 14 of volume 1.

4.3.1 Gunn diodes

Biasing Gunn diodes (Howell, 1987 (bis); Pattison)

A Gunn diode is basically a bulk GaAs device that exhibits negative resistance due to the two-band structure of GaAs when d.c. bias is applied: as the bias is increased from zero, the current increases to a maximum and begins to fall. The rate of fall is determined by contact technology and fabrication processes. Power is generated at about twice the threshold voltage, rises to a peak and then falls off at about four to five times the threshold.

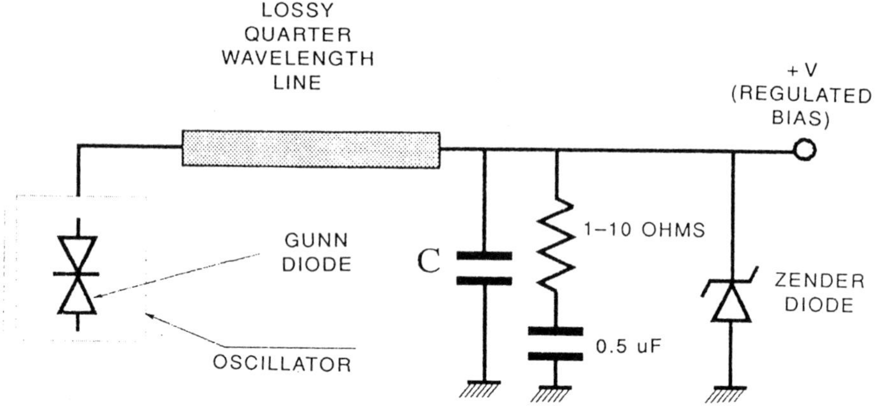

Fig. 4.10 Typical diode biasing circuit.

If allowed to, Gunn diodes can oscillate at many frequencies from VHF to microwave and millimetre range using both microwave and bias circuit resonances. By adding a microwave resonant circuit to the diode so that the total conductance is zero and adding proper bias circuits, it is possible to limit the range of frequency oscillations. However, all microwave circuits such as waveguide, coaxial lines, etc., have more than one resonance, and the art of designing Gunn diode oscillators is to limit the oscillating frequencies to the band of interest while maintaining the required stability.

For biasing, to avoid low frequency oscillations, a lossy low-pass filter can be used to eliminate RF leakage into the bias lines. A typical bypass circuit is shown in Fig. 4.10.

Another problem with Gunn diodes is that once the oscillation has started, if the bias voltage is reduced, there is hysteresis in the power output: as the voltage is decreased, the power remains at a higher value than when the diode was first turned on. This can lead to cold turn-on problems if the power is turned on and the voltage is decreased to give the required level. When the unit is turned off, and temperature reduced, the diode will not start up at a voltage less than the cold start-up voltage.

Using Gunn diodes (Sweet, 1974)

Power can be obtained from Gunn diodes in the frequency range 10 to 110 GHz from either GaAs or InP.

GaAs Gunn diodes operate in the fundamental mode only up to about 70 GHz. Above this frequency, they oscillate at a sub-harmonic of the output frequency and the desired harmonic is coupled from the circuit.

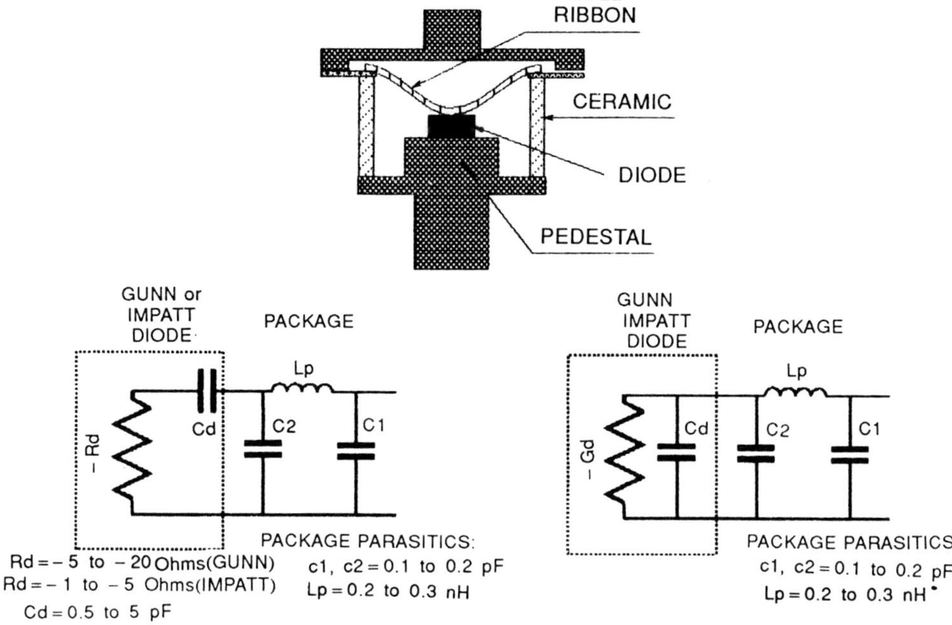

Fig. 4.11 Packaged diode and equivalent circuits.

Major work has been devoted to CW GaAs devices with power levels from 10 to 100 mW, but diodes with high powers (either CW or pulsed) have also been developed with power of more than 2 W at 8 GHz for a simple chip mounted on a diamond heat sink. The efficiency of these devices is in the region of 2 to 12% for the frequency range 10 to 75 GHz. For pulsed GaAs devices, up to 30 W in X-band have been obtained with 10% efficiency at low duty cycle.

InP has higher peak to valley ratios than GaAs leading to efficiencies that are generally twice that of GaAs devices.

These components are small (diameters in the range 50 to 400 μm, thickness 10 to 100 μm), the power density is very high and thus the power to be dissipated is high. It is then necessary to use heat sinking with a good thermal conductor, for example gold, silver or diamond to reduce the thermal resistance as much as possible. This means that Gunn diodes are usually sold in packaged form. A cutaway view with equivalent circuit is shown on Fig. 4.11

4.3.2 Using transistors

Obtaining negative resistance from transistors

An advantage of transistors compared to diodes is that they are three-port devices with gain. They are not, by themselves, negative resistance devices.

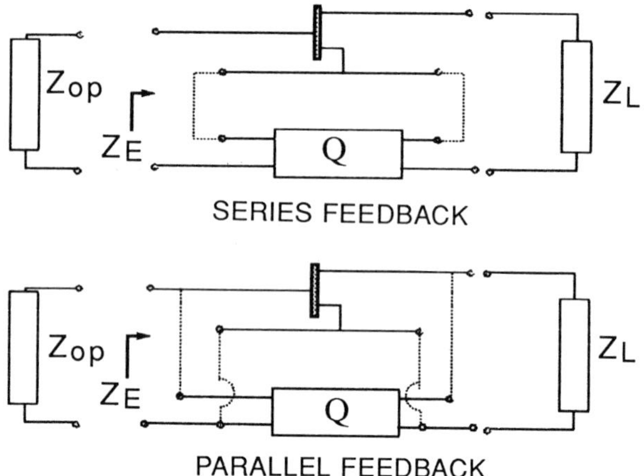

Fig. 4.12 Obtaining negative resistance with feedback on transistors.

It is necessary to add circuits to obtain negative resistance between two ports of the active device in a limited range of frequencies.

As bipolar transistors and FETs have almost the same equivalent circuit (except for the values of capacitance), all discussions on FETs will be directly applicable to bipolar transistors.

The easiest way to analyse negative resistance is to look at the input impedance seen between gate and source. Of course, any other analysis, for example, between drain and source, will lead to the same results.

As shown in Fig. 4.12, there are two possible ways of obtaining negative resistance at the gate: 'series' feedback and parallel feedback. In both cases, when Z_L is connected at the output, the input impedance Z_E has a negative real part and Z_{OP} is the impedance required to oscillate at ω_o such that:

$$\text{Re}(Z_{op}(\omega_o)) + \text{Re}(Z_e(\omega_o)) = 0$$

$$\text{Im}(Z_{op}(\omega_o)) + \text{Im}(Z_e(\omega_o)) = 0$$

These two different circuits can be analysed by using the simplified equivalent circuit of the FET.

Reverse channel FETs

With FETs only, another method to obtain negative resistance is to use a new technique called reverse channel.

In common source, the drain is biased positive with respect to the source and the gate is negatively biased. The electron flow is then from source to drain and the depletion area is larger under the highest field drain region. By reversing the channel, the electron flow is in the opposite direction and the depletion region is larger near the grounded contact which is the drain, result-

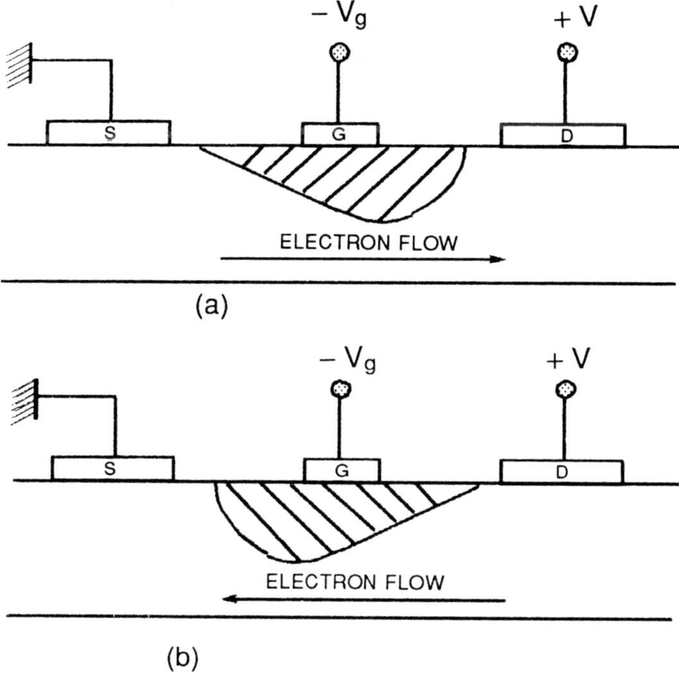

Fig. 4.13 (a) Common source and (b) reverse channel biasing of FETs.

ing in common drain operation which is potentially unstable in a large frequency band (Fig. 4.13). This makes the oscillations possible without any external feedback. Although requiring negative voltage supply, this technique can be used for normally available, common source packaged FETs.

Maximum power of oscillation (Johnson, 1979)

For a FET used as an amplifier, the output power is given by:

$$P_{\text{out}} = P_{\text{sat}}\left[1 - \exp\left(\frac{G_{\text{o}}P_{\text{in}}}{P_{\text{sat}}}\right)\right] \tag{4.29}$$

where P_{sat} is the saturated output power, G_{o} is the small signal gain (or linear gain), and P_{in} is the input power.

For an oscillator, the maximum power occurs at the point where $P_{\text{out}} - P_{\text{in}}$ is maximum, or $\partial P_{\text{out}}/\partial P_{\text{in}} = 1$. By taking the derivative of equation (4.29), the optimum input power can be calculated:

$$P_{\text{inopt}} = P_{\text{sat}}\frac{\ln G_{\text{o}}}{G_{\text{o}}}$$

and then

$$P_{\text{oscmax}} = P_{\text{outopt}} - P_{\text{inopt}}$$
$$P_{\text{oscmax}} = P_{\text{sat}}\left(1 - \frac{1}{G_o} - \frac{\ln G_o}{G_o}\right). \tag{4.30}$$

Thus, for example, an FET having a small signal gain of $G_o = 7.5$ dB with a saturated output power of 1 W would be capable of a maximum oscillator power of 515 mW.

4.3.3 Comparison between diodes and transistors

The main characteristics of each component are summarized in Table 4.1.

Table 4.1 Diode and transistor component's characteristics

	Gunn diode	*IMPATT diode*	*Bipolar transistors*	*Field-effect transistor*
Maximum frequency of oscillation (GHz)	< 70 fundamental mode > 100 harmonics	> 300	commercially: < 8 available Labs: < 20	⩽ 60 (Near future: ⩽ 94)
Output power	2 W X-band CW 200 mW 70 GHz	10–20 W X-band 1 W 200 GHz	50 mW @ 8 GHz	1–3 W X-band 100 mW 40 GHz
Pulsed conditions improvements in power	Yes	Yes	?	?
Efficiency (%)	2–10	10–20	10–15	20–30
Biasing voltages (V)	10	50–80	10–15	5–10
Threshold	Yes	Yes	No	No
Negative resistance	Very broadband	Very broadband	Depends on external circuits	Depends on external circuits
Low frequency problems	Yes	Yes	No	No
Chip/package form	Package	Package	Chip or package	Chip or package
Thermal dissipation	–	+	Best	++
FM noise	Very good	Fair	Very good	Good (can be very good)

4.4 FIXED FREQUENCY OSCILLATORS

These oscillators are required for a variety of applications, including local oscillators, radars, communications transmitters, etc. To be usable, two important characteristics are needed: low frequency drift with temperature and the lowest FM noise possible.

To make such a stable oscillator, two elements are necessary (Section 4.2):

1. negative resistance realized with an active device (diode or transistor)
2. a load.

This load can be created by low Q circuits such as microstrip lines on dielectric substrates. But, due to the low Q, the frequency and power stability are poor. By using impedance matching, the circuit can be calculated to have the good impedance value at the right frequency. Bias filters can also be directly engraved on the substrate.

Apart from the problem of stability, the high thermal impedance of the circuits makes direct insertion of Gunn or IMPATT diodes very difficult without adequate heat sinking of some form. So, to make good fixed frequency oscillators, high Q loads with very low temperature variations must be used in conjunction with a microwave non-linear component. Such a good load can be created by a microwave cavity used either in reflection or in transmission. Two different types of cavities are usually used: metallic cavities or dielectric resonant cavities. With Gunn or IMPATT diodes, both kinds of cavities are used, whereas only dielectric resonators are used with transistors.

4.4.1 Diodes/metallic cavities oscillators (Chang and Ebert, 1980; Sigmon and Ayya Gari, 1987)

Two kinds of metallic cavities can be used: coaxial cavities or waveguide cavities.

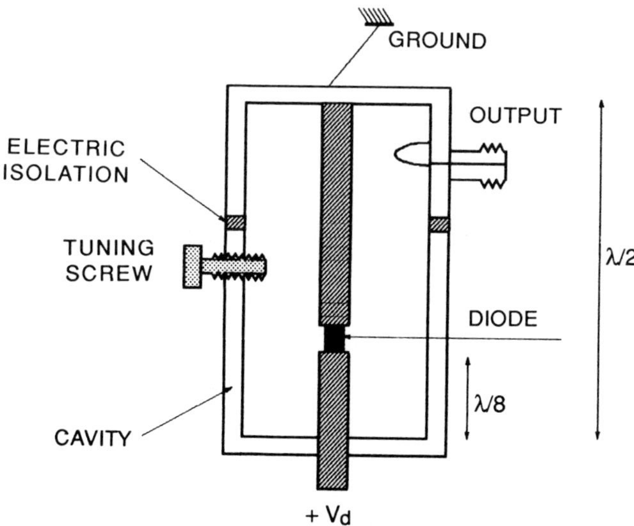

Fig. 4.14 Basic coaxial oscillator.

Coaxial oscillators (Plessey)

The fundamental circuit of such oscillators is presented in Fig. 4.14. In this configuration, the diode is placed between two parts forming the coaxial part of the cavity. The important criteria are as follows.

1. The diameter of the inner coaxial part should be large enough to mount the device with a minimum discontinuity capacitance.
2. The diameter of the outer part should be small enough to eliminate waveguide modes.
3. The unloaded Q factor should be as high as possible.

The diode is mounted close to the end wall of the cavity (very low electrical field) and the power extracted using electric or magnetic coupling.

The frequency is set by the cavity length ($L = \lambda/2$) and adjusted by a screw, preferably at the mid-point where the E-field reaches a maximum. Mechanical tuning can also be obtained by using a plunger modifying the length of the cavity. Power output is adjusted by moving the probe inside the cavity. The simplicity, price and ease of tuning of coaxial fixed frequency oscillations is usually outweighed by their low Q_L, typically 50, and hence low stability for most applications.

In addition, harmonic frequencies can exist at which the diode will produce power, and at high frequencies the losses become severe. To eliminate spurious modes and/or high frequency oscillations and also to improve noise bias, a choke can be directly integrated in the cavity. RF absorber is used to prevent any leakage.

Waveguide oscillators: basic designs

These circuits are the most used due to a reasonable cost and to a relatively high Q factor giving a good frequency stability. The diode is mounted between posts either at $\lambda_g/2$ away from short circuit (Fig. 4.15(a)) or closed to a short circuit, approximately $\lambda_g/2$ from a thin iris which couples into the load (Fig. 4.15(b)). In both cases, frequency and output power are adjusted with screws or variable shorts. In the iris oscillator, the output coupling is set by the iris dimensions and frequency tuning is generally done by a dielectric plunger between diode and iris.

4.4.2 The dielectric resonator

In 1939, R. D. Richtmyer from Stanford University showed that non-metallized dielectric objects, such as spheres or cylinders, could work as microwave resonators. In 1960, some oscillators were made with rutile (TiO_2) as dielectric resonator. But, with this high permittivity material ($\varepsilon_r = 100$), the temperature stability is not very good (~ 100 ppm/°C).

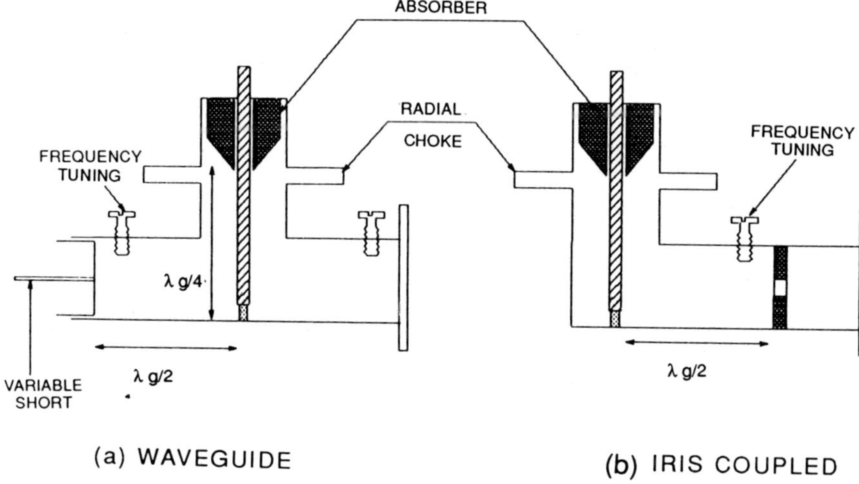

Fig. 4.15 Waveguide oscillators; (a) waveguide, (b) iris coupled.

Raytheon, then Bell Labs, made ceramics based on barium titanate (ε_r is 38, $Q_o > 4000$), very stable in temperature. In parallel, Murata and Thomson-CSF have developed such a ceramic but based on zirconium titanate. Its temperature coefficient can be varied from -4 to $+10$ ppm/°C by adding impurities. The Q factor is about 4000 up to 10 GHz.

The main advantage of this component is its compatibility with hybrid technology. Another advantage is the choice in the temperature coefficient of the resonator in order to compensate the drift in frequency of the oscillator or the filter on a large temperature range, typically from -55 to $+85$ °C.

The main available materials and manufacturers are shown in Table 4.2

Table 4.2 Some materials and manufacturers of dielectric resonators

Material composition	*Manufacturer*	*Permittivity* ε	*Losses* ($\tan\delta$)	*Temperature coefficient* (ppm/°C)
$BaTi_4O_9$	Raytheon Transtech	38	1×10^{-4}	+4
$BaTi_9O_{20}$	Bell Labs	40	1×10^{-4}	+2
$(ZrSn)TiO_4$	Murata Thomson CSF Siemens Transtech MTK	38	1×10^{-4}	adjustable from $-4-+10$
$Ba(Zn_{1/3}Nb_{2/3})O_2$	Panasonic	30	4×10^{-5}	adjustable from $0-+10$
$Ba(Zn_{1/3}Ta_{2/3})O_2$	Murata			

The characteristics of the dielectric resonator are identical to those of a resonant metallized cavity but, due to their high permittivity, the size is much smaller. The volume reduction is by about a factor of six.

Frequency limitations of dielectric resonators are of two types: for $f < 1$ GHz size becomes too big to be usable, and for $f > 100$ GHz, the Q factor becomes too small. Taking into account these limitations gives the useful frequency range of dielectric resonators: from 4 to 40 GHz.

Depending on the diameter to height ratio, two different modes can exist: HE_{111} or $TE_{01\delta}$ which is the most widely used because it is easily coupled to microstrip lines (Konishi, 1976; Kobayashi, 1981).

In effect, as the energy is not completely stored inside the resonator, the magnetic field extending outside the puck can be used either to couple it or to frequency tune it by changing the distance of a metallic grounded piece to the resonator. To avoid radiated emission of the resonator strongly reducing the Q-factor, it is necessary to use it inside a metallic package which, in return, has a great influence on oscillator performance such as temperature drift, FM noise, stability with vibrations, etc. The dielectric resonator is either screwed through a cylindrical hole or attached by using adhesives.

4.4.3 Dielectric resonator oscillators (DROs)

According to the oscillating mechanism, DROs can be classified into three basic types represented in Fig. 4.16 (a), (b), (c).

1. Reflection type: the resonator is used as a matching circuit and an output filter.
2. Shunt feedback: the resonator is used as a feedback element with an active component with enough gain.
3. Series feedback: the resonator ensures the oscillation conditions with a negative resistance (either diode or transistor).

Table 4.3 presents compared performance of the different circuits:

Table 4.3 Performances of different DRO circuits

	Reflection	*Series feedback*	*Shunt feedback*
Mechanical tuning range	Fair	Good	Poor
FM noise	Fair	Good	Best
Output power	Good	Good	Fair
Load pulling	Poor	Best	Good

4.5 ELECTRONICALLY TUNEABLE OSCILLATORS

4.5.1 Main characteristics

These oscillators are used either for ECM systems or for frequency sweepers. Important characteristics are (Kaminsky, 1986):

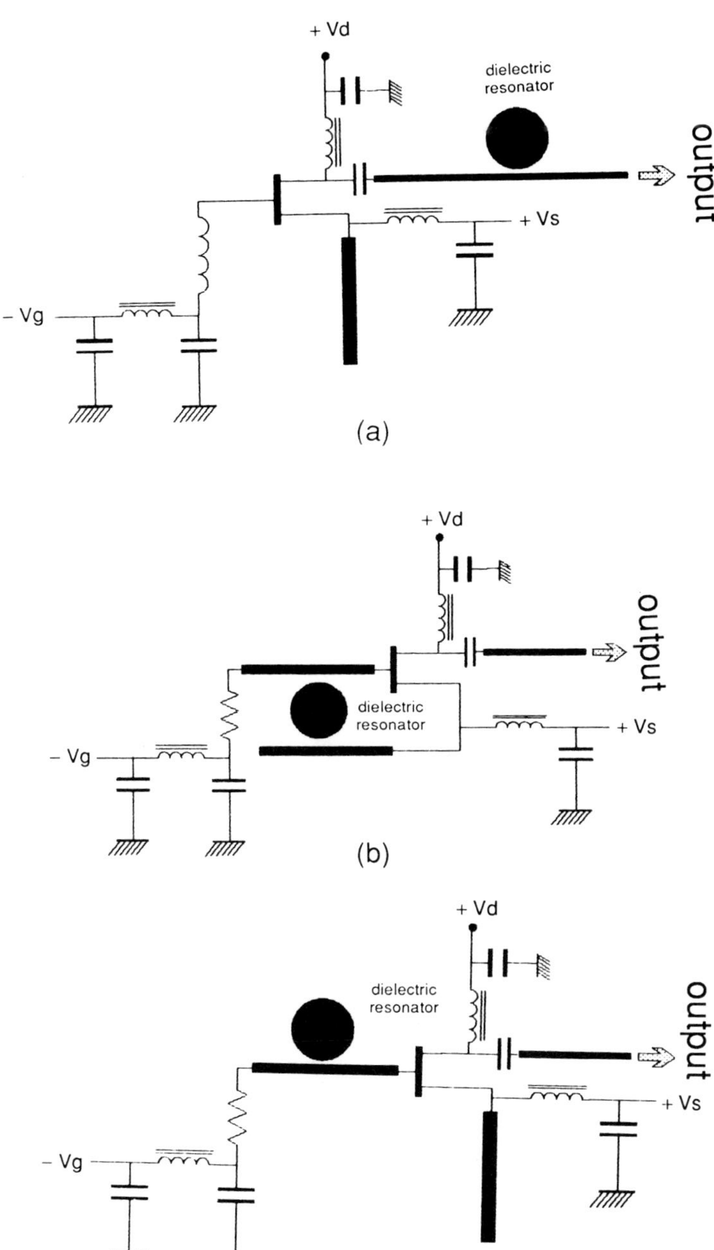

Fig. 4.16 The three basic dielectric resonator oscillator circuits; (a) reflection FET dielectric resonator oscillator, (b) shunt feedback FET dielectric resonator oscillator, (c) series feedback FET dielectric resonator oscillator.

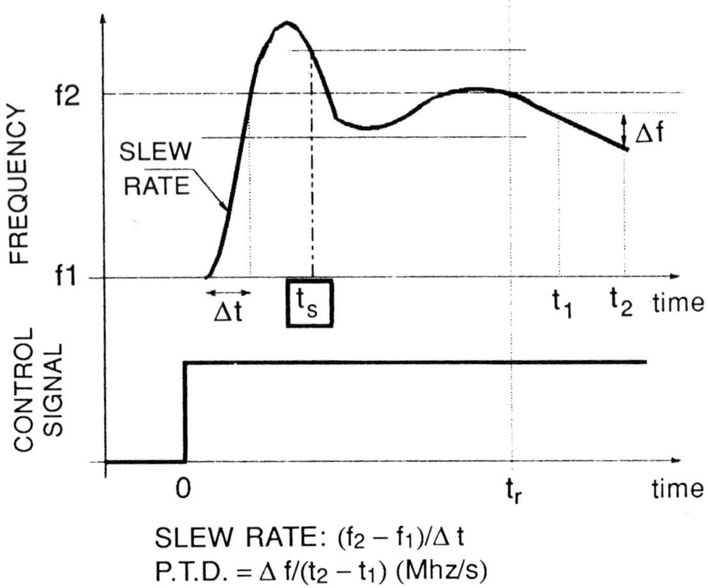

Fig. 4.17 Switching characteristics of tuneable oscillators.

Bandwidth
In general, coverage as broad as possible is needed.

Speed
That is the ability of the source to change (and to obtain) the needed frequency as quickly as possible. Three parameters are used to describe this point.

1. Slew rate expressed in mega or gigahertz per microsecond (GHz/μs). This is the rate at which the frequency of the oscillator can be changed from one end of the tuning range to the other in response to a step change in the control parameter.
2. Settling time (t_s): this is the minimum time after the change of the control parameter to obtain the frequency within a given small band around final frequency measured at reference time t_r.
3. Post tuning drift (PTD) is the drift of the frequency during a time interval large compared to the settling time.

All three parameters are described in Fig. 4.17.

Linearity
For many applications, the curve of frequency versus control signal should be as linear as possible. A measure of this characteristic is the maximum frequency deviation between experimental curve and best linear fit.

Another measure is the sensitivity or the local slope of the frequency versus control signal, expressed in megahertz per volt or per milliampere.

Only two basic concepts are now available to create such oscillators: YIG and varactor tuning. With diodes, only the last one is used now. It is also possible to obtain electronic tuning on the active device itself by modifying the bias voltage. But, as the negative impedance is changed, power, FM noise and stability can be affected. Therefore this technique is not very widely used.

4.5.2 Varactors

These are the most widely used tuning elements (see Vol I, Chapter 9 for complete description). This can take the form of a p-n junction, but the most common varactor is the metal-n type Schottky barrier usually fabricated on GaAs or on silicon. In this component, the reverse biasing causes the depletion barrier to increase, acting as a variable capacitor (Leier and Patston, 1985).

Whether silicon or GaAs, there are two basic types of varactors: abrupt and hyperabrupt. For an abrupt varactor, the concentration of n-type dopant is nearly constant across the depletion region and is non-linear for a hyperabrupt one.

The junction capacitance versus applied voltage is given by:

$$C(V) = \frac{C(0)}{\left(1 + \frac{V}{\phi}\right)^{\gamma}} \tag{4.31}$$

where $C(0)$ is the junction capacitance at 0 V, V is the applied voltage, ϕ contact potential and γ is a constant related to the doping of the device whose value is about 0.5 for abrupt junction, and 1.15 to 1.5 for hyperabrupt junction varactors. As the resonant frequency is inversely proportional to the square root of the capacitance, a doping profile with $\gamma = 2$ will give a linear tuning curve.

The abrupt diode will provide a very high Q and will operate on a very large range of tuning voltages (0 to 50 V). On the other hand, the hyperabrupt diode, due to its linear tuning characteristics, gives a much more linear tuning response and covers a wider frequency range in a smaller tuning voltage range (0 to 20 V). As the Q is much lower than that of abrupt varactors, the phase noise is reduced.

When control voltage is changed, the frequency variation induces changes in the junction temperature of transistor and varactor causing impedance changes and frequency drift. This drift time is directly dependent on the thermal resistance of the device. As the thermal resistance of silicon is better than that of GaAs, settling time and PTD are improved by using silicon varactors. Another effect takes place with GaAs varactors: traps in the junction and impurity build-ups around the junction.

Performance of oscillators as a function of the varactor used are presented in Table 4.4. When the varactor is in packaged form, the inductance of the connections, the capacitance of the package and the losses of the varactor, tend to reduce the capacitance ratio. It is thus more convenient to use chip varactors, when possible, to stretch the bandwidth. As the tuning is controlled by the voltage, oscillators made with these elements are often called voltage controlled oscillators, or VCOs.

Table 4.4 Performance function of varactors

	Si-abrupt	*Si-hyperabrupt*	*GaAs-abrupt*	*GaAs-hyperabrupt*
Linearity	Fair	Good	Fair	Good
Tuning voltage (V)	0–60	0–20	0–50	0–20
Phase noise	Very good	Good	Good	Fair
Temperature stability	Very good	Good	Excellent	Fair
Settling time	Excellent	Excellent	Good	Fair
PTD	Excellent	Excellent	Good	Fair

4.5.3 Diode/varactor oscillators

Available bandwidth is determined by the loaded Q-factor, the capacitance variation of the varactor and the degree of coupling between varactor and circuit. A simple equivalent circuit of the oscillator is shown in Fig. 4.18.

It can be shown that the maximum available tuning range is given by:

$$\Delta F = \frac{-\Delta C_{\mathrm{j}} V^2 F^2 \pi^2}{Q_{\mathrm{L}} P_{\mathrm{T}}} \tag{4.32}$$

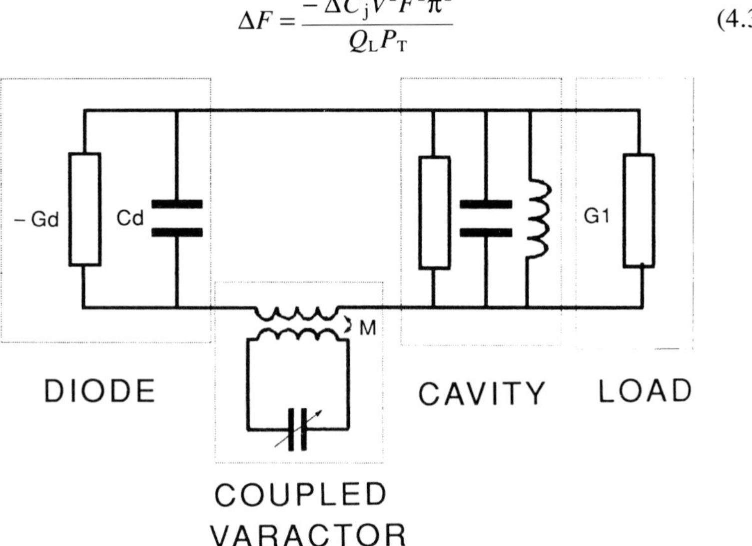

Fig. 4.18 Equivalent circuit of the diode/varactor oscillator.

where ΔC_j is the capacitance variation, V is the RF voltage swing across capacitor, F is the frequency, Q_L is the loaded Q-factor and P_T is the total power dissipated in varactor and circuit. In order to obtain the largest possible bandwidth, Q_L should be minimized. This reduces stability and FM noise. Increasing the voltage swing would, in fact, reduce the capacitance variation. Parasitic elements of the package as well as series resistance of the varactor will also reduce the capacitance variation.

Metallic cavity oscillators

As for fixed frequency oscillators, the same approach can be used to make VCOs by coupling a varactor either by magnetic loop or by a correct positioning of the varactor in the cavity or in the waveguide. A 'correct positioning' is in general obtained by the cut and try method. Different arrangements for coaxial oscillators are presented in Fig. 4.19 and in Fig. 4.20 for waveguide assemblies.

For waveguide VCOs, there are two basic designs: the varactor can be mounted in the same E-plane as the Gunn or IMPATT diode (Fig. 4.20(a)) or mounted near the short circuit (Fig. 4.20(b)).

The coaxial cavities are capable of producing 5 to 10% bandwidth with up to 100 mW (20 dBm) output power with a frequency temperature coefficient of − 1 MHz/°C at X-band.

Waveguide oscillators have higher Q-factor (500 to 1000) leading to good stabilities. They are the best way to obtain low noise narrow band oscillators provided that coupling is light. Typical bandwidths are 0.5 to 1%, with stability of about 0.3 MHz/°C at X-band at up to 300 mW.

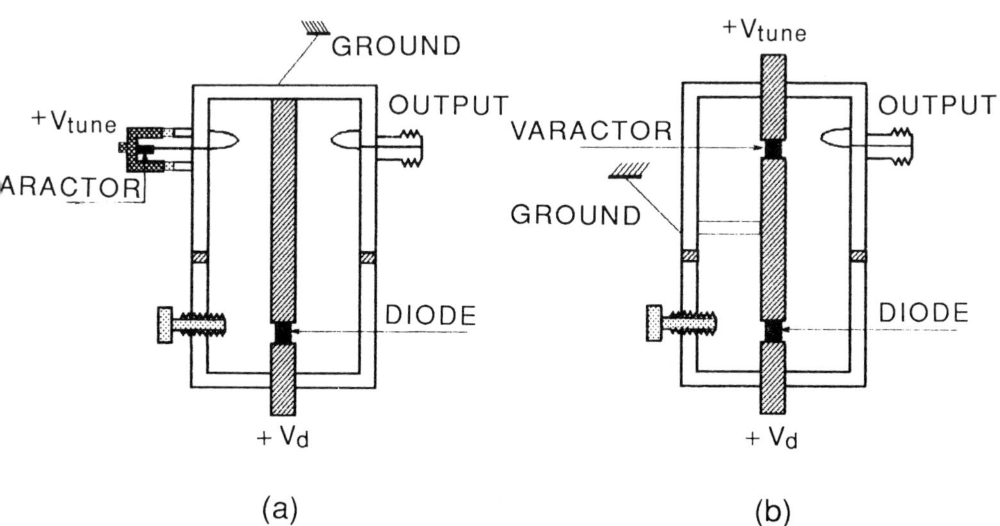

Fig. 4.19 Coaxial varactor tuned basic designs.

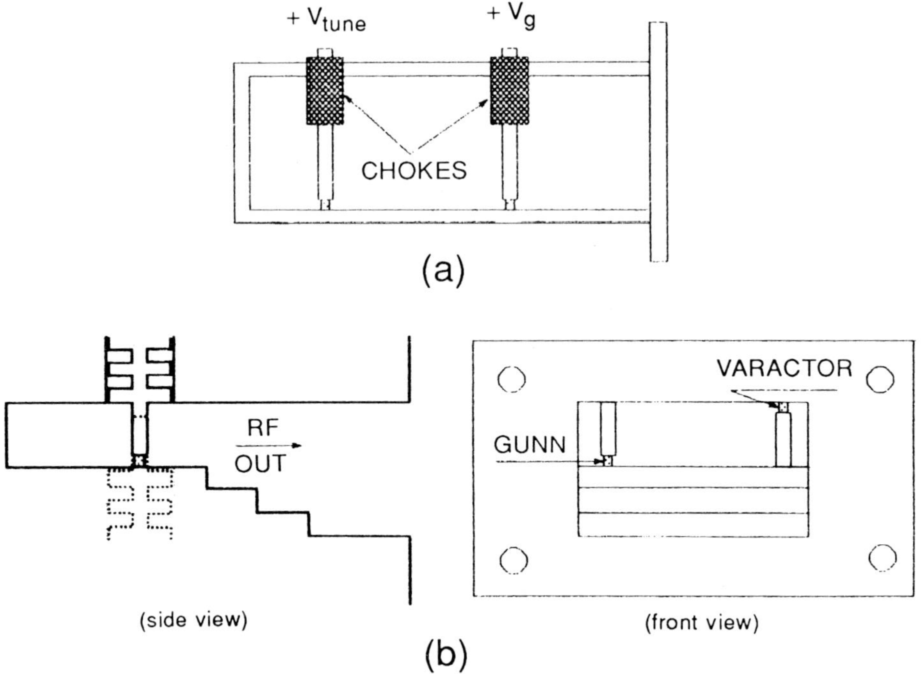

Fig. 4.20 Waveguide varactor tuned basic designs.

Microstrip/diode oscillators

Considerable advantages of microstrip oscillators are size, weight and ease of manufacture as well as low Q-factor. This low Q-factor, unsuitable for low-noise applications, is interesting for obtaining broad-band VCOs. But as Gunn or IMPATT diodes are less and less used under 60 GHz and as microstrip circuits are difficult to use above 40 GHz, these microstrip diode oscillators are disappearing, for the moment.

To oscillate, the condition is $-R_d + \mathrm{Re}Z_{in} \leqslant 0$. As $-R_d$ is of the order of -5 to $-10\,\Omega$, an impedance transformer is needed between the diode and the load, generally $50\,\Omega$. A good impedance transformer is made with coupled microstrip lines. The main advantages of this circuit are as follows.

1. Modification of the imaginary part of the impedance by modifying the length, without changing the real part. The oscillating frequency can be adjusted with same output power.
2. Optimum power can be adjusted by changing distance between two microstrip lines.
3. Such an impedance transformer can be used to obtain a small real part (1 to 5 Ω) needed for example for IMPATT diodes.
4. No decoupling capacitor between bias and RF output.

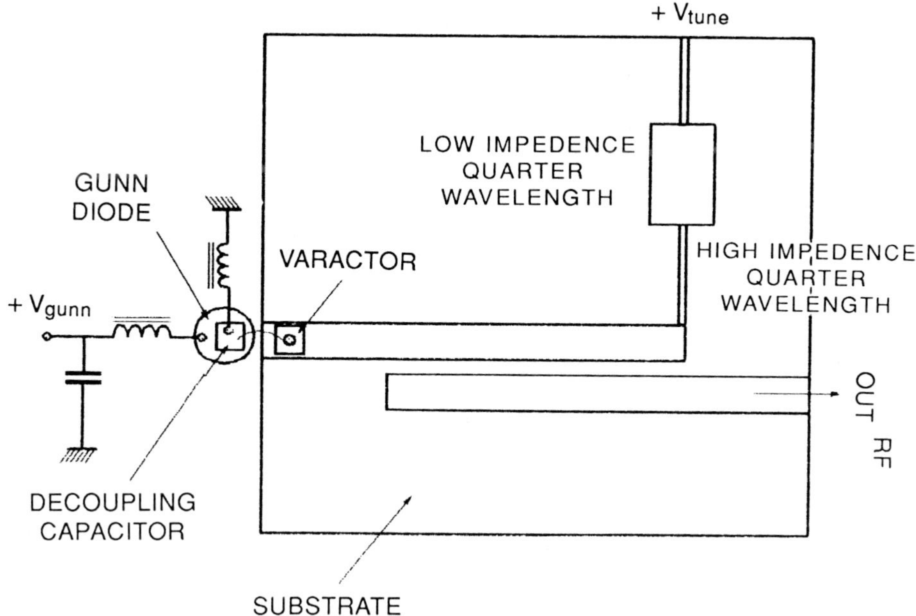

Fig. 4.21 Typical Gunn/microstrip VCO.

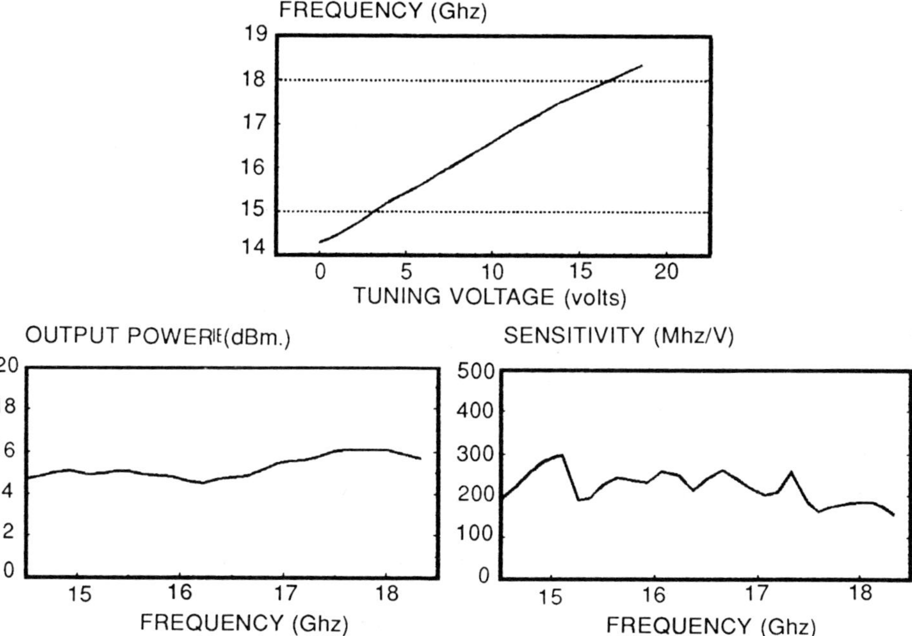

Fig. 4.22 15 to 18 GHz microstrip/Gunn VCO.

A typical circuit using coupled microstrip lines is presented in Fig. 4.21. Results concerning a 15 to 18 GHz oscillator are presented in Fig. 4.22.

4.5.4 Transistor/varactor oscillators (VCOs)

The three main characteristics of VCOs to be usable in ECM systems are:

1. bandwidth;
2. FM noise;
3. settling time and PTD.

Of course, all these points cannot be optimized all together. Compromises must be made. For example, to reduce settling time and PTD, care must be taken with assembly technologies and on structure of the buffer amplifier (Boyd, 1986). But that is not enough. As seen in section 4.5.2, choice of the varactor is very important. Best results are obtained with abrupt silicon varactors but due to low Q, maximum frequency is not very high and bandwidth is small. To increase bandwidth and maximum frequency, GaAs hyperabrupt varactors must be used but settling time is increased. In this case, a better solution is to use two (or more) switched VCOs to cover all the needed bandwidth without sacrificing on settling time.

Another example concerns FM noise: good results can be obtained with bipolar transistors compared to FETs. But it is very difficult to find bipolar transistors above 10 GHz. So, depending on the needs, the choice of the circuit will be different (Bender and Wong, 1983).

In general, the bandwidth and performance of VCOs are a function of:

1. bipolar or field effect transistor;
2. position of the varactor;
3. type of varactors;
4. number of varactors.

To obtain the bandwidth, linear software analysis can be performed. Transistors are described by their scattering matrix and oscillating frequency is found by superposing the negative impedance on the complex conjugated impedance of the load obtained by describing it with negative value of inductances and capacitances. The intersecting point of both imaginary parts ($\mathrm{Im}(Z) = -\mathrm{Im}(Z_c)$) gives the oscillating frequency provided that $-|\mathrm{Re}(Z)| + \mathrm{Re}(Z_c) \leqslant 0$. Typical results of a 9 to 18 GHz, two-varactor buffered oscillator are shown in Fig. 4.23 for the circuit shown in Fig. 4.24.

4.5.5 YIGs (Helszajn, 1985)

The main mechanism is the ferromagnetic resonance which appears when a small magnetic microwave field h is applied perpendicular to a static magnetic field H_o. In microwave frequency ranges (S to Ku-band), spheres of YIG (yttrium iron garnet $Y_3Fe_5O_{11}$) are generally used.

The microwave resonance occurs when the frequency of the RF magnetic field coincides with the natural electron dipole precessional frequency of the material. The dipole resonant frequency for a spherical resonator is:

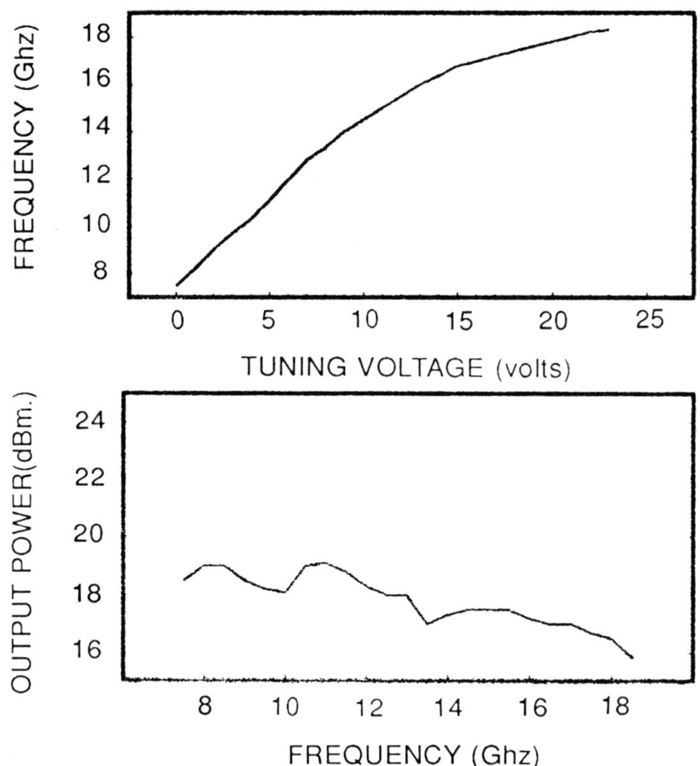

Fig. 4.23 Typical characteristics of a 9 to 18 GHz VCO.

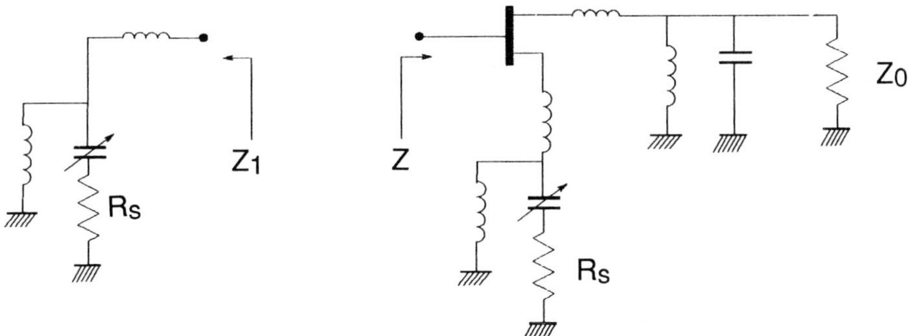

Fig. 4.24 Two varactor FET oscillator.

$$F_r = \gamma(H_o + X h_{an})$$

where γ is the gyromagnetic ratio of an electron (2.8 MHz/Oe), H_o the applied static magnetic field, h_{an} is the internal magnetic field of the crystal and X is a coefficient depending on position of crystal axes of the sphere of the magnetic field ($-1 \leqslant X \leqslant +1$).

There are two consequences. First, resonance frequency is a function of the position of the sphere, and second, the need for a magnetic field implies the use of an electromagnet to modify the oscillating frequency. The structure of such an oscillator is shown in Fig. 4.25.

The coupling between the sphere and the microwave circuit is made by a loop around the sphere (magnetic coupling). By properly orienting the sphere in the static magnetic field, frequency drift with temperature can be strongly reduced (air-gap compensation).

For the magnetic circuit, high nickel content magnetic steels such as supranhyster 36 or 50 are used because they offer the best compromise between the maximum field before saturation (~ 10 000 gauss) and small hysteresis after thermal treatment.

As the inductance of the tuning coil and thus the stored energy is large, the modulation bandwidth is small (some kilohertz). To increase this bandwidth, a small air coil is added. The frequency deviation is small (some megahertz) but the modulation bandwidth can reach 1 to 2 MHz.

YIG oscillators

The YIG sphere, coupled to the circuit, behaves exactly as a resonant cavity having a Q-factor of 1000 to 2000 leading to low FM noise oscillators. As the

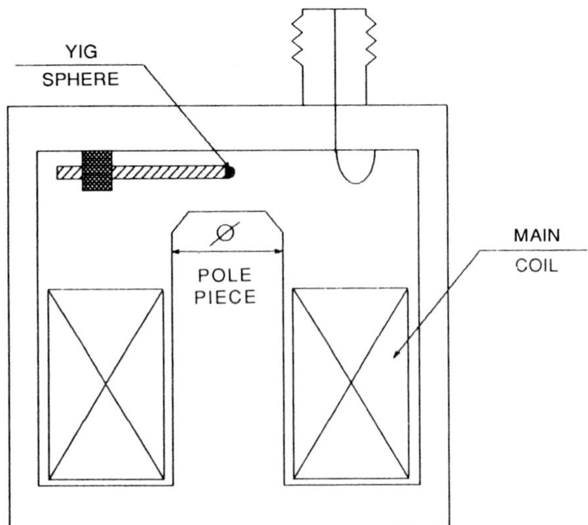

Fig. 4.25 Structure of a YIG oscillator.

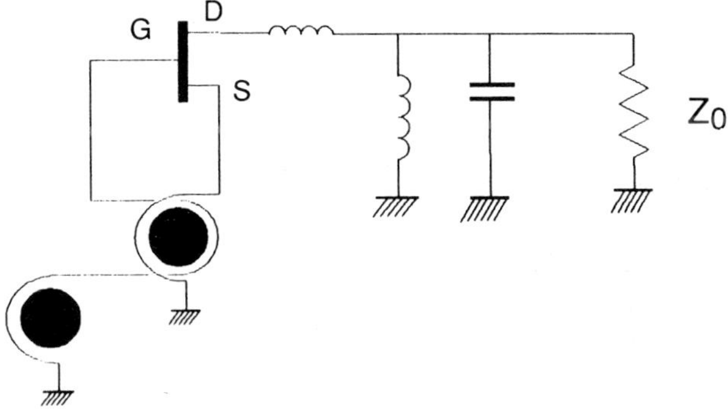

Fig. 4.26 Improved two YIGs oscillator.

imaginary part of its input impedance can be either positive (inductive) or negative (capacitive), the sphere can be put either in the gate (base) or in the source (emitter) or both.

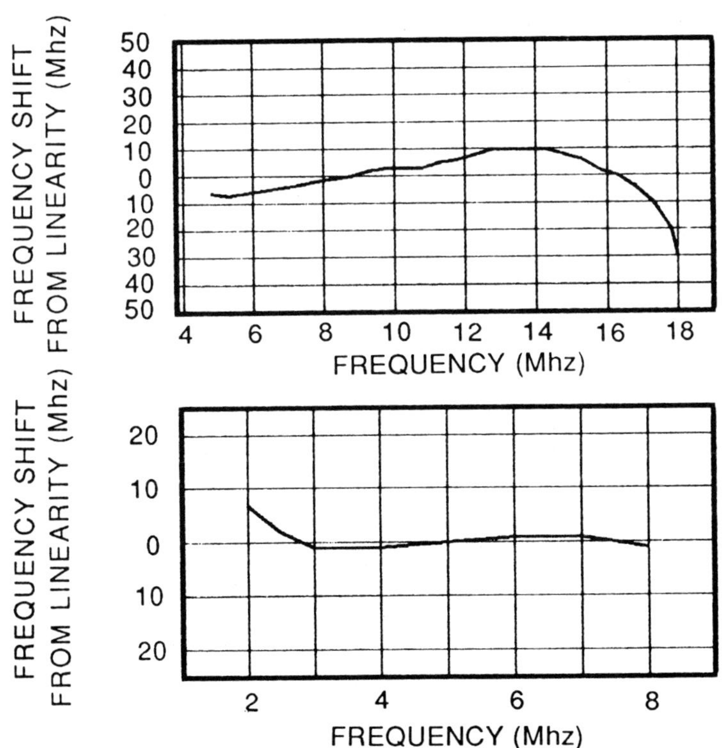

Fig. 4.27 Frequency shift from linearity for 2 to 8 and 6 to 18 GHz YIG oscillators.

Recently a new topology has been presented: a two-sphere oscillator with one double-coupled to gate and source (Fig. 4.26). The main advantages are as follows.

1. For the same inductances, the frequency bandwidth is stretched toward low frequencies.
2. Frequency shift from linearity is improved.
3. Maximum frequency is increased.
4. Efficient temperature compensation is possible.

The results concerning two Thomson-CSF YIG oscillators are presented in Fig. 4.27.

4.5.6 Comparison between VCOs and YIG oscillators

According to the needs (for example, low FM noise versus modulation bandwidth), the system designer will have to choose one or the other. In Table 4.5, a comparison between VCOs and YIGs is presented.

Table 4.5 Comparison between VCOs and YIGs

	VCO	*YIG*
Frequency	Voltage control	Current control
Mechanical tuning	None	None
Bandwidth	1 octave	2 (or more) octaves
Volume	Small	Large (coil)
Pulling	Buffer amplifier	Buffer amplifier
FM noise	Fair	Good
Harmonics	Good	Fair
Linearity	Fair	Excellent
Modulation bandwidth	Good	Main coil: fair Small air coil: better
Settling time	Good (?)	Fair
Weight	Light	Heavy

BIBLIOGRAPHY

Abe H., Takayama Y., Higashisaka A. and Takamizawa H. (1978) 'A highly stabilised low-noise GaAs FET integrated oscillator with a dielectric resonator in the C-band' *IEEE*, **MTT 26**, N° 3.

Anderson K. J. and Pavio A. M. (1983) 'FET oscillators still require modelling, but computer techniques simplify the task' *Microwave Systems News*, **13**, N° 9, 60–72.

Basawapatna G. R. and Stancliff R. B. (1979) 'A unified approach to the design of wide-band microwave solid state oscillators' *IEEE Trans. on Microwave Theory and Techniques*, **MTT 27**, 379–385.

Bender J. R. and Wong C. (1983) 'Push-push design extends bipolar frequency range' *Microwaves and RF*, 91, 98.

Boyd D. A. (1986) 'Design Considerations for Post-tuning Drift Reduction in Modern VCO Subsystems' *Microwave Journal*, 121, 130.

Camiade M., Bert A., Graffeuil J. and Pataut G. (1983) 'Low-noise Design of Dielectric Resonator FET Oscillators' *Proceedings of the 13th European Microwave Conference, Nuremberg.*

Chang K. and Ebert R. L. (1980) 'W-band Power Combiner Design' *IEEE Trans.*, **MTT 28 (4)**, 295–305.

Chen S., Chang L. and Chin J. (1986) 'A Unified Design of Dielectric Resonator Oscillators for Telecommunication Systems' *IEEE MTT-S Digest*, 593, 596.

Edson W. A. (1960) 'Noise in oscillators', Proc. of I.R.E., **48**, N° 8.

Plessey on 'Gunn Effect Technology' Technical publication optoelectronics and microwave.

Hewlett-Packard. 'RF and Microwave Phase Noise Measurement Seminar' handbook.

Helszajn J. (1985) 'YiG resonators and filters' Wiley J. & Sons, editor.

Howell C. M. (1987) 'Gunn Diode Oscillators' (part 1) Microwaves and RF, 125–133.

Howell C. M. (1987) 'Gunn Diode Oscillators' (part 2) Microwaves and RF, 195–207.

Johnson K. M. (1979) 'Large Signal GaAs MESFET Oscillator Design' *IEEE Trans. on Microwave Theory and Techniques*, **MTT 27**, N° 3, 217–227.

Kajfez and Guillon 'Dielectric Resonators' ARTECH HOUSE, Dedham, MA 02026

Kaminsky D. (1986) 'Broad-band microwave oscillators' *Military Microwaves Conference Digest*, 616–621

Khanna A. P. S. and Obregon S. (1981) 'Microwave Oscillator Analysis' *IEEE Trans. on Microwave Theory and Techniques*, **MTT 29**, N° 6, 606–607.

Khanna A. P. S. (1987) 'Review of dielectric resonator oscillator technology' *Proc. 41st Annual Symposium on Frequency Control*, 478, 483.

Kobayashi Y. (1981) 'Resonant modes for a shielded dielectric rod resonator' *Trans. of I.E.C.E.*, **J 64–B**, N° 5, 433.

Konishi Y. (1976) 'Resonant frequency of a TE_{01} Dielectric Resonator' *IEEE Trans. on Microwave Theory and Techniques*, **MTT 2**, N° 2.

Kurokawa K. (1968) 'Noise in synchronized oscillators' *IEEE Trans. on Microwave Theory and Techniques*, **MTT 16**, N° 4, 234–240.

Lan G., Kalokitis D., Mykietyn E., Hoffman E. and Sechi F. (1986) 'Highly stabilized, ultra low-noise FET oscillator with dielectric resonator', *IEEE MTT-S Digest*, 83, 86.

Leeson D. B. (1966) 'A simple model of feedback oscillator noise spectrum' Proc. of the IEEE, **54**, 329–30.

Leier R. M. and Patston R. W. (1985) 'Voltage-controlled Oscillators Evaluated for System Design' *MSN*, 102, 125.

Makino T. and Hashima A. (1979) 'A highly stabilized MiC Gunn oscillator using a dielectric resonator' *IEEE Trans. on Microwave Theory and Techniques*, **MTT 27**, N° 7, 633, 638.

Mamodaly N., Farzaneh F., Bert A., Obregon J. and Guillon P. (1985) 'A fundamental mode InP Gunn Dielectric Resonator Oscillator at 94 GHz' *15th European Microwave Conference Digest, Paris*, 170, 176.

Pattison J. 'Active Microwave Devices: TEDs' in 'Microwave solid-state devices and applications', 14–18, Morgan D. V. and Howes M. J.

Pavio A. M. and Smith M. A. (1985) 'Push-push Dielectric Resonator Oscillator' *IEEE MTT-S Digest*, 266, 269.

Richtmyer R. D. (1939) 'Dielectric Resonators' Journal of Applied Physics, **10**, 391–398.

Sigmon B. E. and Ayya Gari M. (1987) 'A multidiode cavity power combiner using state-of-the-art pulsed Gunn diodes' *1987 MTT Symposium digest*, 871–874.

Sweet A. W. (1974) 'How to build a Gunn oscillator' Micronotes, **11**, N° 2.

Vendelin G., Mueller W., Khanna A. P. S. and Soohoo R. (1986) 'A 4 GHz DRO' *Microwave Journal*, 151, 152.

5

Microwave frequency dividers

Didier Kaminsky

In phase locked loops (PLL) or synthesizers, the microwave signal is compared to a reference signal (generally between 10 and 100 MHz) in order to obtain a stable and accurate microwave frequency. One of the most important elements of the PLL is the frequency divider used to reduce the frequency value. In the microwave field, halvers are generally used. Two divider concepts can be used: digital division using logic circuits, and analogue division generating sub-harmonics.

5.1 DIGITAL DIVIDERS

Basic circuits (Lee and Miller, 1984; Soares, 1988)

These dividers are based on flip-flop designs, thus the output frequency is equal to the clock frequency (the input signal) divided by 2^N where N is the number of divider stages. Figure 5.1 presents three common types of flip-flops.

The maximum frequency is given by:

$$F_{\mathrm{max}} = 1/n\tau_{\mathrm{d}}$$

where τ_{d} is the propagation time and n is a coefficient which depends on the structure of the divider. For structures shown in Fig. 5.1(a), n is between 4.2 to 4.85. The fastest circuit is the flip-flop presented in Fig. 5.1(b), but it requires two complementary signals.

Materials

Two materials are used to manufacture dividers: silicon and gallium arsenide. However, due to their higher electron mobility and higher maximum speed (respectively six and two times that of silicon), GaAs dividers are used more and more. For ECL silicon dividers, the maximum usable frequency is currently 2.5 to 3 GHz, though some results have been presented up to 9 GHz using

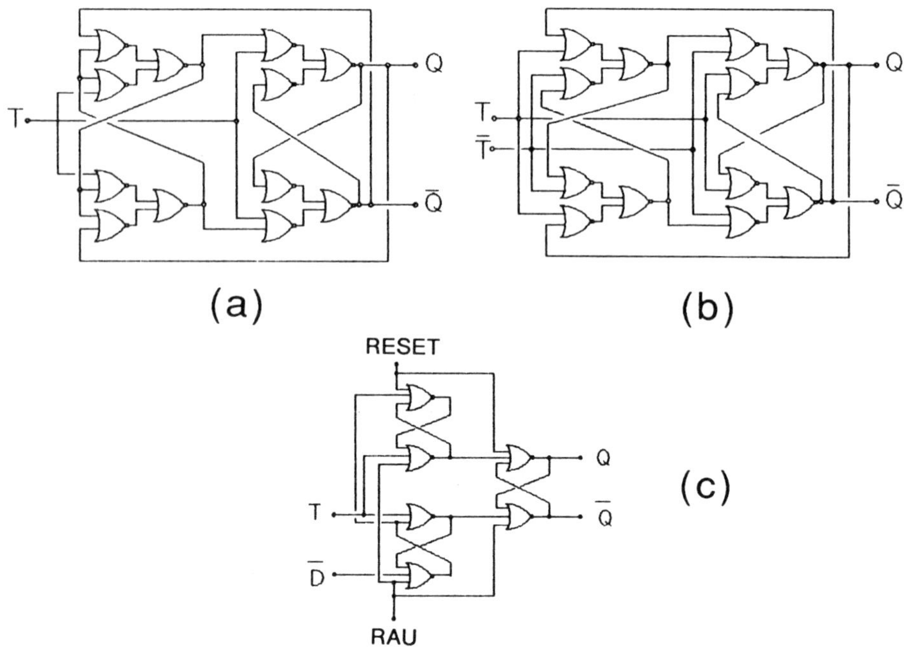

Fig. 5.1 Different structures of flip-flops.

Si bipolar self-aligned process technology. Commercially available GaAs dividers now reach 4.5 to 5 GHz and very soon will be proposed to 10 GHz by using 0.5 μm (or smaller) gate technology.

5.2 ANALOGUE DIVIDERS

Two different circuits are used: parametric frequency dividers, and regenerative frequency dividers.

The first circuit has the broadest bandwidth but requires high input power while the second one needs low input power, may have conversion gain and, moreover, can be more easily manufactured in monolithic form.

Parametric frequency dividers (Harrison, 1983; Harrison and Cornish 1986)

In this type of circuit, an inductive loop contains two varactors. An input at $2f$ excites both varactors in phase. Under specific conditions, balanced loop oscillations occur at a frequency f at which the varactor voltages are 180° out of phase.

In Fig. 5.2., a parametric frequency halver model is presented. The coupling of input signal to varactor (r_s, $C_j(V)$) is represented by L_1 and L_2, and R_L

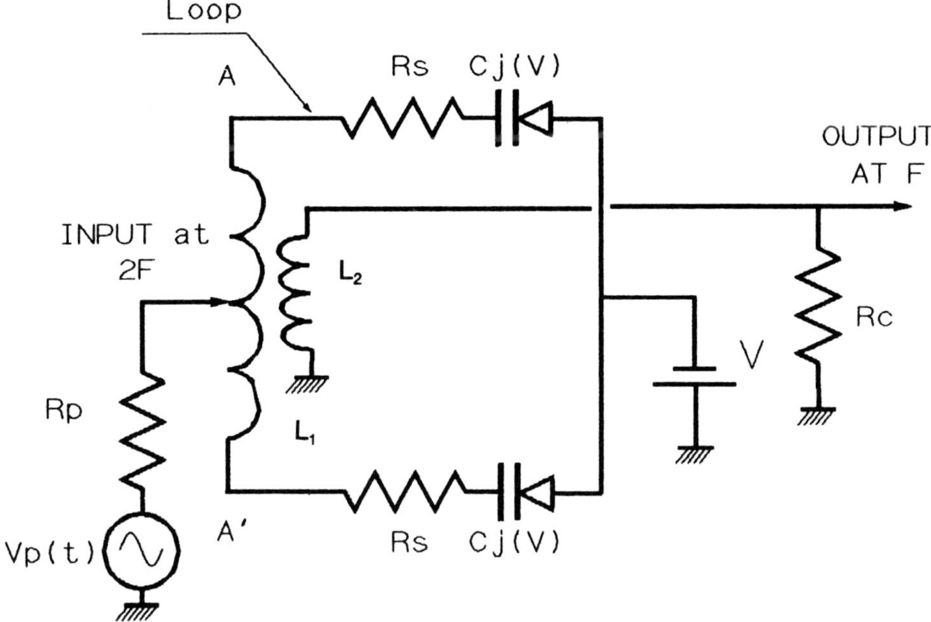

Fig. 5.2 Synopsis of a parametric frequency divider.

represents the coupling of the subharmonic voltage which appears across AA′ to the external load. It can be shown theoretically and experimentally that, for a threshold input level P_{in}, the loop resonates at a particular sub-harmonic frequency.

For input frequency in X-band, typical input powers are between 10 to 20 dBm and losses are between 5 to 15 dB. Up to one octave division bandwidth can be obtained with this type of circuit.

Regenerative dividers (Miller, 1939; Rauscher, 1984; Kaminsky et al.*, 1983)*

The regenerative frequency divider concept, as indicated in Fig. 5.3, involves a mixer element and an amplifying sub-harmonic feedback loop. If some $f/2$ signal appears in the loop, it mixes with the input f signal in the mixer to generate again an $f/2$ signal. The phase condition can adjust itself within the loop over a frequency band since the phase lag between the $f/2$ signal and f signals entering the mixer is not determined. The frequency band of the divider is mainly limited by the condition that the loop gain be greater than unity.

Since both linear and non-linear characteristics are needed for the divider, one FET (or dual-gate FET) can be used to create both the mixing and the amplification processes. For that purpose, the FET (or the dual-gate FET) is

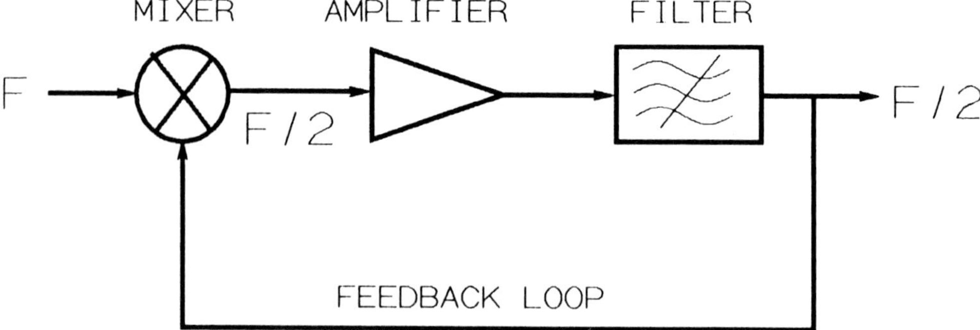

Fig. 5.3 Principle of the regenerative divider.

biased near pinch-off to obtain the maximum non-linearity. In this case, the threshold can be reduced to arbitrarily low levels but at the expense of the bandwidth. With low input power levels (0 to 15 dBm), bandwidth of more than 20% can be obtained (up to 40%) with gains ranging from – 5 to + 5 dB.

BIBLIOGRAPHY

Harrison R. G. (1983) Theory of the varactor frequency halver *IEEE MTT-S Digest*, 203, 205.

Harrison R. G. and Cornish W. D. (1986) Varactor frequency halver with enhanced bandwidth and dynamic range *IEEE MTT-S Digest*, 305, 308.

Kaminsky D., Goussu P., Funck R. and Bert A. G. (1983) A dual-gate GaAs FET analog frequency divider *IEEE MTT-S Digest*, 352–354.

Lee F. and Miller R. (1984) 4 GHz counter bring synthesizers up to speed, *Microwaves and RF*, 113, 124.

Miller R. L. (1939) Fractional – frequency generators utilizing regenerative modulation *Proc. I. RE*, **27**, 446–456.

Rauscher C. (1984) Regenerative frequency division with a GaAs FET *IEEE, Trans. on MTT*, **MTT 32**, (**11**), 1461–8.

Soares R., Graffeuil J. and Obregon J. 'Application of GaAs MESFETs' ARTECH HOUSE, Dedham, MA 02026.

Soares R. A. (1988) *GaAs MESFET Circuit Design* Artech House Books.

6

Microwave frequency multipliers

Jean Anastassiades

6.1 THE NEED FOR FREQUENCY MULTIPLIERS

Frequency multiplication is often a convenient way to increase frequency of electrical signals. Frequency multipliers are particularly needed for applications such as:

1. Very high frequency microwave sources where direct generation may be difficult.
2. Microwave sources with very low level 'coloured noise'. In that case, direct generation by microwave resonator oscillators may be inconvenient because the frequency-dependent noise generally extends to tens or even hundreds of kilohertz from the carrier.
3. Numerous frequency sources. Here frequency multipliers may be of great interest because it is often simpler to generate a large number of discrete frequencies by means of a harmonic generator followed by filters or by a single tuneable filter than to use as many oscillators as needed frequencies.

6.2 BASIC PRINCIPLES

In electronic circuitry, non-linearities often appear and are generally considered as deficiencies or faults. However, that phenomenon is also considered as interesting and even helpful since non-linearities enable electronic engineers to design frequency multipliers. Indeed, if a non-linear impedance is driven by a sinusoidal source, harmonics of the input signal are generated. Non-linear impedance means an impedance whose resistance, capacitance or inductance varies instantaneously with the applied current or voltage.

To illustrate this, let us examine the case of the non-linear resistance of a diode which is a convenient introduction to the various frequency multiplication techniques described later inside this chapter. Figure 6.1(a) shows a typical diode characteristic. Now, let us consider the device of Fig. 6.1(b), which consists in the association of the diode D and a load resistor R_L. A source

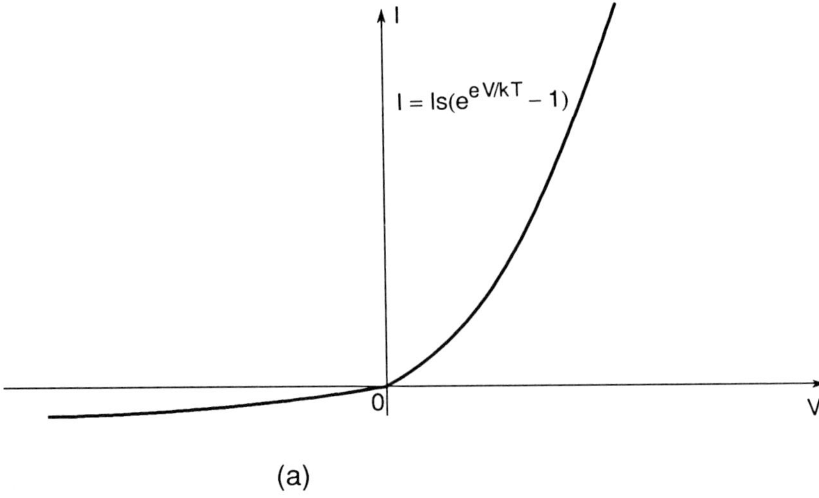

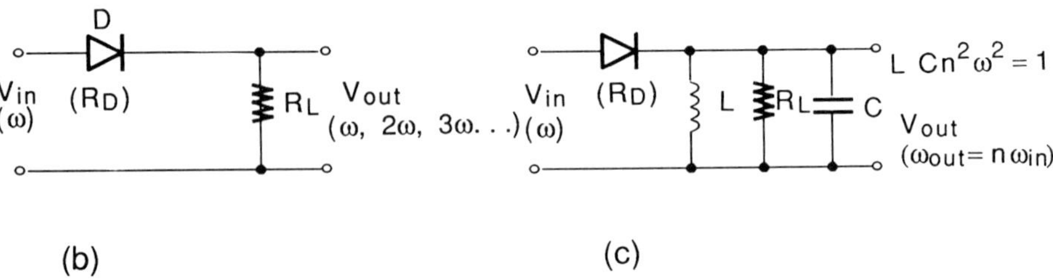

Fig. 6.1 Non-linear circuit and frequency multiplier; (a) diode characteristic, (b) non-linear circuit, (c) non-linear circuit with filter.

delivers a sinusoidal voltage at the input of the device. Then at the output the resulting signal is given by the following relationship:

$$v_{out} = v_{in} \frac{R_L}{R_D + R_L}. \tag{6.1}$$

From this, it is clear there is a non-linear relation between the input and output voltages, since the diode resistance R_D is a non-linear function of v_{in}. Equation (6.1) may also be expanded as follows:

$$v_{out} = a_0 + a_1 v_{in} + a_2 v_{in}^2 + a_3 v_{in}^3 + a_4 v_{in}^4 + \ldots + a_n v_{in}^n. \tag{6.2}$$

Assuming $v_{in} = A \sin \omega t$, equation 6.2. gives the output voltage of the frequency multiplier (Fig. 6.1(b)) as follows:

$$v_{out} = a_0 + \frac{a_2A^2}{2} + \frac{3a_4A^4}{8} + \left(a_1A + \frac{3a_3A^3}{4}\right)\sin(\omega t) - \left(\frac{a_2A^2}{2} + \frac{a_4A^4}{2}\right)\cos(2\omega t)$$

$$- \frac{a_3A^3}{4}\sin(3\omega t) + \frac{a_4A^4}{8}\cos(4\omega t) + \ldots \tag{6.3}$$

Thus, harmonics of the input signal appear and their amplitudes depend not only on the input signal amplitude A, but also on the multiplication factor and the diode and circuit characteristics (coefficients a_n).

Generally the goal is to deliver only one harmonic of the input signal so the device (see Fig. 6.1(b)) must be complemented by an output filter, which cancels the unwanted terms of the multiplication. For example, this may be a simple tuned circuit as shown in Fig. 6.1(c). This circuit is obviously tuned on the desired frequency nF (n = harmonic rank or multiplication factor).

It may be interesting to predict the result of a frequency multiplication following another method. This other method consists first in the determination of the time response of the non-linear circuit when driven by a sine wave and secondly in the derivation of the associated spectrum. The first operation may be done by an analytical derivation or even by means of a graphical method when this is possible. The second operation is to calculate the spectrum by using the Fourier series method.

Application of this to various types of waveforms (parts of sine waves, triangular and rectangular waves, etc.) shows that generation of numerous harmonics with relatively high amplitudes is achievable using rectangular waves, particularly if their duty cycle is low (low ratio between pulse width and period of signal). Conversely, smooth waveforms like rectified sine waves have only a few number of harmonics with high amplitude.

The frequency multiplier shown on Fig. 6.1(b) and (c) is only a simplified model, but it gives an opportunity to notice that the non-linear device is not sufficient to deliver the required signal to the load. In fact, this signal must generally be a sine wave, the frequency of which is an integer multiple of the input signal frequency. As the real generated signal is always complex, it is very often necessary to select one or several harmonics from its spectrum. That function is accomplished by frequency filtering at the non-linear circuit output.

Elimination or reduction of unwanted products are not the only requirements, because efficiency, power level or signal-to-noise ratio are also important considerations in signal generation techniques. Thus, the non-linear device has to be matched to the load and to the source (Fig. 6.2).

Theoretically, when the non-linear device is perfectly matched to the source at fundamental frequency f_0 and to the load at harmonic frequency nf_0, the efficiency could be 100%. Practically, there are losses in every element in the circuits (matching networks, non-linear device, filters, etc.), whatever the

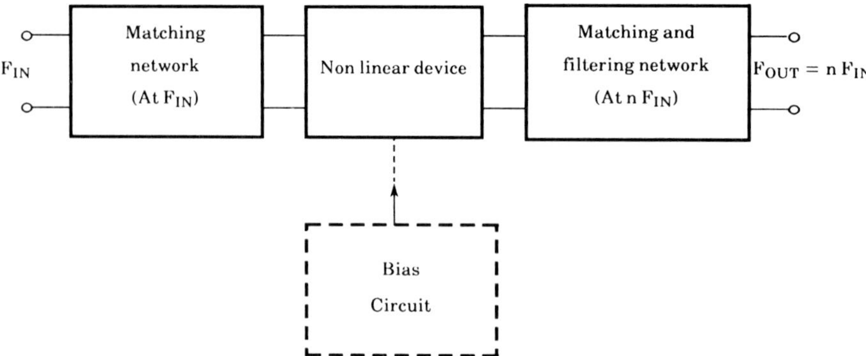

Fig. 6.2 Frequency multiplier block diagram.

technology we use, i.e., distributed elements or lumped circuits, and those losses may greatly reduce the efficiency.

The general diagram on Fig. 6.2 also shows an extra section called bias circuit. This is a part of the multiplier which is often necessary to properly bias the non-linear device in order to generate the required harmonics with a good efficiency.

Design and development of fixed frequency (or very narrow bandwidth) multipliers is not difficult because first, there are generally no major problems in narrow bandwidth circuitry implementation and second, it is always easy to select one harmonic among the others (generally a simple tuned circuit is sufficient to realize a convenient attenuation on the previous and next harmonics). This is illustrated on Fig. 6.3(a).

When the input signal frequency varies around its central value over a larger bandwidth, the design or development of the multiplier may become difficult and even impossible. Let us assume an input sinewave with a frequency $F_{IN} \pm \Delta F/2$ where F_{IN} is the centre frequency, and ΔF its possible variation. The relative frequency bandwidth of this signal is equal to $\Delta F/F_{IN}$.

If we plot the variations of the fundamental and harmonics frequencies on a diagram, we obtain the result shown on Fig. 6.3(b). In this example, the relative bandwidth of variation is equal to 25% and it can be seen that selection of the fourth harmonic is critical and even impossible since the third harmonic at $3(F_{IN} + \Delta F/2)$ is pretty close to the fourth harmonic at $4(F_{IN} - \Delta F/2)$ and above all the fifth harmonic at $5(F_{IN} - \Delta F/2)$ has a lower frequency than the fourth one at $4(F_{IN} + \Delta F/2)$.

If the maximum and minimum frequencies of the Nth harmonic are respectively $NF_{IN.\max}$ and $NF_{IN.\min}$, the condition not to have any overlap between successive harmonics is:

$$NF_{IN.\max} < (N+1)F_{IN.\min}.$$

Since $F_{\text{IN min}} = F_{\text{IN}} - \Delta F/2$ and $F_{\text{IN max}} = F_{\text{IN}} + \Delta F/2$, this unequality becomes:

$$\frac{\Delta F}{F_{\text{IN}}} < \frac{2}{2N+1}. \tag{6.4}$$

From that relation, it is clear that high multiplication factors are not compatible with large relative bandwidth.

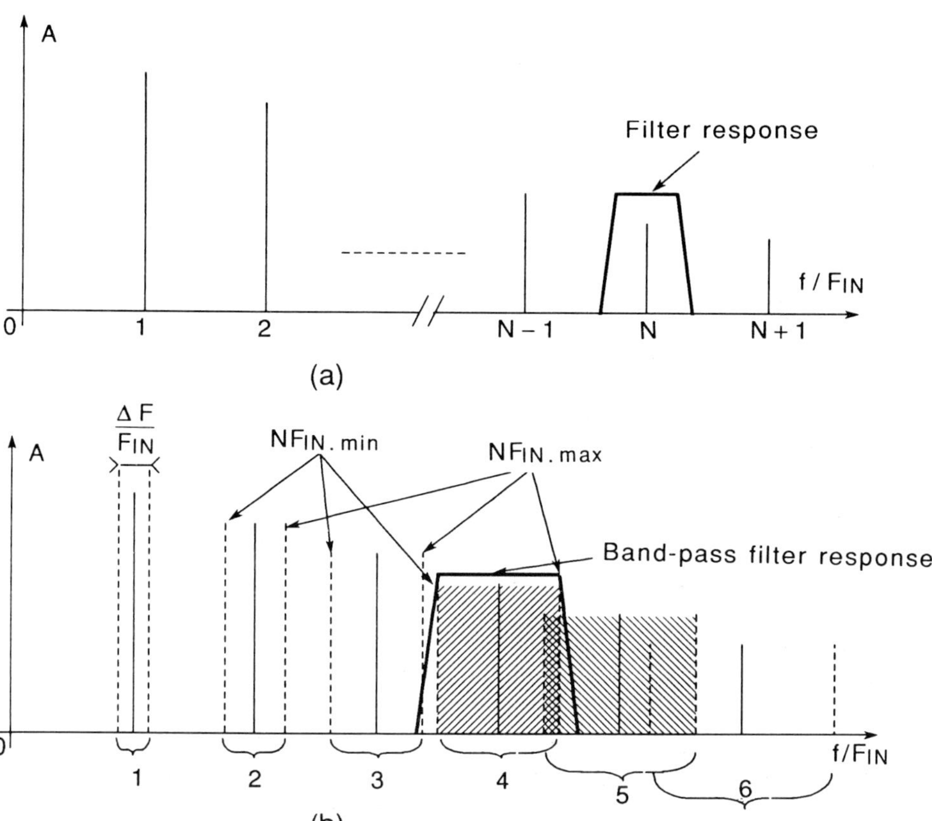

Fig. 6.3 Harmonics generation: spectra; (a) fixed frequency, (b) variable frequency.

The conclusion is that the best way to obtain large relative bandwidth and high multiplication factor is generally to implement multipliers by cascading low factor multipliers. One can also notice this is a good solution to implement frequency multipliers with low additive noise. Another solution using frequency controlled narrow band filters can be used but it often needs temperature compensation and introduces a switching delay.

6.3 TECHNIQUES AND TECHNOLOGIES USUALLY EMPLOYED

6.3.1 Frequency multipliers using transistors (or diodes)

Basic operation

We have already seen a basic example with a diode used as a non-linear element. We can now remember that example to introduce transistors in frequency multiplication because this kind of device may be considered as a diode followed by an amplifier.

In fact, the ability of such devices to generate harmonics of a sine wave is due to several sources of non-linearity:

1. gate-source (or base-emitter) junction non-linear capacitance;
2. output conductance non-linearity;
3. transconductance non-linearity;
4. rectification of the signal due to bias conditions which may be either at pinch-off (or cut-off) voltage or saturation.

Many studies were conducted on this subject during the last 10 years and it has been proven that a large participation in harmonic generation using transistors and particularly MESFET is due to the rectifier effect. Among these studies, essentially devoted to MESFET multipliers, which exhibit better characteristics than bipolar transistors at high frequencies, certain of them present comparisons between theoretical predictions by means of model simulation and experimental results (A. Gopinath and J.B. Rankin, 1982, E. Camargo, R. Soares, R.A. Perichon and M. Goloubkoff, 1983, G.S. Dow and L.S. Rosenheck, 1983).

Some improvements can be brought to the basic multiplier. By using a dual gate FET with appropriate drive instead of a single gate FET, an increased conversion gain may be achieved ($\Delta G \approx 6$ dB). Other improvements are interesting either to enable a transistor to act as an odd multiplier or to increase the conversion gain of $\times 2$ and $\times 4$ multipliers. Examination of a rectangular wave spectrum shows that it is a convenient way to generate odd and even harmonics. So if we need a $\times 3$ multiplier, the solution may be to design a transistor circuit with a bias and a RF input power level such that a quasi-square wave is generated. In other words, if we assume, for instance, that a MESFET is used, the gate bias has to be somewhere between $V_{gs} = 0$ and $V_{gs} = V_p$ (V_p is the pinch-off voltage) and the input swing has to be large enough to cause clipping on both ends. Bias and input level must be adjusted to generate a symmetrical square wave in order to make the third harmonic maximum.

Now, let us have a few words about the other improvement which concerns $\times 2$ and $\times 4$ multiplication. In this case, the main improvement is brought when using a double transistor circuit as in Fig. 6.4(a). This circuit operates as a kind of differential amplifier but with the two drains or collectors connected to a common load. To illustrate this, signals and simplified characteristics of

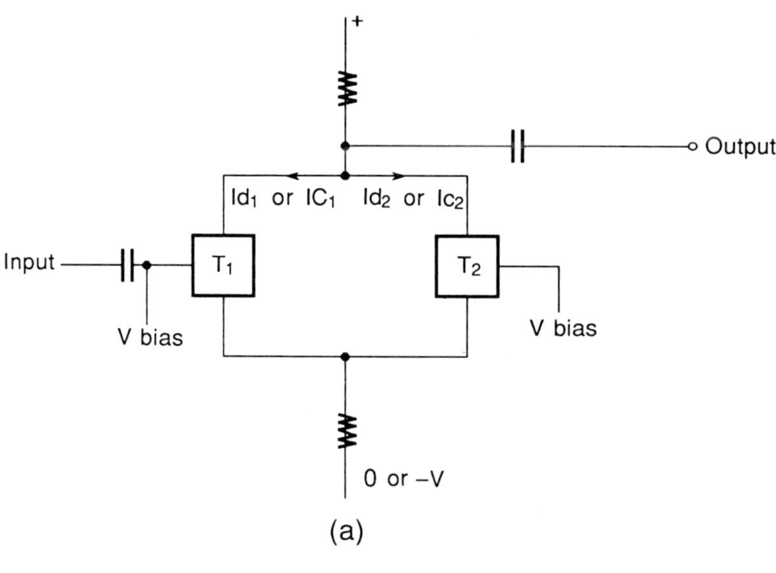

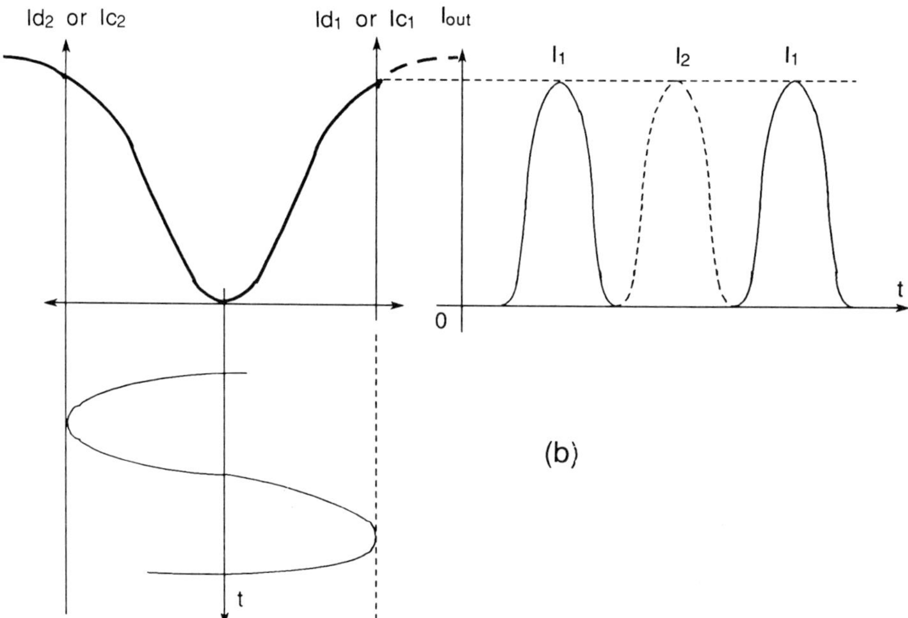

Fig. 6.4 Principle of transistors symmetrical multiplier; (a) basic diagram, (b) operation principle.

the transistors are depicted on Fig. 6.4(b). If we assume that the two transistors are biased close to the cut-off (or pinch-off), there is only a small current flowing through the transistors when the input voltage is equal to zero. Due to the circuit architecture, transistor T_1 conducts during positive alternances of the input sine wave and transistor T_2 remains under cut-off. Conversely, T_1 becomes under cut-off during the other alternance and T_2 conducts. This is depicted on Fig. 6.4(b) where the output currents (I_d or I_c) from transistors T_1 and T_2 are represented. It can be easily shown that the odd harmonics are eliminated in the double transistor multipliers and that the gain conversions of the even harmonics are doubled.

Note that symmetrical diode devices may also be employed in low multiplication factor devices. A typical example would be the use of a double-balanced mixer as a frequency doubler.

Design and examples

The choice of the non-linear element depends essentially on the frequencies and power of operation and it is clear that multipliers from S-band to Ku or Ka-bands will be designed with MESFET or power FET elements. Of course, bipolar transistors cannot handle very high frequencies and they are efficient only up to 1 or 2 GHz.

The design of frequency multipliers using FET or bipolar transistors is relatively simple with techniques very close to those of amplifiers. S parame-

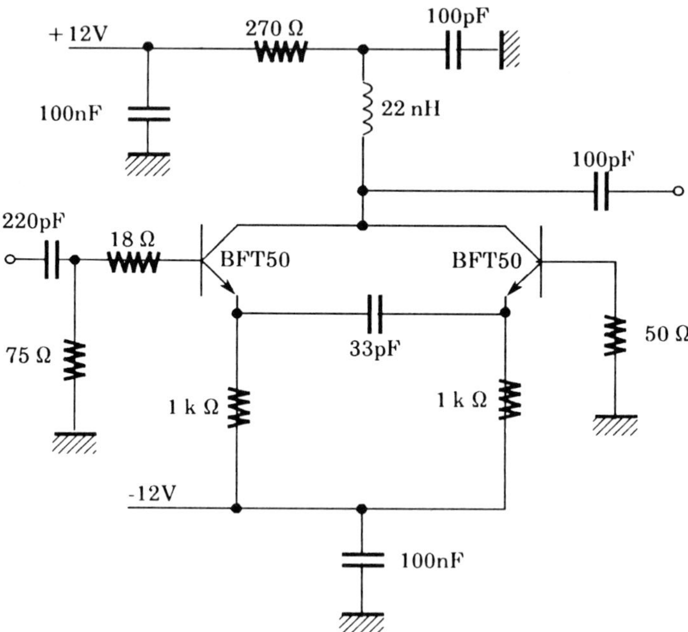

Fig. 6.5 Example of ×4 multiplier (150 MHz/600 MHz) used in Thomson-CSF radar synthesizers.

ters may be used to design such devices (matching sections and stability coefficient essentially) and they may be helpful when accurately known. Nevertheless, it is not always easy to measure the 'equivalent S parameters' at high level, i.e., under non-linear conditions of operation. Anyway, theoretical design often has to be complemented by means of experimental work.

Bipolar and field effect transistors are very convenient for relatively small harmonic numbers and their structure makes the multiplier design easy.

A first example of a frequency multiplier is depicted on Fig. 6.5. It is a bipolar transistor $\times 4$ multiplier biased for class A-B operation with an input frequency of 150 MHz ± 15 MHz (relative bandwidth 20%). It is the first part of a $\times 16$ multiplier and we chose bipolar transistors because of their low noise characteristic at low frequency of operation (flicker noise near the carrier frequency is smaller than in GaAs FET). The output power is equal to −2 dBm ± 1 dB within the 20% bandwidth, with an input power of 10 dBm. The capacitor between emitters was defined for maximum power at the output frequency.

A second example is shown on Figure 6.6. It is a single GaAs FET frequency doubler used, as the previous one, in a synthesizer. The circuit was built on a Duroid 6010 substrate where the input and output matching networks were etched (two microstriplines with a double stub on each). The best conversion gain was obtained with V_{gs} approximately equal to V_p (class B operation) and this was done with self-bias ($R = 22\ \Omega$ and $C = 10$ pF). The two 32 nH inductors are high impedances for decoupling of the self-bias and power supply circuits. There are decoupling capacitors at the input and output of the circuit. For certain applications, a 3 dB attenuator may be added to the input for

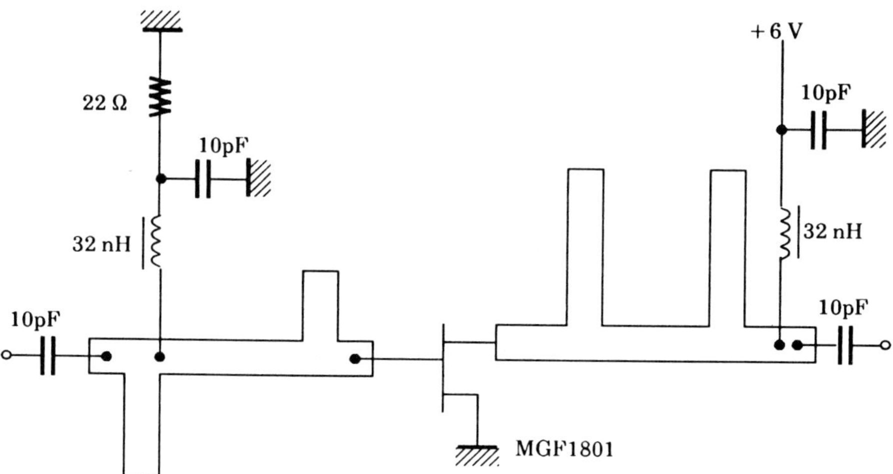

Fig. 6.6 Example of $\times 2$ multiplier (2.4 GHz/4.8 GHz) used in Thomson-CSF radar synthesizers.

matching purposes. From this design, we have obtained the following characteristics. The input operating frequency is 2.4 GHz with a relative bandwidth of ± 5%. The output power is 17 dBm ± 0.5 dB within the output bandwidth (4.8 GHz ± 5%) for a constant input power of 14 dBm (or 17 dBm with the 3 dB attenuator at the input). So the conversion gain is + 3 dB.

6.3.2 Frequency multipliers using varactor diodes

One of the more commonly used devices is the varactor diode which exhibits a non-linear capacitance versus voltage characteristics. Figure 6.7 shows a typical curve of varactor capacitance versus voltage; that characteristic is different from those of the snap varactor or step recovery diode (SRD) (see section 6.3.3).

For convenient frequency multiplication, the varactor diode current must be properly biased, i.e., the bias point is defined to keep the RF signal sweeping only the reverse biased part of the diode characteristic as shown in Fig. 6.7, and also prevent reverse conduction beyond breakdown voltage. If the sum of the bias voltage and the peak RF voltage are such that the diode is conducting during a part of the input signal period (forward or reverse conduction or both), then the losses and thermal dissipation are increased and the signal becomes noisy.

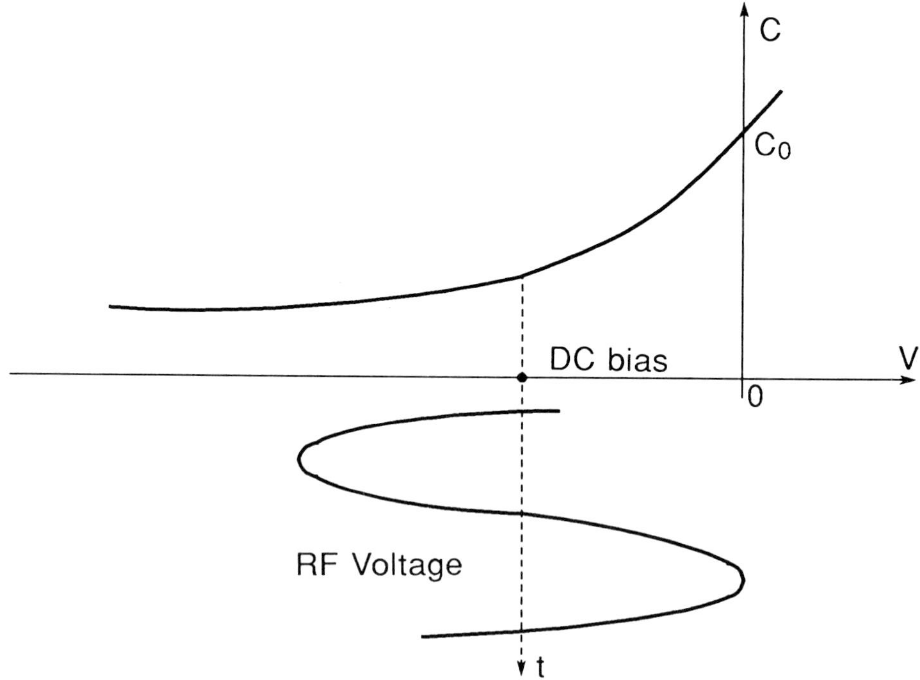

Fig. 6.7 Varactor capacitance versus voltage and basic operating conditions.

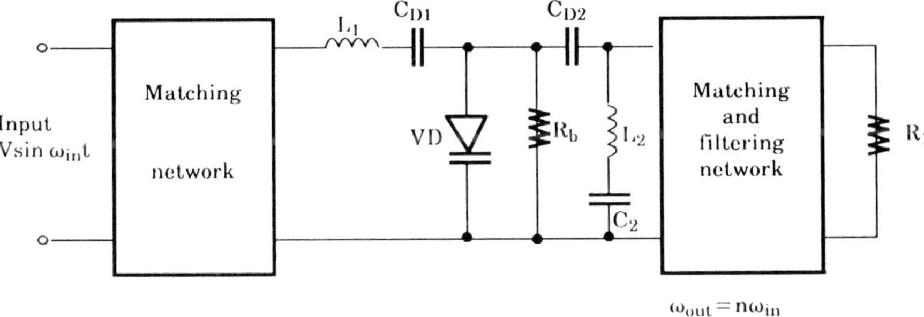

Fig. 6.8 Frequency multiplier with a varactor diode.

The circuits used for varactor harmonics generation are very similar to those of the SRD frequency multipliers (see section 6.3.3). The basic circuit shown in Fig. 6.8 has an input circuit consisting of an inductance L_1 to series-resonate the capacitive input reactance presented by the varactor at the frequency of the input signal. This inductance (or a more elaborate circuit like an LC choke) prevents energy at the output frequency nf_{in} from being lost in the input circuits.

Also on the picture of Fig. 6.8 is a series-resonant circuit $L_1 - C_2$ which acts as a short circuit for one of the unwanted harmonics (several idlers may also be used in order to short several harmonics that are not wanted at the output). The aim of this is to prevent a loss of energy: thermal dissipation of harmonics at the input and output of the multiplier may be replaced by recombination of these harmonics in the non-linear element so as to increase the output energy at the desired frequency.

To conclude, we must also say that varactor diode multiplicators are essentially used for low order, high efficiency (say up to 80 or 90%) broadband applications. Nevertheless, they are now challenged by transistors (bipolar and FET) which bring an additional power gain (for low harmonic numbers).

6.3.3 Frequency multipliers using step recovery diodes*

The step recovery diode is a highly non-linear component because forward stored charge results in low impedance, while reverse stored charge gives high impedance. So the harmonic generation principle is based on impedance variation (depending on the charge) as a function of time when the device is properly driven by means of a sine wave.

A typical step recovery diode characteristic is shown in Fig. 6.9 and the impulse generator circuit in Fig. 6.10 consists of a drive inductance L, the step recovery diode and the load R_L. The input of the circuit is fed with a sine

* Information in this section is partly from S. Hamilton and R. Hall – Hewlett-Packard Application – Notes 913 and 920 – May 1967.

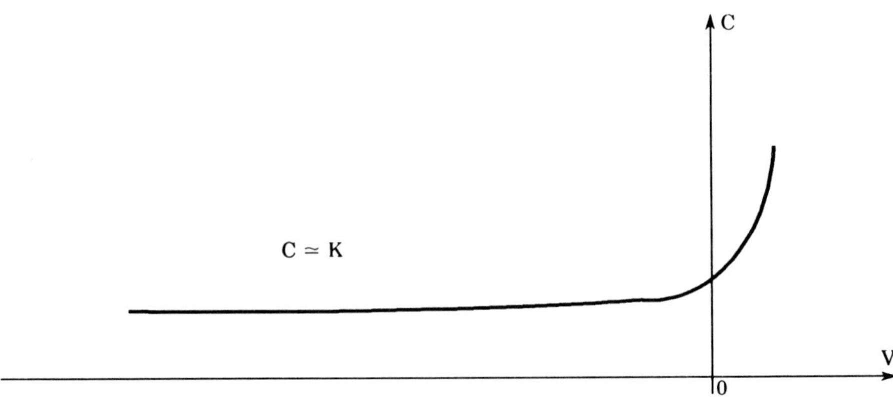

Fig. 6.9 SRD typical capacitance characteristic.

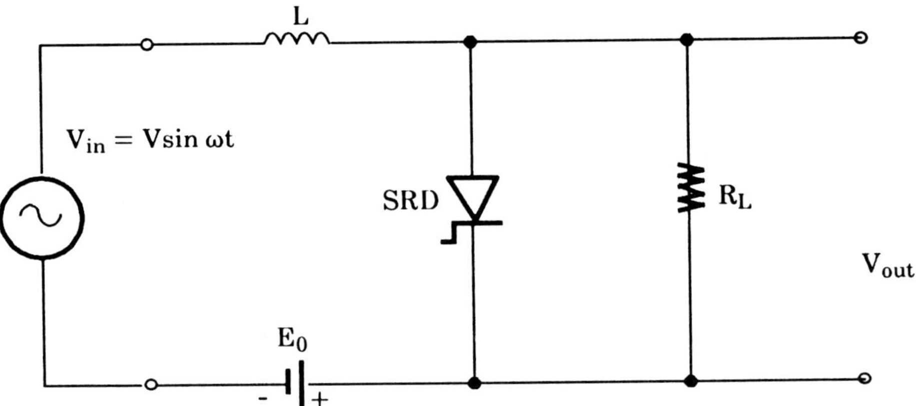

Fig. 6.10 Impulse generator circuit.

voltage from the source and a battery can be used to properly adjust the diode bias. Because they are strictly delimited, the two states of the SRD (short circuit when forward biased and capacitor when reverse biased) make a double linear analysis possible. Then from the impulse generator circuit there are two equivalent circuits to consider: one during the diode forward conducting time and another one during the time when the diode is in its high-impedance state. This is illustrated in Fig. 6.11(a, b), where we can see the two equivalent circuits.

During the conduction interval, the circuit is the drive inductance with the SRD forward impedance (assumed to be negligible, $Z_d = 0$) and the bias battery and the sinusoidal source. As long as the current i_L is positive, the SRD stores an electric charge which is progressively removed from the diode when the current is negative (with the SRD impedance remaining very low).

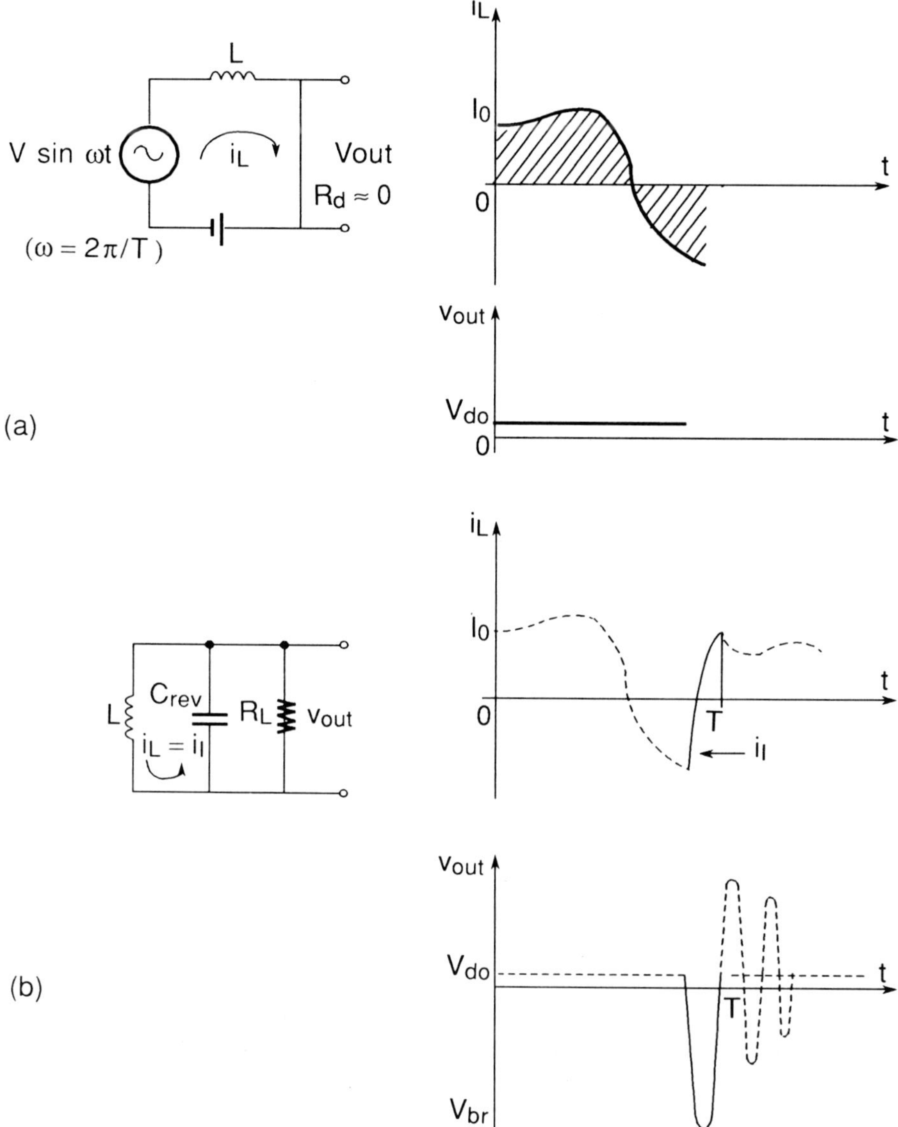

Fig. 6.11 Equivalent circuits and associated current and voltage waveforms; (a) conduction interval, (b) depletion interval.

When there is no more stored charge in the diode, its impedance becomes very high (assumed to be a constant value capacitor C_{rev} for simplification purposes). Note this state occurs when the two areas above and below the $i = 0$ axis are equal (see Fig. 6.11(a). During this interval of conduction, the output voltage is constant and equal to the diode contact potential since the diode resistance was assumed to be zero.

Let us now examine how the circuit works during the impulse (or depletion) interval. At the point where the charge in the diode has been removed, the source and battery voltages are equal and opposite, so the equivalent circuit is the drive inductor L paralleled with the diode reverse capacitance C_{rev} and the load R_L. This is in fact a tank circuit where the energy stored in the inductor generates an impulse current i_I (see Fig. 6.11(b)) flowing through the tank circuit which will normally oscillate until the energy is dissipated. As a matter of fact, only the first half-cycle of this damped oscillation appears across the load because the diode begins to conduct again when the applied voltage goes positive. Then we have only one half sine pulse during the depletion part of the cycle. As the diode is conducting again, a new conduction interval starts and will be followed by an impulse interval and this cycle is renewed at each period of the input sine wave.

The complete SRD frequency multiplier of which the most important part of the device is the impulse generator is represented in Fig. 6.12.

Biasing the diode for optimum efficiency may be achieved by using either a self-biasing resistor or an external source (like the power supply of the overall circuitry). The first solution can be preferred because of temperature compensation, increased bandwidth operation and self-adjustment properties (when the power level at the input varies). Decoupling capacitors C_{D1} and C_{D2} insure correct isolation for the d.c. bias. These capacitors may be a part of the input and output matching network when possible.

This matching network is necessary for impedance matching because the source impedance (generally that of a transistor amplifier) is greater than the impulse input impedance. Similarly an output matching network must be used to transform the low diode impedance into a higher impedance, for instance the 50 Ω normalized impedance of the output filter or the load.

6.3.4 Other types of frequency multipliers

There are many other ways to multiply the frequency of a periodic signal. For instance an avalanche diode, reverse biased so as to reach the avalanche state, is equivalent to a non-linear inductance when its current is varying. This non-linear characteristic $L = f(I)$ can produce harmonics of a sine wave in the same way as the non-linear characteristic $C = f(V)$ of a varactor does.

A second example is the injection locking frequency multiplier that is in fact a stable oscillator which is synchronized by an external signal. The frequency of the oscillator must be an integer multiple of the input frequency (the integer is the multiplication factor).

As a last example, we must also mention the phase locked loop frequency multiplier which is in fact a simple synthesizer whose output frequency is an integer multiple of its input frequency reference (see example of Fig. 6.8).

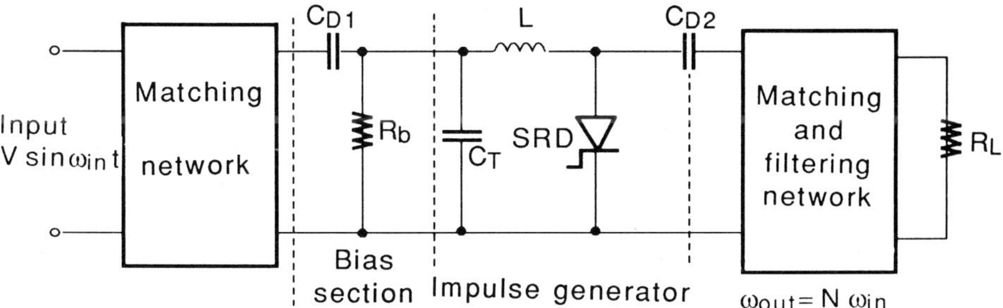

Fig. 6.12 SRD frequency multiplier.

REFERENCES

Copinath A. and Rankin J. B. (1982)

Camergo, E., Soares, R., Perichon, R. A. and Goloubkoff, M. (1983)

Dow, G. S. and Rosenheck, L. S. (1983).

Hamilton, S. and Hall, R. (1967), Hewlett-Packard Application, Notes 913, 920.

7

Frequency synthesizers

Jean Anastassiades and J. P. Aubry

7.1 SYNTHESIZERS: WHAT FOR?

Synthesizers are generally subequipment designed to provide one or several CW reference signals to the equipment or systems into which they are integrated. These signals are used as local oscillators in receivers and often, also to up-convert IF signals before transmission (communications, radars, etc.). Modern equipment and systems operate not only at one single frequency, but rather over a frequency range, so they need sources with several frequencies. These frequencies are generally equally spaced within the overall bandwidth of the equipment and selection of one of them must be made easily and accurately. Among the main areas of application we must at least mention communications, radar, navigation aids and measurements where direct or indirect synthesis or combination of both can be used. (In direct synthesis the frequencies are summed, divided or multiplied to generate the output signal. In indirect synthesis, a voltage controlled oscillator is phase-locked on one or several reference signals.)

In addition to these possible architecture differences and also to physical parameter differences (size, weight, etc.), electrical performances are often different depending on the application. For instance, radars and navigational aids require low or medium relative bandwidths (generally less than 20%) and frequency accuracy (say about 10^{-5} and sometimes 10^{-6}). Also they often require medium frequency resolution (MHz to tens of MHz) i.e., tens to hundreds of different frequencies over the operational frequency range of the equipment. Inversely, 10^4 to 10^6 frequencies may be requested in communications systems over medium to wide relative bandwidths (up to several octaves). The corresponding resolution generally extends from tens of hertz to tens of kilohertz. Also spectral purity and settling time can strongly differ with the application: radars require very low phase noise (about − 125 dBc/Hz at X-band) and short or very short switching time (a few microseconds) while medium performance is often sufficient for communications and navigational aids (e.g., − 110 to − 120 dBc/Hz at C-band or less, and milliseconds or more).

Synthesizers for measurement purposes must exhibit extremely large bandwidths (for instance 100 MHz to 25 GHz) with a resolution between 1 and 100 Hz and a very good frequency accuracy (including long term stability). As a consequence of their complexity, such synthesizers have medium or even poor spectral purities and medium settling time if compared to radar sources that are considerably simpler.

7.2 SYNTHESIZER ARCHITECTURES

There are two major types of synthesis used for microwave signal generation. They are called direct (analogue) synthesis and indirect synthesis. Nevertheless, direct digital synthesis may be associated with the two analogue techniques. In this part, typical architectures of the two principal kinds of synthesis will be examined so as to give the necessary basis for further study and design.

7.2.1 Direct analogue synthesis

This is of course the oldest type of synthesis used to generate high frequency signals. It generally consists in the use of one or several reference sources associated with frequency multipliers and dividers, mixers, amplifiers, switches and filters. Generally the reference sources must be accurate, and stable sources with very high quality resonators are used.

The simplest way to design a synthesizer, i.e. a device that generates many frequencies from one or several reference sources, is to make a combination of sources (S_1, S_2, . . . , S_n) with their output being selected by means of a switch. The sources can be stable microwave oscillators using dielectric resonators, or HF, VHF or UHF oscillators using crystal resonators, etc., and selection of one or the other type depends on the required characteristics. This type of device remains convenient only when low numbers of frequencies are requested since size and cost increase linearly with number.

Synthesizer architectures using several banks of sources successively associated by means of mixers are used where high numbers of frequencies are requested. As a matter of fact, if there are N_a, N_b, . . . , N_n frequencies delivered by banks *a*, *b*, . . . , *n*, the total number of frequencies available at the synthesiser output is:

$$N_t = N_a \times N_b \times N_c \times \ldots \times N_n \tag{7.1}$$

or, with an equal number *N* of frequencies in each bank:

$$N_t = N^n \tag{7.2}$$

Design and manufacture of synthesizers according to this principle may be done by means of crystal oscillators (XO or surface acoustic wave oscillators) that are switched in order to deliver the desired frequencies at the synthesizer output. Of course, banks of resonators can be associated with oscillators to reduce cost and size of the device. In that case, the switching time between two frequencies is greatly increased (oscillator switching takes a few microseconds while resonator switching requires a few milliseconds).

A very flexible synthesizer using non-coherent direct synthesis has been made, essentially for radar application purposes, i.e. with a relatively small number of available frequencies quickly switchable. The block diagram of Fig. 7.1 shows this synthesizer. The basic synthesis was created by means of three banks of five high spectral purity crystal oscillators. Frequency selection is performed by very efficient diode switches properly controlled to obtain the desired frequency at the synthesizer output. Quartz crystal resonator frequencies lie approximately from 50 to 200 MHz, according to the oscillators bank. The first bank is designed to produce the coarse frequency steps whereas the second is devoted to the fine frequency steps and the third one creates the intermediate frequency steps. The second frequency mixing acts as a frequency subtractor so as to maintain a low spurious level. Each of the mixers is followed by an amplifier driving a bandpass filter used to reduce out-of-band spurious (in-band spurious are minimized by proper selection of the frequencies and levels of the mixed signals).

Amplifiers are also used to drive mixers and particularly the frequency multiplier chain. This chain depends on the frequency range desired at the output of the synthesizers so several modules were designed to produce signals at S, X and Ku-bands. These modules are (2×3) for S-band, $(2 \times 3 \times 3)$ for X-band and $(2 \times 3 \times 7)$ for Ku-band. We can also notice that the multiplication

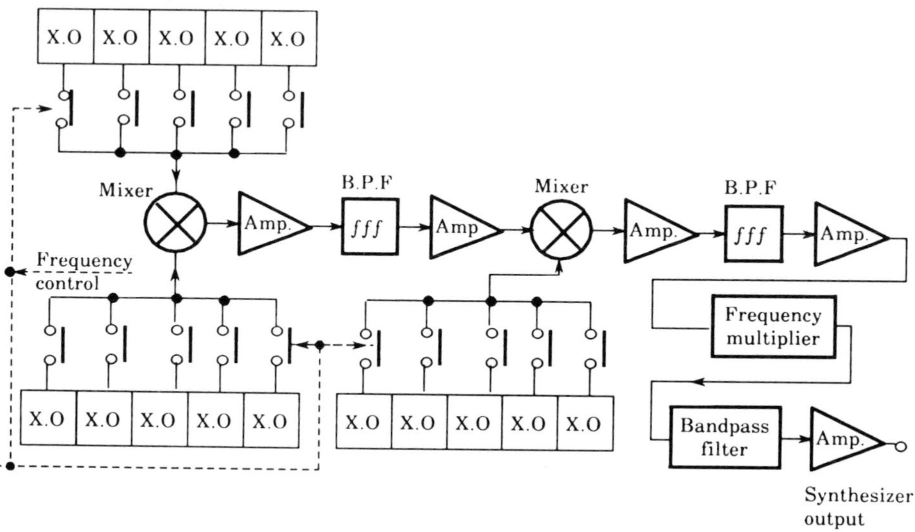

Fig. 7.1 Example of non-coherent synthesis used in a Thomson-CSF radar system.

factor was split into several parts to ensure convenient (spurious free) frequency multiplication (see Chapter 6). SSB signal-to-phase noise ratios of these three versions of the synthesizer is about – 135 dBc/Hz at S-band, – 125 dBc/Hz at X-band and – 118 dBc/Hz at Ku-band for an offset frequency of about 100 kHz from the carrier. The point where the slope of the 'coloured noise' meets the 'white noise' asymptotic curve is located approximately at 5 kHz from the carrier. Spurious elements largely depend on the relative bandwidth (BW_{rel}) of the synthesizer: they reach only – 50 dBc for the S-band device with a 20% relative bandwidth but they are between 65 and 70 dBc for a 10% relative bandwidth. For the X-band they reach – 55 to 60 dBc ($BW_{rel} = 10\%$) and for the Ku-band they are around – 50 dBc ($BW_{rel} = 6\%$). In all cases, the switching time to ensure proper operation of radar (phase stability) is less than 5 μs whatever the frequency (i.e. for any change from one frequency to another among the 125 available frequencies).

The main advantages of direct incoherent synthesis are a relatively low cost and a good flexibility due to the fact that the sources utilized in the synthesizer are independent so their number is chosen to define the number of frequencies available. Nevertheless, this type of synthesis also has its own inconveniences such as limited frequency stability (contribution of several sources) and a limited number of available frequencies (cost increases with the number of sources used in the device). Sometimes, the non-coherence of all the sources of a synthesizer can also induce disturbances such as signals beating in receivers. The importance of this kind of phenomenon depends on the type of signal processing used in the receiver of the equipment so each case has to be carefully examined.

When non-coherent direct synthesis appears not to be convenient for a particular application, a direct (or indirect) coherent synthesis may be used. We will first examine the case of the direct coherent synthesis that may be considered as an evolution of the previous case. As a matter of fact, and while many approaches may be used to achieve direct coherent synthesis, the most popular principle is to use a chain of mixers, filters and amplifiers but with all the signals that drive the mixers being derived from a single source.

Practically, this basic principle is often improved by using the so-called mix and divide system in which a frequency divider is associated to each mixer of the chain (Fig. 7.2). The main advantage of this type of synthesis is that very small frequency increments can be generated by cascading the adequate number of stages since the basic increment is divided by D^n at the output if D is the division factor and n the number of stages. This greatly simplifies the problem of filtering that occurs when very narrow increments are needed. Let us now come back to Fig. 7.2 to explain briefly how this type of synthesizer operates. As shown in the diagram, there are two different parts that are the divide and mix part and the reference generator section. The latter was represented here as a chain constituted with a stable reference oscillator driving a comb generator followed by a bank of narrow band filters that feed a switching array.

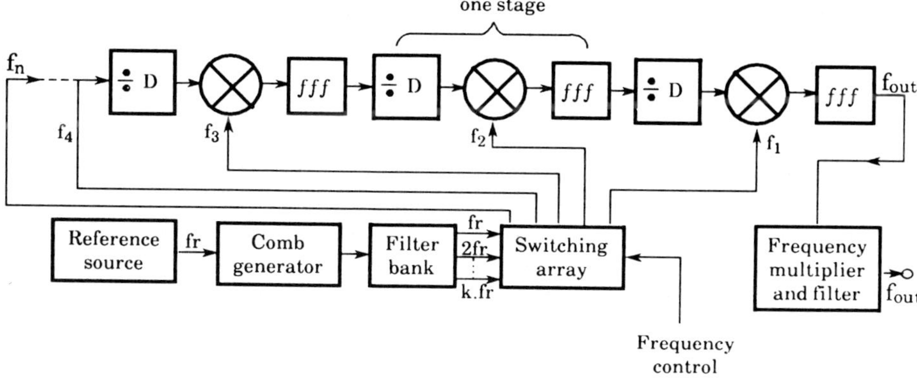

Fig. 7.2 Block diagram of a typical coherent direct synthesizer.

As in non-coherent synthesis, the reference source may be a crystal or surface acoustic wave (SAW) oscillator or even a microwave source but in this case, the designer will keep in mind there may be a difficult problem to solve when implementing the switching array. This is the reason why the reference sources generally used are often XO or SAW oscillator devices. Making comb generators is not particularly difficult whatever the frequency is and the most interesting way is to use a step recovery multiplier that allows easy generation of large harmonic spectra. As already mentioned just before, the main difficulty may arise from the implementation of the selection matrix since each mixer must receive one after the other, each of the harmonics from the comb generator. This leads to the use of multiple switches with high isolation and to arrange all the circuits conveniently in the matrix so as to avoid leakages, coupling between mixer input connections and so on. Note also there may be some filtering problems when very high orders of harmonics are used, but these orders generally do not exceed ten, so filters can easily be implemented.

The mix and divide section of the synthesizer consists of a sequence of quasi-identical stages containing a frequency divider followed by a mixer and a bandpass filter. For simplification purposes, necessary amplifiers are not represented in Fig. 7.2. At the output of the last stage, a frequency multiplier may be used to increase the frequency of the synthesized signal. This is obviously the case when the synthesis utilizes XO or SAW oscillators and the output must deliver a microwave signal. The second input of each mixer is connected to the switching matrix so as to be fed with the signals generated in the reference generator section. For easier comprehension, mixer input frequencies are indexed f_1 for the last (output) stage to f_n for the first stage. Any frequency $f_1, f_2, \ldots, f_n$ can be equal to f_r or to one of its harmonics $k \cdot f_r$ (k is an integer), according to the selection made through the switching matrix.

Thus, the total number of frequencies available at the synthesizer output is as follows:

$$N_t = k^n = k^{m+1} \tag{7.3}$$

m being the total number of mixers used in the device and $n = m + 1$ the total number of reference signals applied to the mix and divide chain. As the number of harmonics cannot be increased beyond a reasonable value (because of filtering problems), the total number of frequencies will be essentially determined by the number of stages used to make the synthesizer.

If D is the division factor of each stage, the expression of the output frequency is:

$$f_{out} = f_1 + \frac{f_2}{D} + \frac{f_3}{D^2} + \ldots + \frac{f_{n-1}}{D^{n-2}} + \frac{f_n}{D^{n-1}}. \tag{7.4}$$

The output frequency may also be expressed as a function of the reference frequency f_r (i.e. the XO frequency) since $f_x = k_x \cdot f_r$. So equation (7.4) may be written:

$$f_{out} = \left(k_1 + \frac{k_2}{D} + \frac{k_3}{D^2} + \ldots + \frac{k_{n-1}}{D^{n-2}} + \frac{k_n}{D^{n-1}} \right) f_r \tag{7.5}$$

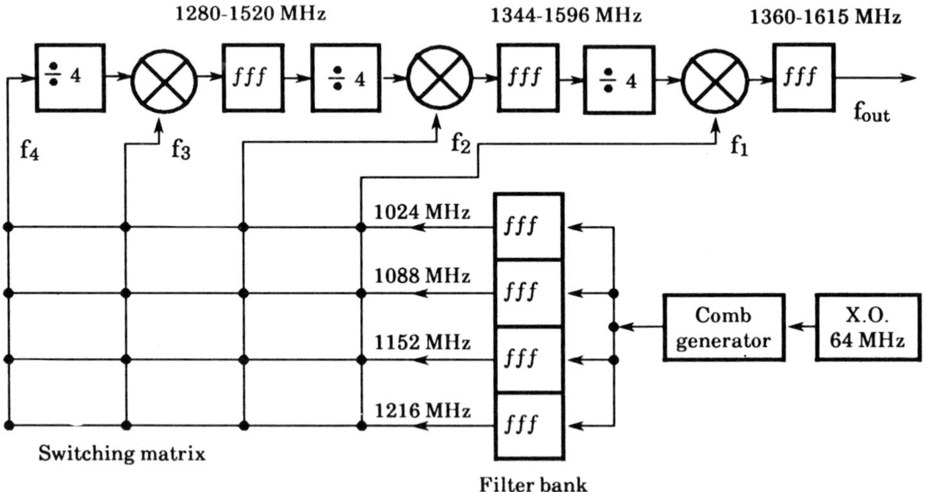

$$f_{out} = f_1 + \frac{f_2}{4} + \frac{f_3}{16} + \frac{f_4}{64}$$

Fig. 7.3 An example of coherent direct synthesis used in a Thomson-CSF communication system.

with k_x varying from 1 to $k_{x\,\max}$. For regularly spaced frequency within the overall bandwidth, the number of reference frequencies $k_{x\,\max}$ must be (at least) equal to the division factor D.

An example of coherent direct synthesis is represented in Fig. 7.3. This design was done by Thomson-CSF for a satellite communications project. The reference source delivers four coherent signals generated from a single crystal oscillator operating at 64 MHz. The switching matrix consists essentially of four SP4T (single pole 4 throw) switches that receive the four reference signals, each of the swiches having its output connected to one of the four inputs of the mix and divide chain. The divide ratio is equal to the number of reference frequencies (i.e. $D = 4$). Out-of-band spurious elements are rejected by means of filters located at the mixer outputs. This synthesizer operates in the 1360 to 1615 MHz bandwidth with a minimum frequency increment of 1 MHz (the total number of channels is 256). The switching time is lower than 2 µs to reach the phase steady state with an error of less than 5 °. The noise level is characterized by a PSD of less than – 110 dBc/Hz at a 100 kHz offset frequency, – 135 dBc/Hz at 3 MHz and – 142 dBc/Hz at 15 MHz and the spurious elements are less than 55 dBc. This performance was measured over the temperature range of – 5 to + 70 °C.

7.2.2 Indirect synthesis

Indirect synthesis is a very common type of frequency generation that utilizes an oscillator to generate its output frequency. That oscillator is generally controlled by a phase locked loop (PLL).

Brief theoretical study

The following few pages are intended to give the reader some basic concepts that are helpful to understand PLL synthesizer operation and to go thoroughly into this study. Details on PLL synthesizers theory (and applications) may be found in numerous specialized articles and books (V. Manassewitsh, 1980, N.F. Egan, 1981, A.J. Viterbi, 1966).

The general block diagram of a typical microwave synthesizer is shown on Fig. 7.4. Two reference signals at low and high frequencies (f_{RL} and f_{RH}) are generated from the reference source (at f_{RO}) by frequency division and multiplication. The output signal at f_{OUT} is delivered by a voltage controlled oscillator (VCO). A microwave mixer is used to generate a signal, the frequency of which is the difference between f_{RH} and f_{OUT}. The resulting frequency is divided by N in a variable ratio divider. A phase comparator delivers an error signal which varies with the phase difference between the signal at f_{RL} and the signal at $|f_{OUT} - f_{RH}|/N$. That signal is amplified and filtered to slave the frequency of the VCO.

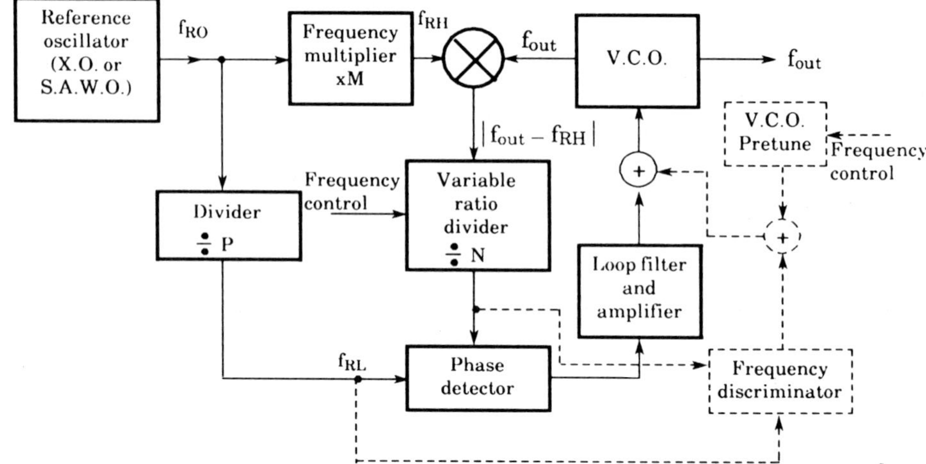

Fig. 7.4 Basic PLL microwave synthesizer with XO or SAW oscillator source.

Figure 7.5 shows a block diagram which is the mathematical representation of the basic PLL synthesizer depicted in Fig. 7.4. The variables at the input and output are frequencies but it is possible to replace them by phases.

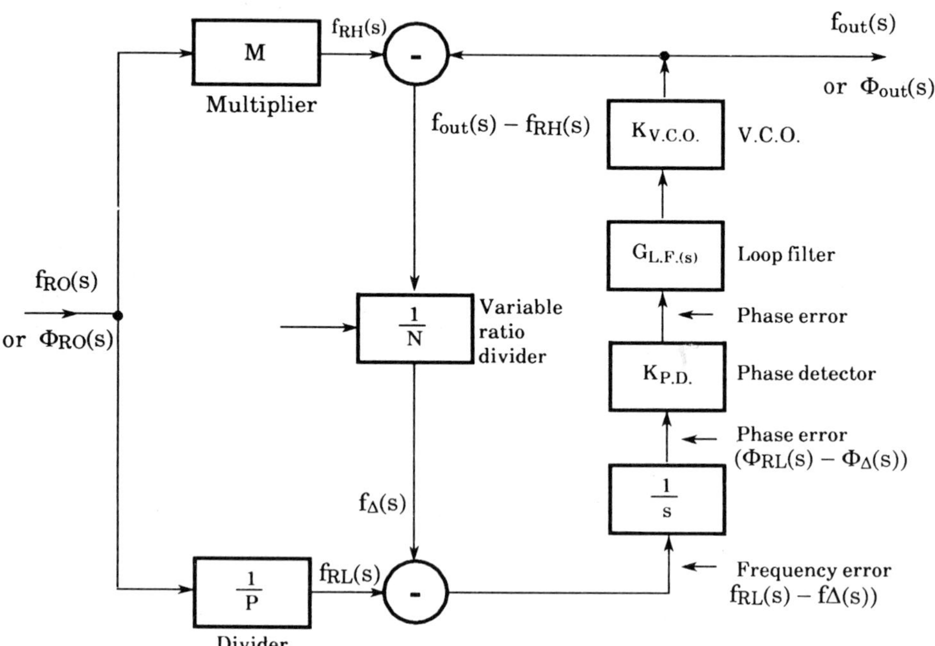

Fig. 7.5 Mathematical representation for basic PLL synthesizer of Fig. 7.4.

Mathematical relations hereafter established are derived for the overall device, i.e. it is the ratio f_{OUT}/f_{RO} that was considered but separate relations between f_{OUT} and f_{RH} or f_{OUT} and f_{RL} could be established in the same way in order to examine independently the behaviour of the output versus each of the two input references.

It is assumed that the system operates with small signals, i.e. under linear conditions, and the Laplace transform is used to make the calculation. The VCO slope, K_{VCO}, i.e. the ratio of fiequency to control voltage, is supposed to be a constant and also the phase detector slope K_{PD} which is the ratio of voltage to phase. (Even if the phase detector is a mixer, its sinusoidal characteristics may be assimilated to its tangent at the zero volt point.) Other constant parameters with respect to the Laplace variables are the multiplication factor M producing the high frequency reference f_{RH} and the fixed and variable division ratios P and N used respectively to generate the low frequency reference f_{RL} and the frequency variation of the synthesizer. As the phase detector delivers a voltage and this resulting voltage is applied to the VCO input so as to control its frequency, transformation between phase and frequency must be translated in Laplace notation. This corresponds to the $1/s$ function introduced in the block diagram which means that frequency difference is integrated to give phase difference. To close the loop, the loop filter is added so as to filter spurious signals and noise, to amplify the phase error and to ensure loop stability. That function is noted $G_{LF}(s)$ as it is a frequency dependent quantity. The frequency and phase differences are simply materialized by two subtractors in the block diagram.

Assuming, for instance, $f_{OUT} > Mf_{RO}$, the closed loop transfer function is:

$$F(s) = \frac{f_{out}(s)}{f_{RO}(s)} = \left(M + \frac{N}{P}\right) \frac{K_{VCO}\, G_{LF}(s)\, K_{PD}/N}{s + K_{VCO}\, G_{LF}(s)\, K_{PD}/N} = \frac{\Phi_{out}(s)}{\Phi_{RO}(s)} = \left(M + \frac{N}{P}\right) H(s). \tag{7.6}$$

This may also be found by means of feedback theory, that is by application of the general rule: the transfer function from any point in the loop to the output is equal to the ratio between the forward gain from that point to the output (here $K_{VCO}\, G_{LF}(s)\, K_{PD}/s$ from the phase detector inputs) and the sum [1 + (open loop gain)], (here: $1 + K_{VCO}\, G_{LF}(s)\, K_{PD}/Ns$).

Examination of equation (7.6) where $s = j\omega$, shows that the output signal at frequency f_{OUT} is slaved to the reference frequency f_{RO} (or to f_{RL} and f_{RH} together). At low offset frequencies, $K_{VCO}\, G_{LF}(s)\, K_{PD}/N$ is generally large and constant so $j\omega$ may be neglected and the spectrum of the output signal is that of the reference signal multiplied by $(M + N/P)$.

On the other hand, at far offset frequencies, the loop gain $G_{LF}(s)$ decreases and $j\omega$ becomes larger and larger so that input and output frequencies are finally related by a complex function and this means the output spectrum is no longer an image of the input spectrum. In other words, the phase locked loop is a low-pass filter that can be used in frequency synthesis to copy the spectrum of one (or several) reference signals. Efficiency of copy can be limited to the lower part of the reference spectrum by selection of the loop cut-off frequency.

Let us now consider the simplest PLL that can be made. It is a loop with a constant gain ($G_{LF}(s) = K_{LF}$) with respect to the offset frequency. The open loop transfer function is then $-jK/N\omega$ so the phase between input and output is $-\pi/2$. Then it is clear that whatever the gain, the loop will be stable, since at the frequency where the open loop gain is 1 (or 0 dB) there is still a $\pi/2$ phase margin that prevents the loop from oscillating, as illustrated in Fig. 7.6. Also represented in this figure is the asymptotic closed loop transfer function. In addition to the theoretical Bode plot, influence of spurious elements has been represented by means of dashed curves. It was supposed here that there were two spurious low-pass filters and for example they may be one of the loop amplifiers whose gain decreases at high frequencies and one of the filters at the VCO input (high input impedance and parasitic parallel capacitor). Their influence is not only to lower the loop gain, but also to introduce a phase shift that quickly reduces and even cancels the phase margin. This is the case in Fig. 7.6 where it can be seen that the actual closed loop is quite unstable since the open loop gain is greater than unity when the phase is equal to 180 °. As a consequence, the designer has to define a loop filter to avoid loop instabilities. Numerous solutions can be used and the convenient solution depends on certain characteristics of the synthesizer. As an example, the case of the lag-lead filter (Fig. 7.7), which is particularly suitable for wide acquisition and hold in ranges, is briefly described hereafter. The transfer function of this filter is:

$$G_{LF}(s) = K_{LF}\frac{1+\tau_2 s}{1+\tau_1 s} = K_{LF}\frac{1+(s/\omega_2)}{1+(s/\omega_1)}$$

so the open loop gain is:

$$G_{OL}(s) = \frac{K_{VCO}K_{LF}K_{PD}}{Ns}\frac{1+\tau_2 s}{1+\tau_1 s} = \frac{K}{N}\frac{1+\tau_2 s}{s(1+\tau_1 s)}. \tag{7.7}$$

Figure 7.7 shows the phase shift increases at low frequencies then decreases at higher frequencies so it can be seen that proper location of the cut-off frequencies of the lag-lead filter with respect to the parasitic cut-off frequencies may ensure a stable operation of the loop. As a matter of fact, when the decreasing part of the phase shift (from about $-\pi$ to $-\pi/2$) is properly chosen, it counteracts the spurious phase shift, thus increasing the phase margin around the point where the open loop gain is equal to unity. Of course, the d.c. gain (gain at zero frequency) is also of importance for loop stability and the designer has to properly combine its actions on cut-off frequencies and overall gain to achieve a good compromise. The transfer function of the loop when the loop filter is a lag-lead circuit is expressed by the following relation:

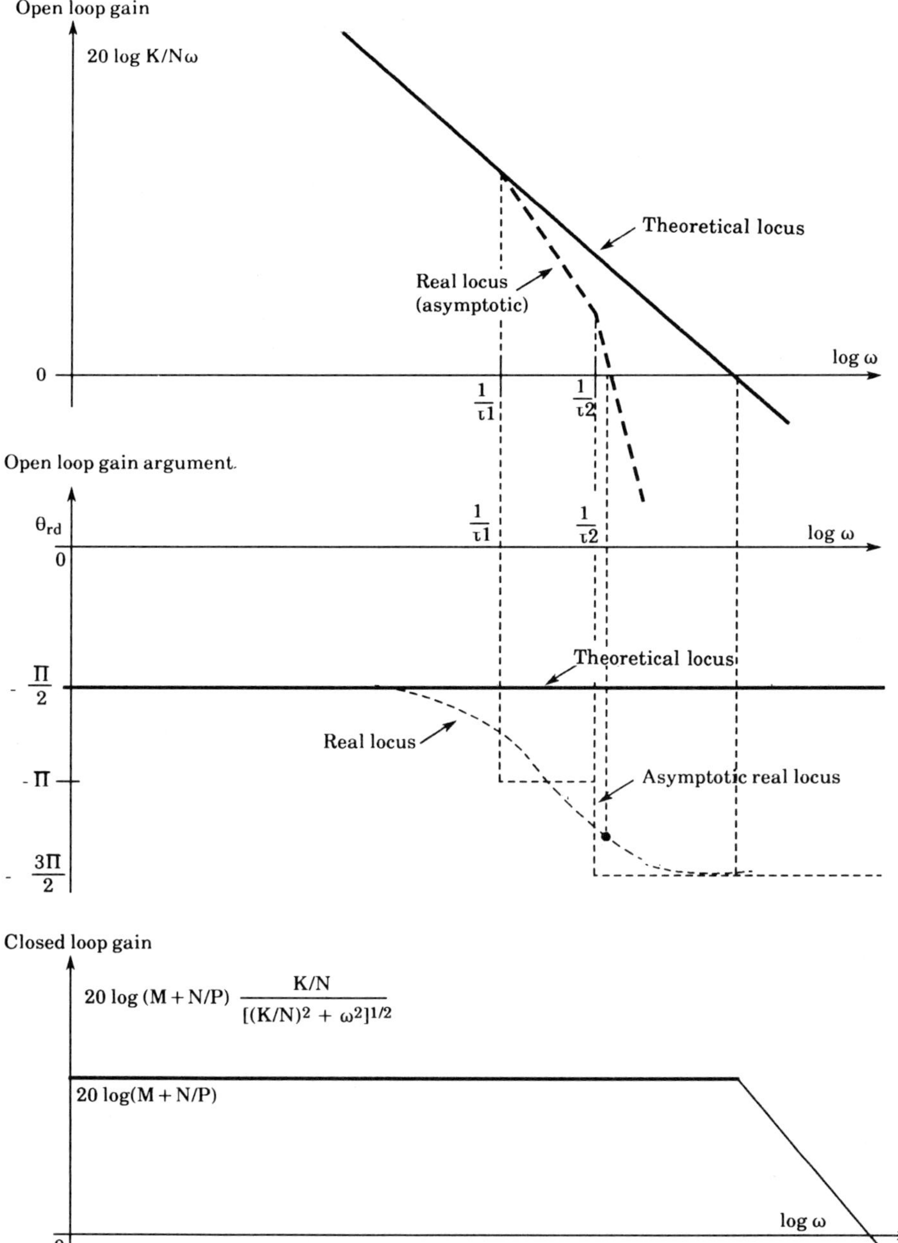

Fig. 7.6 Bode plot for a first order loop.

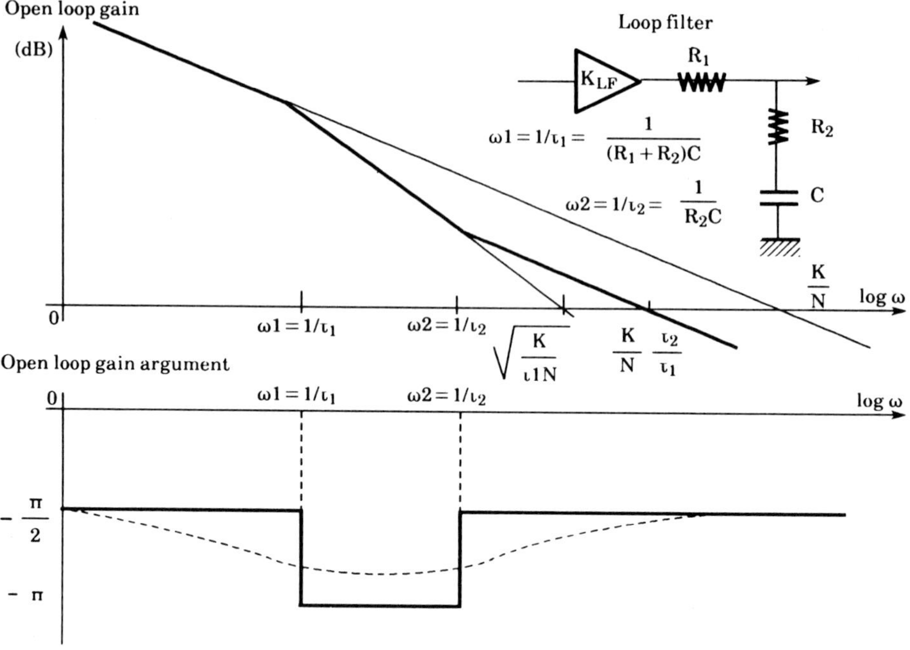

Fig. 7.7 Bode plot for a second order loop with lag-lead filter.

$$F(s) = \frac{f_{out}(s)}{f_{RO}(s)} = \left(M + \frac{N}{P}\right) \frac{\frac{K}{N}(\tau_2 s + 1)}{\tau_1 s^2 + [1 + (K/N)\tau_2] s + \frac{K}{N}} = \frac{\Phi_{out}(s)}{\Phi_{RO}(s)}. \tag{7.8}$$

Note that all the previous relations are only valid under linear operation which is not the real case during loop transients.

PLL synthesizer examples

To illustrate the previous sections, two examples are now briefly described. The first one is a single reference PLL synthesizer used in a Thomson-CSF DME (distance measurement equipment) whose block diagram is represented in Fig. 7.8. Its operation bandwidth is 962 to 1213 MHz with a 1 MHz frequency increment. This equipment was designed to be as cheap as possible and without severe constraints on its settling time. This, and also the fact that only medium spectral purity was required, allows the use of a narrow loop bandwidth. The synthesizer was manufactured using only off-the-shelf components: most of them are integrated circuits but the VCO is a hybrid component. As no variable ratio divider working at L-band was available, an ECL fixed ratio divider was used to lower the frequency at the variable divider input thus reducing the phase detector reference frequency to 125 kHz. So the frequency of the stable reference crystal oscillator is divided by 32 to generate the 'step reference' of the synthesizer. The main characteristics of the device (over the temperature range – 10 to + 55 °C) are a long term frequency stability of ± 10. 10^{-6}, a PSD of – 110 at a 100 kHz offset frequency and spurious elements at less than 60 dBc. The 251 frequencies (with a 1 MHz step) are manually controlled.

The second example (Fig. 7.9) is a double loop radar synthesizer from Thomson-CSF operating between 4560 and 5060 MHz with a 10 MHz frequency increment. The main requirements for this device were low noise and (relatively) fast switching time. The double loop architecture was chosen in order to minimize noise. The first loop generates a stable signal whose frequency varies by 10 MHz increments (fine step) while the second has its frequency changed by 100 MHz steps (coarse step).

Also, to reduce noise, the division ratio of the second loop is only four or five, the complementary variation being generated by means of a divider ($P = 3$, 4, 6 or 12) located out of the loop so there is no multiplication effect (P^2) on the loop noise. The low noise reference is delivered by a hybrid crystal oscillator whose frequency is multiplied by 35 (7 × 5) to provide the high frequency reference of the first loop. A frequency control system converts the frequency and switching data into signals that drive the programmable dividers and the VCO pre-tunings. A similar synthesizer has also been made

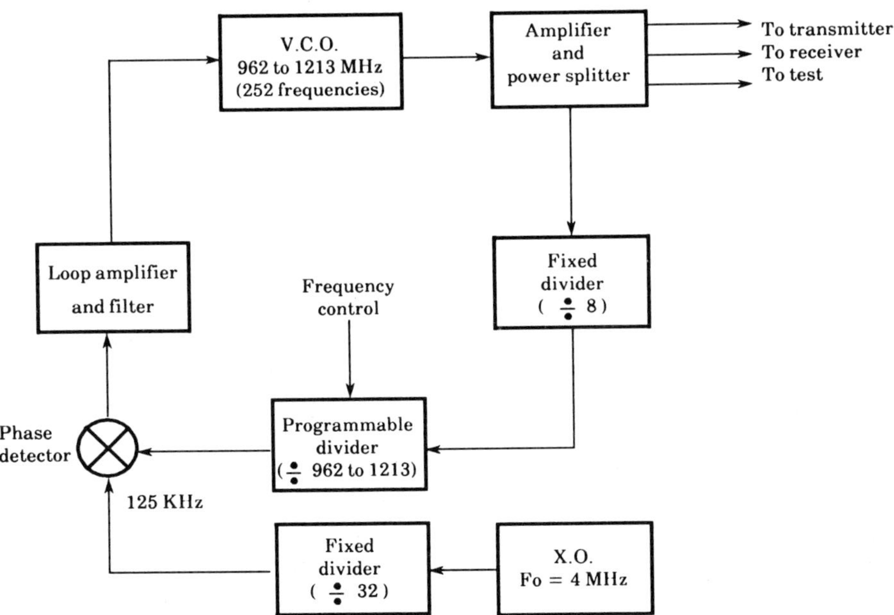

Fig. 7.8 An example of simple PLL synthesizer in a Thomson-CSF DME application.

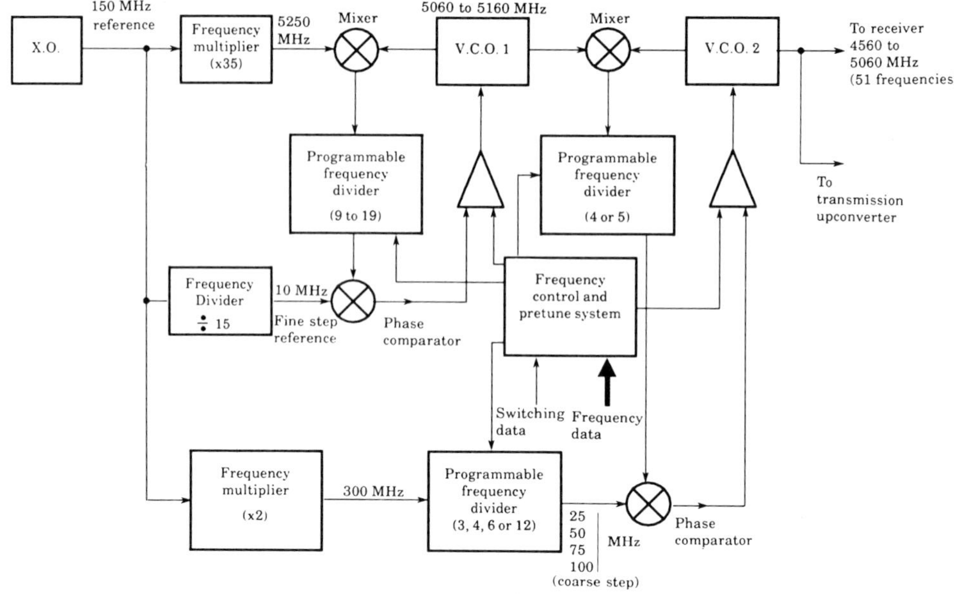

Fig. 7.9 An example of Thomson-CSF radar PLL synthesizer.

for radars operating over 450 MHz at S-band with a 5 MHz step. The main characteristics of the C-band synthesizer are a settling time of less than 35 μs (for the Doppler filtering compatibility) or less than 10 μs without phase stability condition and a PSD of – 115 dBc at 100 kHz from the carrier. Spurious emissions lie at less than 50 dBc.

7.3 PIEZOELECTRICITY FOR SYNTHESIZERS

Since the discovery of the piezoelectric effect, at the end of the last century, and its first professional application to control the frequency of a radiobroadcast in New York in 1930s, progress in the field was directed toward two main areas: improved stability, and increased frequency.

Piezoelectricity (from the greek *piezo* which means *pressure*) describes an electrical polarity that appears on some crystals submitted to mechanical stresses. This property is anisotropic and appears only in non-centrosymmetric crystals (classes 32, 3 m, etc.).

Piezoelectric resonators present a voltage polarity associated with an acoustic standing wave of motion in a piezoelectric material at given frequencies (called resonant frequencies).

Piezoelectric devices (i.e. filters and oscillators) are used in various equipment such as radars, communication networks, positioning, etc.). They can be used as time or frequency references, HF source, local oscillators or filters in modulation/demodulation or synthetizers. These applications cover a wide range of frequencies (1 MHz to 1 GHz) using various types of oscillators (clock, TCXO, VCXO, OCXO, ...).

Digital communications networks call for specific requirements such as low phase noise oscillators and wide band linear phase filters. Increasing intermediate frequencies (up to 70/100 MHz) call for high frequency fundamental mode resonators. Recent developments toward high frequency, wide band devices are mostly based on ion etched (or chemical etched) resonators and on the introduction of new piezoelectric materials ($LiTaO_3$, $AlPO_4$) which offer stronger piezoelectric coefficients than quartz.

7.3.1 Quartz material and cuts

Quartz is the α phase of crystalline SiO_2 around ambient temperature. A β phase of SiO_2 also exists at temperatures above 540 °C.

Natural quartz was used, up to the 1970s, primarily to build resonators. Most of the mines were in Brazil and Arkansas. Synthetic quartz is now the most widely used. Growing is performed in high temperature, high pressure autoclaves, in an aqueous solution of $NaCO_3$ or NaOH.

Raw material is placed in the bottom, seeds in the upper area, and a small temperature difference is maintained between these two areas. Raw quartz dissolves in the high temperature, lower liquid and recrystallizes, preferably on the seeds in the low temperature, upper side of the autoclaves.

For radiation hardening or aging purposes, chemical purification can be performed by 'sweeping': an electromigration process of free ions in the crystalline matrix under high voltage and high temperature. High crystalline quality, with low dislocation content, is available for chemical etching of high frequency devices.

The low frequency resonators were longitudinal (compression) or flex modes of motion in plates close to the 'X' cuts (a cut is identified by the direction of the normal of its main face).

Today's resonators, which cover the 1 MHz to 1 GHz range, use thickness shear waves of AT, BT or doubly rotated cuts shown in Fig. 7.10.

7.3.2 Piezoelectric resonators

Propagation of acoustic waves in a piezoelectric medium

Acoustic waves in a piezoelectric medium are solutions of a differential equation of motion which is established from the relations:

$$T_{ij} = C_{ijkl} S_{kl} - e_{mij} E_m$$

$$D_n = \varepsilon_{mn} E_m + e_{nkl} S_{kl} \qquad (7.9)$$

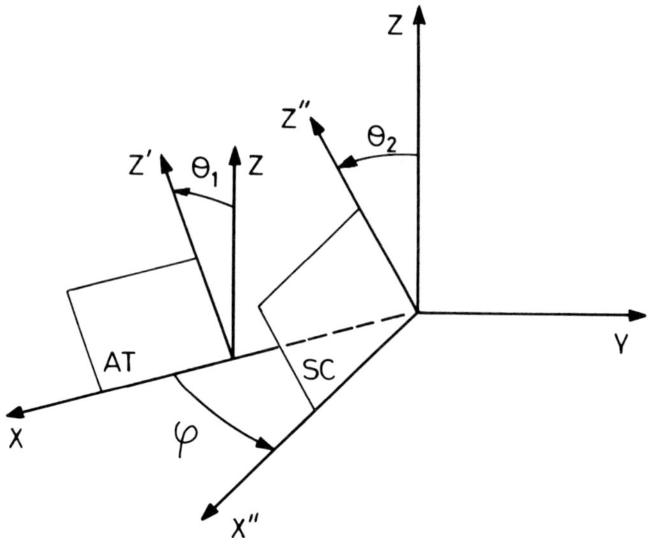

Fig. 7.10 Thickness shear geometry in quartz piezolectric resonators.

T_{ij}, is the stress tensor, E_m and D_n are the electric field and electric displacement; C_{ijkl}, e_{mij}, ε_m are the elastic, piezoelectric and dielectric parameters. S_{kl} is the strain tensor described on first order of displacement, by:

$$S_{kl} = \frac{1}{2}(U_{k,l} + U_{l,k}). \tag{7.10}$$

Equations of dynamic and electromagnetism give:

$$T_{ij,i} + F_j = \rho\, \ddot{U}_j \tag{7.11}$$

Plane waves propagating along an S-direction are

$$U_j = a_j \exp\left[j\omega\left(t - \frac{s}{V}\right)\right]$$

$$\phi = a_4 \exp\left[j\omega\left(t - \frac{s}{V}\right)\right] \tag{7.12}$$

with $S = n_l x_l$ and $V = \omega/k$.

Equations (7.9), (7.10), (7.12) in (7.11) lead to a determinant relation

$$[\Gamma'_{ijk} - \rho v^2 \delta_{jk}] \cdot a_j = 0 \tag{7.13}$$

where

$$\Gamma'_{jk} = \Gamma_{jk} + \frac{\gamma_j \gamma_k}{\varepsilon_s}$$

with

$$\Gamma_{jk} = c_{ijkl}\, n_i\, n_l \qquad \gamma_j = e_{mij}\, n_m\, n_i \qquad \varepsilon_s = \varepsilon_{mn}\, n_m\, n_n\,.$$

System (7.13) has non zero solutions in a_j if the determinant of the matrix vanishes.

There are then three solutions of plane waves which satisfy the propagation equations, given by propagating speed and acoustic polarization $U1^{(r)}$, $U2^{(r)}$, $U3^{(r)}$ where $r = 1,\ 2,\ 3$ for the three modes of vibration. One solution is longitudinal waves (polarization parallel to the propagation direction), 2 modes are transverse.

Resonant frequency and equivalent circuit

A resonator is created by a thin disk, metallized on both faces, the direction of the normal to the major faces is then the propagating direction 'S'. A stationary wave is given by:

$$U^{(r)} = a^{(r)} \exp j_\omega \left(t + \frac{S}{V} \right) + b^{(r)} \exp j_\omega \left(t - \frac{S}{V} \right) \tag{7.14}$$

takes place between both faces, for one of the 3 modes 'r'. Mechanical and electrical conditions on each face:

$$T_{ij} = 0 \qquad \text{(for stress free faces)}$$

$$\phi = \pm \frac{V_o}{2} \exp j\omega t \tag{7.15}$$

allow a complete description of the stationary wave.

After some algebric manipulation whose main steps are changing the coordinate base to diagonalize the Γ matrix and applying relations (7.15) in this new base, we can reach the relation

$$K^2(r) \tan \frac{\omega h}{v(r)} = \frac{\omega h}{v(r)} \tag{7.16}$$

where $K^2(r)$ is the piezoelectric coupling coefficient of the mode (r) along direction S, h being half the thickness of the plate.

Frequencies given by $\dfrac{\omega h}{v(r)}\,(2n+1)\,\pi/2$ correspond to stationary waves whose amplitude goes to zero (antiresonant frequencies), those given by (7.16) are stationary waves of maximum amplitude (resonant frequency). These frequencies are related by

$$r_r^{(n)} = r_a^{(n)} \left(1 - \frac{4K(r)^2}{N^2 \pi^2} \right) \tag{7.17}$$

where $N = 2n + 1$ is the overtone rank.

The calculation of the current intensity across the plate ($i = Q$, $Q = OS$, $O = D_m\, N_m$) allows us to write the impedance of the resonator as:

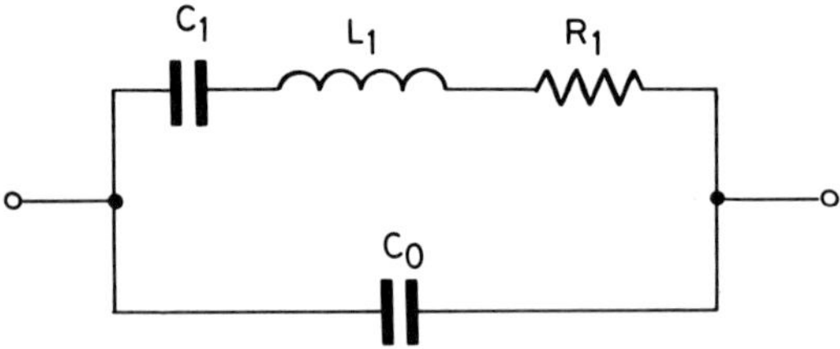

Fig. 7.11 Equivalent circuit of a piezoelectric resonator.

$$Z(\omega) = \frac{1}{jCo\,\omega}\left[\frac{1 - K^2(r)\tan\dfrac{\omega h}{v(r)}}{\dfrac{\omega h}{v(r)}}\right]. \tag{7.18}$$

The resonator can be seen as a capacitor for $\omega < \omega_r$ or $\omega > \omega_a$, and as an inductor between ω_r and ω_a. This impedance is described by an equivalent circuit given in Fig. 7.11. From this circuit, we can write:

$$\frac{\omega_a - \omega_s}{\omega_s} \simeq \frac{1}{2}\frac{C_1}{C_0}$$

or

$$\frac{C_1}{C_0} = \frac{8K^{(r)2}}{\pi^2} \cdot \frac{1}{N^2}. \tag{7.19}$$

This equation gives the main relation between acoustic behaviour and electrical equivalent circuit and is valid at least on first order for large lateral dimensions of piezoids.

7.3.3 Technological trends for VHF resonators

Communication networks require higher and higher frequency generation or filtering. Digital modulation leads to special requirements for filtering such as wide band and linear phase for signal integrity. Equation (7.19) shows that wideband filters call for high frequency fundamental mode resonator and/or high coupling coefficient material.

Classical lapping methods are limited to around 40 to 50 MHz on fundamental AT cut plates (thickness $\simeq$ 30–40 μm). Higher frequency fundamental modes require special techniques to reach a thickness thinner than 30 μm. Great technological efforts were made on this topic and two basic processes were established: chemical etching and ion beam etching.

The first one is strongly material sensitive. It requires high quality, low dislocation quartz plates and it does not apply to some other materials such as lithium tantalate. The second one is described on Fig. 7.12.

The beam of accelerated argon ions (C) is driven onto a 50 μm thick quartz plate (E) through a glass mask (D). The etching rate is approximately 7 to 10 μm per hour to keep a good surface quality. The ion etched resonator is shown on Fig. 7.13. A central area only was etched and the 50 μm thick surrounding ring allows mechanical strength for manipulation.

These resonators range up to 300 MHz (≃5 μm) on a fundamental mode and higher than 1 GHz on overtones. Measurements must be performed by S parameter methods to take into account parasitic capacitances (holder, pin, ...).

7.3.4 Filters

Filters are widely used in communications, synthesizers, radars, etc. to select given frequencies or bands in a complex spectrum. Crystal filters take advantage of resonator technologies such as high fundamental frequency or high coupling material, to cover many different fields.

To reduce cost and size, monolithic coupled mode resonators (i.e. two or more resonators on the same plate) technology is used. Design criteria take into account lateral dimensions (metallization, interelectrode gap) and electrode thickness. Energy trapping by electrodes allows control of acoustic active modes to minimize spurious resonances. Lithium tantalate allows wide band filters to be made, such as IF (Fo 10, 7 MHz, BW 3 dB = 200 kHz) and ion etched monolithic allows VHF filters (Fo 110 MHz, BW 3 dB = 200 kHz); both being used in digital communications.

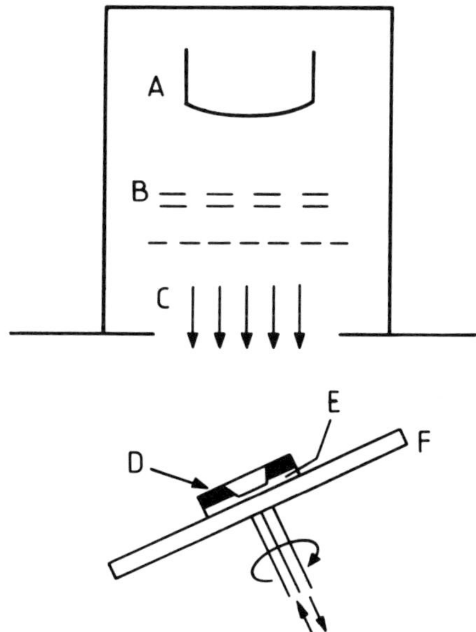

Fig. 7.12 Argon ion beam etching apparatus.

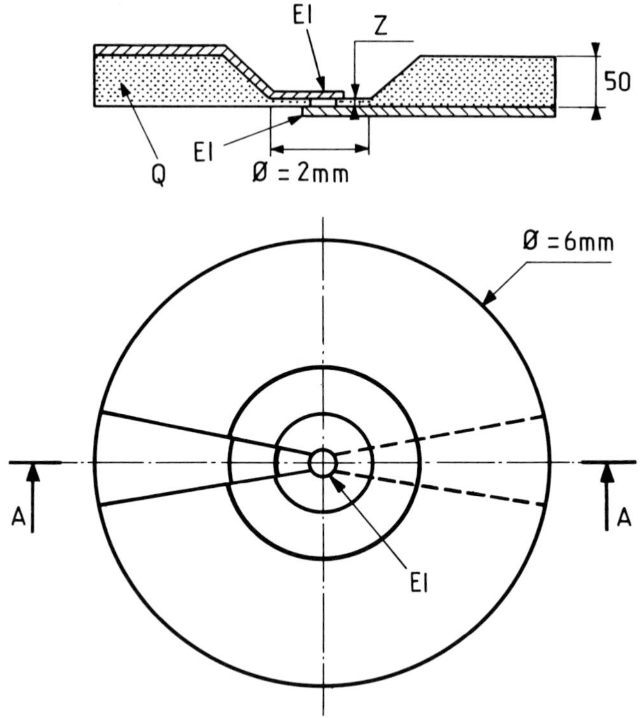

Fig. 7.13 Argon ion etched resonator.

7.3.5 Oscillators

Various kinds of oscillators can be built around a piezoelectric resonator. Depending on the required stability, we may want

1. clock oscillators: some 10^{-5} in temperature range – 20 to + 70 °C;
2. temperature compensated crystal oscillators (TCXO) either analogue ($\simeq 10^{-6}$) or digital ($\simeq 10^{-7}$);
3. Ovenized oscillators (10^{-8} to 10^{-10} in temperature).

These oscillators may have frequencies in the kHz to GHz range, the later ones using ion etched resonators described in section 7.3.3 of this chapter.

In addition to frequency and frequency vesus temperature stability, the main criteria a system designer must specify are:

1. aging;
2. phase noise;
3. consumption, supply;

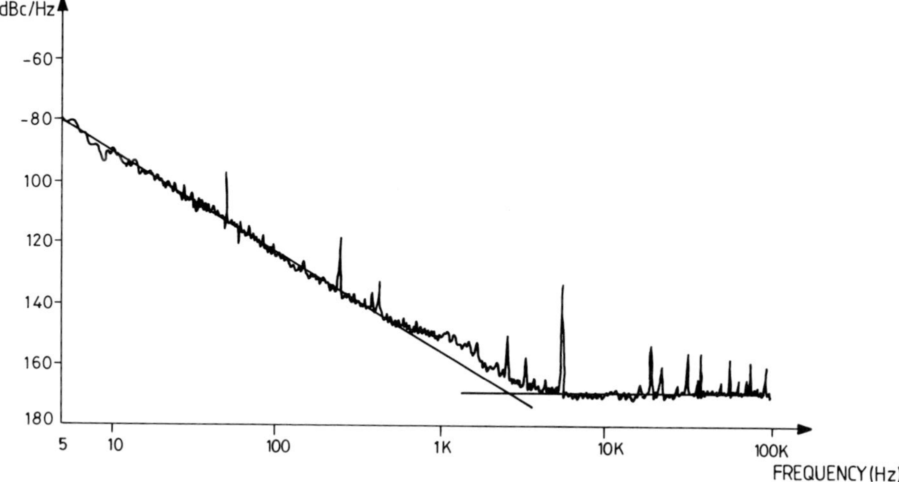

Fig. 7.14 Measured performance of a reference oscillator for airborne radar applications.

4. environmental sensitivity, among which sensitivity to accelerations can affect the phase noise spectrum under vibrations.

This last criterion is mostly sensitive in airborne radar or communication systems and in positioning equipment. Figure 7.14 gives, for example, the performance of an up-to-date reference oscillator for airborne radar.

The phase noise performance ($\simeq -170$ dBc/Hz at 5 kHz offset from the carrier) is very lightly affected by random vibrations up to 0.2 g^2/Hz. In this

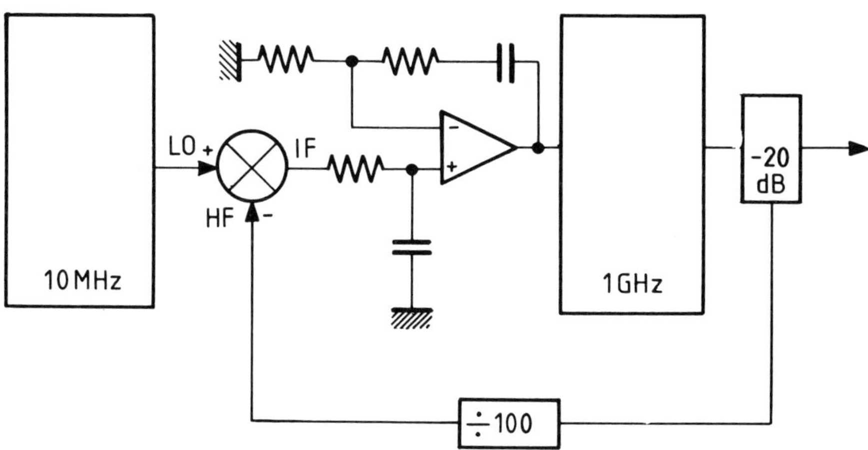

Fig. 7.15 Typical phase locked oscillator configuration for radars, synthesizers, or high frequency sources.

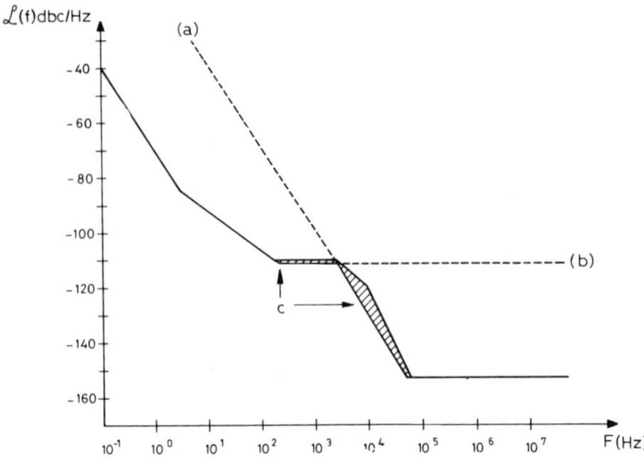

Fig. 7.16 Phase noise spectrum of the feedback controlled 1 GHz direct oscillator.

oscillator, g sensitivity of the resonator is lower than 5×10^{-10}/g. Figure 7.15 gives typical phase locked configuration which can be used either in radars, synthesizers or high frequency sources. In this example, a 1 GHz bulk wave oscillator is locked on to a high stability 10 MHz one.

Curve (a) on Fig. 7.16 gives the phase noise spectrum of the 1 GHz direct oscillator. The noise floor is very low (– 155 dBc/Hz) but the spectrum close to the carrier is affected by the relatively low Q of the 1 GHz resonator ($Q \simeq 30.000$ at 1 GHz for BT cut 5th overtone quartz). Curve (b) gives the spectrum of a 10 MHz × 100. Due to high Q ($Q \simeq 1.3 \cdot 10^6$ at 10 MHz), the spectrum decreases quickly close to the carrier, but, because of the multiplication rank of 100, the noise floor is poor. Curve (c) gives the spectrum of the 1 GHz output locked by division on the 10 MHz reference. Such a configuration allows the benefit of the main advantage of each separated oscillator: low noise floor, low noise close to the carrier, and low ageing.

REFERENCES

Egan, N. F. (1981)
Manassewitsh, V. (1980)
Viterbi, A. J. (1966)

Part Two
Microwave Hybrid and Integrated Circuits

8

Microwave solid-state amplifiers

Yves Charlet

8.1 INTRODUCTION

What are the main characteristics of microwave solid state amplifiers?

Currently, the frequency range covered by the microwave solid-state amplifiers goes from about 1 GHz to more that 40 GHz. But, as RF components are now working from 1 MHz to more than 4000 MHz, this chapter will also describe these amplifiers. The RF range will be devoted to the silicon transistor amplifiers and the microwave range to the gallium arsenide transistor amplifiers.

In terms of output power, generally speaking, it is from 10 to 100 mW (+ 10 to + 20 dBm). The medium power amplifiers (in solid state) begin at 0.5 W, with an upper limit of 50 W—higher power is obtained by the combination of many elementary amplifiers.

For the noise figure (Fig. 8.1), the boundaries are about 2 dB in the L-band (1 to 2 GHz), 3 dB in X-band (8 to 12 GHz) and 6 dB in Ka-band (26 to 40 GHz). Below these limits are the low-noise amplifiers (LNA) and above are the others. This boundary changes with the relative bandwidth: for a wideband amplifier, say 6 to 18 GHz, an LNA will have a noise figure less than 5 dB.

The gain of solid-state amplifiers is generally not critical, because of the modular aspect of their construction. The gain will go from 5 dB up to 70 dB, only by cascading elementary stages, whose minimum gain values are fixed by the transistor.

The last important characteristic is the size. Often, microwave solid-state amplifiers are small compared to the tubes, but it depends on the technology used by the manufacturer. It could be monolithic microwave integrated circuits (MMIC), microwave integrated circuits (MIC) or common printed circuits, then the size will go from a few hundred square micrometres to a few square inches.

With such differences of frequency, power, noise and size, the applications of microwave solid-state amplifiers are numerous: radiolinks, TV transmission and reception, radar, electronic counter-measures (ECM), test equipment, etc.

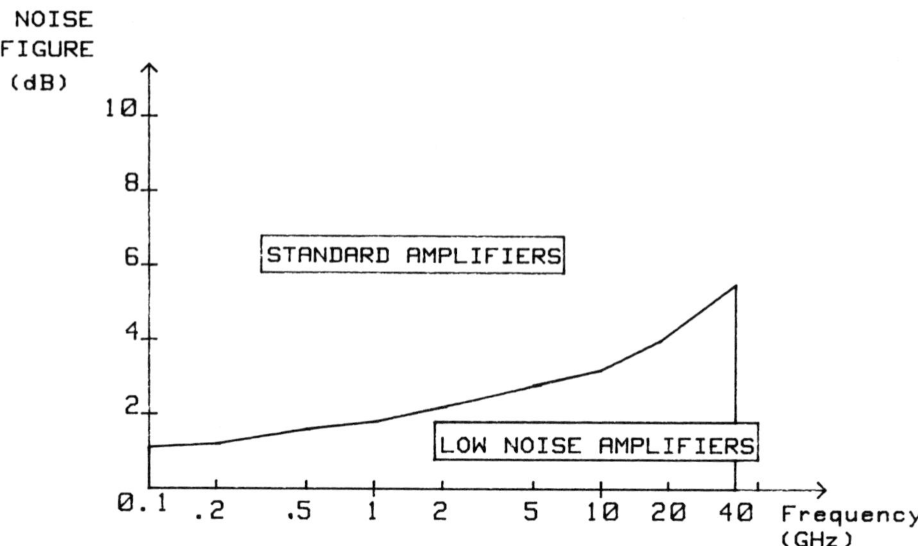

Fig. 8.1 Noise figure of standard and low noise amplifiers.

What are the definitions of the main electrical parameters?

1. The bandwidth defines the frequency range where the different parameters are guaranteed. In general, the amplifier response exceeds this range.
2. The output power. The 1 dB gain compression value defines the linear operating zone of the amplifier. It is the output power obtained when the gain has decreased by 1 dB as compared to its low level value. The saturation power is the maximum output power of the amplifier. The power P_o is given in dBm (10 log P_o/1 mW) or in dBW (10 log P_o/1 W) and of course in watts, or milliwatts.
3. The noise figure of a device defines signal-to-noise ratio degradation between input and output. It is expressed in dB (the reference noise power is the thermal noise of a 50 Ω load, at a 290 K temperature).
4. The gain is defined as measured in a 50 Ω system for low power levels ensuring linear operation of the amplifier. It is the ratio between the output level and the input level, expressed in dB (10 log P_{out}/P_{in}).
5. The voltage standing wave ratio (VSWR) expresses the mismatch of microwave ports of the device with respect to 50 Ω.
6. The group delay is the instant angular phase slope versus aneous frequency. It is expressed in nanoseconds.

8.2 SILICON TRANSISTOR AMPLIFIERS

As stated above, these amplifiers are used from 1 MHz (or less) to 4 GHz, even 10 GHz in the latest development areas. The design is mainly related to performance requirements. If we need only a narrow band amplifier, without particular requirements on the gain parameter, we will use a single ended configuration. If we need large bandwidth operation (i.e. 200 to 500 MHz) then we can also try the single ended configuration with the constant gain circle calculations; but probably instability and poor VSWR will drive us to the feedback structure.

8.2.1 Single ended amplifier

The structure of a single ended amplifier is shown in Fig. 8.2. The two matching networks, at the input and the output of the transistor, allow the best power transmission between the generator and the transistor, and between the transistor and the load. The best way to design such an amplifier is to use the scattering parameters of the transistor.

$$\begin{matrix} S_{11} & S_{21} \\ S_{12} & S_{22} \end{matrix}$$

Note that the parameters to match S'_{11} and S'_{22} depend on the load and the source as shown in Fig. 8.3. In this case, if Γ_L and Γ_S are the reflective coefficients of the load and of the source, respectively:

$$S'_{11} = S_{11} + \frac{S_{12}\,S_{21}\,\Gamma_L}{1 - S_{22}\,\Gamma_L}$$

and

$$S'_{22} = S_{22} + \frac{S_{12}\,S_{21}\,\Gamma_S}{1 - S_{11}\,\Gamma_S}\,.$$

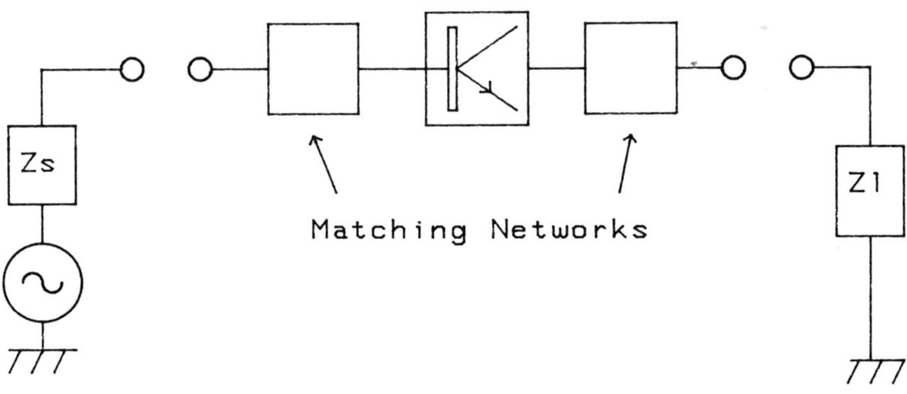

Fig. 8.2 Structure of a single-ended amplifier.

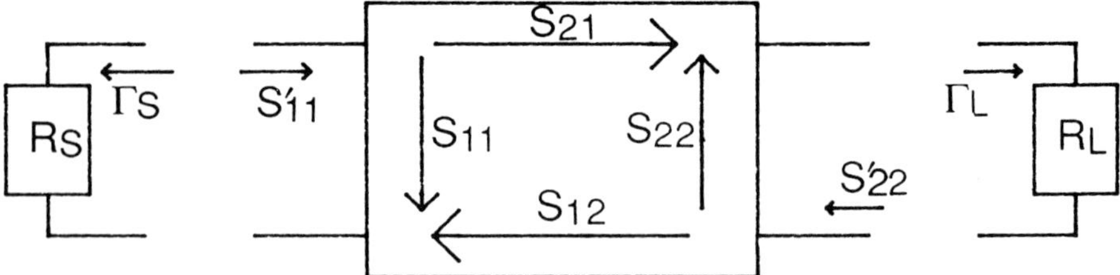

Fig. 8.3 S-matric parameters of a single-ended amplifier.

To match these values to 50 Ω considering the conjuguated values of S'_{11} and S'_{22}, Γ_L has to be transformed in $(S'_{22})^*$ and Γ_S in $(S'_{11})^*$. This gives two equations, with two unknowns S'_{11} and S'_{22}, but don't forget they are vector variables (amplitude + phase). So, with such a simple problem, you already get 4 equations. Fortunately, software exists which can very easily calculate the matching networks (e.g.: ESOPE from Thomson, see Volume 1, Chapter 20). The exact solution will give R_L. Corresponding to Γ_L by the relation:

$$\Gamma_L = \frac{R_L - 50}{R_L + 50}$$

and R_S corresponding to Γ_S by the relation:

$$\Gamma_S = \frac{R_S - 50}{R_S + 50}\cdot$$

Typical matching networks are shown in Fig. 8.4.

In the above example, the amplifier will be a narrow-band amplifier, with a gain value very close to the maximum available gain, G_{max}, where

$$G_{max} = 20 \log |S_{21}| + |10 \log (1 - |S_{11}|^2)| + |10 \log (1 - |S_{22}|^2)|\ .$$

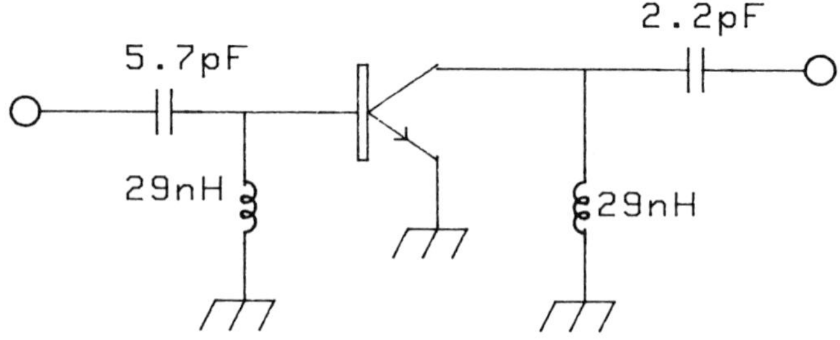

Fig. 8.4 Matching networks calculated for the example of Fig. 8.2.

This formula comes from the general expression of the transistor gain which is:

$$G = |S_{21}|^2 \cdot \frac{1 - |\Gamma_S|^2}{|1 - S_{11}\Gamma_S|^2} \cdot \frac{1 - |\Gamma_L|^2}{|1 - S_{22}\Gamma_L|^2}$$

where $\Gamma_S = S_{11}^*$ and $\Gamma_L = S_{22}^*$

We can, of course, make an amplifier with lower gain and larger bandwidth by choosing different values for Γ_S and Γ_L. The locus of Γ_S for a given value of $1 - |\Gamma_S|^2 / |1 - S_{11}\Gamma_S|^2$ on a Smith chart is a circle, called a constant gain circle. The same can be done at the output with Γ_L. Thus, one can adjust the matching circuits to compensate for the natural gain decrease with frequency of the transistor and obtain a flat gain amplifier in a larger bandwidth. Unfortunately, this mismatching can significantly degrade the input and output VSWR.

8.2.2 Resistive feedback amplifier

As the bandwidths approach a decade of frequency, gain compensation based on matching networks is very difficult. With resistive feedback configuration, we can achieve very large bandwidths such as 10 to 2000 MHz, or even 100 to 4000 MHz, the problems coming from the polarization circuits and not from the RF part.

A common design (without coupling capacitors and bias network) is shown in Fig. 8.5. This structure can accept variations in the S parameters from transistor to transistor (with minimum values of course), reduces the gain variation with the temperature, and gives flat response and good VSWR.

The series feedback resistor R_S will be generally associated with a capacitor C_S to add extra correction, i.e. in the high frequency range, R_S is shunted and

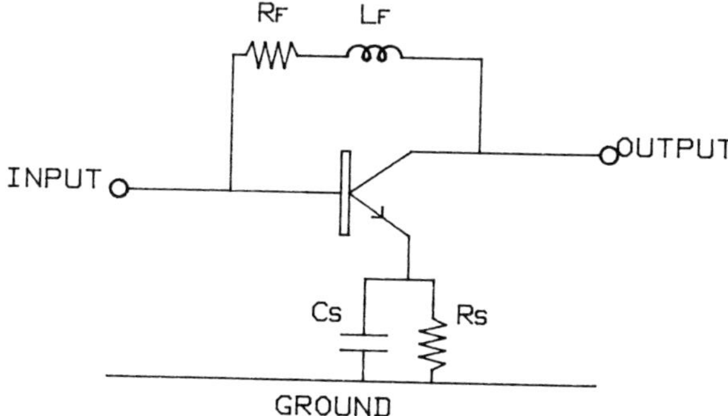

Fig. 8.5 Simple configuration of a resistive feedback amplifier.

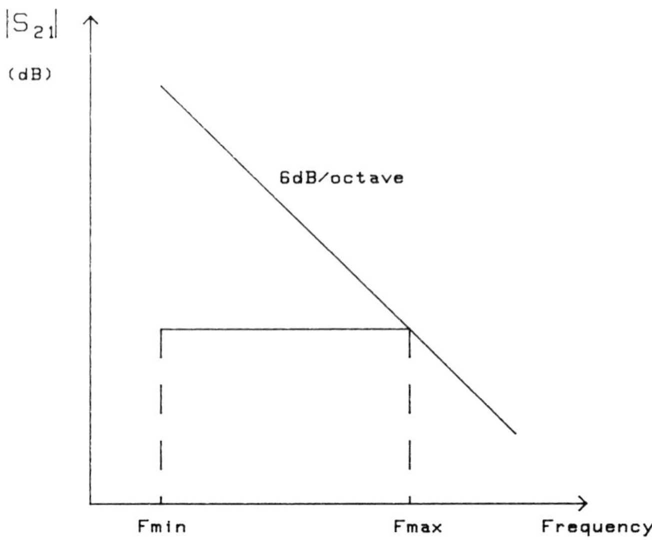

Fig. 8.6 Typical transistor gain $|S_{21}|$ vs frequency is – 6 dB per octave; corrected by the feedback network, gain is flat between F_{min} and F_{max}.

then extra gain is obtained to compensate the gain reduction of the transistor. Figure 8.6 shows the natural gain variation $|S_{21}|$ of the transistor: – 6 dB per octave. The feedback network acts to keep the gain flat from F_{min} to F_{max}. In the same way, an inductance L_F is placed in series with the parallel resistor R_F, to reduce the negative feedback at the higher frequencies.

Then, step by step, the circuit in Fig. 8.7 is obtained. Such an amplifier can be optimized by a computer program as ESOPE™, COMPACT™, TOUCHSTONE™, etc. L_{in}, C_{in} and L_{out}, C_{out} are used to improve the matching and reduce to the VSWR below 2:1 which is an acceptable level.

Example: A transistor has S parameters as given in Table 8.1.

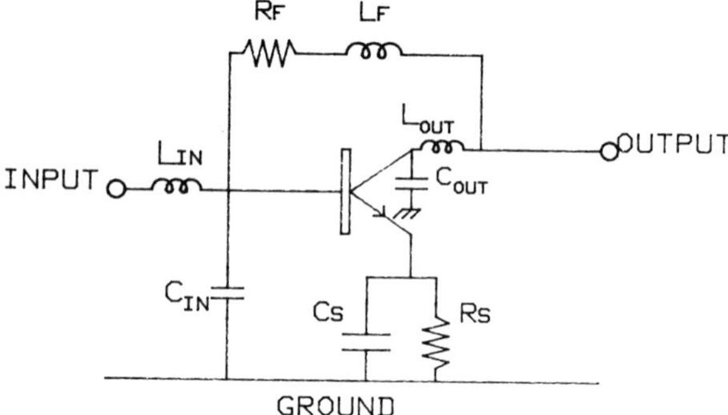

Fig. 8.7 Resistive feedback amplifier with coupling capacitances and inductances.

Table 8.1 Transistor *S* parameters

	S_{11}	S_{21}	S_{12}	S_{22}
500 MHz	0.31–135 °	9.45 (+ 19.5 dB) + 100 °	0.03 + 73 °	0.57–14 °
1000 MHz	0.28–164 °	5.5 (+ 14.8 dB) + 84 °	0.056 + 78 °	0.56–15 °
1500 MHz	0.30–175 °	3.7 (+ 11.4 dB) + 72 °	0.084 + 78 °	0.56–12 °
1000 MHz	0.25–182 °	2.8 (+ 9 dB) + 68 °	0.112 + 80 °	0.56–20 °

With a design of fig. 8.7 the following gain response is given (VSWR < 2:1, gain flatness = ± 1 dB). With $R_S = 7.5\ \Omega$

$R_F = 160\ \Omega$ Gain = 8 dB up to 2300 MHz

$R_F = 200\ \Omega$ Gain = 10 dB up to 2000 MHz

$R_F = 300\ \Omega$ Gain = 13 dB up to 1500 MHz

$R_F = 420\ \Omega$ Gain = 15 dB up to 1000 MHz

The same transistor with $R_S = 2.2\ \Omega$ and $R_F = 420\ \Omega$ exhibits 18 dB gain up to 600 MHz.

As seen above, the resistive feedback amplifier exhibits good performance in terms of gain, VSWR, bandwidth, and even in power output if one uses a powerful transistor; but on the minus side, in some cases the stability could be hazardous (when the gain per stage is too high), and especially the noise figure is rather mediocre compared to other solutions.

Figure 8.8 gives the noise figure degradation due to the feedback resistors. One observes that to reduce the noise figure, one can increase the R_F resistance, which means increasing the gain. So a large bandwidth amplifier with only 8 dB gain up to 2300 MHz as shown before, will exhibit poor noise figure (5 dB for instance), even at the low frequency edge (100 MHz). A particularity of the resistive feedback amplifier is that the noise figure is nearly flat across the band. The only way to achieve good noise performance is to use a low noise transistor with the minimum feedback circuit, but then the VSWR will be bad and the output power low. The solution is to take away the resistors, which are 'noisy', from the feedback network, to obtain the lossless feedback amplifiers.

8.2.3 Lossless feedback amplifier

The transistor is placed in a non-dissipative network, which means that all the power delivered by the transistor is transmitted to the load. As a result, there is no source of noise except the transistor itself, and no extra power dissipation, so the noise figure is a minimum and the dynamic range maximum. Figure 8.9 describes such a structure.

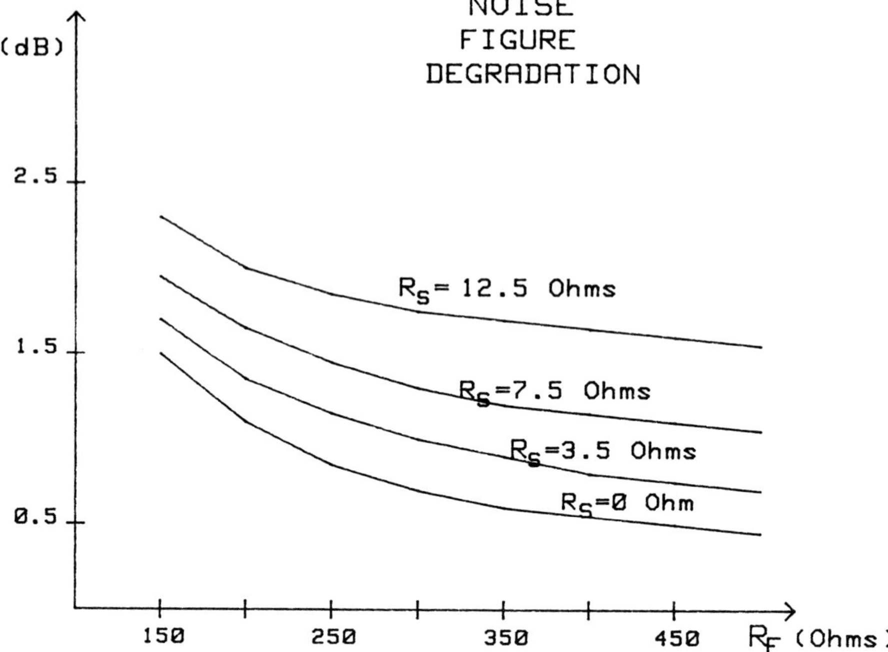

Fig. 8.8 Noise figure degradation due to feedback resistors.

In practice the directional coupler is as shown in Fig. 8.10. It is wound on a ferrite; the coupling ratio between c and b is: 10 log $(N^2 + 1)$. As the amplifier in Fig. 8.9 is an inverting one (common emitter) there must be no phase shift between ports c and d to get the feedback operation.

An amplifier with 13 dB gain up to 500 MHz and such a coupler has a noise figure of about 1.8 dB instead of 3 dB with a resistive feedback. Another advantage is that you can reduce the gain without noise degradation only by

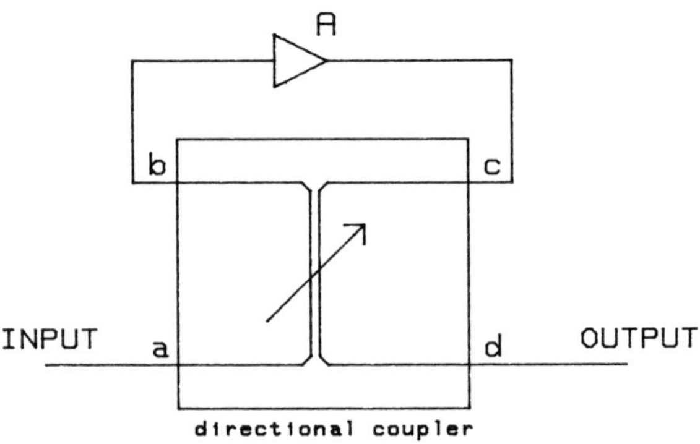

Fig. 8.9 Schematic configuration of a lossless feedback amplifier.

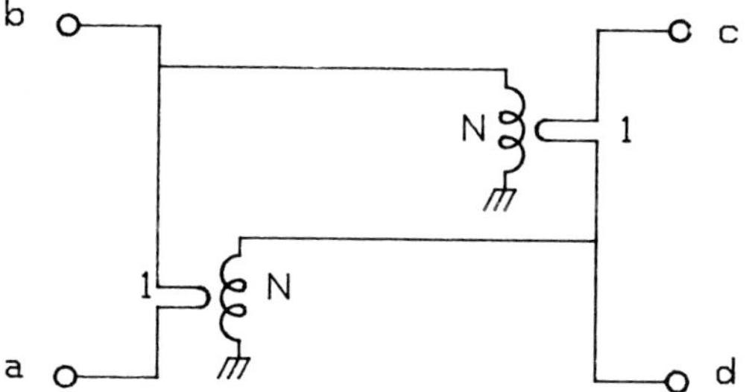

Fig. 8.10 Practical directional coupler for lossless feedback amplifier.

changing the coupling value. The limit of these structures comes from the wound component, which is unable to reach frequencies higher than 1000 MHz. So the range covered is about 1 to 500 MHz.

8.2.4 High isolation structure

The coupler, when used with the appropriate load impedances, exhibits a rather good isolation between ports d and b (for instance 35 dB), but not between ports d and a which is the same as between c and b. However, in some cases we need good isolation between the output and the input of the amplifier. Then we use an extra component, to get this isolation. The circuit is shown in Fig. 8.11. This component is a power divider.

A simple divider is as shown in Fig. 8.12.

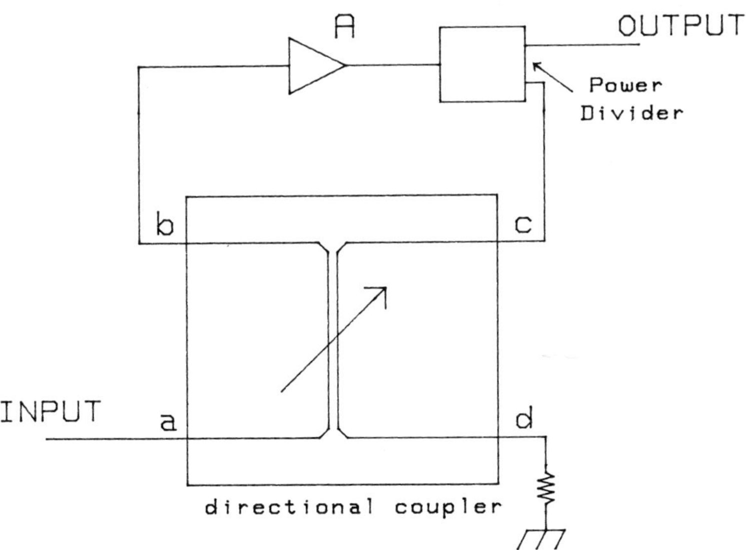

Fig. 8.11 Schematic of a high isolation lossless feedback amplifier.

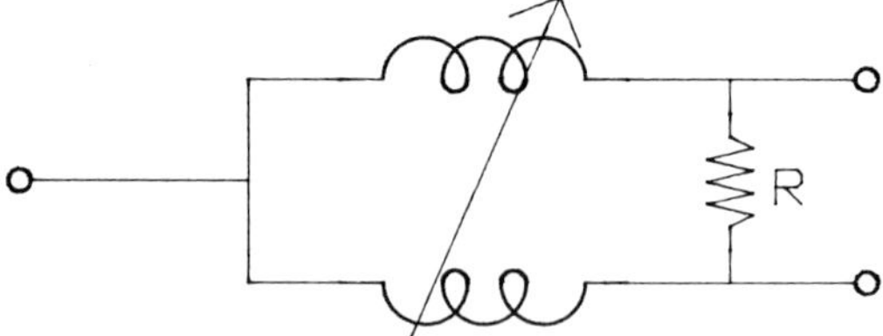

Fig. 8.12 Simple configuration of a power divider used to obtain high isolation.

R allows the good matching at ports 2 and 3. The isolation between 2 and 3 can reach 20 dB, with a proper resistance value. So, if the coupling value of the feedback coupler is 12 dB, one obtains 20 + 12 = 32 dB of isolation between output and input, which is in the range of the intrinsic S_{12} parameter of the transistor around 500 MHz. This isolation is very useful for buffer amplifiers behind a frequency source for instance, where a load variation has a direct influence on the frequency.

8.2.5 Biasing and temperature operation

The simple biasing network is shown in Fig. 8.13. This is already a stabilized network: if i_C increases, V_C goes down, so i_B goes down and the new i_C goes down. But the current value i_C is very dependent on the β value of the transistor ($\beta = h_{fe}$).

An improved circuit is given in Fig. 8.14. One chooses the emitter current i_E (depending on the transistor characteristics and the performance desired), so V_E is determined by the R_E value; then one calculates the $R_1 - R_2$ voltage divider to perform $V_B = V_E + V_{BE}$ ($V_{BE} \simeq 0.8$ V for silicon transistor). To get a linear operation, one chooses R_C to get:

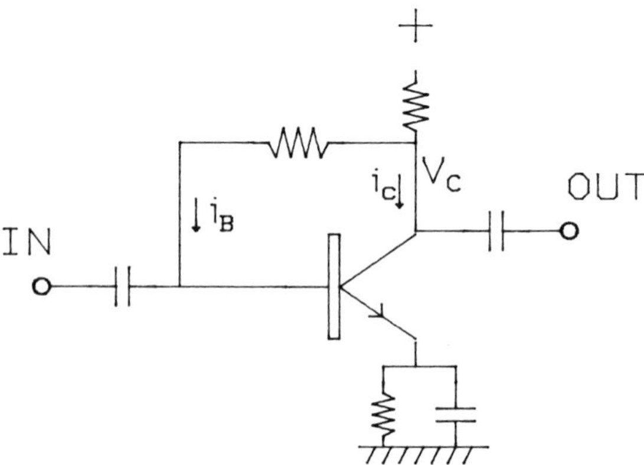

Fig. 8.13 Simple transistor amplifier biasing network.

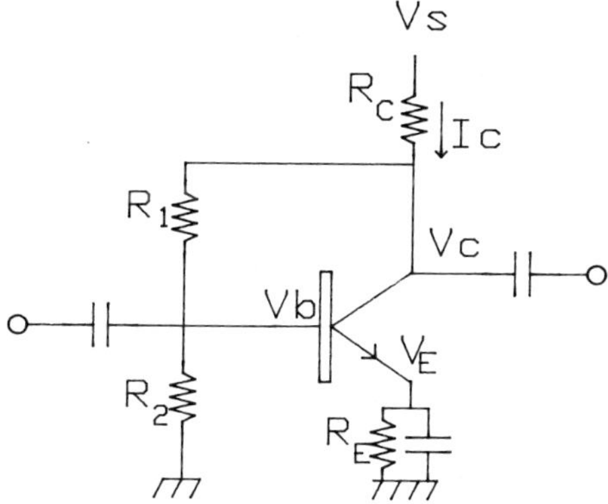

Fig. 8.14 Improved transistor amplifier biasing network.

$$R_C * I_C \simeq V_{CE}$$

(for instance, $V_S = 12$ V, $V_E = 2$ V, then $V_{CE} = 5$ V).

The current in the divider $R_1 - R_2$, must be at least five times the i_B value (which is i_E/β) to have a well stabilized network, rather independent of the transistor. Nevertheless, the values obtained for R_C, R_1 and R_2 could be wrong with respect to the RF operation (in resistive feedback configuration for instance): one needs a R_1 value matched to the polarization requirements, and matched to the RF circuit. So, the universal biasing network is as shown in Fig. 8.15.

The principle is to settle the i_C current by a current generator, which controls the i_B value. In this current generator, T_1 acts to compensate the temperature variation of T_2. If V_{EB1} and V_{EB2} are the emitter-to-base voltages of the T_1 and T_2 transistors, and β the gain current of the microwave transistor T_3, then:

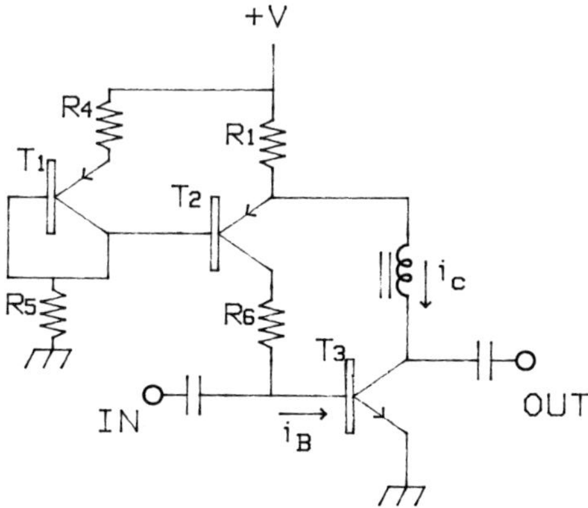

Fig. 8.15 Universal transistor amplifier biasing network.

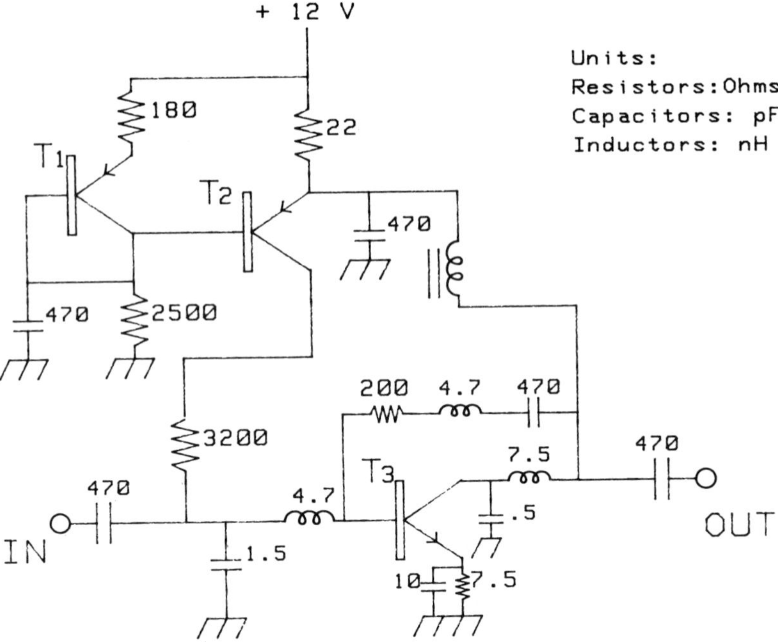

Fig. 8.16 Practical example of a universal transistor amplifier biasing network.

$$I_C = \frac{1}{R_1}\left[V \cdot \frac{R_4}{R_4 + R_5} + V_{EB1}\frac{R_5}{R_4 + R_5} - V_{EB2}\right]\frac{\beta}{\beta + 1}.$$

A typical value for R_5 would be about $10 \times R_4$, thus the temperature variations of V_{EB1} and V_{EB2} are annihilated, and as $\beta >> 1$, one obtains:

$$I_C \simeq V \cdot \frac{1}{R_1} \cdot \frac{R_4}{R_4 + R_5}.$$

This is independent of the microwave transistor, and independent of the temperature (temperature variation of resistors could be assumed to be negligible).

This biasing network is the most 'secure' one. An illustration of the biasing network, and of the negative feedback is given in Fig. 8.16. A photograph of this amplifier, constructed with thin film technology on alumina substrate is given in Fig. 8.17. The electrical characteristics (at 25 °C) are given in Fig. 8.18. This amplifier will operate between − 54 and + 110 °C without any problem. It could be used in the IF section of microwave receivers.

8.3 GaAs FET AMPLIFIERS

These amplifiers operate in a microwave frequency range from about 0.1 GHz to more than 40 GHz. We find the standard topology used with silicon transistor amplifiers, i.e. single-ended, resistive feedback amplifiers, but also other designs well-matched to the microwave field such as balanced configuration and distributed circuits.

Fig. 8.17 Photograph of the amplifier of Fig. 8.16.

8.3.1 Single-ended, resistive feedback amplifier

All the comments in sections 8.2.1 and 8.2.2 apply to GaAs FET amplifiers as well. The main differences are of course, the values of the S parameters, which are extended into higher frequency ranges.

The software for computer-aided design has to be much more powerful than for RF amplifiers. In fact all the commercially available software offer these capabilities (ESOPE™, TOUCHSTONE™, COMPACT™). They include a library of element models including microstrip, stripline, coplanar circuits; they also provide tools for modelling the active components.

The single-ended structure is used in low cost technology (with printed soft-board material), and also for low-noise amplifiers, both for narrow bandwidths.

The resistive feedback GaAs amplifier is, in fact, an extension of the silicon transistor amplifiers. In that case upper frequency ranges with microwave integrated technology can reach 8 GHz. The difficulty will be in terms of gain stability (GaAs transistors exhibit positive gain up to 18 GHz), and also in the choice of the transistor to achieve a minimum gain value.

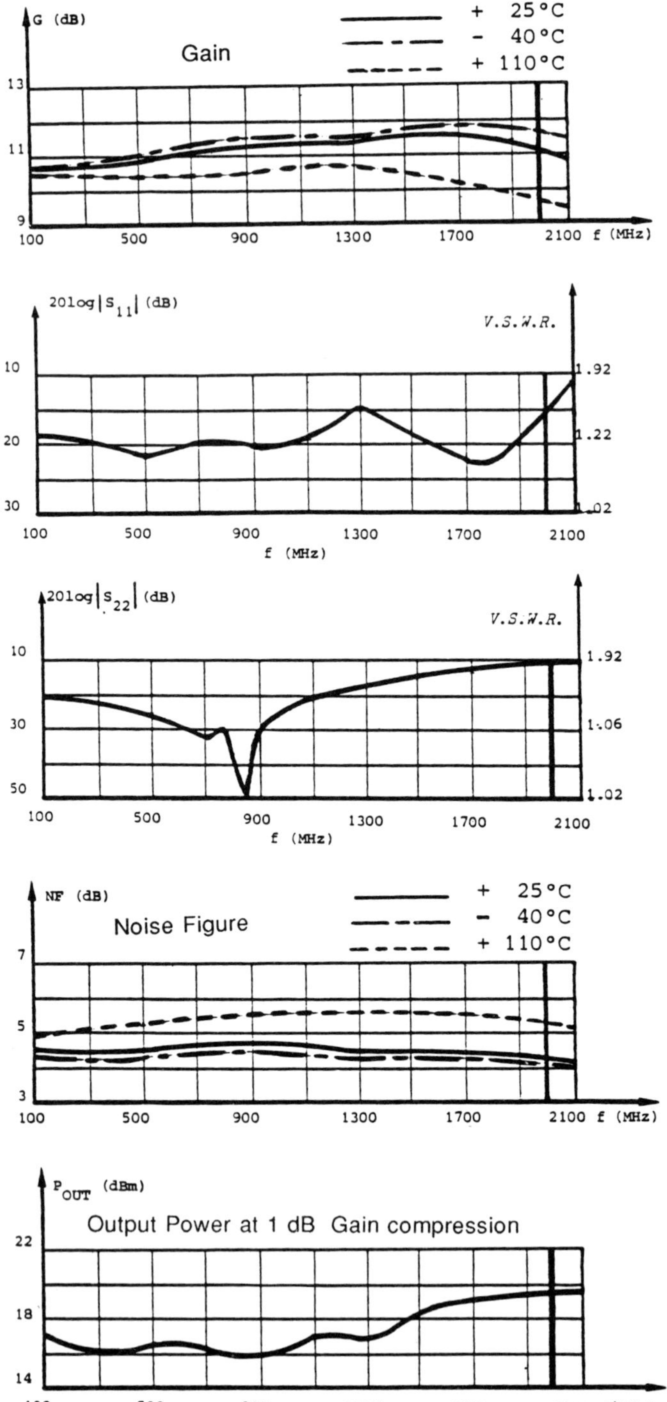

Fig. 8.18 Electrical characteristics of the amplifier of Fig. 8.16.

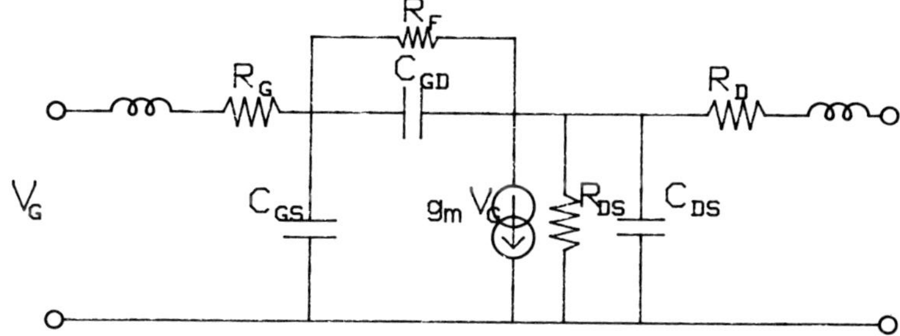

Fig. 8.19 Equivalent circuit of a single-ended, resistive feedback amplifier.

The equivalent circuit is shown in Fig. 8.19. As the amplifier must have a flat gain response, we can determine the gain at low frequency with the reduced scheme in Fig. 8.20. Then one transforms the admittance parameters into scattering parameters. With a few calculations, and some usual rules as $|S_{11}| < 0.33$ (VSWR < 2), one finds $R_F = F\ (g_m,\ R_{DS})$.

For instance for a 0.5 μm gate length MESFET with 450 μm gate width, $g_m = 48$ mS $\rightarrow R_F \leqslant 250\ \Omega$, then $|S_{21}| \leqslant 6.4$ dB, which is very low, but with a MESFET having 900 μm gate width, and $g_m = 110$ mS one obtains $R_F \leqslant 400\ \Omega$, and $|S_{21}| \leqslant 12$ dB. Thus, it is important to choose a transistor having a high g_m value. Then one must add matching networks at the input and the output, to get both the good input and output VSWR, and the gain flatness. This will be done by simulation.

8.3.2 Balanced configuration

This design separates the VSWR parameters from the others such as: gain flatness, optimum noise figure, or optimum output power. To get this, two 3 dB

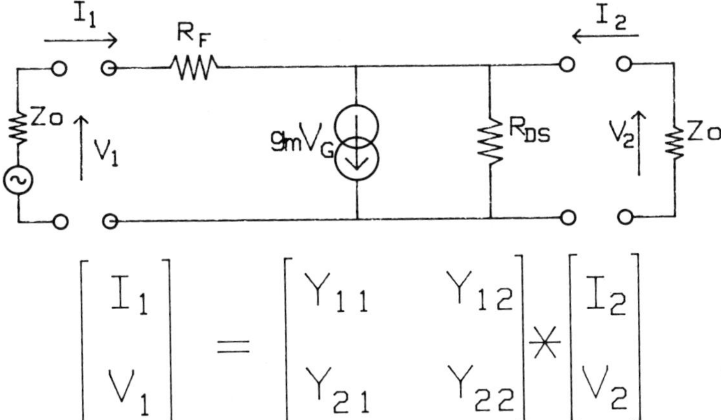

Fig. 8.20 Equivalent circuit of Fig. 8.19 in a low frequency approximation.

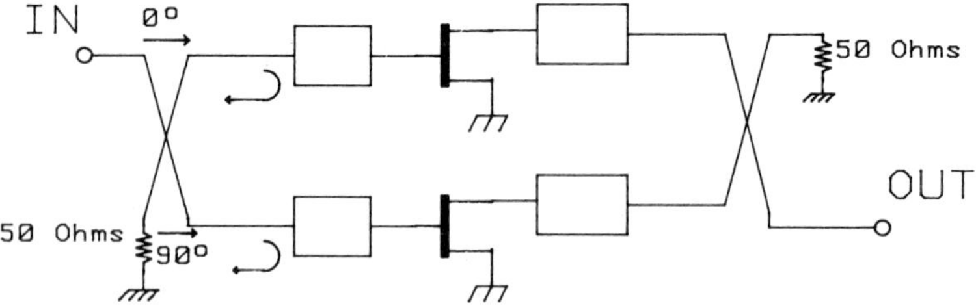

Fig. 8.21 Balanced amplifier configuration.

hybrid couplers are used, as shown in Fig. 8.21. The two paths between the couplers must be identical. So if A is the input signal, it is divided into A/2 (0 °) on path 1 and A/2 (90 °) on path 2. The signal reflected by each amplifier, will be $\rho \cdot A/2\ (\phi)$ and $\rho \cdot A/2\ (\phi + 90\,°)$, $\rho(\phi)$ being the reflective coefficient of each single amplifier. These signals are again divided by the coupler and one obtains:

at the input: $\rho \cdot A/4\ (\phi) + \rho \cdot A/4\ (\phi + 180\,°) = 0$

in the load: $\rho \cdot A/4\ (\phi + 90\,°) + \rho \cdot A/4\ (\phi + 90\,°) = \rho \cdot A/2\ (\phi + 90\,°)$.

That means that all the reflective power is delivered to the load, and nothing appears at the input, so the VSWR is minimum. The same phenomena exist at the output.

As the VSWR can be controlled with this structure, one can design single amplifiers to achieve the required performance: gain flatness, optimum noise figure, or optimum output power. The gain of a balanced stage is equal to the

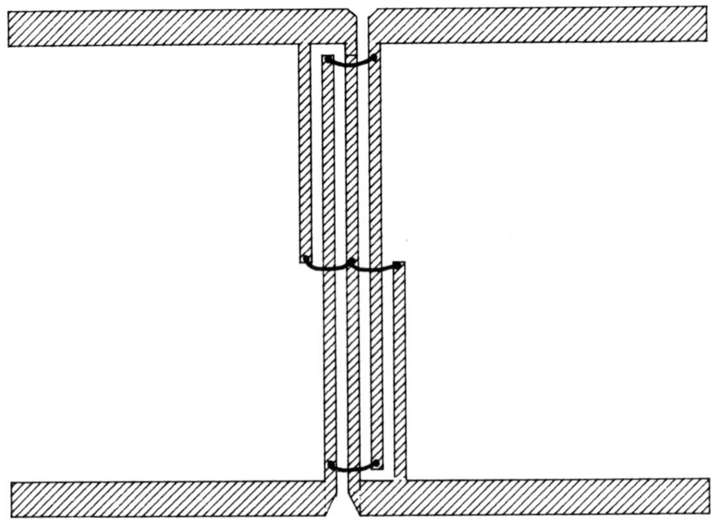

Fig. 8.22 Lange 3 dB coupler in microstrip technology.

gain of a single amplifier, and the output power is twice the output power of the single. If one transistor fails, the gain is 6 dB less and the output power 3 dB less, but the amplifier 'works'; it will not be completely faulty, which could be very useful.

To achieve the 3 dB coupler, a microstrip uses the interdigitated structure known as the Lange coupler (Fig. 8.22). The coupling value between the lines determines the achievable bandwidth. The maximum ratio between the upper and the lower frequency is about 4 (e.g. 2 to 8 GHz), so that, on alumina substrate, with a thickness of 0.635 mm, the width is 30 μm for the digits and only 14 μm for the interdigits. The length of the coupler is about a quarter of the wavelength of the mid-band frequency. The loss introduced by the coupler is about 0.4 dB at 18 GHz. The isolation of such a stage is rather good, and reaches about 15 dB. Typical gain values for low power stages in large bandwidth operation are set out in Table 8.2.

Table 8.2 Typical gain values for low power stages in wide band operation

Bandwidth (GHz)	*Gain* (dB)
2 to 4	13
2 to 8	10
6 to 18	7
18 to 40	4.5

These good performance figures explain why the balanced configuration is very popular in large bandwidth applications, combined with the excellent cascadability due to the low VSWR of each stage. Commercial amplifiers used in military applications employ this topology. Nevertheless, the balanced structure is limited in large bandwidth area (e.g. 2 to 18 GHz), compactness for low frequency (due to the coupler), and cost (you need two single amplifiers per stage). For these reasons, there is a tendancy to use the distributed structure, especially well suited to MMIC technology (see Chapter 10).

8.3.3 Distributed structures

These combine matching and travelling wave distribution. Two propagation lines of equal velocity are created using the input and output capacitances of the FET (see Fig. 8.23). The principle is to transfer the power from the input line to the output line through the transistors and then benefit from the gain of the FET. The basic structure is similar to a directional coupler, with two transmission lines terminated by their characteristic impedances (50 Ω). These lines are connected by the periodically placed transistors. The gates have to be excited in such a way that the drains deliver signal in phase with the progressive

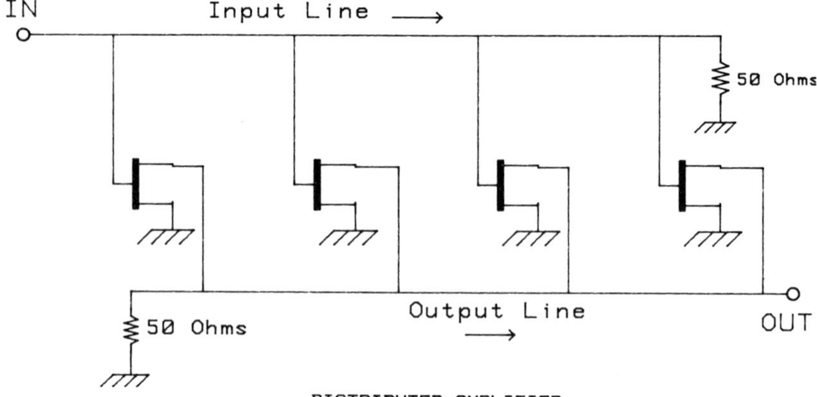

Fig. 8.23 Distributer amplifier configuration.

wave on line 2 toward the output and in opposite phase toward R_C. A more complete description of the distributed structure is given in Section 10.4.4.

With micro-electronic technology, the distributed design is limited by the parasitic capacitances, and also by the residual inductances of the bonding wires, which limit the bandwidth. Nevertheless, hybrid amplifiers covering 2 to 40 GHz have been made. In that case other difficulties come from the biasing circuits.

As shown in Fig. 8.23, the distributed amplifier consists of four or five transistors between two progressive lines. The overall gain is equivalent to the gain of one transistor. The noise figure is rather good with respect to the bandwidth (e.g. 4 dB from 2 to 18 GHz). This structure could be used for power combination.

8.3.4 Medium power amplifiers

FET characterization

Of course the scattering parameters are needed to determine the linear equivalent circuit of the transistor, but it is also necessary to know the large signal characteristics. Figure 8.24 shows the equivalent circuit model used to design medium power amplifiers.

The main non-linear elements of a GaAs MESFET are (Fig. 8.25):

the input Schottky diode I_{gs} representing gate current in the gate–source junction;
the gate–source capacitance C_{gs};
the drain current source I_{ds};
the drain–gate voltage controlled-current source I_{gd} describing the drain–gate avalanche current.

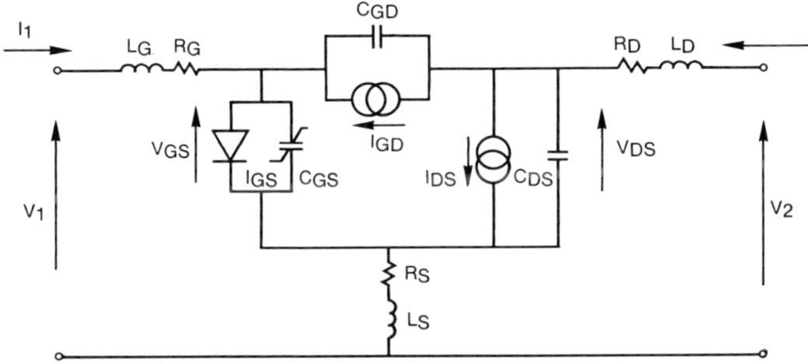

Fig. 8.24 Non-linear equivalent circuit model for large signal FET behaviour.

Several methods are used to obtain enough parameters to design a power amplifier. A well-known technique is the load-pull measurement. It consists of producing a variable and controlled source impedance in order to evaluate the performance of a microwave device under source impedance variable conditions. Variable and controlled loads can be a passive network with variable elements or an active load. For a highly mismatched device, the losses of passive networks don't allow measurement of the necessary reflection coefficient. To overcome these difficulties, methods based on active loads are used to simulate all of the impedances.

The parameters we can determine are:

1. Optimum input and output impedances for a maximum output power or a minimum intermodulation under multicarrier operation.
2. Gain and efficiency at different classes of operation.

A more general technique to characterize the non-linearities of the FET is a pulsed measurement set-up. To minimize thermal and trapping effects, a short

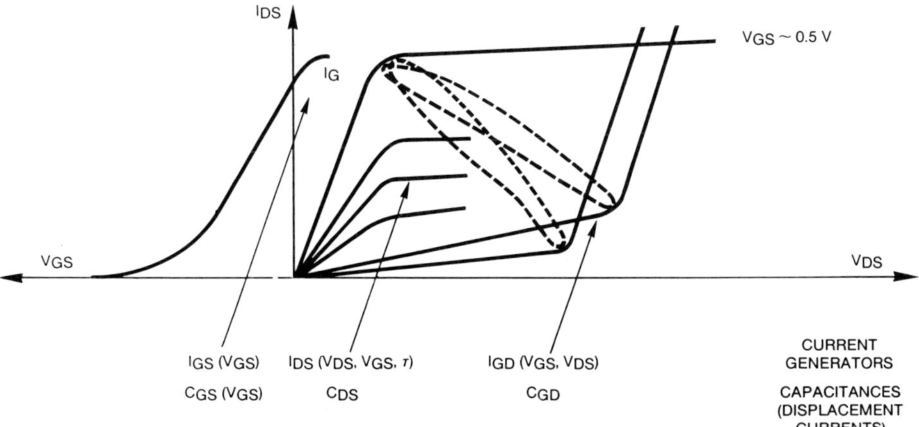

Fig. 8.25 Main non-linear elements of a GaAs MESFET.

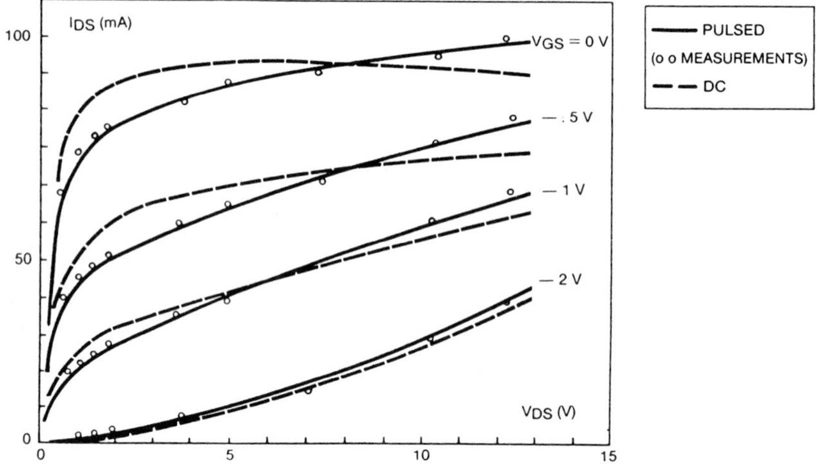

Fig. 8.26 Measured field effect currents under DC and pulsed conditions.

pulse voltage is superposed on the d.c. voltage around the operating point. A comparison between several repetition rates and widths shows that a 200 ns pulse width and 10 to 100 kHz repetition rates are the best one. However these values depend on the FET. Figure 8.26 shows the difference between pulsed measurement and d.c. measurement for the I_{ds} (V_{gs}, V_{ds}) characteristics. This set-up allows independent measurement of the different current sources, as functions of the external applied voltages:

$I_{ds}(V_1, V_2)$: field effect current

$I_g(V_1)$: Schottky current

$I_{gd}(V_1, V_2)$: avalanche current.

These experimental curves are fitted to an analytic expression and are transformed into internal current sources depending on V_{gs} and V_{ds}. Linear elements, measured by the S parameters, and non-linear current sources give a very good equivalent circuit model of the FET. This method allows the use of non-linear computer aided design and many types of circuits can be designed: power amplifiers, mixers, oscillators, frequency multiplier or divider, etc.

Design of microwave power FET amplifiers

The main FET output power limitations are the forward gate current and the drain–gate avalanche current. The first limitation gives the maximum channel current ($I_{ds\,max}$) and the second gives the voltage between drain and source ($V_{ds\,max}$). Figure 8.27 shows the limitation mechanisms considering an ideal FET where the channel current can be written by the simplified expression:

$$I_{ds}(V_{gs}, V_{ds}) = g_m(V_{gs} - V_p) = G_{ds}V_{ds}.$$

The maximum available output power will be:

$$P_{max} = \frac{1}{8} V_{ds\,max} \times I_{ds\,max}.$$

To obtain this power the optimum load is $g_{opt} = I_{ds\,max}/V_{ds\,max}$. Generally g_{opt} is greater than g_{ds} where g_{ds} is the optimum load for maximum gain. If the load g_l is greater than g_{opt} the power limitation is the forward gate current and $V_{ds\,max}$ will be reduced. If g_l is lower than g_{opt} the power limitation is the breakdown voltage and I_{ds} will be reduced. A large breakdown voltage decreases the difference between g_{opt} and g_{ds}. For a given output power and FET technology it is necessary to adjust $I_{ds\,max}$ by a good choice of the size of the FET. The output power is directly proportional to the total periphery gate length. The difficulty consists in recombining each elementary cell without losses.

After this simplified analysis, it is necessary to take into account every non-linearity and parasitic effect of the GaAs FET. The objectives in designing one stage of power amplifier are the following:

1. Input and output reflection coefficients must be compatible with the cascadability between different power stages. Generally, the VSWR must be lower than two.
2. A minimum gain value at 1 dB gain compression (for instance 6 dB between 6 to 18 GHz).
3. The output matching circuit must be optimized for the maximum output power.

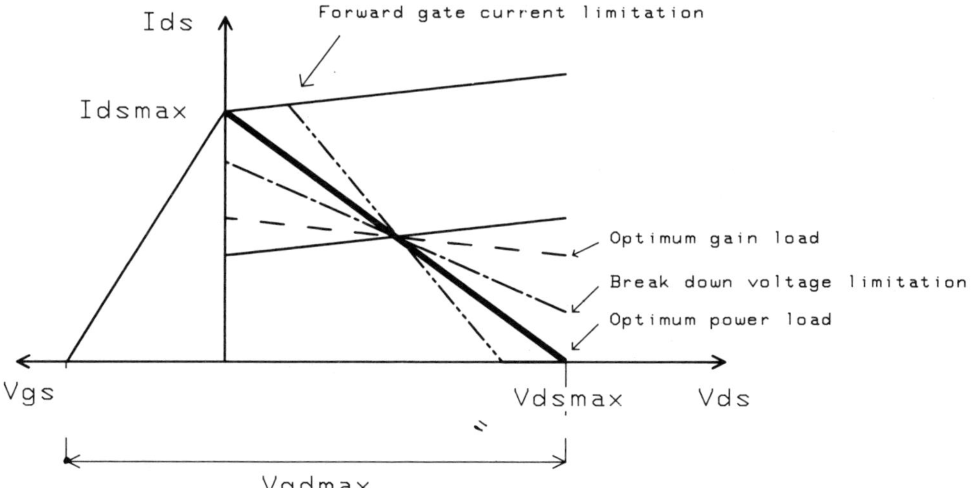

Fig. 8.27 Various limiting factors in power FET design.

4. Some applications require a good power added efficiency and not necessarily the maximum output power.

The synthesis of input/output matching networks can be done with the linear CAD using the load-pull measurement. It is the simplest and fastest method but cannot predict the performance versus power input nor the d.c. operating point. The use of this method is limited to a simple matched input/output stage and is not usable to design feedback amplifiers or distributed amplifiers.

A large signal FET model can predict large signal characteristics such as power saturation and distortion at arbitrary input levels. A non-linear circuit may be defined by at least one semiconductor device and a linear embedding circuit. Optimization of these circuits is usually done by first fixing a topology, and then optimizing the values of the linear elements. Another method consists of finding the upper limits that can be achieved by a given active device. The impedances presented to the FET are optimized independently of the topology chosen for their realization: they are then synthesized by usual methods of linear circuits. The bias voltages are also optimized by this method. For example if the main objective is the optimum power added efficiency it will be possible to search for the optimum class AB operating point.

Every analysis or optimization using a non-linear FET model needs a non-linear circuit analysis routine. These non-linear routines can be divided into two groups: time domain analysis and hybrid analysis that iterates between the frequency and time domain. The time domain approach requires excessive computation time when the transient time of the circuit under analysis is long. In the hybrid analysis approach, the circuit is divided into a non-linear sub-circuit and a linear one. They are, respectively described by linear equations in the frequency domain and by non-linear differential equations in the time domain. The goal is to stir-fry both sets of equations simultaneously. The hybrid analysis method most often used is the harmonic balance method. (see Volume 1, Chapter 20)

Wide band medium power amplifiers and combination techniques

The previous conditions concerning the output power must now be realized in a wide band. The passive circuit coupled with the FET is a low-pass or band-pass filter for the output. The FET is also a low-pass amplifying device. The gain has a slope of 6 dB/octave decreasing with the frequency. So the input filter must have a 6 dB/octave slope transmission increasing with frequency. The main conditions to make a wideband power amplifier are:

1. The capability of the FET to have gain at high frequencies of operation. This is mainly related to the technological state of the art.
2. The small losses and ripple of the passive input and output filters. This is a limitation concerning the size of the elementary FET cell.
3. The use of a good combination technique with several FET cells.

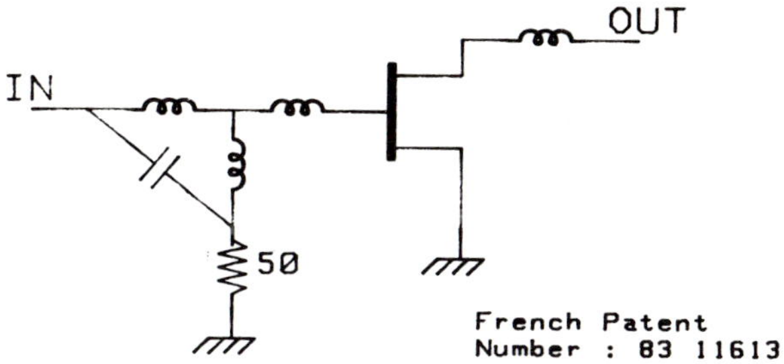

Fig. 8.28 Resistive input matching using a lossless filter.

The simplified equivalent circuit of the FET is a R C series and R C parallel input and output impedance over the whole bandwidth. An input voltage dependent current source at the output port gives the field effect current. The different matching circuits are:

1. Reactive matching. It is generally a band-pass filter. The limitation is the general lossless filters properties depending on the bandwidth and on the technology used. So to obtain a constant gain the input filter must reflect back a part of the low frequency input signal.
2. Resistive matching. This circuit is only an input lossless filter. A frequency selective resistive circuit within the matching circuit absorbs the reflected signal at the low frequency (Fig. 8.28).
3. Resistive feedback. The input and output are connected by a resistive filter which contributes to the flat gain and the input/output matching.

Table 8.3 summarizes the relative advantages and disadvantages of the different circuits.

Table 8.3 Comparison of different circuits

Circuit matching	*Advantages*	*Disadvantages*
Relative matching	Low losses Efficiency Noise	Poor input VSWR at low frequencies Bandwidth
Resistive matching	Good VSWR Easy to bias Small size Bandwidth (lowpass)	Losses Noise
Resistive feedback	Good VSWR Small size Noise Bandwidth (lowpass)	Losses High frequency limitations

The power density of a Ku-band FET is today about 0.3 to 0.5 W/mm of gate width. The size of one elementary cell varies between 0.3 and 3 mm. It is generally very difficult to match directly more than 1 mm in a wide frequency band. Therefore large power values (0.5 to 3 W) require wideband combining circuits.

Some combining circuit configurations are:

1. Paralleling. This is the usual way to connect FET cells. We note that power FET chips are a paralleling combination of elementary gates. The difficulty of this method results from the low impedance level at input port.
2. Balanced configuration. See section 8.3.2.
3. Distributed structures. See section 8.3.3 and Fig. 8.29.
   They combine matching and travelling-wave distribution. Two propagation lines of equal velocity are created using the input and output capacitors of the FET. The main feature is the line losses due to the FET's resistors and the quality factor of the circuit elements. Thus, the input power decreases

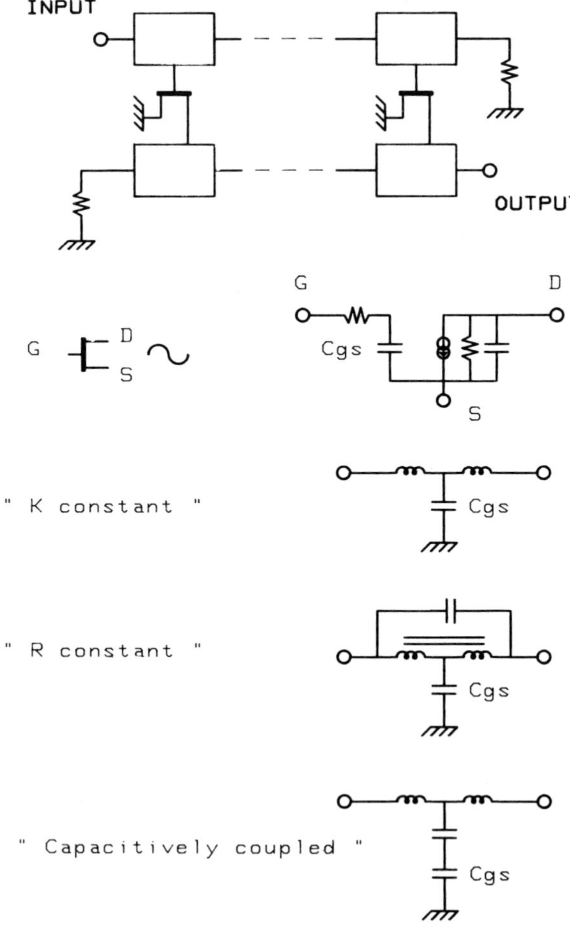

Fig. 8.29 Distributed structures for circuit matching.

along the line and the last FET doesn't work at high frequency. To reduce this problem, different circuits are used (Fig. 8.29).

To maintain the wideband frequency response with a large gate periphery a 'constant R network' can replace the more traditional 'constant K network'. Theoretically this principle allows one to obtain constant real input impedance independently of frequency. Practically, according to the FET parameters, the cut-off frequency of the amplifier can be multiplied by $\sqrt{2}$.

To extend the cut-off frequency, a capacitor can be inserted in series with each gate FET forming a voltage divider to ground. Thus the gate line capacitors can be used to equalize the FET voltages along the line at any frequency, resulting in significant improvement of output power and efficiency.

Gate line and drain line can be tapered to reduce the power loss in the terminations.

4. Travelling wave divider/combiner (Fig. 8.30). The divider results from cascading in phase several power splitters connected by sections of propagation lines in such a way that incident waves have different times of arrival at the output ports of successive power splitters. Impedance values at the outputs of each power splitter are chosen proportional to the reciprocal of the circulating power. Resistors are introduced between the outputs of each power splitter to absorb the power reflected by transistors no longer matched. The combiner is the same as the divider and allows an in-phase recombination. This combination technique has the following advantages:

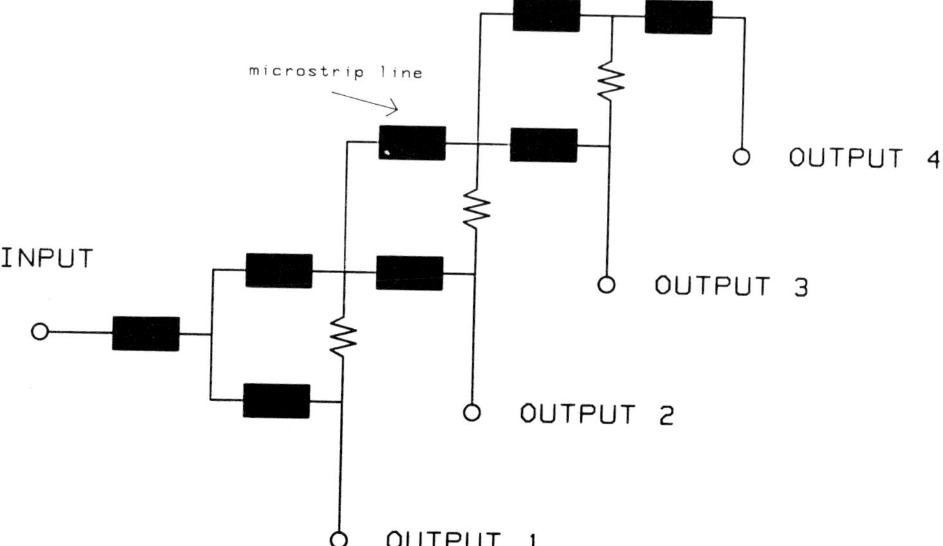

Fig. 8.30 Travelling wave divider/combiner.

wideband frequency operation (e.g., 2 to 18 GHz), possibility to combine mismatched FETs, and low sensitivity to FET dispersion.

Today, with this combiner and four power transistors, it is possible to obtain about 15 W in X-band, but each FET needs about 1.2 A and 11 V respectively for the drain current and the V_{DS} voltage, hence the importance of the technological solution able to dissipate such a power level.

8.3.5 Complementary circuits

To improve some characteristics of the FET amplifiers, we can use: temperature compensation, power limitation and voltage regulation.

Temperature compensation

GaAs field effect transistors exhibit a gain variation of about − 0.012 dB/°C. That means that a 6 dB gain stage (6 to 18 GHz), has about 1.5 dB gain variation between − 55 and + 85 °C, which is a rather high value. Two solutions are used to compensate the gain variation: active biasing and variable attenuator.

Active biasing
This changes the drain current I_D with the temperature, so it changes the gain value S_{21} of the transistor. The phenomena are described in Fig. 8.31 and 8.32. Nevertheless, one must select the FET offering monotonic gain variation with I_D, in a relatively large area, in order to achieve enough gain control. Unfortunately, in a large temperature range, the current will change by such a big

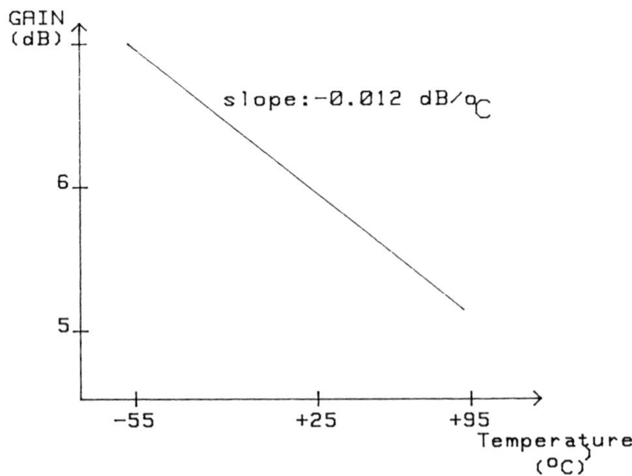

Fig. 8.31 Gain variation of a GaAs FET vs temperature correction.

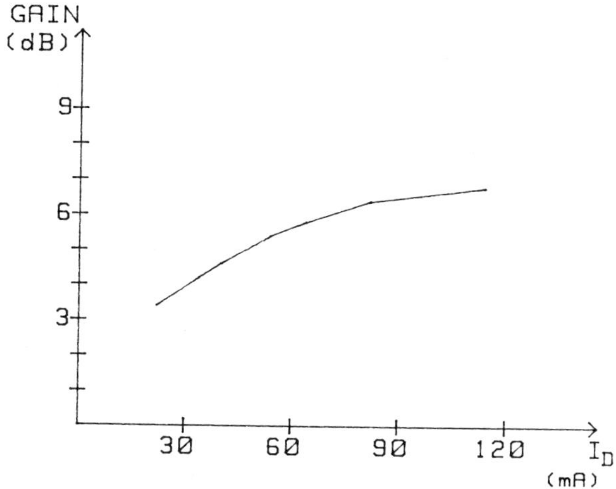

Fig. 8.32 Active biasing for gain vs temperature before correction.

ratio (e.g., 1 to 3), that the general characteristics will be completely modified, especially the output power and the noise figure. That means that this active biasing is reserved for small temperature variations, or for non-critical gain stages.

The active biasing could be achieved by a negative temperature coefficient resistor in the source path, or by a gate-source voltage control (Fig. 8.33 and 8.34). Another way of achieving the gain variation by an external control is the use of dual-gate FET (Fig. 8.35). One gate is for the microwave input and the other one for gain control. Such FETs are available up to 18 GHz, but not for applications such as low noise, or high output power.

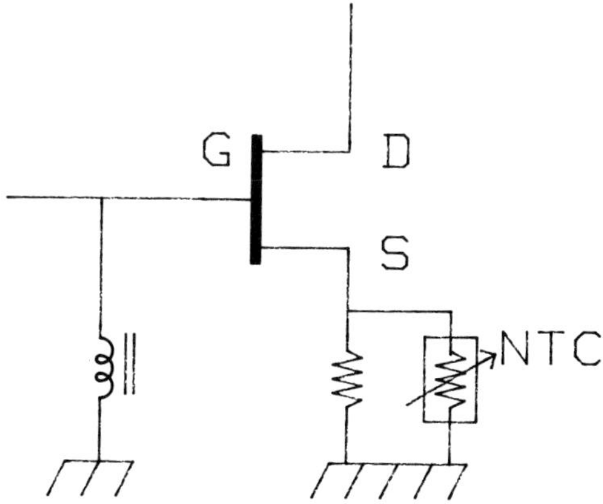

Fig. 8.33 Active biasing circuit using a Negative Temperature Coefficient (NTC) resistor.

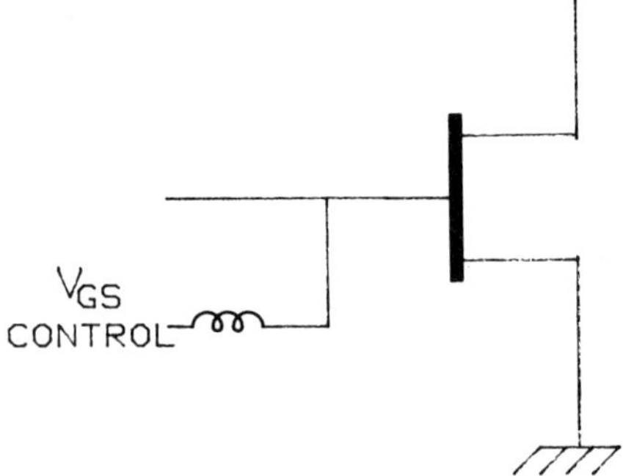

Fig. 8.34 Active biasing circuit using a gate-source voltage control.

Variable attenuator

The most common way to compensate gain variation, is the insertion of a p-i-n diode attenuator in line with the gain stages (Fig. 8.36). This attenuator is non-reflective, and driven by a negative temperature coefficient resistor which achieves the insertion loss variation with the temperature. This solution can be well matched to the amplifier performance, nevertheless the position of the attenuator in the amplifier chain must be determined to avoid noise figure degradation (when placed close to the input) or output power degradation (when placed close to the output).

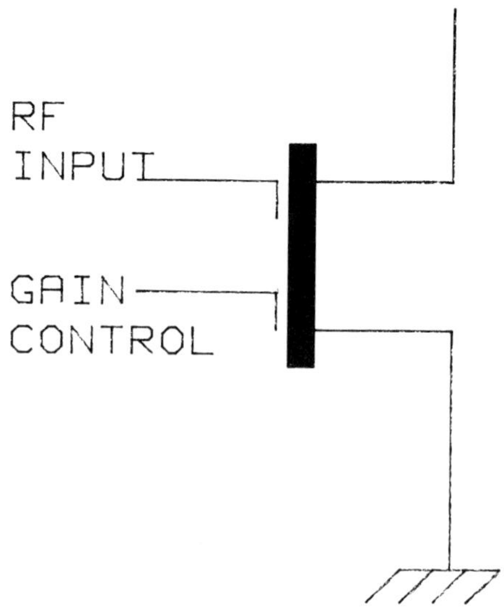

Fig. 8.35 Use of a dual-gate FET for external control of gain variation.

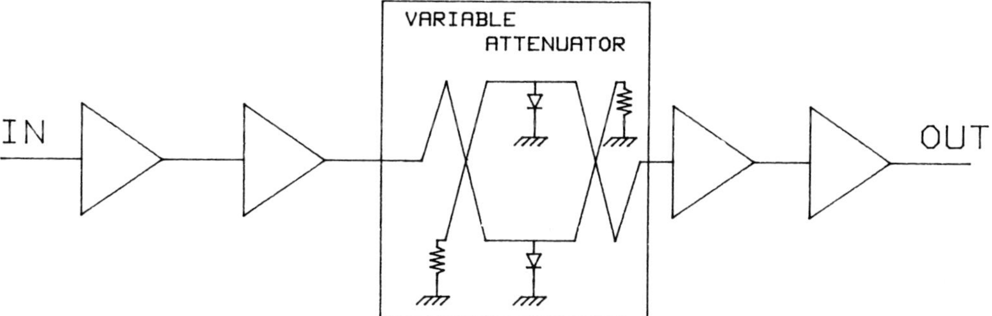

Fig. 8.36 Use of a p-i-n diode variable attenuator for amplifier chain gain control.

Power limiting

Two types of microwave power limiting are sometimes requested: at the input for protection and at the output for levelling.

A microwave amplifier in Ku-band can withstand about 100 to 500 mW of continuous wave power (CW). With a p-i-n diode limiter, it can reach 5 W (CW) and a few kW peak. This limiter will be placed in front of the first stage. For levelling, the limiter has to be placed at the output. It will also be a p-i-n diode limiter or a Schottky diode limiter for smaller output levels.

Voltage regulation

Simple biasing circuits with self-polarization in the source are often used (Fig. 8.37). To achieve voltage protection, and also to deliver the voltage level

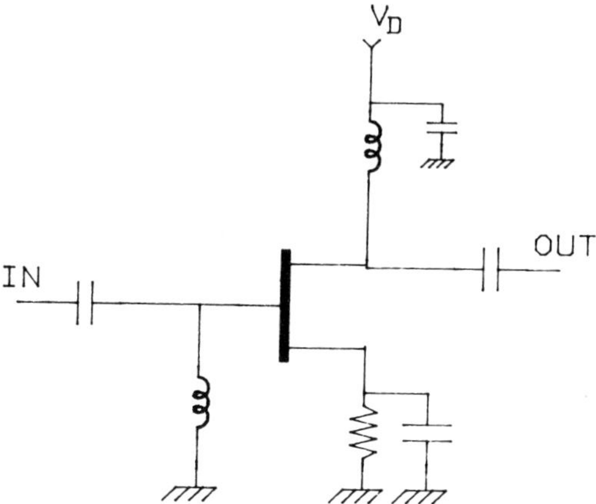

Fig. 8.37 Simple self-polarized, source biasing circuit.

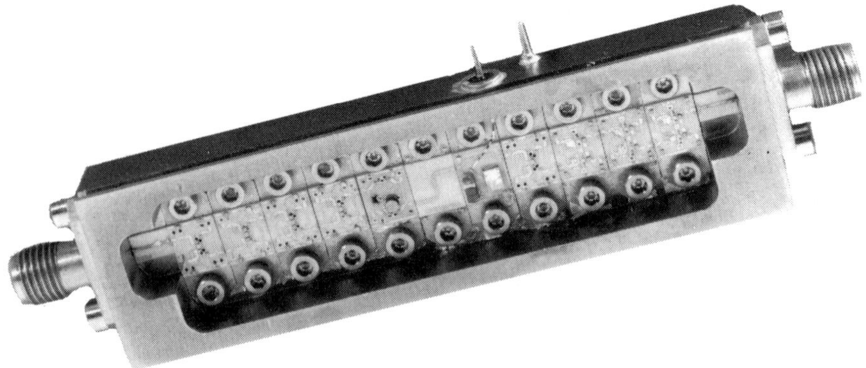

Fig. 8.38 Photo of a 6–18 GHz amplifier built in a balanced configuration with a p-i-n attenuator module for temperature compensation.

required by the FET, we need to drop the voltage issued from the equipment or from the power supply. We use a standard voltage regulator, for example, 10 to 18 V becomes 5 V regulated.

The decoupling components such as capacitors and chokes are chosen with respect to the microwave frequency range. For instance, in the 6 to 18 GHz band the capacitances are around 10 pF.

We should note that for power transistors, the source is already grounded, thus we need a separate negative power supply for the gate.

Figure 8.38 shows a 6 to 18 GHz amplifier made with MIC technology. The gain stages are in the balanced configuration, the temperature compensation is achieved by a p-i-n attenuator module. The main characteristics of this amplifier between – 55 and + 85 °C are:

frequency range:	6 to 18 GHz
gain:	36 to 41 dB
noise figure:	< 7.5 dB
power output:	> + 20 dBm at 1 dB gain compression
power supply:	Voltage: + 12 V
	Current: 500 mA
size:	69.8 × 11 × 20 mm.

8.4 TECHNOLOGY

This chapter gives only a glance at the technologies used for the manufacture of microwave amplifiers. They are particularly suitable for the most common transmission media, which is the 'microstrip'.

8.4.1 Printed circuit boards (PC boards)

The material used must have good microwave performance. Up to 500 MHz, standard epoxy boards can be used; above we use 'soft materials' as glass – PTFE with a dielectric constant, ε_r, of 2 to 2.5, or ceramic – PTFE with $\varepsilon_r \simeq 10$. The metallization is copper, or copper-gold.

These circuits are associated with encapsulated components (transistors, diodes, etc.), soldered by standard processes. They are easily drilled, and are cheap. Of course, due to the surface of such boards, it is not possible to create 3 dB interdigitated couplers on them. These PC boards are reserved for narrow band applications where the size is not critical.

8.4.2 Microwave integrated circuits (MICs)

The substrate is alumina with $\varepsilon_r = 9.7$. The thin-film metallizations are chromium–copper–gold or nickel–chromium–gold. With thick films, it will be gold type conductors. This MIC technology permits the integration of resistors. The other components (semiconductors, capacitors, etc.) are in chip form, and attached by epoxy or by solder. With the proper etching process, the 3 dB Lange couplers are easily achieved. The connections are by gold wires attached with specialized equipment. The wire diameters could be as small as 12 μ m.

The MICs are very common in military applications. An extension is the MHMIC (miniaturized hybrid MIC). These include: metallized holes, dielectric film deposition for capacitors, and air bridges. The only extra components needed to perform a microwave active function are the semiconductors (diodes, transistors). This technology gives very high performance amplifiers, and is used especially above 18 GHz up to 40 GHz.

8.4.3 Monolithic microwave integrated circuits (MMICs)

This technology is detailed in Chapter 10, and brings a new type of development for microwave amplifiers. If today the performance obtained by MMICs is not yet equivalent to MICs, this gap will be progressively reduced. Microwave designers are thinking of new concepts with MMICs, which could not be achieved any other way. For example, amplifier modules in phased array antennas, very large bandwidth distributed amplifiers, etc. Nevertheless, even with MMICs, the connections between functions remain necessary, and the PC boards or MICs will be always useful.

8.5 CONCLUSION

The evolution of microwave solid-state amplifiers continues and performance is improving every year.

In 1980, commercial amplifiers had a maximum output power of 10 mW for large bandwidth applications such as 8 to 18 GHz, and the maximum operating frequency was about 20 GHz. In 1988, the same large bandwidth amplifiers (8 to 18 GHz) reached 2 W and the maximum operating frequency for commercial products reached 50 GHz.

But the most important evolution is with the MMIC development, as it was for data processing with the LSI components during the 1970s. These electrical and technological changes bring the microwave amplifiers progressively into the civil market: direct broadcasting, microwave detectors, sensors, and as yet unthought of applications that will appear with the beginning of mass production.

9

Low noise microwave amplifiers

Yves Charlet and M. Depré

9.1 INTRODUCTION

These amplifiers are especially important in the telecommunications field. This means that emphasis is given to very low noise performance, linearity, and to the cost when they are used in the semi-public area. The particularities are: narrow bandwidth, good VSWR, well optimized transistors, sometimes cooled stages, but with standard technology and without special size requirements.

9.2 LOW NOISE AMPLIFIER DESIGN

9.2.1 Noise parameters

The noise figure (*NF*) is the signal-to-noise ratio degradation between the output and the input of the amplifier.

$$NF = \frac{P_{in}/N_{in}}{P_{out}/N_{out}}$$

where P_{in} is the signal power at the input, P_{out} is the signal power at the output, N_{in} is the noise power at the input, N_{out} is the noise power at the output, and where

$$N_{out} = G \times N_{in} + N_a$$

G being the amplifier gain $= P_{out}/P_{in}$ and N_a the noise contribution of the amplifier.
So

$$NF = 1 + \frac{N_a}{N_{in} \times G}$$

it is often expressed in dB, $NF_{dB} = 10\log(NF)$. As $N_a = GkT_eB$, and $N_{in} = kT_{in}B$ we obtain

$$NF = 1 + \frac{T_e}{T_{in}}$$

where k is the Boltzmann constant (1.38×10^{-23} J/K), T_e is the noise equivalent temperature of the amplifier, T_{in} is the noise temperature at the input (290 K) and B is the bandwidth. For very low noise amplifiers, the noise figure is expressed in degrees Kelvin. For instance an amplifier with $NF = 3$ dB, has a noise equivalent temperature of 290 K.

When two amplifiers with respectively NF_1, NF_2 and G_1, G_2 noise figures and gains, are cascaded, the resulting noise figure is given by the Friis equation:

$$NF = NF_1 + \frac{NF_2 - 1}{G_1}.$$

This formula is true only with perfect impedance matching between stages. That means that the noise contribution of the second stage is reduced by the gain ratio G_1 of the first stage. This is why all the emphasis must be placed on the first stage for the noise and gain performance.

The designer of a low noise amplifier needs the transistor characteristics NF_{min} (minimum noise figure), R_n (noise resistance) and $Y_{opt} = G_{opt} + jB_{opt}$ (optimal noise admittance). These parameters permit the noise figure calculation by the following relationship:

$$NF = NF_{min} + \frac{R_n}{G_s} \times |Y_s - Y_{opt}|^2$$

with $Y_s = G_s + jB_s$ (source admittance connected at the input of the transistor).

Unfortunately these parameters are not always given by the transistor manufacturer and need to be 'measured'. This will be done by measurements of the noise figures of the transistor for different values of Y_s. With a minimum of seven values, we are able to calculate all the noise parameters.

9.2.2 Noise optimization

Circuit design

The noise figure expression given above can be drawn on a Smith chart, with Y_s as variable, then one obtains constant NF circles as well as constant gain circles. So the noise optimization of an amplifier will consist of determination of the circuit which will present the good Y_{opt} to the transistor (Fig. 9.1).

In practice, first we optimize the output network to obtain the maximum gain, then we achieve Y_{opt}. This could easily be done at a fixed frequency, but

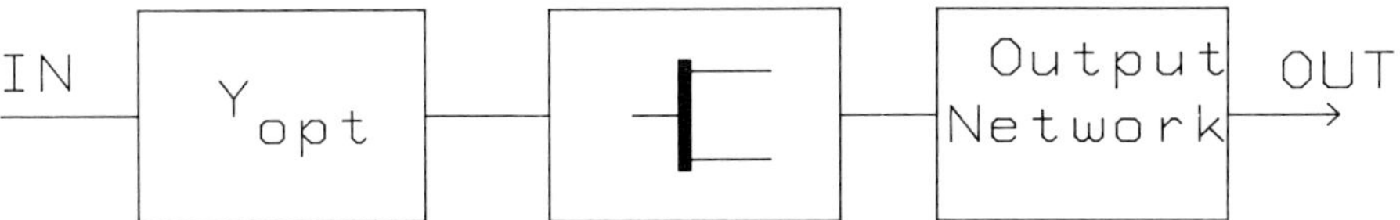

Fig. 9.1 Block diagram circuit design for optimum noise performance.

is much more difficult in a given bandwidth. Also, there will be a compromise between the noise figure, and the maximum gain due to the input matching. Indeed, the constant gain circles at the input, and constant noise circles are generally very different, as seen in Fig. 9.2.

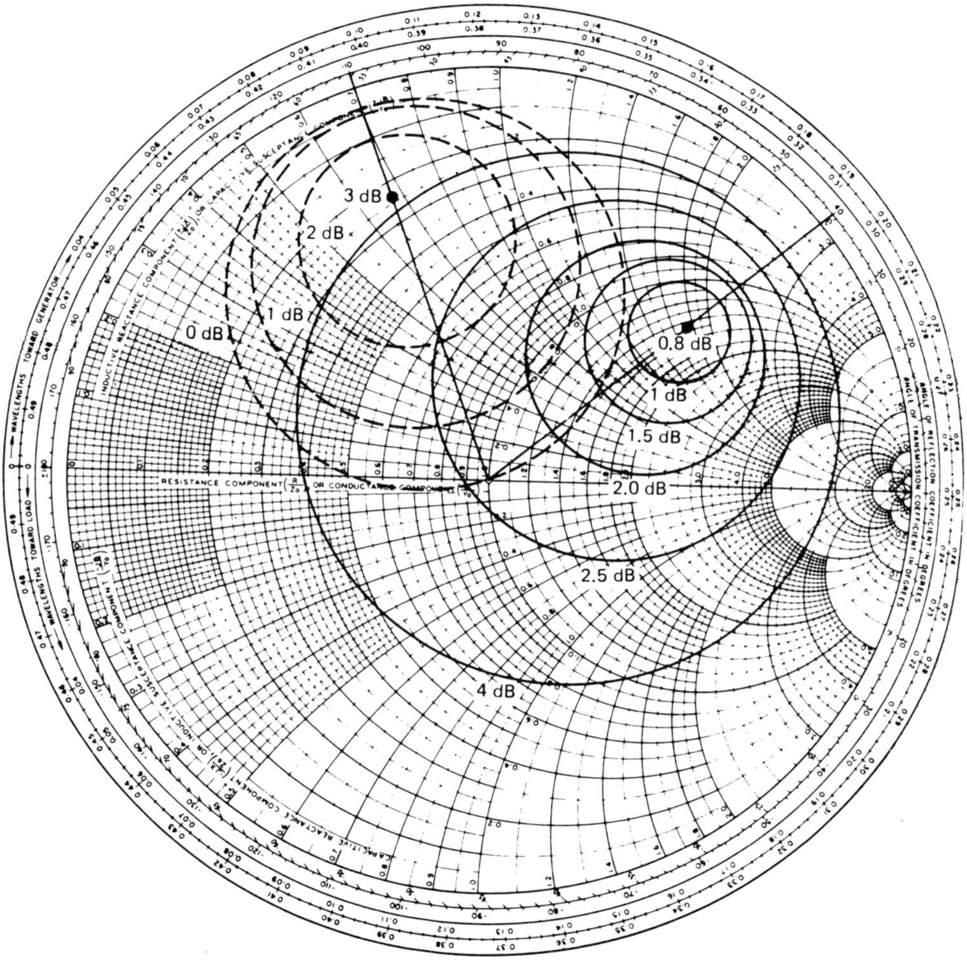

Fig. 9.2 Noise figure circles (solid) and constant gain circles (dashed) on a Smith chart.

As you remember that the gain value of the first stage is very important in the global noise performances of the amplifier, you will have to choose! Fortunately, now many software products are able to optimize the noise figure and the gain simultaneously.

Nevertheless other parameters are always critical: they are the input and output VSWR of the amplifier. In fact in low noise applications, the problem is solved by an isolator at the input. For the output, the VSWR could be good even with noise degradation of the last stage as we know that its noise contribution is low.

The use of an isolator at the input is acceptable because of the low insertion loss, and the good VSWR and isolation of such devices. Typical parameters for a waveguide isolator in the 7.1 to 7.7 GHz band are:

insertion loss:	< 0.15 dB (some high performance devices can decrease to 0.07)
VSWR:	< 1.06
isolation:	> 30 dB.

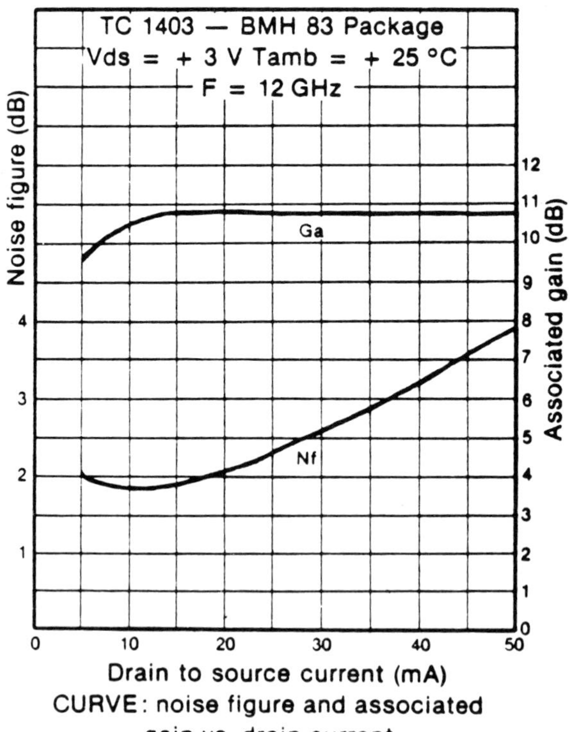

Fig. 9.3 Noise figure and associated gain VS. drain current.

The problem is the size: $80 \times 80 \times 50$ mm. In coaxial form it will be only $12.7 \times 17 \times 12.7$ mm but the insertion loss will be about 0.25 dB, and the VSWR 1.20.

Transistor

The noise performance of an amplifier first comes from the transistor. As explained in Vol. I, ch. 13, the noise parameters of a transistor depend on the technical aspects involved (semiconductor material, topology, gate or emitter size, etc.). Nevertheless a common feature is the current influence on the noise figure. For instance with a 0.5 μm gate length and 250 μm gate width, the noise figure and gain variations of a GaAs transistor with the drain current are as shown Fig. 9.3. Again, one finds a compromise between the maximum gain and the minimum noise figure.

For the GaAs transistor, the noise being thermal, a cooled mounting can reduce the noise figure by a large amount. For example, a value of 0.5 dB at 290 K becomes 0.2 dB at 77 K (liquid nitrogen). Without going up to the cryogenic cooling which is very expensive and complicated to drive, we increasingly use the thermoelectrical properties of a Peltier element. The first stage of the amplifier can then be cooled to – 40 °C without any refrigerating equipment, using only a current power supply.

9.2.3 Parametric amplifier

There is not enough space here to describe parametric theory fully (for that, see *Parametrics Electronics* by Löcherer Brandt), only a typical parametric amplifier (Fig. 9.4).

The principle is to 'pump' energy from an external source via a non-linear device, and transfer it to the signal circuit in a way to increase the signal

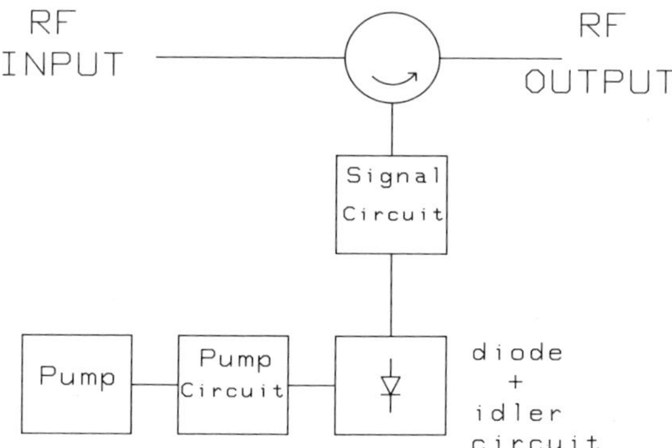

Fig. 9.4 Schematic diagram of a typical parametric amplifier.

power. The pump frequency F_p must be higher than the source frequency F_s. An example of a parametric oscillating system is the swing which pumps energy from the child who periodically bends and straightens his legs and body.

For a parametric amplifier, generally the non-linear circuit is a varactor diode. The pump source could be a Gunn oscillator, or any other oscillator able to deliver enough power (+ 10 to + 20 dBm) at very high frequency (40 GHz or more).

The key circuit is the varactor diode which must have a very high cut-off frequency, that means a very good quality factor (low series resistance, low junction capacitance). The varactor diode will see its capacitance 'modulated' by the pumping power. In fact this circuit is a 'three-frequency' one as shown in Fig. 9.5. Z_s is a series resonant circuit on the signal port, Z_p is a series resonant circuit on the pump port, and Z_i is a series resonant circuit on the idle port. The idle port will be terminated with a reactive load. The signal is fed into port 1 and taken off just after amplification.

Gain considerations

The impedance of the 'pumped' varactor can be calculated and expressed through the diode characteristics and the idle and pump circuits. The equivalent circuit at the signal frequency is shown in Fig. 9.6. $R_n + jX_n$ includes the influence of the idler circuit. Then one can determine the reflective gain of the amplifier when the varactor is matched at Ω_s.

At the resonance, the gain is:

$$G = \frac{\Omega_s}{\omega_p - \Omega_s} \cdot \frac{4\,\alpha_i}{(1 - \alpha_i)^2}$$

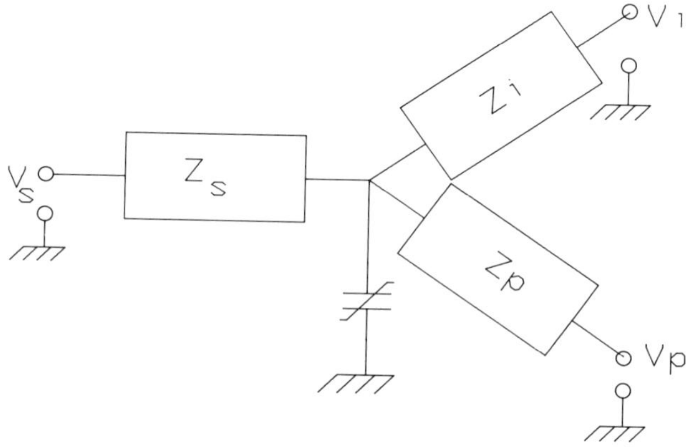

Fig. 9.5 Equivalent diagram of a varactor diode non-linear circuit for a parametric amplifier.

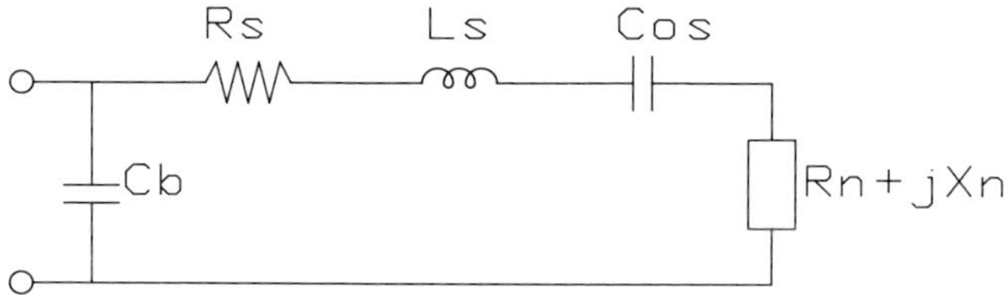

Fig. 9.6 Equivalent circuit of a pumped varactor.

Where Ω_s is (signal frequency) $\times 2\pi$, ω_p is (pump frequency) $\times 2\pi$, and, α_i is $f(S(1), \Omega_s, \omega_p, R_{11}, R_{22})$, with $S(1)$ being the first term of the Fourier transform of the diode reactance, R_{11} the real component of the signal circuit including the generator impedance and the bulk resistance of the diode, and R_{22} the real component of the load impedance including the idle impedance and also the bulk resistance of the diode:

$$\alpha_i = \frac{|S(1)|^2}{\Omega_s \cdot (\omega_p - \Omega_s) R_{11} \cdot R_{22}}.$$

That means we can achieve gain with a sufficient value of α_i. With particular conditions, the maximum available gain would be

$$\frac{\omega_p + \Omega_s}{\Omega_s}$$

which shows the importance of the high value for the pump frequency.

Noise considerations

The thermal noise coming from the loss resistances in the circuits, lines, and diode are not correlated between signal, idler and pump. The shot noise coming from the diode is negligible, because it is operating in the backward region. Also the pump generation noise may be neglected, because the major one being the phase noise, it has no influence on the signal performance of parametric circuits.

The minimum noise temperature is achieved with an optimum ratio Ω_i/Ω_s ($\Omega_i = 2\pi *$ idler frequency). This minimum noise is related to the quality factor Q of the diode:

$$T_{min} \simeq T_0 (1 - 1/G_0)(2/Q - 1)$$

where T_0 is the operating temperature, G_0 is the gain and $Q = \alpha/(R_s C_0 (1 - \alpha^2) \Omega_s)$, and with R_s the series resistance, C_0 the junction capacitance, and α the capacitance ratio of the pumped varactor.

To obtain a minimum value for R_s and C_0 one must choose a varactor diode with the maximum cut-off frequency F_c (2000 GHz is now achievable). α is related to the pump power, but also to the non-linearity parameter of the diode described by

$$\Delta N_j = \frac{C_{j0} - C_{j-6v}}{C_{j0}}$$

which must be a maximum.

Example 1. With a diode having a cut-off frequency of 600 GHz, a 4 GHz parametric amplifier can have the characteristics: $T_{min} \simeq 55$ K, with $F_p = \Omega_p/2\pi = 61$ GHz.

Example 2. With a diode having $F_c = 320$ GHz, a 9 GHz parametric amplifier can exhibit a $T_{min} \simeq 120$ K, with $F_p = 45$ GHz.

These characteristics explain why the parametric amplifier was often used for low noise applications. Nevertheless we have to notice that now, with an HEMT, without any cooling a $T_{min} = 100$ K is achievable at 7.5 GHz, the design being much simpler than those of the parametric amplifier, and thus the cost much lower.

9.3 TYPICAL APPLICATIONS

9.3.1 A 30 GHz low noise amplifier

With MESFET, chosen for their high performance (*NF* = 3.5 dB at 40 GHz and associated gain = 6 dB), the following amplifier was realized:

bandwidth:	27.5 to 30.5 GHz
gain:	19 dB
noise figure:	4.6 dB
output power at 1 dB:	> + 3 dBm.

The elementary stage on an alumina substrate is shown in Fig. 9.7. The size is only 2.5 × 2.6 mm. One can notice the matching networks: at the input for noise reduction, and at the output for gain optimization. The source is grounded, and the biasing delivered separately to the gate and the drain. The pads near the lines permit adjustments of the electrical parameters by connecting them with gold ribbons to the microstrip.

This stage has a gain of 7.5 dB and a noise figure of 3.9 dB. With three equivalent stages one obtains the global performances given above, including the losses of an input isolator, and of a waveguide – microstrip transition. This same construction with an HEMT in place of the MESFET will present a noise figure of about 0.5 dB less, such as 4.0 dB at 30 GHz.

9.3.2 X-band amplifiers

With similar technology, on alumina substrate, we are able to achieve a noise temperature of about 100 K between 7.25 and 7.75 GHz.

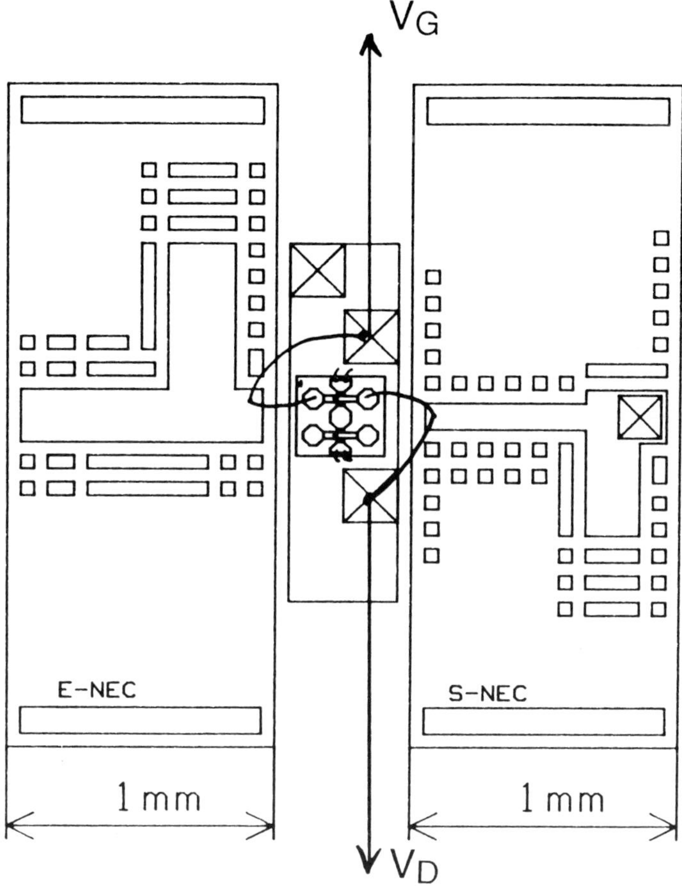

Fig. 9.7 Layout of an elementary stage of a 30 GHz LNA.

9.3.3 C-band amplifiers

With HEMT transistors on printed circuit boards, the following results were obtained:

bandwidth:	3.625 to 4.20 GHz
noise temperature:	< 45 K (40 K typical)
temperature range:	− 20 to + 45 °C

In this case the first transistor is cooled by double Peltier cells.
Same as above except:

noise temperature:	< 65 K (60 K typical)

In this case, there is no cooling circuit.

These two amplifiers use a very specific isolator at the input with less than 0.1 dB loss. A particular design is chosen for the cooled stage to reduce the thermal conductivity. The power necessary to cool the transistor is about 30 W.

9.3.4 Ku-band amplifiers

Bandwidth:	10.95 – 12.75 GHz
noise temperature:	< 130 K with Peltier cells
	< 150 K without Peltier cell

These C and Ku Band amplifiers are used in the telecommunications field, which means that cost reduction is very important. So, the technology used is the printed circuit board on soft material (PTFE glass), with packaged transistors which permit noise or gain screening before mounting, to achieve the best performance.

9.4 CONCLUSION

The availability of very low noise transistors such as the HEMT, allows the design of amplifiers much simpler than the parametric amplifiers, so it will

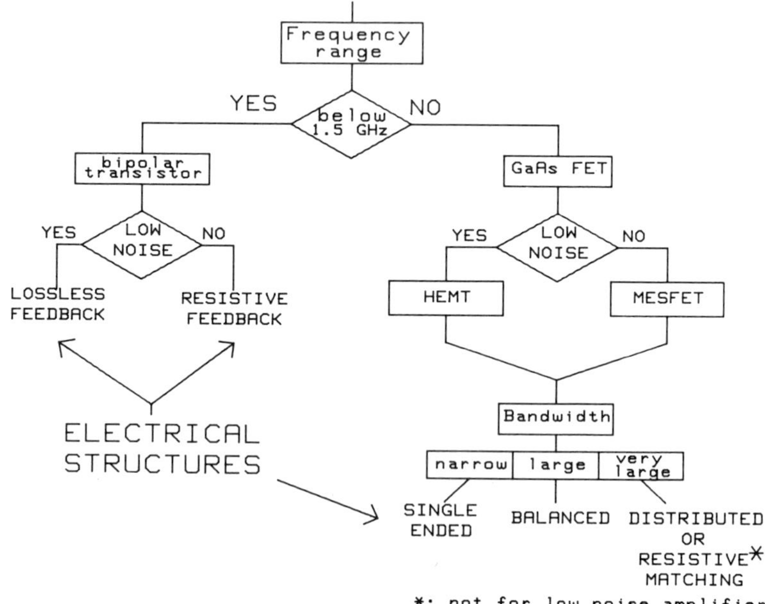

Fig. 9.8 Design flow chart for microwave solid state amplifiers.

increase their use, for instance, in direct broadcasting (DBS), small radio links, very low noise receivers, and industrial fields, etc. The technology involved in such products, is connected to the customer needs. For military equipment, it will be microwave integrated circuits (MIC) for small sizes and reliability; and for professional or industrial applications, the printed circuit board is the only solution in terms of cost reduction.

As a summary, Fig. 9.8 gives the decision chart for the electrical characteristics of microwave solid state amplifiers. This decision chart must be understood in the light of the preceding chapters. The resulting design choices are connected to the evolution of the basic components: transistors and microwave monolithic integrated circuits. Nevertheless, these electrical structures will remain as standard, only the frequency and noise limits will progress.

10

Monolithic microwave integrated circuits

Christian Rumelhard

10.1 INTRODUCTION

10.1.1 History

The first attempts to make microwave integrated circuits were made with mixing or p-i-n diodes on silicon. Then, the arrival of field effect transistors led to the development of hybrid microwave circuits which were made of an alumina or quartz substrate and additional passive (inductances, resistances, capacitances) or active elements (Gunn diodes, transistors, etc.). These circuits were wrongly called microwave integrated circuits. The first truly integrated circuit was presented in 1975: it was an X-band amplifier with a transistor and a few passive elements. To indicate that these circuits were really integrated they were called monolithic microwave integrated circuits while microwave integrated circuits based on alumina or quartz were named hybrid microwave integrated circuits. A number of monolithic circuits were shown in the following years: low noise or power amplifiers, mixers, fixed or voltage controlled oscillators, analogue or digital phase shifters. All these circuits were created with MESFET on GaAs.

In 1980, an new heterojunction device was simultaneously patented by Thomson-CSF which called it two dimensional electron gas field effect transistor (TEGFET), and Fujitsu which called it high electron mobility transistor (HEMT) (see Volume 1, Chapter 15). A first monolithic circuit based on this transistor was made in 1983.

Then, several circuits were associated to create subassemblies on the same chip: a four bit phase-shifter, a DBS (direct broadcast satellite) receiver (Kermarrec *et al.*, 1984), a transmit/receive T/R module for phased array antenna, but these presentations served mainly as demonstrations; the present subassemblies projects based on monolithic circuits are made of several chips which associate functions of the same type like low noise amplification, power amplification, signal control, etc.

At the present time, the monolithic circuit activity is shared between foundries which offer to make circuits starting from more or less complex elementary cells, and equipment teams which design systems containing full custom monolithic circuits.

10.1.2 Physical aspects

The basic material of monolithic circuits is GaAs. The main physical properties of this material are now recalled.

The electron velocity as a function of electric field is given on Fig. 10.1 in comparison with other materials such as InP, GaInAs or Si. The saturation velocities are about the same for all these materials, but for III-V compounds, this velocity passes by a maximum which is two or three times greater than for the Si. Under the gate of a MESFET, the saturation velocity is reached only at the end of the length and therefore, the main part of the path is covered at a higher velocity. This result leads to faster devices. Contrary to Si, GaAs substrate is semi-insulating and its resistivity is $10^7\ \Omega$ instead of 100 to 1000 Ω for Si. Therefore, propagation lines made in GaAs have much lower losses.

GaAs is also less sensitive to ionizing radiations and high temperature. Because of this, it can be used in harsh environmental conditions. GaAs emits light in the visible spectrum which allows the coupling on a same chip of

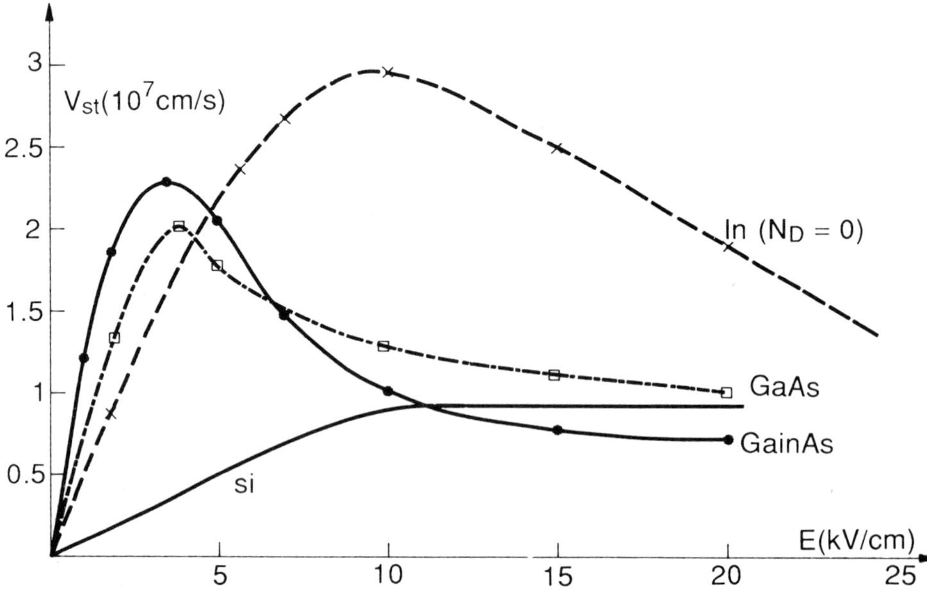

Fig. 10.1 Comparison of electron velocities as a function of electric field in Si, GaAs, GaInAs and InP.

electronic circuits with light emitting or receiving diodes. GaAs is piezoelectric which allows coupling with surface acoustic wave devices. This last aspect has seen few realizations for the moment. Finally, the electro-optic properties of GaAs could lead to original test methods.

10.1.3 Hybrid and monolithic circuits

The fields of hybrid and monolithic integrated circuits are complementary. The use of monolithic circuits is justified in the case of large quantities because of their potential low cost, their better reliability (decreasing number of connections), their low weight, their better reproducibility or bandwidth.

But, on the other hand, a monolithic chip can't be used alone. To make the connections with other chips, to bring the biasing voltages, to associate the logic command circuits, to ensure hermeticity, to drain heat, and to avoid electromagnetic coupling, it will always be necessary to use hybrid circuit techniques. And once the monolithic circuit techniques are stabilized, the points which have been mentioned above concerning the hybrid aspects will appear as the main problems to solve.

10.1.4 Necessary tools for the design and fabrication of MMICs

The design of MMICs calls for specific CAD tools like linear or non-linear electrical simulators, layout software, design rule check software, or programs that allow the coupling of layout and simulation to check the circuits from the point of view of microwaves.

A monolithic microwave technology is based on a library of passive and active components which must be developed and modelled electrically or described as graphic cells to be used in CAD software. All these models constitute the circuits design rules. The circuits are manufactured in a technological line and then tested by static and microwave methods. This is done by specific and generally automatic equipment.

In the next sections, the technology will be described first and then the active and passive components will be presented. The last part will be dedicated to the description of some typical monolithic circuits.

10.2 TECHNOLOGY AND PROCESS CONTROL

10.2.1 Description of a technological process

Figure 10.2 summarizes a technological process used to manufacture MMICs. In this process, different layers are created: active layers in the semiconductor, metallic layers, dielectric layers, etc. These layers must have particular

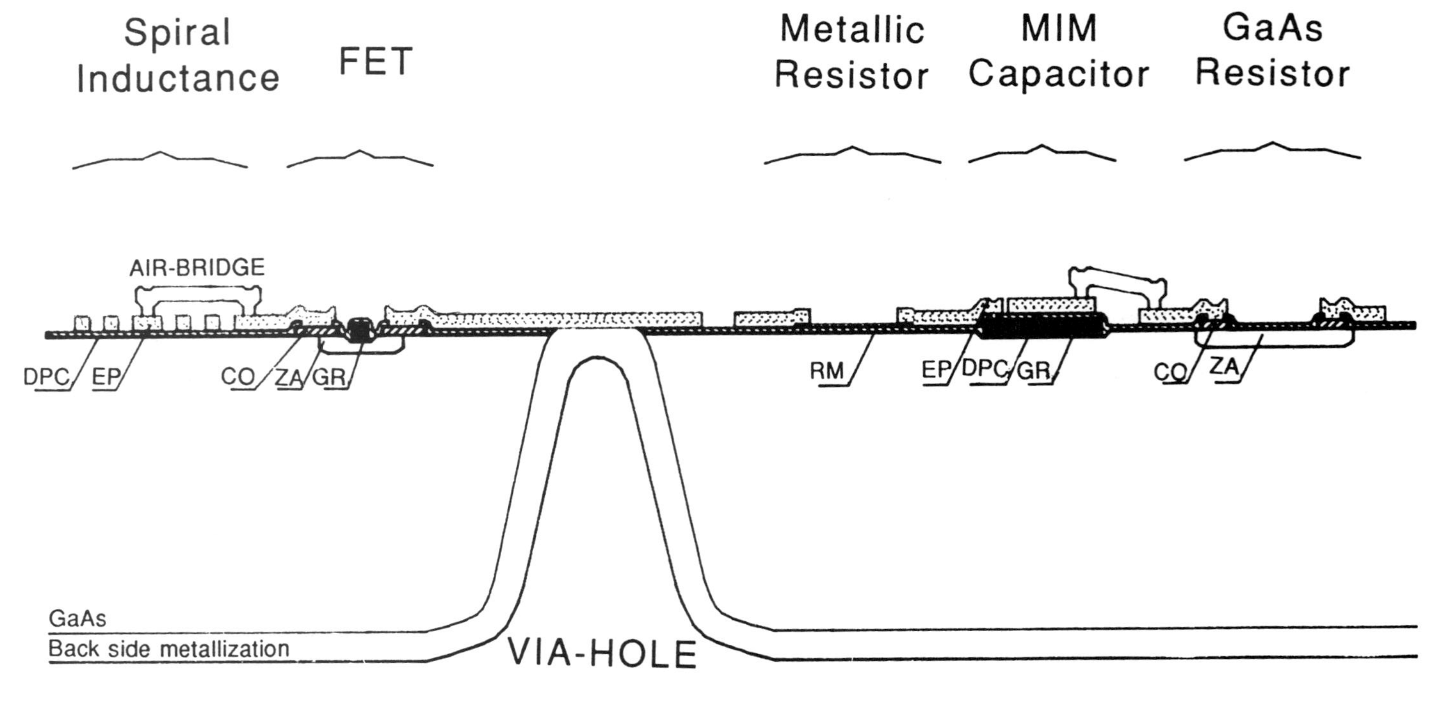

Fig. 10.2 Cross-section of typical MMIC components.

properties (insulation, conduction, resistivity, etc.) and must be correctly aligned. The different layers are now described.

Active layer and isolation

Starting from a GaAs semi-insulating substrate, the active layer can be made in different ways. A first way consists of growing a layer by epitaxy (by a vapour phase or molecular beam process). A first layer is used as a buffer and is made of intrinsic GaAs. This layer is followed by an active layer doped with Si atoms. Each of the different active layers corresponding to the transistors, diodes or resistances must be insulated from the others. This can be obtained by the fabrication of mesa by etching (old method) or by the implantation of Bo atoms which break up the crystalline structure and thus bring an insulation.

A second way consists in the implantation of ionized Si atoms (ionic implant) either on the whole surface, or selectively on the different locations of active layers. As mentioned above, an effect of this implantation is the partial break up of the crystalline structure. To the recover the original crystalline structure, the implant must be activated by an anealing at 800 °C. If the implantation is made on the whole surface of a wafer, the isolation between the different components can be obtained with one of the techniques recalled in the preceding section. The selective implantation introduces a natural insulation which can be enhanced by boron implantation, but the main advantage of this technique is to allow the creation of different active layers to get different components like small signal transistors (low energy implantation) associated to varactors (high energy implantation).

Ohmic contacts

The ohmic contacts (source or drain electrodes of FETs) are obtained by depositing successively an alloy of AuGe, then Ni, then Au, and proceeding with anealing at 450 °C. These operations give a reprocical or 'ohmic' metal semiconductor contact.

Schottky gates

Transistor or diode gates are created by successive deposit of Ti, Pt and Au. The Ti deposit gives the Schottky contact with the active layer; the objective of the gold deposit is to decrease the resistance; and the Pt is put between Au and Ti to avoid the migration of gold into the active layer.

All these deposits are obtained by optical lithography except for the gates having a length equal or less than 0.5 μm. In this case, it is necessary to use electron beam lithography. The deposit of the gate is preceded by a recess in the active layer made by wet or ion etching. The object of this recess is to avoid possible problems due to surface charges and to pass through a small highly doped layer which is located near the surface to decrease the resistance of the active layer.

First metal

A deposit of Ti, Pt, Au gives the lower electrodes of capacitors and of certain connections. The thickness of this layer is 0.4 μm.

Dielectric for passivation and capacitances

A coating of Si_3N_4 is deposited to protect the active layer of transistors and diodes and to make the dielectric of the capacitors.

Second metal

A second gold layer, relatively thick (2 μm), is deposited to constitute the upper electrode of capacitances, spiral inductors and connections.

Air bridges

To make crossing with a parasitic capacitance as low as possible, air bridges are fabricated by the sputtering of gold followed by electroplating. These bridges are also used in spiral inductors and to connect the upper electrodes of capacitors (Fig. 10.3).

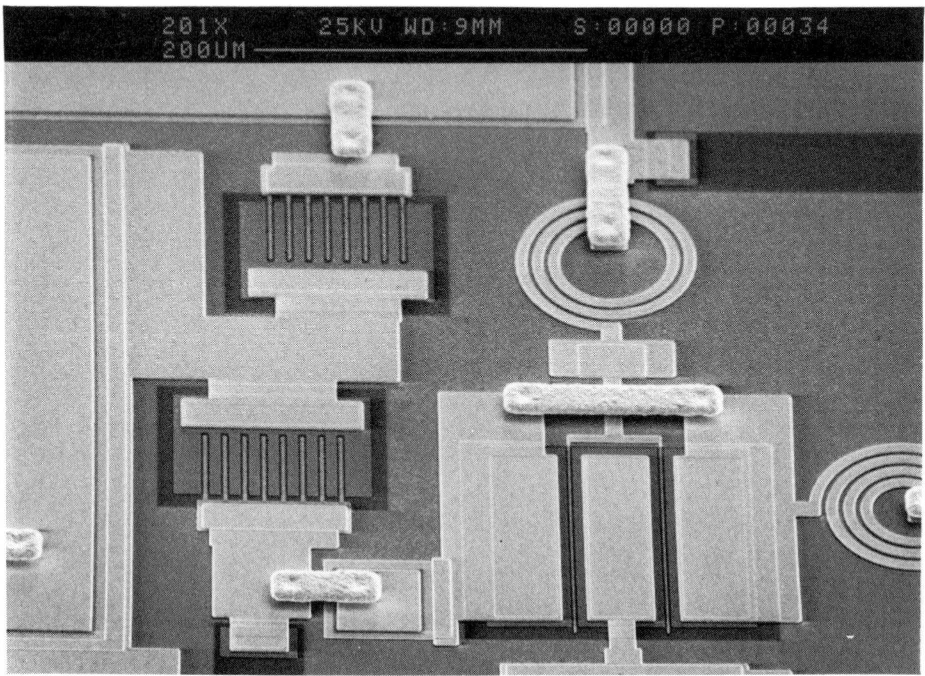

Fig. 10.3. Use of air bridges for conducting crossovers and spiral inductors.

Backside metalization and via holes

After the process described above, the GaAs wafer is thinned to a thickness of 100 μm. Holes are etched through the substrate starting from the backside of the wafer. In this case, the alignment between the back side and the upper surface of the wafer is made with an infrared microscope because GaAs is transparent to infrared light. Then the via holes are metallized by sputtering and electroplating of gold.

The result of all this technology is to introduce constraints in the design of circuits such as minimum distances between the different metal, dielectric or active layers or minimum dimensions of elements. All these constraints form a set of technological rules which must be respected for designing monolithic circuits ('design rules')

10.2.2 Process control monitor

The process described in the above section must be checked either during the process (in line tests) or at the end of the process by a final test. In order to do this, a process control pattern is added to the layout of the circuit (from 5 to 40 times on a wafer). This process control monitor is shown on Fig. 10.4. The patterns which are used during the manufacturing are located at the centre of the figure and are the following:

drain of ohmic contacts to measure the square resistance of the active layer;
isolating curve which gives indications on the isolation between active elements;
transistor to check the recess: the measurement of saturation current between ohmic contacts during the gate recess operation allows to adjust the characteristics of the final transistor;
gate pattern to test the gate resistance.

The patterns which are tested at the end of the process are located on the periphery:

fat FET to know the doping profiles;
'Greek key' pattern of gate to measure the gate resistance;
chain of air bridges for optical and electrical check of the quality of bridges;
metallic and active layer resistance;
capacitor to monitor the capacitance density per surface unit area and breakdown voltages;
two via holes located in series electrically to test them by a simple resistance measurement;
a piece of line to test the resistance of the second level thick metallization.

All along the process and at the final test, limit values are fixed and are used as acceptance criteria. For instance, the acceptance or rejection of wafers is decided on the measurement of five parameters: ohmic contact resistances, square resistance of active layer, threshold voltage of transistor, gate resistance, MIM capacitance and metallic resistance. These measurements give an

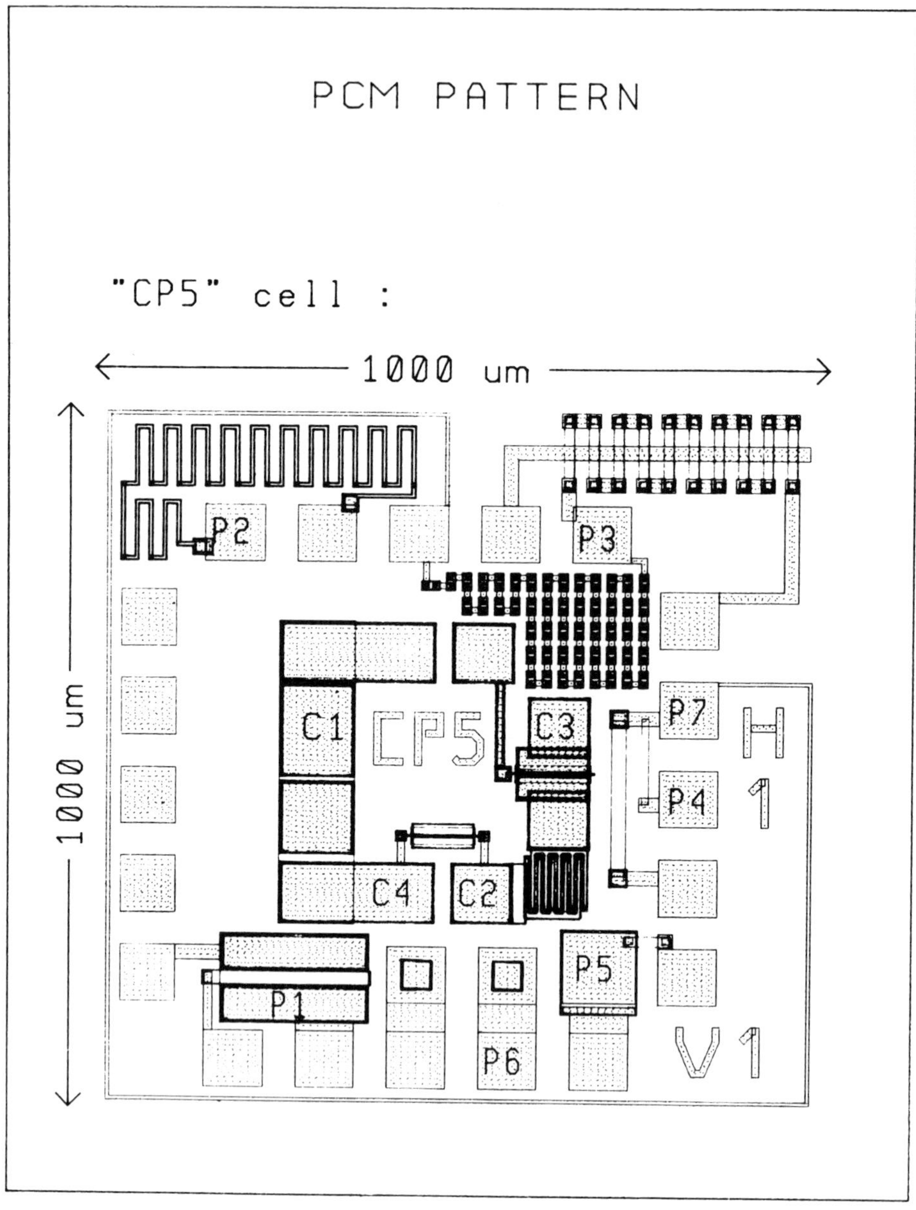

Fig. 10.4 Process control monitor.

indication of the success of the process, independently of the circuits which are on the wafer.

10.3 PASSIVE AND ACTIVE COMPONENTS

Generally, once it is finished, a monolithic circuit can't be adjusted or trimmed. This fact leads to a need to get a comprehensive component library to avoid many repetitions of the design, manufacture and measurement cycles. Some examples of these component models are now described.

10.3.1 Passive components

Resistances

The resistances are obtained by using a piece of active layer or a specific metallic deposit such as Ti or NiCr or with a more complex deposit like TaN (tantalum nitride).

Capacitances

There are two main configurations to design capacitances. The first type (interdigitated capacitance) gives very low value capacitances but with a very good accuracy. This pattern is preferably used above 20 GHz.

The second type (MIM capacitance) is by far the most used in monolithic circuits. The dielectric can be taken from the materials listed in Table 10.1. To obtain compact circuits, the permittivities are as high as possible. But, the breakdown voltage must be simultaneously large. Therefore the value which best characterizes a dielectric material is the dielectric efficiency given by the relationship:

$$D = \frac{1}{2} \varepsilon_0 \varepsilon_r E_2^2$$

where ε_c is the breakdown electric field and ε_r is the relative permittivity. This efficiency can be computed from Table 10.1.

Table 10.1 Some dielectrics for monolithic capacitors

Materials	*Permittivity*	*Breakdown electric field* (V/μm)	*Losses* (tan δ) *at* 20 °C, 100 kHz
SiO_2	4–6	400	0.01–0.04
Polymide	3–5		0.001–0.005
Si_3N_4	5.5	200	
Al_2O_3	8.8	250	0.008
TiO_2	55	50	0.04–0.09
Ta_2O_5	22–28	200	0.02–0.01

Inductances

The drawing of a spiral inductor is given on Fig. 10.5(a) and its equivalent circuit on Fig. 10.5(b). The values of inductance, resistance and parasitic capacitances are given by electromagnetic models (Parisot *et al.*, 1984) which are then checked by systematic measurements of *S* parameters.

Starting from these computations verified by the measurements, it is possible to extract functions which fit the different values of the equivalent circuit of Fig. 10.5(b). These values are given by the coefficients of a quadratic polynomial expansion $y = A_0 + A_1 x + A_2 x^2$ where x is the external diameter in μm. Then the values of Q, R, C_{10}, C_{20} and F_c (cut-off frequency) are given as a function of the inductance represented by x.

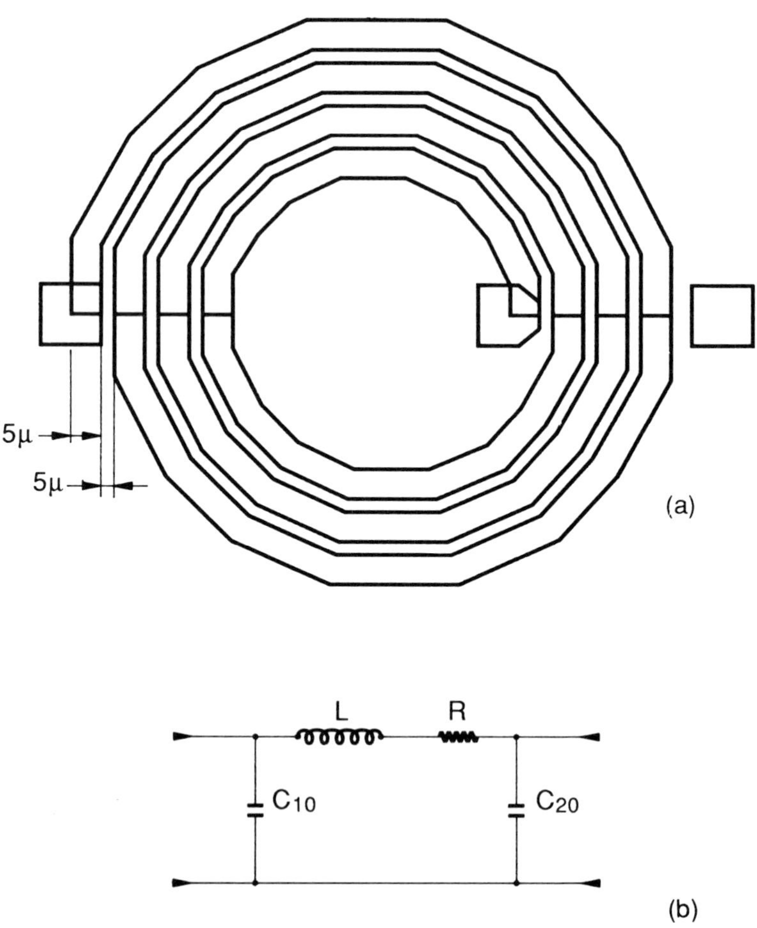

Fig. 10.5 Layout of a spiral inductor and its equivalent circuit.

Table 10.2 Elements of equivalent circuit for a mixing diode

D.C. bias (V)	C_j (pF)	R_s (Ω)	R_p (Ω)	L_1 (pH)	L_2 (pH)	F_c (GHz)
0.4	68	108	228	—	—	22
0.5	77	101	232	—	—	20
0.6	127	91	127	—	—	14
0.7	163	44	44	30	40	22
0.8	113	21	22	35	32	66
0.9	101	14	14	30	30	111

Lines

In a monolithic circuit, the lines which ensure the connections between elements or which create inductances are always approximately in the configuration described in Fig. 10.6. A part of the ground is on the back side of the wafer which gives a microstrip line and another part of ground is coplanar. The curves of Fig. 10.7 give characteristic impedance Z_0 and relative permittivity ε_r for a line of monolithic circuit. To compute the inductance and spurious capacitance of a piece of line, it is possible to use the relationships given here:

$$Z_0 = (L/C)^{1/2} \qquad V_p = 1/(LC)^{1/2} \qquad V_p = c/\sqrt{\varepsilon_{\text{eff}}}$$

where L is the inductance per unit length, C is the capacitance per unit length, V_p is the phase velocity and c is the velocity of light.

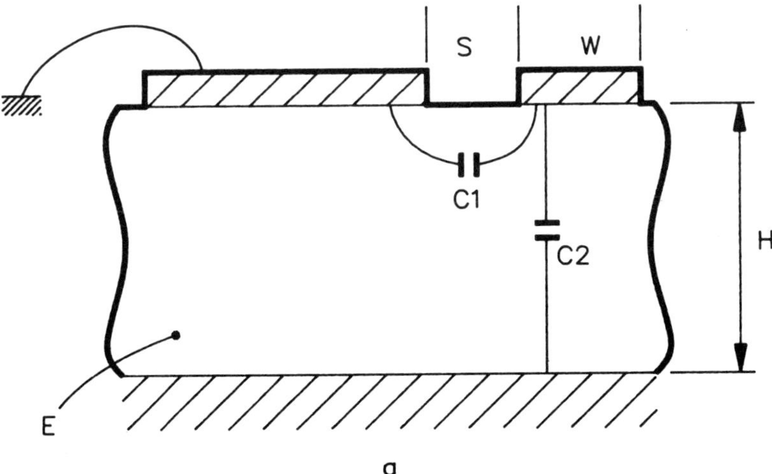

Fig. 10.6 Diagram of a monolithic circuit connection line.

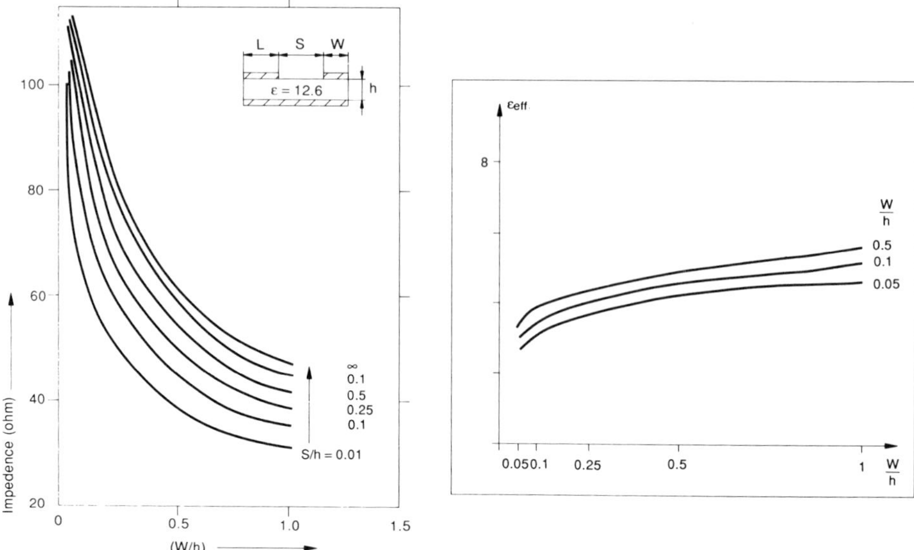

Fig. 10.7 Characteristic impedance and equivalent dielectric constant of a monolithic circuit line.

10.3.2 Diodes

In monolithic circuits, diodes can be used either as mixers or as varactors. Contrary to hybrid circuits, the diodes encountered in monolithic circuits always have a planar structure, which increases their losses. Although they are used in very different conditions (direct bias for mixing diodes and inverse bias for varactor diodes), the equivalent circuit is identical and is shown on Fig. 10.8.

Mixing diode

Table 10.2 gives an example of values of the equivalent circuit for a mixing diode made of two fingers of Schottky contact having a width of 5 μm, a length of 1 μm and interdigitated between three cathodes themselves spaced by 3 μm.

Varactors

Table 10.3 gives an example of the elements of the equivalent circuit for a varactor diode made up of four fingers of Schottky contact having a width of 50 μm and a length of 4 *μm* in a space between successive cathodes of 8 μm. The pinch-off voltage of such a diode is – 7 V.

Table 10.3 Elements of equivalent circuit for a varactor diode

D.C. bias (V)	C_j (pF)	R_s (Ω)	R_p (Ω)	L_1 (pH)	L_2 (pH)	F_c (GHz)
0.1	1.34	2.38	4	22	2.83	93
0.2	0.91	3.01	4	22	2.19	93
0.3	0.72	3.28	4	22	2.54	94
0.4	0.60	3.65	4	22	3.00	94
0.5	0.51	4.13	4	22	3.30	94
0.6	0.43	5.05	4	22	3.65	91
0.7	0.25	6.24	4	21	4.08	72
0.8	0.16	2.06	4	21	3.19	85

10.3.3 Transistors

Single-gate field-effect transistor

A first possibility to model a transistor is to define an equivalent circuit extracted from S parameter measurements made between 0.1 and 26 GHz. Figure 10.9 presents such an equivalent circuit in which each of the elements can vary as a function of biasing voltages and of geometrical dimensions of the transistor. Figure 10.10 shows the variation of the elements for three different biasing voltages.

Concerning geometrical dimensions, the expressions given below enable us to get the elements of the equivalent circuit of a transistor constitued of m fingers in parallel, having a width W', starting from the equivalent circuits of a transistor having n fingers in parallel and having a width W.

$$R_i' = R_i\, W/W' \qquad G_O' = G_O\, W/W'$$

$$R_s' = R_s\, W/W' \qquad C_{GS}' = C_{GS}\, W/W'$$

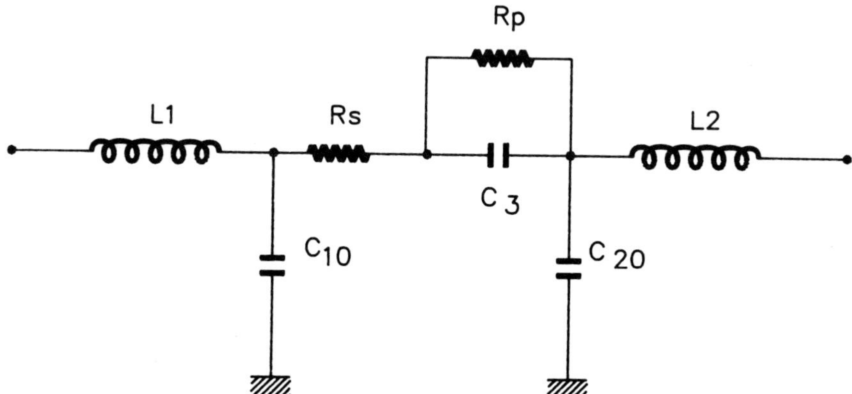

Fig. 10.8 Microwave small signal equivalent circuit for varactor and mixing diodes.

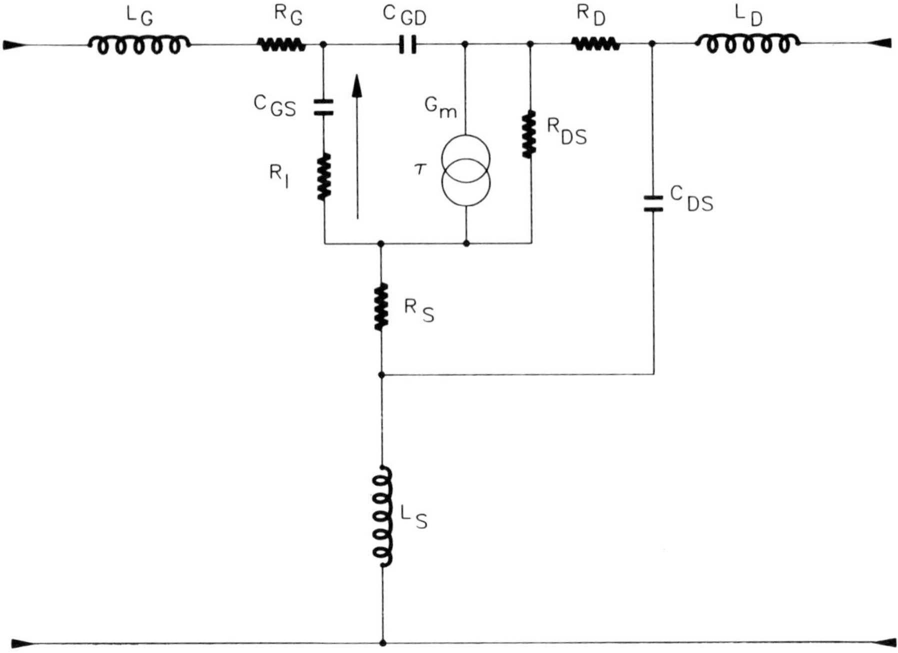

Fig. 10.9 Small signal equivalent circuit of a FET.

$$R'_D = R_D\, W/W' \qquad C'_{GD} = C_{GD}\, W/W'$$

$$R'_{DS} = R_{DS}\, W/W' \qquad C'_{DS} = C_{DS}\, W/W'$$

$$\tau' = \tau$$

$$R'_G = (W'/W)\, R_G\, (n^2/m^2)$$

For a transistor, a non-linear equivalent circuit can also be given. The elements are given for a particular d.c. bias. Figure 10.11 shows an example of a non-linear diagram. Values of the elements are given by the following relationships:

$$I_{ds} = (A_1 + A_2\, V_{gs} + A_3\, V_{gs}^2)\tanh[(A_4 + A_5\, V_{gs})V_{ds}] + (A_6 + A_7\, V_{gs})V_{ds}$$

$$I_{gs} = I_{gso}\,(\exp(kV_{gs}) - 1).$$

The expression for C_{gs} is:

$$C_{gs}\,(\mathrm{pF}) = C_1\, V_{gs}^2 + C_2\, V_{ds}^2 + C_3\, V_{gs} + C_4\, V_{ds} + C_5.$$

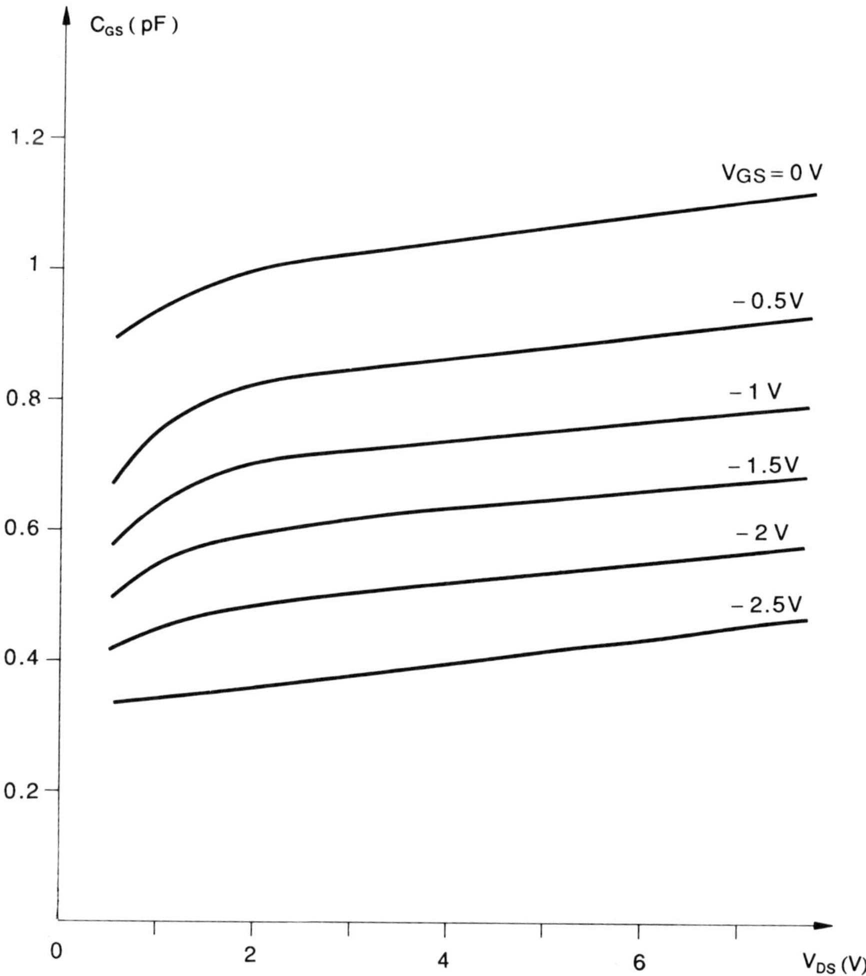

Fig. 10.10 C_{GS} as a function of V_{DS} and V_{GS} for transistor of $4 \times 150 \times 1$ μm.

Dual gate field effect transistor

The same type of equivalent circuit as for a single gate transistor can be given. Figure 10.12 presents such a circuit and Table 10.4 gives an example of values of elements for a dual date transistor having gates of 150×0.5 μm. During the extraction of the elements, the following values are fixed:

$$R_{G1} = R_{G2} = 8\ \Omega \qquad L_{G1} = L_{G2} = 100\ \text{pH}$$

$$R_{D1} = 3.5\ \Omega \qquad L_{S1} = 0\ \text{pH}$$

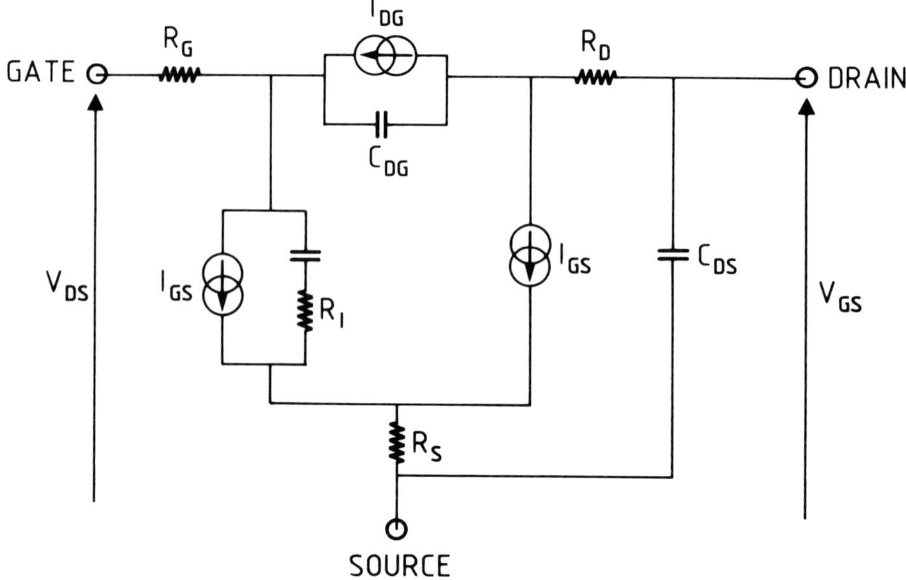

Fig. 10.11 Non-linear equivalent circuit for a transistor.

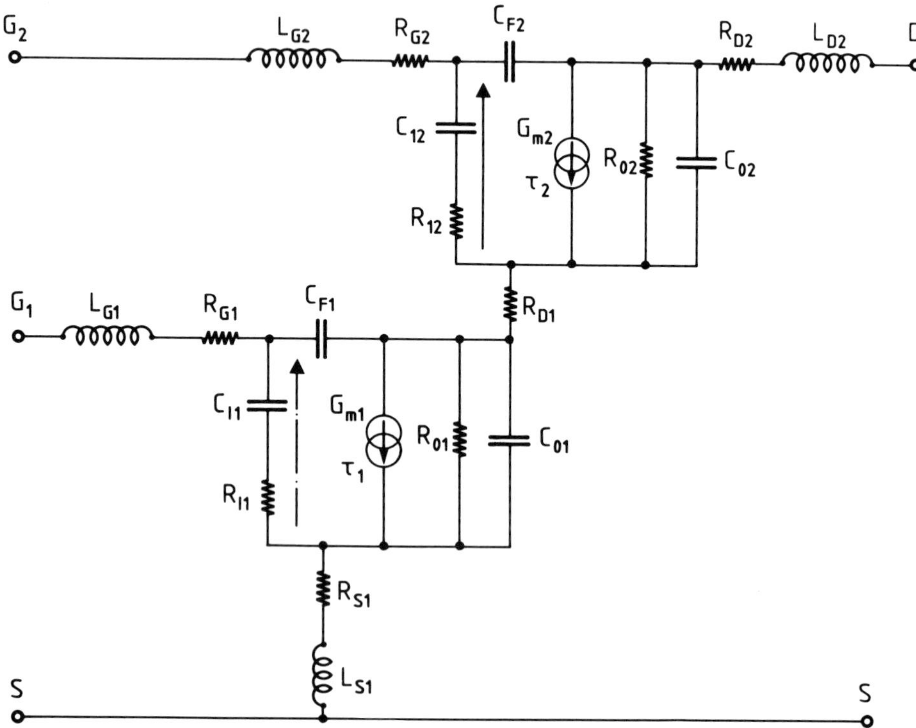

Fig. 10.12 Equivalent circuit of a dual gate transistor.

$R_{D2} = 2\ \Omega \qquad L_{D2} = 95$ pH.

$R_{S1} = 2\ \Omega$

10.4 AMPLIFIERS

10.4.1 Topologies of monolithic amplifiers

We recall the most common way to get an amplifier is to use a common source transistor with matching circuits at the input and the output. With only resistances in the matching circuits or with a resistive feed back, the curve A_1 (Fig. 10.13) can be obtained.

Table 10.4 Equivalent circuit elements for a dual gate transistor: $150 \times 0.5\ \mu$m; Vds = 4 V; $V_{GS1} = 0$ V

V_{G2S}	V	−1	2
I_{DS}	mA	0.6	12.3
C_{I1}	pF	131	117
R_{I21}	Ohms	7	6.1
C_{FI}	pF	47.1	20.3
G_{m1}	mS	0.2	22.6
τ_1	ps	2.8	3.3
C_{O1}	pF	36.5	9.8
R_{O1}	Ohm	37.6	404
C_{I2}	pF	82	140
R_{I22}	Ohm	10.8	16.7
C_{F2}	pF	24.6	18.7
G_{m2}	mS	5.3	26.6
τ_2	ps	3.7	3.6
C_{O2}	fF	26.9	16.8
R_{O2}	Ohm	1094	298

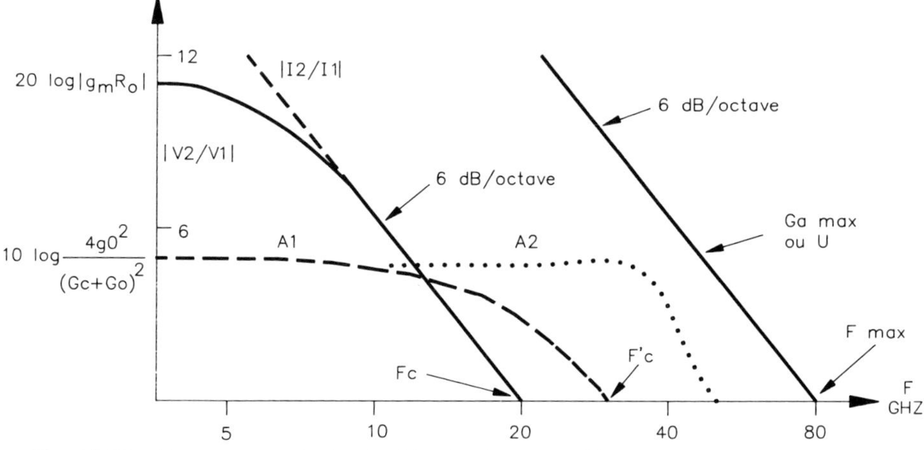

Fig. 10.13 Gain curves obtained with a FET mounted in a common source.

With a perfect matching of input and output by reactive circuits (and as far as the transistor is stable for these conditions), an envelope of all the possible gains for each frequency can be plotted. These gains can be obtained with a narrow band circuit. By combining resistive and reactive matching, the curve A_2 can be plotted. It shows a typical gain for a large bandwidth amplifier.

Different topologies can be used as a function of the type of amplification which is wanted (Rumelhard *et al.*, 1987). Using reactive matching gives amplifiers with a bandwith of less than an octave. The introduction of dual gate transistors introduces the possibility of varying the gain without any change in the input or output matching and without any change in phase. An example of such a circuit will be shown below. The distributed structure is well fitted to monolithic circuits; two examples of such circuits will be described. For amplifiers supposed to deliver a certain power (more than 1 W), another topology is used: the arborescent (or tree) structure. Two examples will be given.

10.4.2 Biasing circuits

Before showing some monolithic amplifiers, the different ways to bring d.c. biasing voltages to the transistor must be reviewed. The constraints to consider are:

The different biasing circuits must not interfere between themselves, which means they must be isolated with decoupling capacitors.

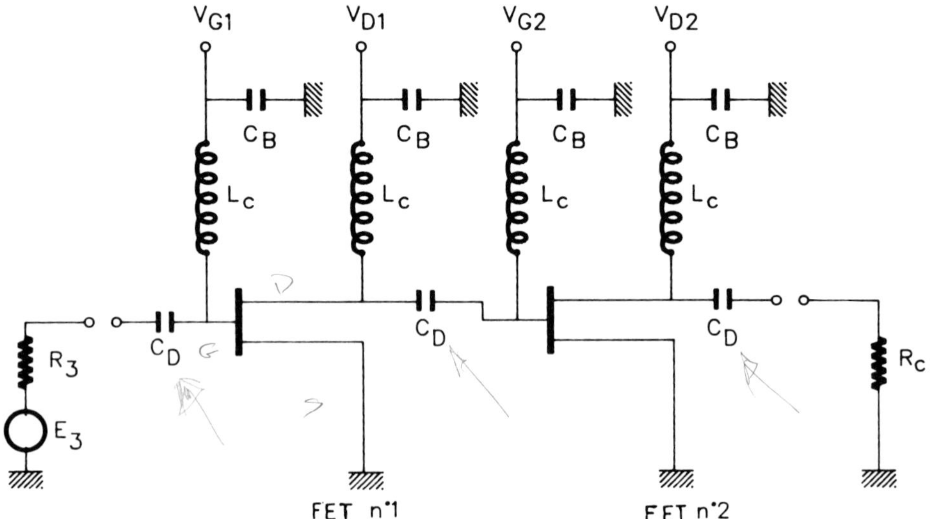

Fig. 10.14 Basic diagram for biasing circuits.

The biasing circuits must not perturb the matching or feedback microwave circuits.
The biasing circuits must not introduce spurious coupling at the working frequencies of the circuits at any other frequency which could possibly induce oscillations.

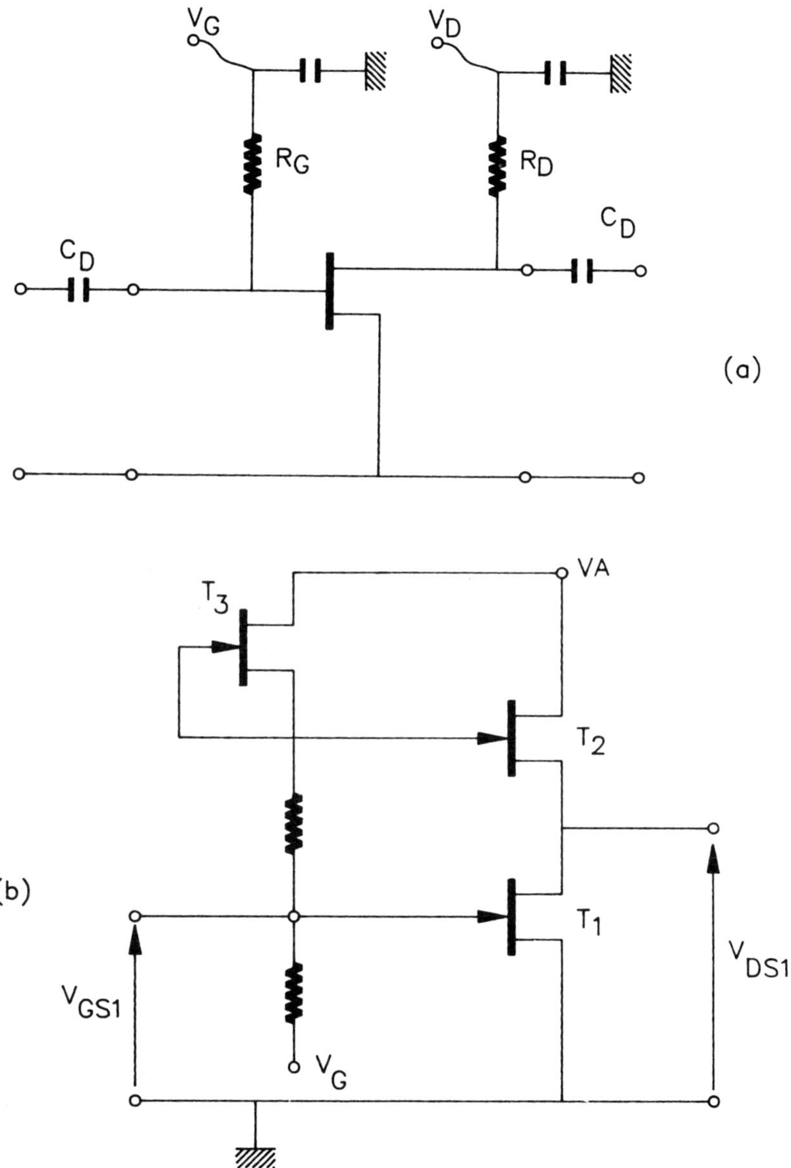

Fig. 10.15 Biasing through (a) resistances; (b) an active load.

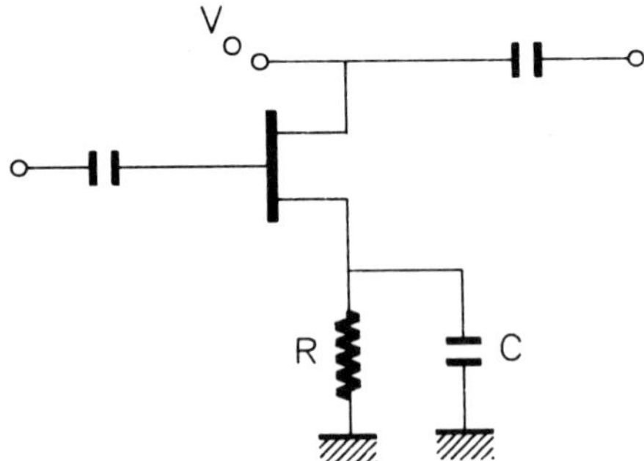

Fig. 10.16 Self-biasing circuit for the gate.

Figure 10.14 shows the basic diagram for the biasing circuits of two cascaded transistors mounted in common source. The biasing circuits comprise capacitances C_D to isolate the d.c. voltages, capacitance C_B to bypass the microwave signal and choke inductances L_C to isolate the microwave circuits from the biasing supplies. In fact, the choke inductances must have very high values which are most often impossible to create in monolithic circuits. These inductances can then be replaced by quarter-wavelength lines with a high characteristic impedance. But this solution can lead to too large dimensions for the circuits: a quarter-wavelength on GaAs is about 2.2 mm at 10 GHz. Another solution consists of using resistances (Fig. 10.15 (a)). On the gate side, this solution is very useful; on the other hand, on the drain side, the resistances see the biasing current of the transistor, which introduces a large

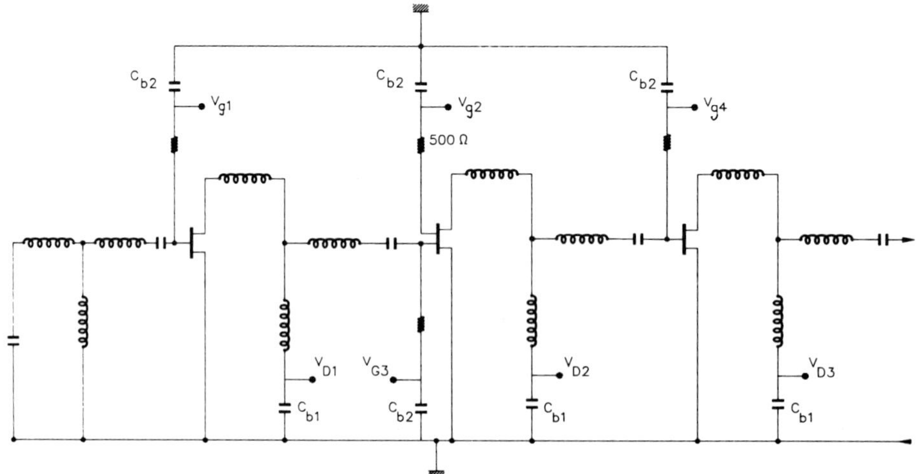

Fig. 10.17 Diagram of a gain controlled amplifier in X-band.

voltage drop and strong dissipation of power because the resistance must be sufficiently large not to perturb the microwave circuit.

In Fig. 10.15(b), a biasing circuit with an active load can be seen (Rumelhard *et al.*, 1985), which allows a limit to the disadvantages due to the losses of power and the voltage drop in the drain resistance. The particularity of this circuit is to give the possibility of varying the d.c. gate voltage, while at the same time keeping the advantages of the active load as it is used in logic circuits. In all the preceding diagrams, the gate is biased by an independent voltage. It is possible to avoid this by using the diagram of Fig. 10.16 showing a self-biasing circuit.

Some particular configurations of the microwave matching circuits allow applying the biasing voltages with a minimum of added elements. Some examples of such circuits are visible in the power amplifier described in a following section.

10.4.3 Variable gain amplifier

A variable gain can be obtained in an amplifier with a dual-gate transistor. Figure 10.17 depicts such an amplifier where reactive elements create the matching at the input and at the output. The biasing of the gates is made through high value resistances while the drain biases are applied through the output matching circuits with the addition of capacitances to short the microwave signal (C_{b1}). The biasing resistance of the second gate of the dual gate ensures at the same time that the phase of this amplifier is kept constant during the variation of gain. The photograph in Fig. 10.18 shows the amplifiers which comprises two single gate transistors and a dual gate transistor. The curves of Fig. 10.19 presents the gain variation as a function of the voltage of the second gate of the dual gate transistor between – 1.6 and + 1 V by steps of 0.2 V.

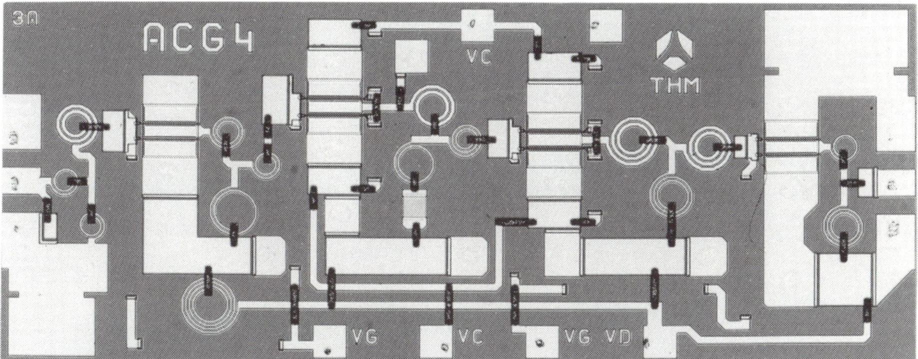

Fig. 10.18 Gain controlled amplifier in X-band.

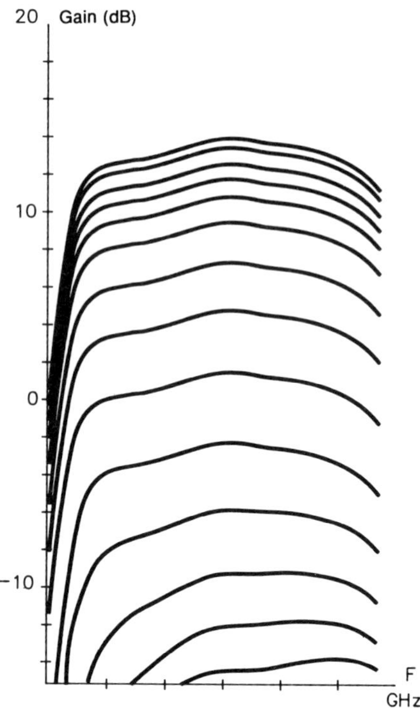

Fig. 10.19 Gain of the X-band amplifier as a function of second gate voltage.

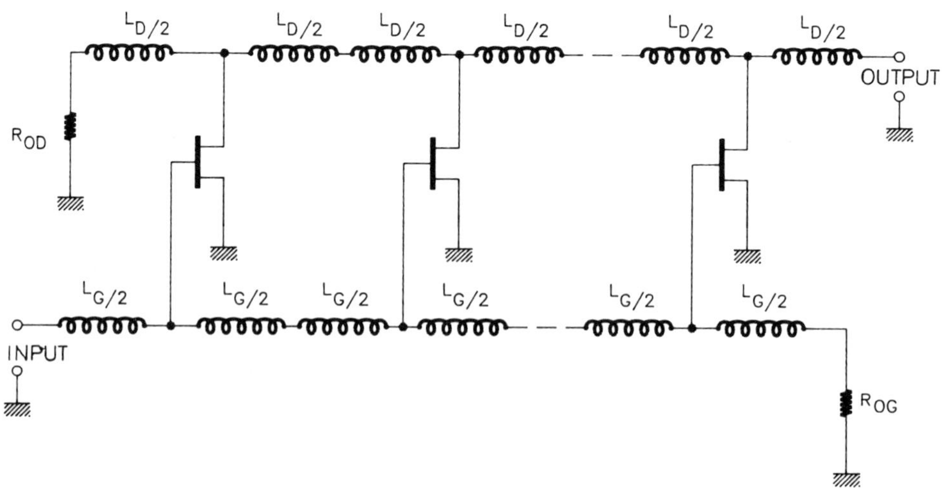

Fig. 10.20 Basic diagram of a distributed amplifier.

10.4.4 Distributed amplifier

The basic diagram of a distributed amplifier can be seen in Fig. 10.20. This type of circuit was first made in a monolithic circuit (Ayasli *et al.*, 1982) because there is in principle no difficulty in putting five transistors on the same circuit instead of one. A gate line is made up of the inductances L_G and of the input capacitances of the transistors while a drain line is made with inductance L_D and the output capacitances of the transistors. The values of inductances are designed to have the same propagation constant in the two lines which are coupled through the transistors. This configuration procures a very wide band amplification. The gain in this type of amplifier is limited by the losses in the gate line which are mainly due to the input resistances of the transistors. This type of structure gave circuits with very wide bandwidth, for instance from

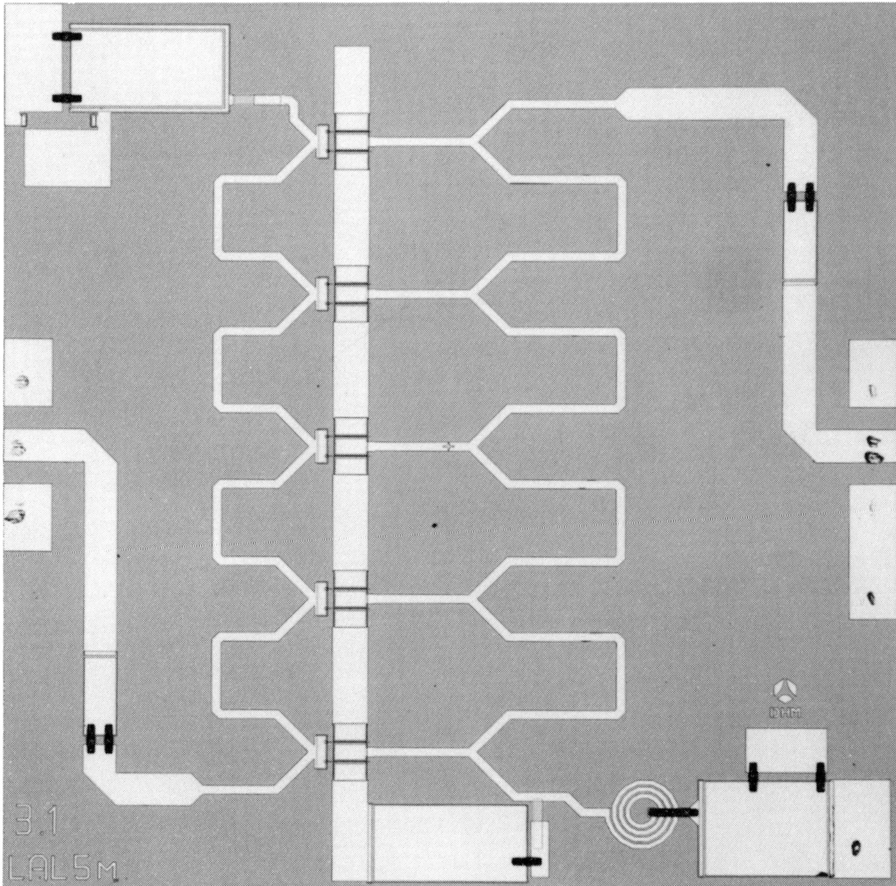

Fig. 10.21 Distributed amplifier 2–18 GHz.

2 to 40 GHz, and were made with MESFETs on GaAs or with TEGFET or HEMTs.

On the photograph in Fig. 10.21, a distributed amplifier is shown. It works between 2 and 18 GHz with a gain of 6 dB, a noise figure of 6 dB until 14 GHz and 8 dB until 18 GHz. The power at 1 dB of compression of such an amplifier is 15 dBm. It is made up of five transistors of $200 \times 0.5\ \mu$m.

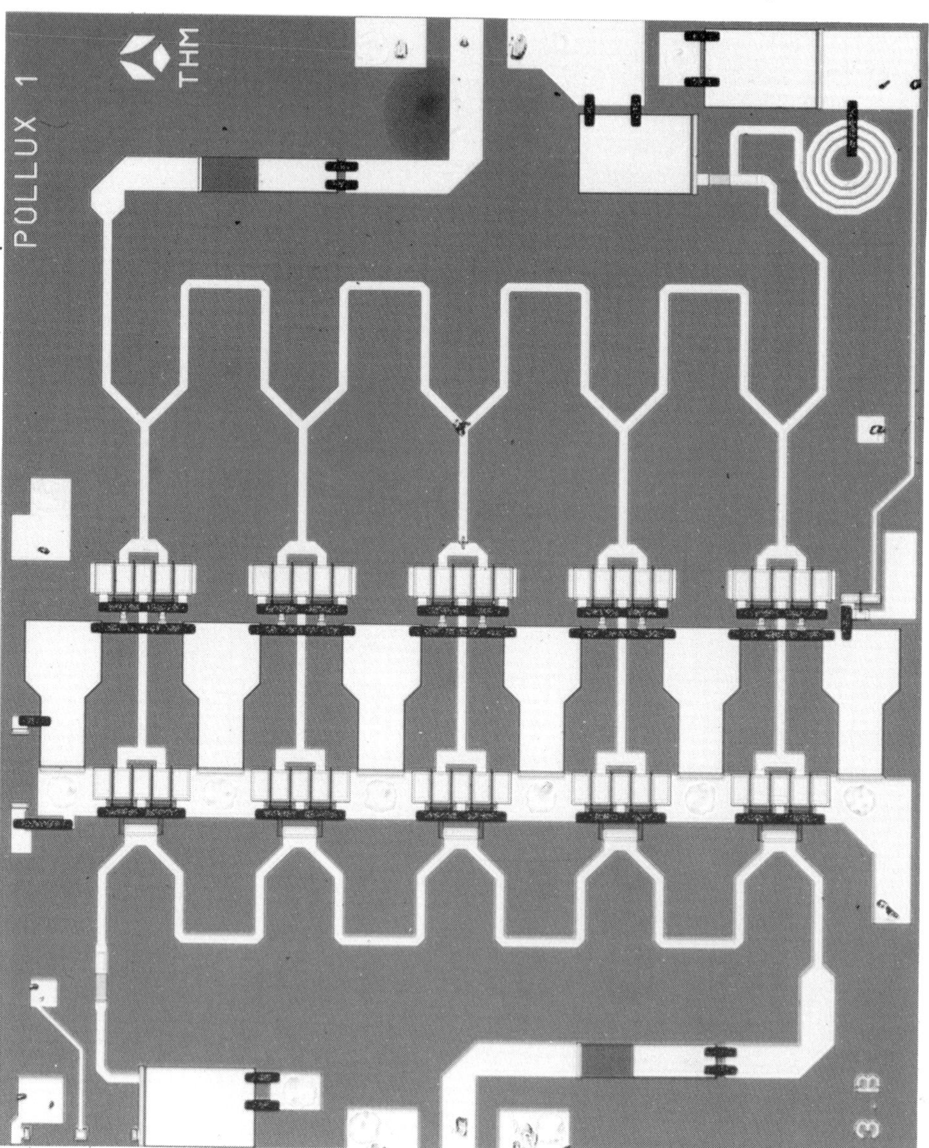

Fig. 10.22 Medium power distributed amplifier delivering 100 mW between 2 and 18 GHz.

Starting from this distributed structure, many variants were developed. A first solution is to use two transistors mounted in cascade instead of one. This diagram increases the gain of one stage (8 dB instead of 6 dB) with the same biasing current. With such a diagram or with dual gate transistors, it is also possible to get more power. As an example, Fig. 10.22 shows a distributed amplifier delivering 100 mW between 2 and 18 GHz.

10.4.5 Power amplifiers

Before considering the design of a narrow band power amplifier with several stages and having to deliver a certain power at 1 dB of compression, the following remarks must be made.

Starting from the desired output power for the amplifier and an estimation of the output circuit losses, it is possible to deduce the total gate width of the last stage from the power per unit gate width of the power transistor which will be used. To optimize the electrical efficiency of the global amplifier, the total gate width of the penultimate stage must be minimum. To do this, an optimum impedance must be shown at the output of this penultimate stage at the maximum frequency. If the transistors were perfectly matched at each frequency, a two stage amplifier would have a slope of 12 dB/octave. A flat gain is obtained by introducing a progressive mismatch when the frequency decreases. This last condition requires one to know perfectly the load pull chart of the transistor used for the design.

It is also to be recalled that a transistor associated with its output impedance corresponding to a maximum power gives a compression curve of the type shown on Fig. 10.23. The linear gain which appears on this figure is different from the maximum gain in small signal. To be in a linear region of this curve the output power must be at – 2 dB lower than the power at 1 dB of compression. In order that the global amplifier compression doesn't exceed 1 dB, the stages preceding the output stage have to work linearly and deliver a sufficient power.

An example of a simple three-stage power amplifier is shown in the diagram of Fig. 10.24.

A power of 1.4 W corresponds to 31.5 dBm. The output circuit having a loss of 0.8 dB, the power at the drain of the output stage must be 32.3 dBm. With the assumption that the transistor delivers a power of 25.5 dBm/mm, 4800 μm of total width will be necessary (for instance, four transistors of 16×75 μm). Such a transistor has a linear gain of 7 dB. To know the power at 1 dB of compression of the transistors of the penultimate stage the following subtractions must be made:

– 1 dB of compression for the output stage

– 1 dB of losses in the matching circuit at the input of the last stage

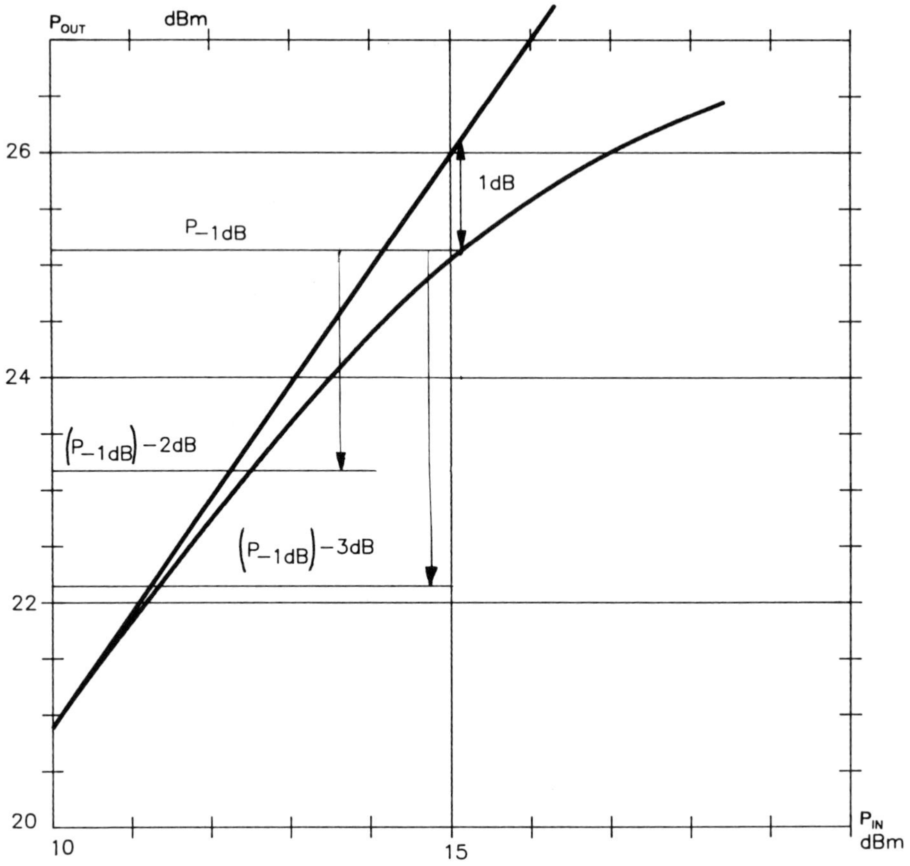

Fig. 10.23 Typical compression curve for a power FET.

– 1 dB of mismatch of the penultimate stage

– 2 dB to be in the linear region of the curve.

Therefore, the output power at 1 dB of compression of the transistors of the penultimate stage must be 32.3 dBm – 2 dB = 30.3 dBm corresponding to a

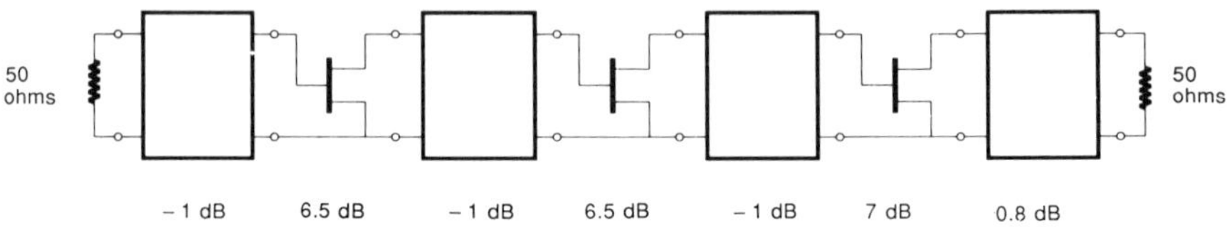

Fig. 10.24 Basic diagram for the design of a three stage power amplifier.

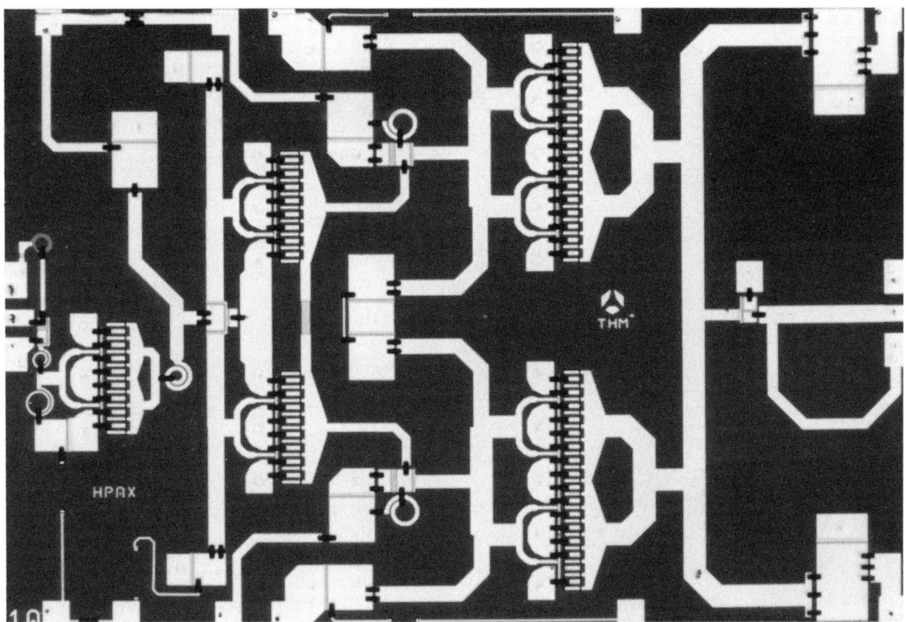

Fig. 10.25 Example of a 3 stage 1.5 W amplifier in X-band.

width of about 3 mm which can be obtained with two transistors of $16 \times 90\ \mu m = 2880\ \mu m$.

This type of transistor having gate fingers wider than those used for the output stage, has linear gain of only 6 dB. When applying the same method to know the preceding stage, for instance the first stage of a three stage amplifier, the following subtractions must be made:

– 1 dB of losses in the matching circuit of the output of the second stage

– 2 dB of mismatch in the output of the first stage.

Then, the first stage must have a width of 1440 μm given by a transistor having 16 fingers of 90 μm.

In this presentation, it was supposed that the different stages were made up of only one transistor. Actually, to get the optimal matching, a special configuration is better suited to monolithic circuits: the arborescent structure (Pavlidis, 1984, Rumelhard, 1987). A circuit which was designed with this approach is shown on the photograph of Fig. 10.25. Its dimensions are $3 \times 4\ mm^2$. It delivers a power of 1.5 W at 1 dB of compression between 8 and 12 GHz and its linear gain is 18 dB. It is supplied with a voltage of 10 V and its efficiency is 20%.

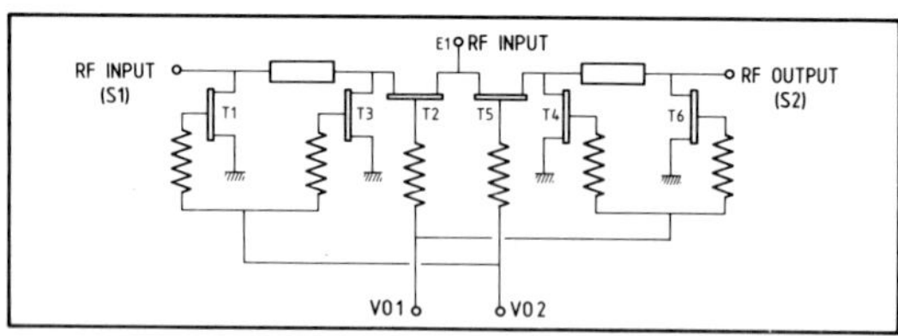

Fig. 10.26 Diagram of a SPDT switch.

10.5 SIGNAL CONTROL CIRCUITS

10.5.1 Introduction

In many applications and for instance in phased array antenna modules, the amplitude and phase of the signal must be adjusted and the signal has to be switched in different ways. In hybrid circuits these functions are made with a basic component: the p-i-n diode. But to work correctly, this component must be made with a vertical structure which is not convenient for the planar technology used in monolithic circuits. Therefore, it was necessary to find components matched to the monolithic technique. Among many possible diagrams, an example of a switch and an analogue phase shifter will be described. As to the amplitude control, it can be executed by an amplifier with a dual gate transistor like one which was shown in section 10.4.3.

10.5.2 Switches

The basic component of monolithic switches is a transistor working with no drain-source voltage. This 'transistor' is 'passing' or 'blocked' according to

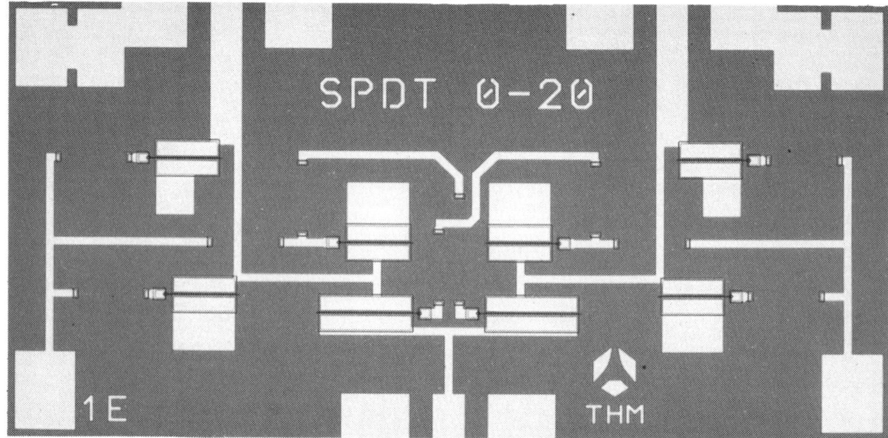

Fig. 10.27 SPDT 0–20 GHz.

the voltage applied on the gate. Such a device is placed in series, then it acts like a standard switch. It can also be placed in parallel: its effect is to bypass the signal when it is passing. Figure 10.26 shows the diagram of an SPDT switch which adds the two functions. When V_{01} is negative, V_{02} equals zero, T_2, T_4 and T_6 are blocked and T_1, T_3 and T_5 are passing: the signal enters in E_1 and goes out in S_2 when V_{01} equals zero and V_{02} is negative, the microwave signal goes out in S_1.

The switch of Fig. 10.26 is shown on the photograph on Fig. 10.27. The circuit of 2×1 mm^2 is made of four transistors of 300×1 µm. In the passing way, the attenuation is lower than 2 dB between 0 and 20 GHz and in the blocked way, the attenuation varies from -25 to -30 in the same frequency range. The SWR is lower than 1.5 at the input and lower than 2 at the output. To decrease the losses in the passing way, the transistors must be larger by increasing the number of fingers. Another possibility is to suppress the series transistors. But the result is to decrease the attenuation in the blocked way.

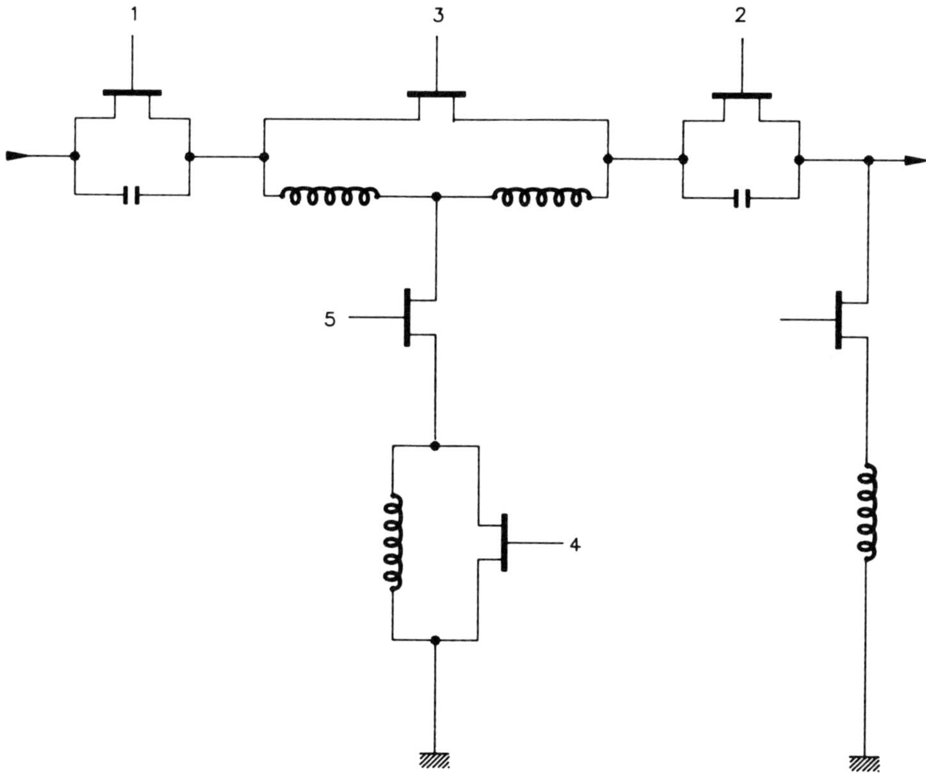

Fig. 10.28 Diagram of a digital monolithic phase shifter.

10.5.3 Phase shifters

Phase shifters used in monolithic circuits are of two different types. They are either digital or analogue. In the first case, a biasing voltage applied to one or several switches puts them in a blocked or passing state which introduces two different paths by which the microwave signal passes. In one path a cell can introduce a delay in phase and in the second path, another cell can introduce an advance. Thus, two positions in phase are created. In the second type of phase-shifter, one or several voltages make the phase vary in a continuous way. Many diagrams of digital or analogue phase shifters have been proposed.

Figure 10.28 shows an example of a digital phase shifter elaborated from 'cold' transistors (without drain-source voltage). When the transistors 1, 2 and 4 are blocked and the transistors 3 and 5 are passing, the cell introduces an advance in phase of + 45 °, when the transistors 1, 2 and 4 are passing and the transistors 3 and 5 blocked, the cell introduces a phase delay of – 45 °. Several cells of the same type or of different configurations can form pads of 180 °, 90 °, 45 °, etc.

The diagram of Fig. 10.29 shows an analogue phase-shifter of the vectorial type. First of all, the signal is divided by two and shifted by + 90 or – 90 °. A second division by two is followed by a variable attenuation and of phase shifts of + 45 or – 45 °. Thus, four orthogonal vectors have been created and

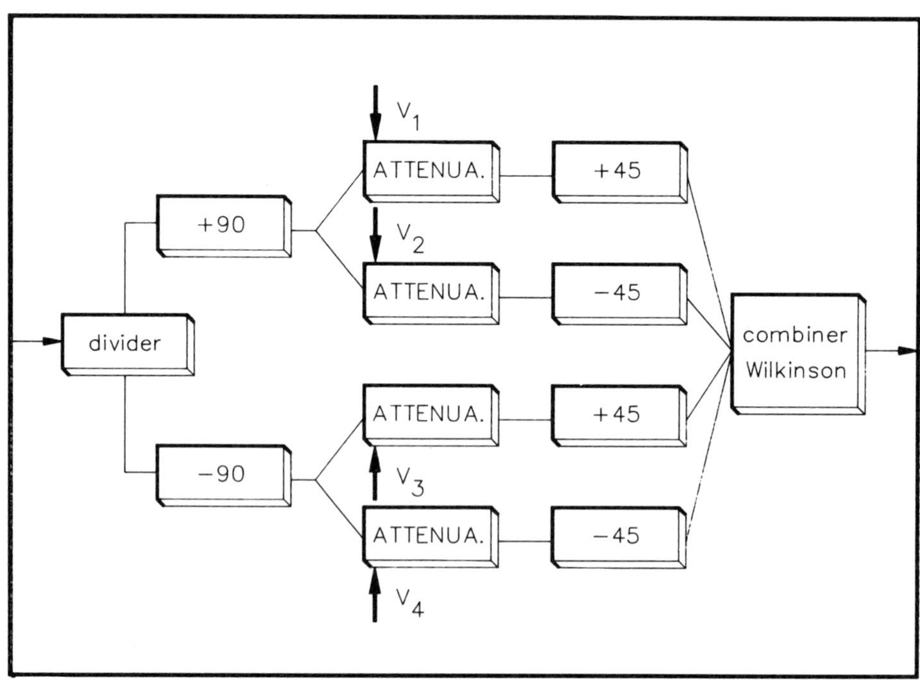

Fig. 10.29 Vectorial analogue phase shifter.

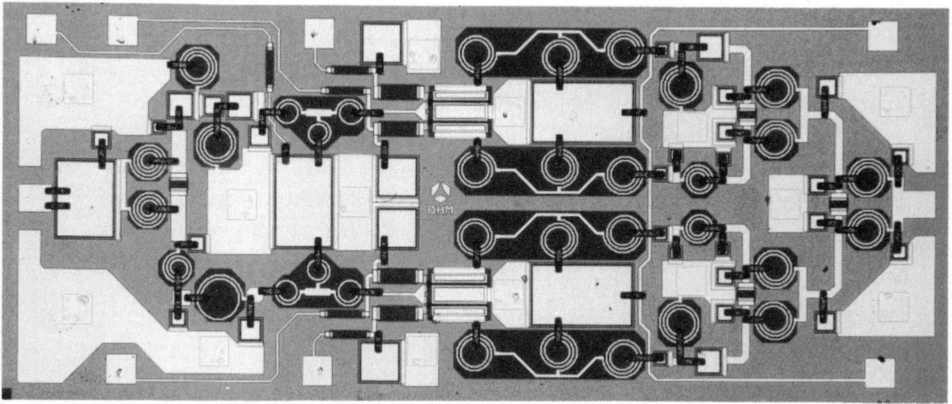

Fig. 10.30 0–360 ° monolithic vectorial analogue phase shifter.

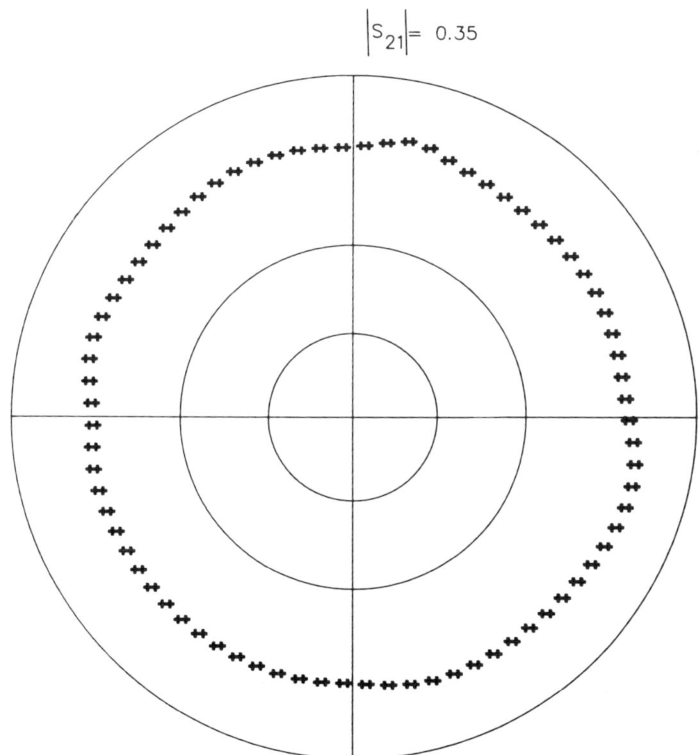

Fig. 10.31 Phase excursion between 0 and 360 ° when applying linear voltages on second gates of dual gate FETs number 1 and 2, then 2 and 3, then 3 and 4, then 4 and 5.

can be combined two by two with a variation of their amplitude (for instance with a dual gate transistor). The variations of voltages V_1, V_2, V_3, V_4 applied on the second gates of the corresponding transistors allow obtaining a phase excursion between 0 and 360 °. Figure 10.30 shows such a phase shifter working in X-Band and constituted of four dual gate transistors of 150×0.5 μm. The dimensions of the circuit are 2×1 mm^2. Figure 10.31 shows plots of the phase excursion at 10 GHz when applying voltages varying linearly on the second gates of the transistors taken successively two by two.

10.6 OSCILLATORS AND MIXERS

10.6.1 Voltage controlled oscillators

A voltage controlled oscillator in a monolithic circuit comprises a FET to produce the negative resistance and a varactor for the variable capacitances. The diagrams of Fig. 10.32 give an idea of the frequency range which can be obtained as a function of the position of the varactor with respect to the transistor (the results come from simulations) (Pataut and Pavlidis, 1988). The two best solutions are to put the varactor in the gate or in the source circuit. Another possibility is to use two varactors (one in the gate circuit and the other in the source). The photograph of Fig. 10.33 corresponds to a monolithic oscillator with two varactors (eight fingers of 50×0.5 μm). This oscillator is isolated from the 50 Ω load by a buffer amplifier which has three functions:

Isolate the oscillator from the load to decrease the frequency pulling.
Introduce an impedance matching by the presentation of a 15 Ω impedance at the oscillator in order to increase the tuning range.
Act as a filter to decrease the harmonics level.

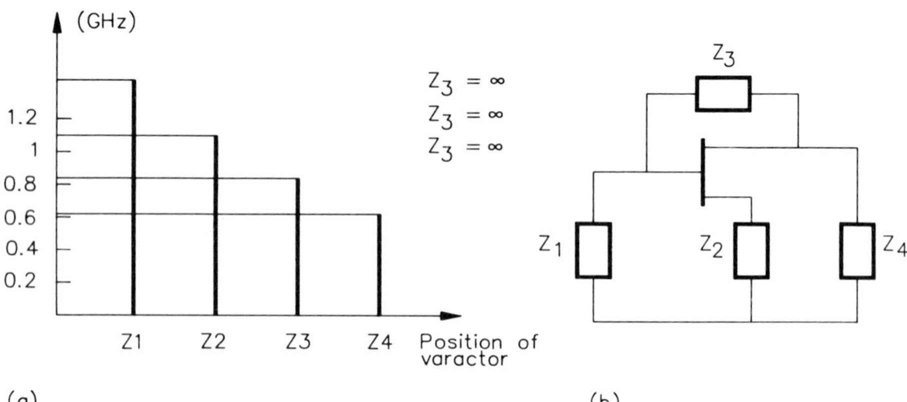

Fig. 10.32 Different configurations of monolithic voltage controlled oscillators.

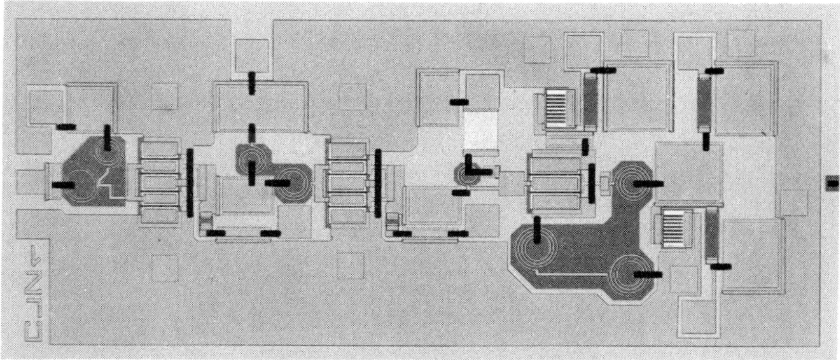

Fig. 10.33 VCO + buffer amplifier.

With such a circuit followed by a two stage amplifier, the frequency tuning range is 2 GHz around 10 GHz and the output power is 10 mW.

10.6.2 Mixers

In monolithic circuits, single, balanced or doubly balanced mixers can be designed with diodes, single FETs, two FETs mounted in cascade or dual gate FETs. The simulation of these circuits is still very rough. In Fig. 10.34 can be seen a diode balanced mixer working from 5.9 to 8.5 GHz with an IF from 0 to 100 MHz and a conversion loss of 7 dB. In this circuit, the local oscillator signal is phase shifted by 180 ° by applying it to the diodes through cells with a phase shift of + 90 ° on one side and – 90 ° on the other side.

10.7 A SUBASSEMBLY: A T/R MODULE

An example of a subassembly with several MMIC chips is shown on Fig. 10.35. It is a transmit receive module for phased array antenna working in X-Band. Such a module comprises a transmit path supposed to deliver a power of 1 W and receive path having a noise factor lower than 4 dB.

Figure 10.36 indicates the different circuits which are necessary to make this module and points out the circuits which were made in MMIC. The hybrid circuits on one part and the monolithic circuits on the other part are associated in two packages to create the microwave connections of the three-port or two-port types or to apply the biasing voltages. The hermeticity is ensured separately for each package. Figure 10.37 is a view of the package containing the signal control circuits, namely a SPDT switch, two gain blocks, two phase-shifters and two circuits to control the amplitude. The phase shifter and the amplitude control circuit were described in a preceding section. The curves

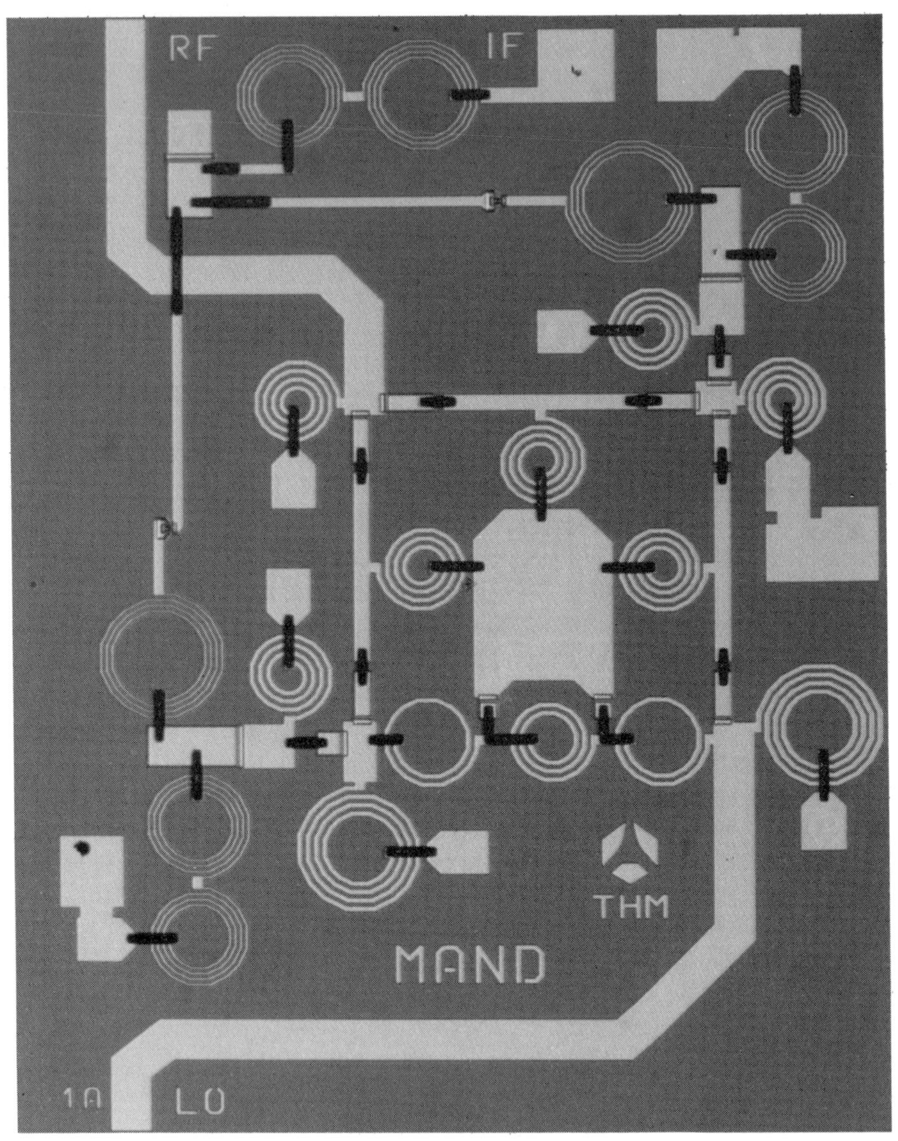

Fig. 10.34 Balanced mixer.

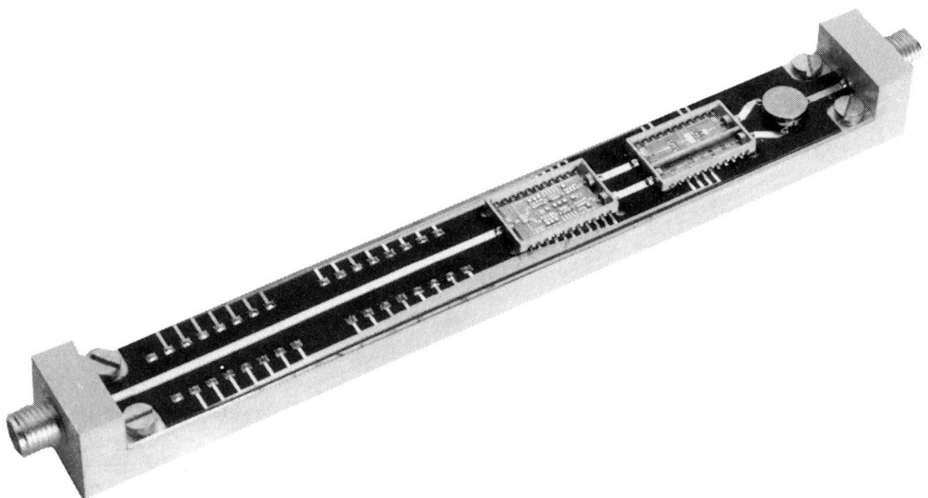

Fig. 10.35 T/R module.

of Fig. 10.38 give an idea of the phase excursion which is possible with this type of module.

This type of experience points out the necessity for a global design of the monolithic circuits and of their environment by taking into account the different aspects like biasing circuits, logic circuits for the command of microwave circuits, the hermeticity, the evacuation of heat, the choice of the optimal number of chips (in this example, the switch and the gain blocks could be associated on the same chip).

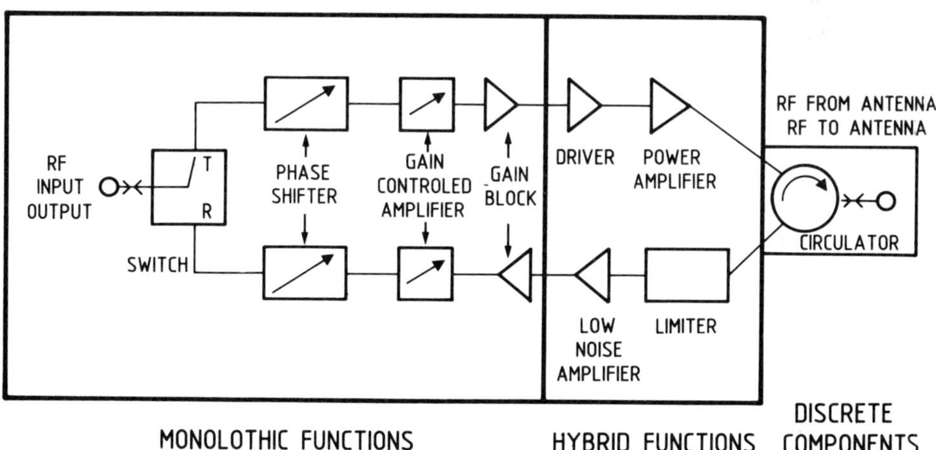

Fig. 10.36 Diagram of the T/R module of Fig. 10.35.

Fig. 10.37 Signal control.

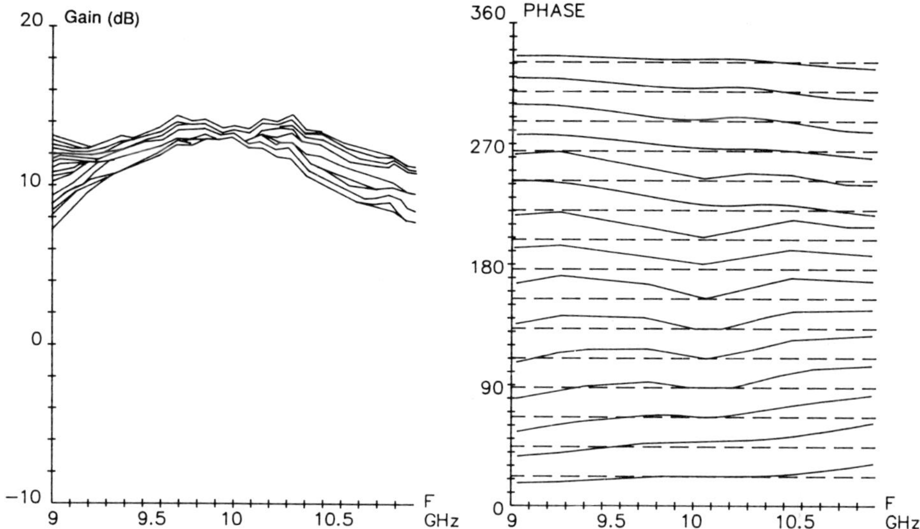

Fig. 10.38 Phase of the T/R module of Fig. 10.35.

10.8 FUTURE PROSPECTS

Many monolithic circuits remain to be explored such as multipliers, analogue dividers, circuits for active filtering, etc. As the development proceeds, these circuits will be available as predesigned cells with guaranteed performance. The use of these cells will allow design of various subassemblies for different applications.

The introduction of new devices like HEMTs, pseudomorphic HEMTs or HBT will give the possibility of affecting some chips of a subassembly to a particular function like low noise amplification or high power. It will be possible to associate a part of the logic command circuits directly on the microwave chips to decrease the number of connections (Versnaeyen, 1988). The decreasing of the gate lengths and the appearance of new devices will allow going to higher frequencies (60 GHz and 94 GHz). At these frequencies it will be possible to use quarter- or half-wavelength elements.

BIBLIOGRAPHY

Allamando E., Salmer G., Constant E., Radhy N. E., Cappy A. and Carnez B. (1983) 'A New Noise Model of Submicrometer Dual-Gate MESFET', *7th Inter. Conf. On Noise in Physical Systems*, Montpellier.

Allamando E., Radhy N.E. and Constant E. (1985) 'Broadband High Order Frequency Multipliers By Using Dual-Gate MESFET's', *Proc. 15th European Microwave Conference*, PARIS.

Andrade T. (1985) 'Manufacturing technology for GaAs monolithic microwave integrated circuits', *Solid State Technology*, 199–205.

Ayasli Y., Miller S. W., Mozzi R. and Hanes L. (1984) 'Capacitively coupled travelling-wave power amplifier', *IEEE microwave and Millimeter-Wave Monolithic Circuits Symposium*, 52–54.

Ayasli Y., Reynolds L. D., Mozzi R. L. and Hanes L. K. (1984) '2–20 GHz GaAs travelling-wave power amplifier', *IEEE Trans. on Microwave Theory and Technics*, **MTT-32**, n. 3.

Ayasli Y., Mozzi R. L., Vorhaus J. L., Reynolds L. D. and Pucell R. A. (1982) A monolithic GaAs 1–13 GHz travelling-wave amplifier, *IEEE Trans. on Electron, Devices*, **ED-29**, 7.

Ayasli Y., Reynolds L. D., Vorhaus J. L. and Hanes L. (1982) Monolithic 2–20 GHz GaAs travelling-wave amplifier, *Electronic Letters*, **18**, n. 14.

Camacho-Reñalosa C. (1983) 'Numerical steady-state analysis of non linear microwave circuits with periodic excitations', *IEEE Trans. on Microwave Theory and Techniques*, **MTT-31**, September 1983, n. 9, 724–730.

Danto Y., Barriere A. S., Girma T., Pichon J., Jay P. R. and Rumelhard C. (1983) 'Physical characterisation of different dielectrics for thin films capacitors in GaAs MMIC's', *Proceedings of the First International Conference on conduction and breakdown in solid dielectrics*, Toulouse, France.

Dao T. H. and Ha T. T. (1981) 'Explicit formulas for GaAs FET amplifier interstage matching networks', *IEEE Proc.*, **128**, Pt. G., n. 1.

Delagebeaudeuf D., Delescluse P., Laviron M., Chaplart J. and Linh N. T. (1980) *Electron. Letters*, **16**, 667.

Finlay H. J., Jenkins J. A., Pengelly R. S. and Cockrill J. (1983) 'Accurate coupling predictions and assessments in MMIC networks' – *GaAs IC Symposium Technical Digest*, 16–19.

Freeman J., Weingarten W., Rodwell M., Diamond S., Fong H. and Bloom D. (1985) 'Direct Electro Optic Sampling of Analog and Digital GaAs Integrated Circuits', *GaAs IC Symposium Technical Digest* , 147–150.

Gamand P., Deswarte A., Wolny M., Meunier J. C. and Chambery P. (1988) '2 to 42 GHz Flat Gain Monolithic HEMT Distributed Amplifiers' *GaAs IC Symposium Technical Digest*, 109.

Hara S., Tokumitsu T., Tanaka T. and Aikawa M. (1988) 'Broad-Band Monolithic Microwave Active Inductor and Application to a Miniaturized Wide Band Amplifier', **IEEE**, *Microwave and millimeter-wave monolithic circuit symposium*, 117.

Higgins J. (1988) 'Heterojunction Bipolar Power Transistors For High Efficiency Power Amplifiers', *GaAs IC Symposium Technical Digest*, 33.

Ho W., Sovero E., Deakin D., Stein R., Sullivan G., Higgins J., Trinh T. and August R. (1988) 'Monolithic Integration of HEMT's and Schottky diodes for millimeter wave circuits' *GaAs IC Symposium Technical Digest*, 301.

Kermarrec C., Faguet J., Vancon B., Mayousse C., Collet A., Kaikati P., and Beaufort D. (1984) The first GaAs fully integrated microwave receiver for DBS applications at 12 GHz, *14th European Microwave Conference*, 749–754.

Jay P. R. and Wallis R. H. (1981) 'Magnetotransconductance mobility measurements of GaAs MESFET's', *IEEE Electron, Device Letters*, **EDL-2**, n. 10.

Laviron M., Delagebeaudeuf D., Rochette J. F., Jay P. R., Chevrier J. and Nguyen T. Linh (1984) 'Ultra low noise and high frequency operation of TEGFETs made by MBE, ESSDERC.

Le Brun M., Jay P. R. and Rumelhard C. (1981) 'Coupling and impedance between line and ground electrodes on GaAs: implications for MMIC design', *Proceedings of the 11th European Microwave Conference*, Amsterdam.

Le Brun M., Jay P. R. and Rumelhard C. (1981) 'Strip cuts coupling on crowded MMIC chips, Microwaves'.

Le Brun M., Jay P. R., Rumelhard C., Rey G. and Delescluse P. (1983) Monolithic microwave amplifier using a two-dimensional electron gas FET, a comparison with GaAs, *IEEE* gallium arsenide IC Symposium, Phoenix, 20–23.

Le Brun M. and Perez F. (1983) 'Graphical design of feedback amps', Microwave & RF, 78–86.

Mimura T., Hiyamasu S., Fuji T. and Nanbu K. (1980) Appl. Phys., n. 19, L.225.

Mishra U. and Brown A. (1988) 'In GaAs/AlInAs HEMT Technology for Millimeter Wave Applications', *GaAs IC Symposium Technical Digest*, 97.

Niclas K. (1980) 'GaAs MESFET feedback amplifier-Design considerations and characteristics', *Microwave Journal*, 39–85.

Niclas K. (1984) 'Reflective match, lossy match, feedback and distributed amplifiers; a comparison of multioctave performance characteristics', *IEEE MTT Symposium Digest*, 215–217.

Parisot, M., Archambault, Y., Pavlidis, D. and Magarshack, J. (1984) 'Highly accurate design of spiral inductors for MMICs with small size and high cut-off frequency characteristics, *IEEE Microwave and Millimeter Wave Monolithic Circuits Symposium Digest*, 91–94.

Pataut, G. and Pavlidis, D. (1988) 'X-Band varactor tuned monolithic GaAs FET oscillators', *Int. J. Electronics*, **64**, 5, 731–751.

Pavio A. M., McCarter S. D. and Saunier P. (1984) 'A monolithic multi-stage 6–18 GHz feedback amplifier', *IEEE Microwave and Millimeter-Wave Monolithic Circuits Symposium*, 45–48.

Pengelly R. S. and Turner J. A. (1976) 'Electronics Letters', **12**, 251.

Pengelly R. S. (1980) 'GaAs monolithic microwave circuits for phased-array applications', *IEEE Proc.*, **127**, pt. F. n. 4.

Pengelly R. S. (1983) 'Hybrid vs monolithic microwave circuits-A matter of cost', MSN, 77–114.

Pucel A. R. (1985) 'Monolithic Microwave Integrated Circuits' IEEE, PRESS.

Rumelhard C. and Carnez B. (1985) 'A new biasing circuit for high integration density GaAs MMIC's', *15th European Microwave Conference.*

Rumelhard C., Le Brun M. and Quentin P. (1987) 'A Comparison between Different Configurations of Monolithic Amplifiers on GaAs', 1987 *SBMO International Microwave Symposium Proceedings*, 355–359.

Sangiovanni–Vincentelli (1981)'Circuit Simulation in Computer Design Aids for VLSI Circuits' Edited by Antognetti P., Pederson D. O., De Man H., Sijthoff and Noordhoff.

Scott B. N., Wurtele M. and Gregger B. B. (1984) 'A family of four monolithic VCO MIC's covering 2–18 GHz', 1984 *IEEE Microwave and Millimeter-Wave Monolithic Circuits Symposium*, 58–61.

Soares R., Graffeuil J. and Obregon J. (1984) 'Application des transistors à effet de champ en arséniure de gallium', *Éditions Eyrolles, Collection Technique et Scientifique des Télécommunications.*

Strid E. W. and Gleason K. R. (1982) 'A DC-12 GHz monolithic GaAs FET distributed amplifier', *IEEE Trans. on Electron Devices*, **ED-29**, n. 7

Tajima Y. and Millet P. (1984) 'Design of broad band power GaAs FET amplifiers', *IEEE MTT-32*, n. 3.

Telliez I. 'Modèlisation Grand Signal de TEC pour la Simulation de Circuits Intégrés Monolithiques Hyperfréquences en GaAs', Thèse de Doctorat, Université de Lille, à paraître.

Tsironis C. and Meierer R. (1982) 'Microwave Wide Band Model of GaAs Dual Gate MESFET's', *IEEE Trans. On MTT*, **30**, n°3, 243–251.

Turner T., Lemnios Z., Moghe S., Podell A., Korgell C. and Genin R. (1988) 'Application Specific MMIC: A Unique and Affordable Approach to MMIC Development', IEEE, *Microwave and millimeter-wave monolithic circuit symposium*, 9.

Versnaeyen C., Gavant M., Ruggeri S. and Le Brun M. (1988) 'GaAs Fully Monolithic X-Band SPDT Switch with Integrated TTL Compatible Driver' *Conference Proceedings of 18th European Microwave Conference*, 409–414.

Wallace P., Bayruns J., Schneingerg N., Rachlin M. Krautrer H. and Park S. (1988) 'A Temperature Compensated L-Band Variable Gain Amplifier with Eight Bit Digital Control', *GaAs IC Symposium Technical Digest*, 281.

Wisseman W. R., Witkowski L. C., Brehm G. E., Coats R. P., Heston D. D., Hudgens R. D., Lehmann R. E., Macksey H. M. and Tserng H. Q. (1987) 'X-Band GaAs Single Chip T/R Radar Module', *Microwave Journal*, 167–173.

Yuen C., Nishimoto C., Glenn M., Pao Y. C., Bandy S. and Zdasiuk G. (1988) 'A Monolithic 3 to 40 GHz HEMT Distributed Amplifier', *GaAs IC Symposium Technical Digest*, 105.

Zamansky P. (1973) 'Calcul des lignes hyperfréquences plates (microstrip recouvert, microstrip suspendu, triplaque bifilaire) et de leurs discontinuités', *Revue Technique Thomson-CSF*, **5**, n. 3.

PART THREE
PROPAGATION, ANTENNAS AND MEASUREMENTS

11

Microwave propagation

Michel-Henri Carpentier

11.1 PROPAGATION IN FREE SPACE

11.1.1 Field in free space created by an isotropic antenna

Let us consider first the case of a transmitter, transmitting a microwave power P via an isotropic antenna, even if we know that such an antenna is not physically feasible. It will create at a distance D a spherical wave carrying a power density (in watt/m^2) given by

$$\frac{P}{4\pi D^2}$$

This power density is equal to the amplitude of the Poynting vector $\boldsymbol{V} = \boldsymbol{E} \times \boldsymbol{H}$. If we are far enough from the antenna the vectors $\boldsymbol{E}$, $\boldsymbol{H}$ and $\boldsymbol{V}$ are a trirectangular direct trihedron with the scalar relationship

$$\frac{E}{H} = \sqrt{\frac{\mu_0}{\varepsilon_0}}$$

where ε_0 (the permittivity of free space) $= (1/36\pi) \times 10^{-9}$ F/m and μ_0 (the permeability of free space) $= 4\pi \times 10^{-7}$ H/m. So

$$\sqrt{\frac{\mu_0}{\varepsilon_0}} = 120\pi = 377\,\Omega$$

$$\frac{P}{4\pi D^2} = \frac{E^2}{\sqrt{\mu_0/\varepsilon_0}} = \frac{E^2}{120\,\pi} = \frac{E^2}{377}$$

$$E^2 = \frac{30\,P}{D^2} \qquad E = \frac{\sqrt{30\,P}}{D}$$

E is in V/m, P is in W and D in metres.

Example: $P = 1$ kW, $D = 1$ km, and $E = 0.173$ V/m. This is valid for a linear polarization. If the polarization is elliptic:

$$E = E_V + E_H$$

E has to be replaced by $\sqrt{E_V^2 + E_H^2}$. If the polarization is circular:

$$E_V = E_H = E$$

and E has to be replaced in the former formulae by $E\sqrt{2}$.

11.1.2 Field in free space created by a practical antenna

If we call G_T the gain (in power, i.e. its value in decibels is $10 \log_{10} G_T$) of the transmitting antenna, the antenna will create at a large enough distance D a spherical wave carrying a power density given by:

$$\frac{PG_r}{4\pi D^2}$$

and, in linear polarization, an electrical field the amplitude of which is given by:

$$E = \frac{\sqrt{30\,PG_T}}{D}\cdot$$

For instance, if the antenna is a circular dish of diameter d, at a wavelength λ, uniformly illuminated, the gain in the direction of the axis of the dish, at a distance large enough, is:

$$G_T = \frac{4\pi S_T}{\lambda^2}$$

where S_T is the surface of the dish:

$$S_T = \pi\frac{d^2}{4}\cdot$$

A distance 'large enough' means a distance $D >> d^2/\lambda$, where

$$G_T = \frac{\pi^2 d^2}{\lambda^2}\cdot$$

Example: $\lambda = 0.1$ m (S-Band), $d = 10$ m and $d^2/\lambda = 1000$ m. The gain G_T is (at a distance $D >> 1$ km) equal to 98 700 (around 50 dB): at a distance $D = 10$ km, the electrical field is $E = 5.44$ V/m for $P = 1$ kW.

11.1.3 Power received by a far enough receiving antenna

If we now consider (at a far enough distance D) a receiving antenna whose gain (in power) is G_R, in the direction of the transmitting antenna, it will collect the power density at that place multiplied by the effective area S_R such as

$$G_R = \frac{4\pi S_R}{\lambda^2} \quad \text{or} \quad S_R = \frac{G_R \lambda^2}{4\pi}.$$

The power collected by the receiving antenna (in line-of-sight) will be:

$$\frac{PG_T}{4\pi D^2} \times S_R = \frac{PG_T G_R \lambda^2}{(4\pi D)^2}$$

In the case of two dishes used on transmission and reception

$$\frac{d^2}{\lambda} = \frac{G_T \lambda}{\pi^2}$$

'far enough' means that $D >> 0.1\, G_T \lambda$ and $D >> 0.1\, G_R \lambda$.
Example: $G_T = 50$ dB (10^5), $G_R = 50$ dB (10^5), $D = 10$ km ($D >> 0.1\, G_T \lambda$ and $D >> 0.1\, G_R \lambda$), $P = 1$ kW and $\lambda = 0.1$ m. The power received (collected) by the receiving antenna is 6.3 W. Note that applying the above formula for $D = 500$ m would give 2500 W for the received power, which is obviously quite wrong.

11.1.4 The radar equation

Let us now consider a radar system, detecting a target in a direction where the transmission gain is G_T and the receiving gain is G_R, the target being at a distance D far enough ($D >> 0.1\, G_T \lambda$ and $D >> 0.1\, G_R \lambda$), the power density at the distance D is

$$V = \frac{PG_T}{4\pi D^2} \ (\text{W/m}^2).$$

By definition of the radar cross-section (RCS = σ) of the target, everything is as if, at the time of the detection, the target would collect $V\sigma$ watts and re-radiate it isotropically. Hence, at the location of the radar system, the power density (in watts/m^2) is

$$\frac{V\sigma}{4\pi D^2} = \frac{PG_T \sigma}{(4\pi D^2)^2}.$$

The receiving antenna collects

$$\frac{PG_T \sigma}{(4\pi D^2)^2} S_R \ \text{(in watts) with}$$

$$G_R = \frac{4\pi S_R}{\lambda^2} \qquad S_R = \frac{G_R \lambda^2}{4\pi}.$$

The power collected on reception is then

$$\frac{PG_T G_R \lambda^2 \sigma}{(4\pi)^3 D^4}.$$

Example: $P = 10\,\text{MW}$, $G_T = 48$ dB (63 000), $G_R = 38$ dB (6300), $\sigma = 1\,\text{m}^2$ and $D = 400\,\text{km}$ ($0.1\, G_T \lambda = 600$ m and $0.1\, G_R \lambda = 60$ m). The receiver power is 7.8×10^{-13} W.

11.2 EXAMPLES OF RADAR CROSS SECTIONS

11.2.1 Flat metallic circular dish

Let us consider a flat metallic dish, large when compared to the wavelength (diameter d, wavelength λ, with $d >> \lambda$) perpendicular to the direction radar dish (Fig. 11.1). The dish collects a total surface

$$S = \frac{\pi d^2}{4}$$

of the incident microwave power and re-radiates it as an antenna which is uniformally illuminated, i.e. with a gain

$$G = \frac{4\pi S}{\lambda^2} = \frac{\pi^2 d^2}{\lambda^2}$$

so that is similar to a target collecting SG watts but isotropically re-radiating it. Hence the radar cross-section σ is

$$\sigma = SG = \frac{4\pi S^2}{\lambda^2}.$$

Example: $d = 2$ m, $S = 3.1\,\text{m}^2$, $\lambda = 0.03$ m and $\sigma = 140\,000\,\text{m}^2$.

Of course if we move the dish just a little, the gain in the direction of the radar will change steeply: it comes to zero when the axis of the dish makes an angle of around $0.6\,\lambda/d$ with the direction of the radar (in the former example, if the axis of the dish moves by about 0.009 radian, i.e. half a degree,

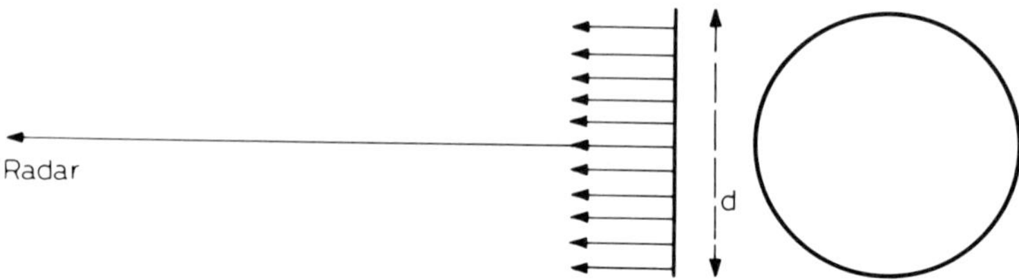

Fig. 11.1 Re-radiation of a flat metallic dish.

the RCS comes to zero). The RCS fluctuates widely if the dish has very slight angular vibrations. Of course this notion of RCS is valid only if the notion of gain G is valid, i.e. if the radar is far enough from the dish (at a distance $D >> 0.1\,(4\pi S/\lambda^2)\,\lambda \approx S/\lambda$ (S surface of the dish).

Generally if the maximum dimension of a target is d_M, there is no physical meaning in speaking of its radar cross-section when the target is at a distance not large enough compared to d_M^2/λ. (For example: $d_M = 100\,\text{m}$ $\lambda = 0.1\,\text{m}$, $d_M^2/\lambda = 100\,\text{km}$. $d_M = 0.1\,\text{m}$ $\lambda = 1\,\mu\text{m}$, $d_M^2/\lambda = 10\,\text{km}$.) The notion of RCS is meaningless for a large building perpendicular to the direction of an S-band radar, and also without meaning for a small plane dish of 10 cm diameter perpendicular to the direction of a lidar (a radar in micrometre wavelength) when the dish behaves exactly like a mirror. The radar equation is then in D^2 instead of in D^4.

The power density at the location of the radar (or lidar) is

$$\frac{PG_T}{4\pi(2D)^2}$$

and the received power is

$$\frac{PG_T\,G_R\,\lambda^2}{(8\pi)^2\,D^2}.$$

But obviously too, what has been said about the RCS of the flat dish assumes that it is really flat (the reflection is 'specular') i.e. that the irregularities of its plane surface are very small compared to the wavelength λ. If the dish has a surface which is rough when compared to the wavelength, the reflection is diffuse instead of specular, the dish re-radiates more or less according to Lambert's law and the radar cross-section is then, in a rough approximation, close to the surface S of the dish (a few times larger on average). Finally it is clear that if the dish is not metallic, it could absorb more or less of the impinging radiation, changing its effective RCS.

11.2.2 Metallic sphere

If we consider a metallic sphere (Fig. 11.2), the diameter d of which is very large compared to the wavelength λ, it is clear that only the region close to the point N (where the tangent plane is perpendicular to the radar) re-radiates toward the radar. The sphere captures $\pi d^2/4$ of the incident microwave power but it re-radiates it in all directions, which allows us to know that the cross-section of the sphere is

$$\pi\frac{d^2}{4}$$

(or πR^2 if $R = d/2$ is the radius of the sphere).

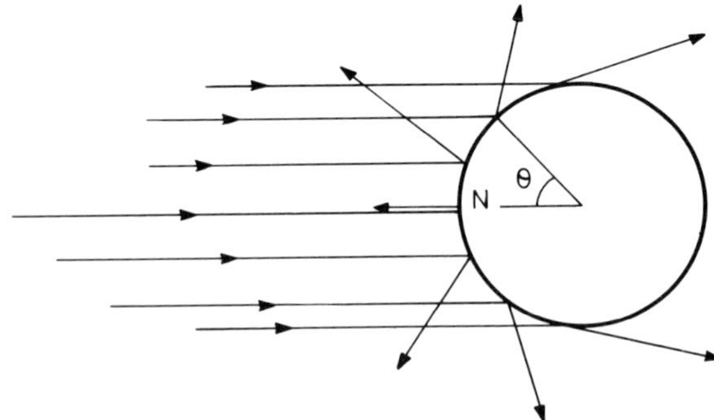

Fig. 11.2 Re-radiation of a metallic sphere.

Computation proves that result: consecutive Fresnel's zones (depth: $\lambda/4$ (see Fig. 11.3)) contribute to the re-radiation, compensating each other in such a way that the total re-radiation is eventually the same as a portion of the first zone. This is no longer valid if the sphere contains only a limited number of Fresnel's zones. If there is only about one ($R \approx \lambda/4$), the cross-section will be close to a maximum, if there are about two ($R \approx \lambda/2$) it will be close to a minimum, etc. This explains the curve represented in Fig. 11.4.

Now if we consider a sphere the diameter of which is very small when compared to the wavelength, Rayleigh's laws for diffraction of small objects are applicable

$$\sigma \approx 900 \left(\frac{d}{\lambda}\right)^4 \cdot \frac{\pi d^2}{4}.$$

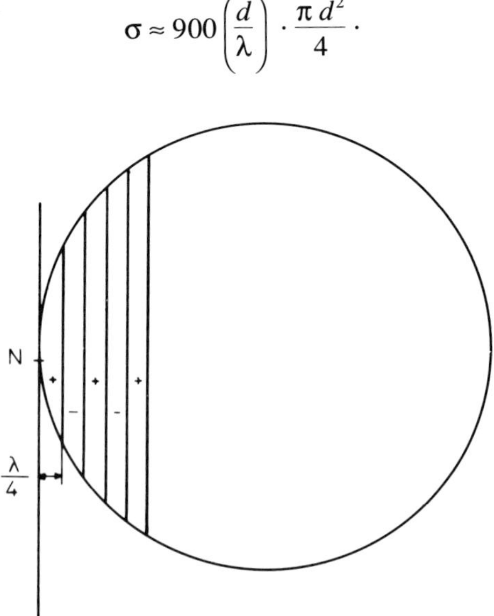

Fig. 11.3 Fresnel's zones of the re-radiation of a metallic sphere.

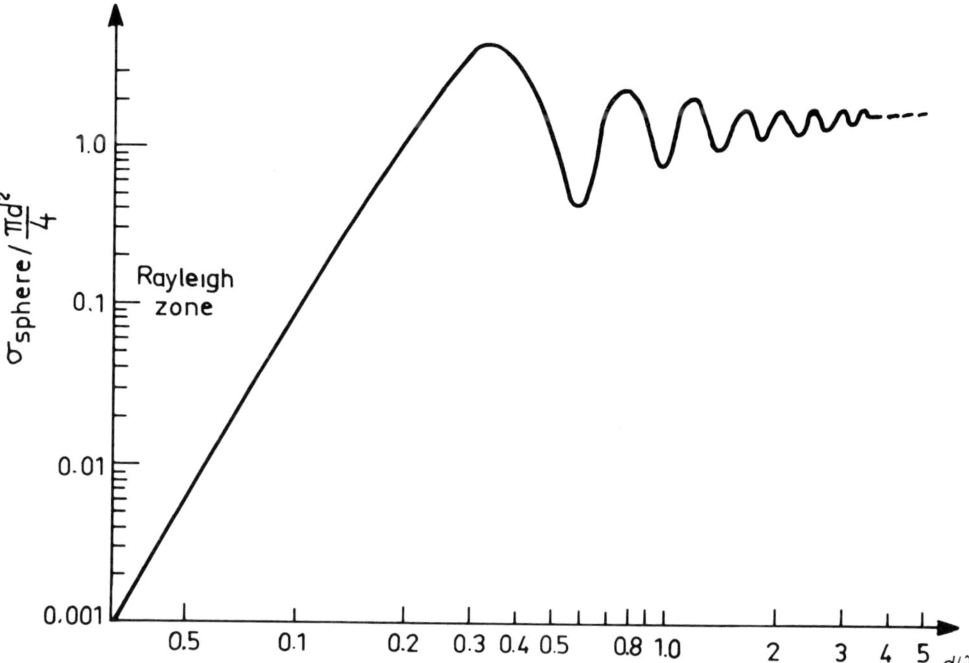

Fig. 11.4 RCS versus d/λ of a metallic sphere.

This indicates why rain drops are not detected by radars using a large wavelength.

Obviously the radar cross-section of a metallic sphere does not change if we move the sphere around its centre: a sphere is an isotropic target.

11.2.3 Irregular sphere or 'potato'

Let us consider what could be called an irregular sphere, or a metallic potato, i.e., a metallic surface, convex in all points, where the radii of curvature are always very large when compared to the wavelength (Fig. 11.5). As for the metallic sphere, only the zone around the 'nose' N (the point of the target the closest to the radar) is re-radiating microwave power in direction of the radar. Similarly the radar cross-section is given by

$$\sigma = \pi R_1 R_2$$

where R_1 and R_2 are the principal radii of curvature of the surface at N. Obviously since R_1 and R_2 are not constant along the surface, the radar cross-section fluctuates when the orientation of the target changes, i.e. a potato is not an isotropic target.

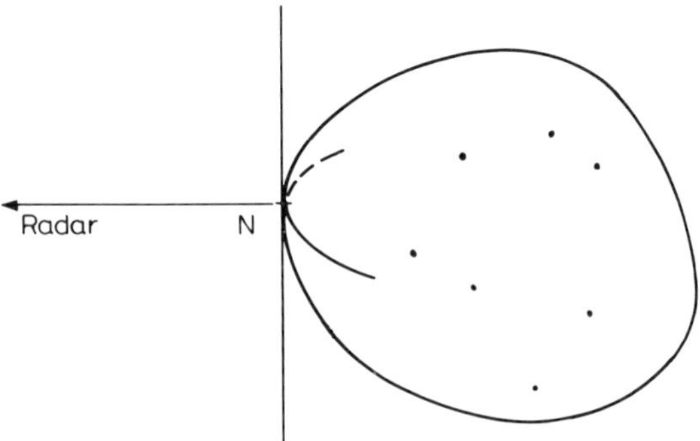

Fig. 11.5 A 'potato'.

11.2.4 Very complex target

Let us consider a very complex metallic target such as represented in Fig. 11.6 which contains, seen from the radar, several 'noses' where the tangent plane is perpendicular to the direction of the radar. If the various noses (N_1, N_2, . . .) cannot be separated (in angle or in distance) then they combine in amplitude and in phase to give a final result. If the radar operator is lucky, combination is mainly with similar phases and the resulting radar cross-section is large. If the radar operator is not lucky, the combination gives a small result and the resulting radar cross-section is small.

Of course, if the distance between noses is large when compared to the wavelength λ, any slight change in frequency or in orientation completely changes the result: the target RCS is strongly fluctuating versus frequency or angular orientation.

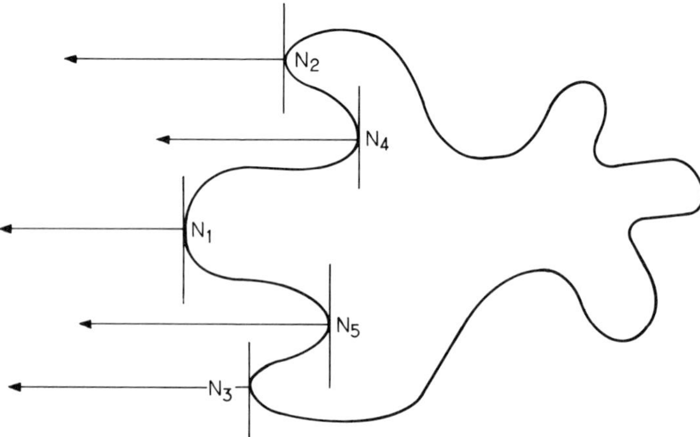

Fig. 11.6 A very complex target.

11.2.5 Radar cross-section of a cone

If we consider now a conical target whose axis of revolution is directed towards the radar, it re-radiates towards the radar only by diffraction, and if its dimensions are very large compared to the wavelength, its radar cross-section is given by:

$$\sigma = 0.08\,\lambda^2 \tan^4 \alpha$$

where α is the cone half-angle.
Example: $\lambda = 0.1$ m, $\alpha = 20\,°$ and $\sigma = 1.4 \times 10^{-5}$ m^2.
Of course if the orientation changes by 70 ° the radar direction becomes perpendicular to the surface of the cone and the radar cross-section becomes significantly large.
These former examples of very large targets (compared to λ) are summarized in Table 11.1.

Table 11.1 Reflective properties of different types of targets

Target	*RCS*	*Wavelength dependence*	*Angular dependence*
Sphere	$\sigma = \pi R^2$	not depending on $\lambda :: \lambda^0$	Isotropic not depending on θ
Plane dish	$\sigma = \dfrac{4\pi S^2}{\lambda^2}$	$:: \lambda^{-2}$	strongly fluctuates with θ
Potato	$\sigma = \pi R_1 R_2$	$:: \lambda^0$	slightly fluctuating with θ
Very complex target	no direct relation with any surface	deeply fluctating with λ	strongly fluctuating with θ
Cone	$\sigma = 0.08\,\lambda^2 \tan^4 \alpha$	$:: \lambda^2$	fluctuates with θ for very large motions

11.3 TROPOSPHERIC REFRACTION

11.3.1 The refractive index of the atmosphere

The refractive index n of the air above the surface of the earth is very close to 1, that is the reason why it is often replaced by the coindex $N = (n - 1)10^6$, which is more convenient.

For microwaves the coindex N depends on the atmospheric pressure p (in millibars or hectopascals), the absolute temperature T (in K), and the specific humidity S (in grams of water per kilogram of air); according to the formula

$$N = p\left(\frac{77.6}{T} + 600\,\frac{S}{T^2}\right).$$

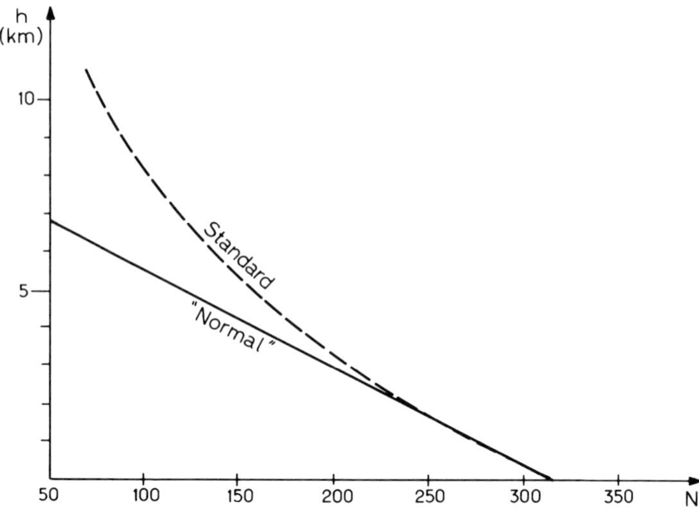

Fig. 11.7 N versus h for 'normal' and 'standard' atmospheres.

Example: $p = 1013$ millibars, $T = 300$ K, $S = 11$ (around 50% relative humidity); $p\,\dfrac{77.6}{T} = 262$, $p \cdot 600 \cdot \dfrac{S}{T^2} = 74$ and $N = 336$.

When the altitude increases, on the average N decreases and goes to zero: the average law comes mainly from the diminishing of the pressure versus increasing altitude while irregularities around the average law come from the irregularities of the temperature and humidity.

A standard variation of N versus h the altitude in kilometers has been defined to represent that average law which is given by $N(h) = 315 \exp(-0.136\,h)$. For that law $\dfrac{\mathrm{d}N}{\mathrm{d}h} = -43$ for $h = 0$ but $\mathrm{d}N/\mathrm{d}h$ comes to zero if h is very large (-22 for $h = 5$ km).

People generally call 'normal atmosphere' an hypothetical atmosphere for which $\mathrm{d}N/$"d$\mathrm{d}h = -39$éfor all values of h (see Fig. 11.7). Clearly N decreases versus h much more rapidly if $\mathrm{d}T/\mathrm{d}h > 0$ (temperature inversion) or if $\mathrm{d}S/\mathrm{d}h << 0$ (dry air above humid air).

11.3.2 Propagation in normal atmosphere

Let us consider the earth as spherical, with a refractive index depending only on the altitude h. If we have for instance (see Fig. 11.8) three successive layers of respective index n_1, n_2 and n_3. Then we have, according to Snell's law:

$$n_1 \sin i_1 = n_2 \sin i_2'$$

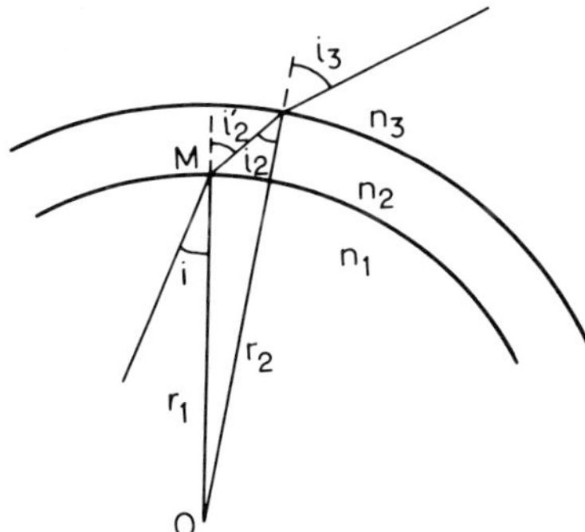

Fig. 11.8 Propagation around the earth when the index depends only on the altitude.

and

$$n_2 \sin i_2 = n_3 \sin i_3$$

but we also have

$$\frac{r_1}{\sin i_2} = \frac{r_2}{\sin i'_2} = \frac{n_2 r_2}{n_1 \sin i_1}$$

and so

$$n_1 r_1 \sin i_1 = n_2 r_2 \sin i_2 .$$

If now the index n regularly varies, and if the thickness of the layer n_2 is very small, we obtain the Bougueur's law. Along the 'path' of the light

$$n r \sin i = \text{constant}.$$

If we call $i = \pi/2 - \phi$

$$n r \cos \phi = \text{constant}.$$

It comes from that (See Fig. 11.9):

$$\frac{dn}{n} + \frac{dr}{r} + \frac{\cos i \, di}{\sin i} = 0 \tag{11.1}$$

with

$$\cos i = \frac{dr}{ds} \qquad \sin i = \frac{r d\theta}{ds} \qquad \tan i = \frac{r d\theta}{dr} .$$

The radius of curvature of the trajectory (Fig. 11.9) at M is given by the well-known formula:

$$\rho = \frac{ds}{d\alpha}$$

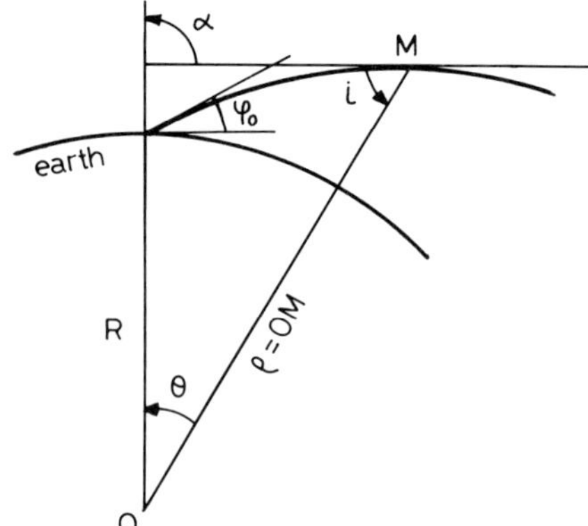

Fig. 11.9 Propagation around the earth when the index depends only on the altitude.

with $\alpha = \theta + i$

$$\mathrm{d}\theta = \frac{\mathrm{d}r}{r}\tan i \qquad \text{or} \qquad \frac{\mathrm{d}r}{r} = \frac{\mathrm{d}\theta}{\tan i}$$

Equation (11.1) becomes

$$\frac{\mathrm{d}n}{n} + \frac{\mathrm{d}\theta}{\tan i} + \frac{\mathrm{d}i}{\tan i} = 0$$

or

$$-\frac{\mathrm{d}n}{n} = \frac{\mathrm{d}\,\alpha}{\tan i} \Rightarrow \mathrm{d}\alpha = -\frac{\mathrm{d}n}{n}\tan i$$

$$\mathrm{d}s = \frac{\mathrm{d}r}{\cos i}$$

We obtain

$$\rho = \frac{\mathrm{d}s}{\mathrm{d}\alpha} = -\frac{\mathrm{d}r}{\cos i}\cdot\frac{n}{\mathrm{d}n}\cdot\frac{\cos i}{\sin i}$$

$$\frac{1}{\rho} = -\frac{\mathrm{d}n}{\mathrm{d}r}\cdot\frac{\sin i}{n}$$

$$\frac{1}{\rho} = -\frac{\mathrm{d}n}{\mathrm{d}h}\cdot\frac{\cos\phi}{n}$$

If ϕ is small $\cos\phi \approx 1$ and since $n \approx 1$

$$\frac{1}{\rho} = -\frac{\mathrm{d}n}{\mathrm{d}h}\,.$$

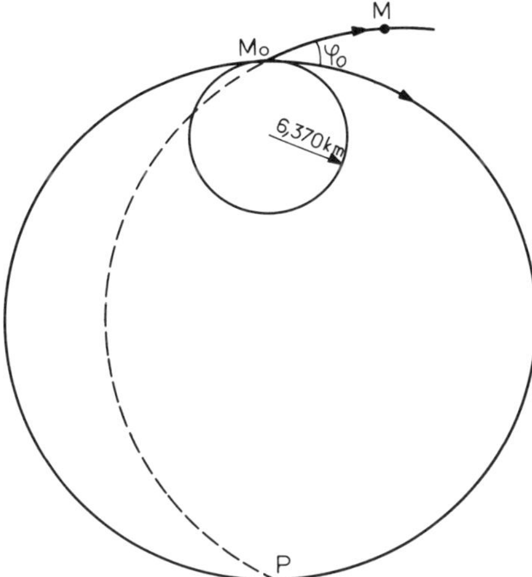

Fig. 11.10 Propagation around the earth when the index depends only on the altitude.

If ϕ is small in normal atmosphere the 'light path' has a constant radius of curvature $\rho = -1/(dn/dh)$ and $\rho = 25\,640$ km.

The path is a portion of circle of radius 25 640 km in normal atmosphere if ϕ is small. If ϕ is not so small ρ becomes $25\,640/\cos\phi_0$, and the path is a portion of another circle, which crosses the former one in P (see Fig. 11.10).

We may transform the figure by transforming M into N according to

$$PM \cdot PN = PM_0^2$$

(the older reader will remember this as a reversal): the earth becomes a sphere with a radius of 8470 km instead of 6370 km and the paths become straight lines, with the same elevation angle in M_0 (Fig. 11.11). This representation is often used to represent radar coverages, except the fact that generally the scale for altitude is changed (see Fig. 11.12).

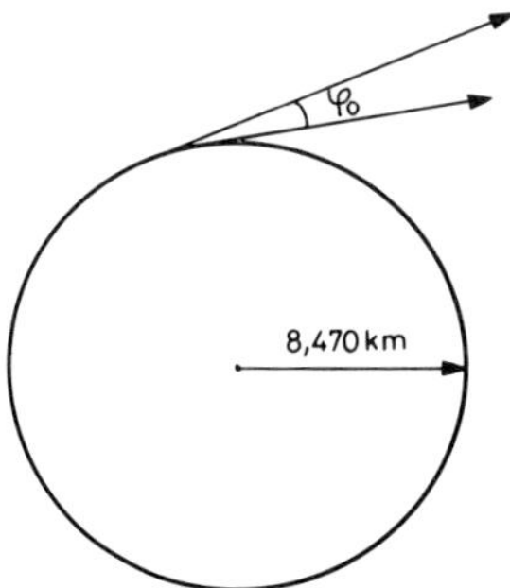

Fig. 11.11 Replacement of the actual earth by an earth with a larger radius.

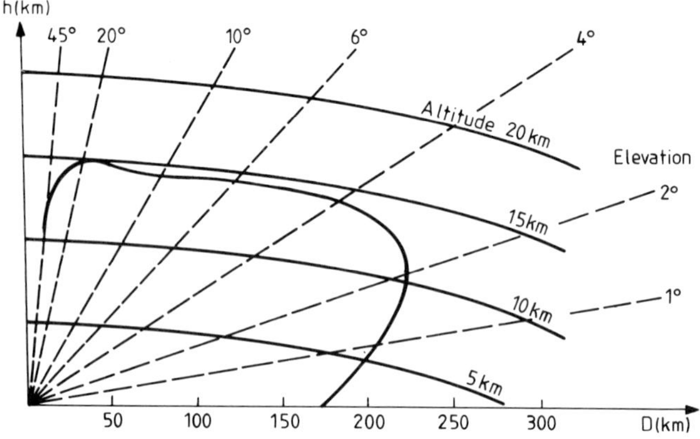

Fig. 11.12 Coverage of a surveillance radar.

For instance the path corresponding to $\phi_0 = 0$ corresponds, for a distance d, to an altitude of

$$\frac{d^2}{2 \times 8\,470\,000} = 5.9 \times 10^{-8}\, d^2 = h$$

which gives

$$\approx 600\ \text{m} \qquad \text{at } d = 100\ \text{km}$$

$$\approx 150\ \text{m} \qquad \text{at } d = 50\ \text{km}$$

$$\approx 2400\ \text{m} \qquad \text{at } d = 200\ \text{km}$$

instead of about 800 m at $d = 100$ km for an earth of 6370 km of radius. Thus in normal atmosphere, we are able to slightly see below the horizon.

Similarly if a radar is on an aircraft flying at an altitude h over the earth, it is able, in normal atmosphere, to detect targets at low altitude or on the earth up to a distance d given by

$$d = (2 \times 8\,470\,000\, h)^{1/2}$$

$$d = 4100 \cdot h^{1/2}$$

Examples: obviously $h = 600$ m corresponds to $d = 100$ km and $h = 5000$ m corresponds to $d = 290$ km.

Now if a transmitter is at an altitude h_1 and a receiver at an altitude h_2, the maximum range between them in normal atmosphere is

$$4100\,(h_1^{1/2} + h_2^{1/2}).$$

In fact it is necessary to have a margin, for instance equal to10 λ for h_1 and h_2 and that maximum range becomes

$$4100\,[(h_1 - 10\,\lambda)^{1/2} + (h_2 - 10\,\lambda)^{1/2}].$$

11.3.3 Anomalous propagation

If R is the radius of the earth (6370 km), r (of the Bouguer's law) is $r = R + h$ and the relevant formula becomes

$$nR\left[1+\frac{h}{R}\right]\cos\phi = \text{constant}$$

$$R\left[1+10^{-6}N\right]\cdot\left[1+\frac{h}{R}\right]\cos\phi = \text{constant}.$$

Calling the modified refractive index

$$n+\frac{h}{R}$$

and the relevant coindex

$$M=\left(n+\frac{h}{R}-1\right)10^{-6}=N+10^{-6}\frac{h}{R}$$

we find

$$(1+10^{-6}M)\cos\phi = \text{constant}$$

and if ϕ is small

$$(1+10^{-6}M)\left(1-\frac{\phi^2}{2}\right)=\text{constant}$$

$$10^{-6}M-\frac{\phi^2}{2}=\text{constant}.$$

If $M = M_0$ and $\phi = \phi_0$ at the point of departure of the path

$$\phi^2=\phi_0^2+2(M-M_0)\,10^{-6}.$$

To understand easily the various phenomena, it is convenient to represent the path in rectangular coordinates, altitude h versus $R\,\theta$ (refer to Fig. 11.9) where the earth is a flat plane.

The first interesting case is the normal atmosphere. Figure 11.13 represents on the left the curve of M versus h (in fact h versus M) in case of normal propagation.

$$\frac{\mathrm{d}M}{\mathrm{d}h}=-39+\frac{10^6}{R}=118$$

(h in km) and on the right a relevant path ($\phi = 0$) is in the virtual part of the trajectory corresponding to

$$M-M_0=-\frac{\phi_0^2}{2}\,.$$

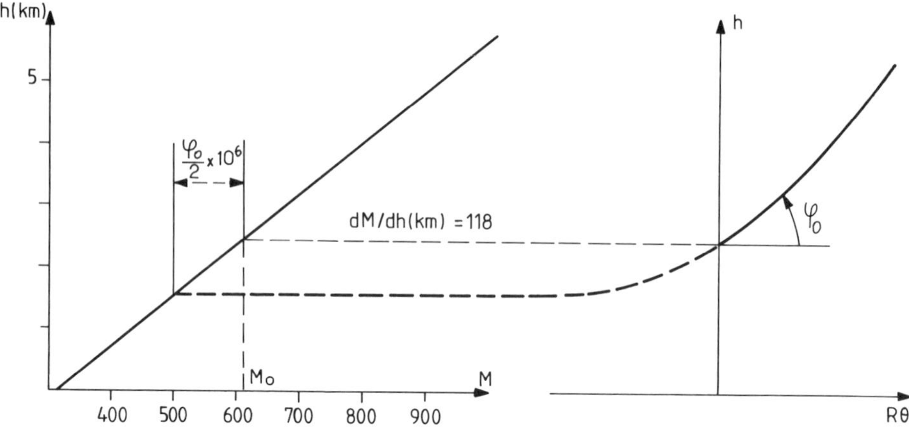

Fig. 11.13 'Normal' propagation.

The path appears to be up-convex because of the choice of the coordinates while it is actually down-convex (and straight with a modified radius of earth of 8470 km).

The second interesting case corresponds to the curves of Fig. 11.14 (infrarefraction). In the representation of flat earth the radius of curvature of paths is smaller than in the former case. The range of systems at low elevation angle is reduced, while coverage is increased towards the zenith. In the representation of an earth with a modified radius (8470 km), the paths from the ground are up-convex instead of being linear. With the real earth radius, the paths are less bent towards the earth and are possibly up-convex.

A third interesting case is represented in Fig. 11.15; in the representation with flat earth, paths beginning with $\phi_0 = 0$ (horizontally) remain horizontal indefinitely.

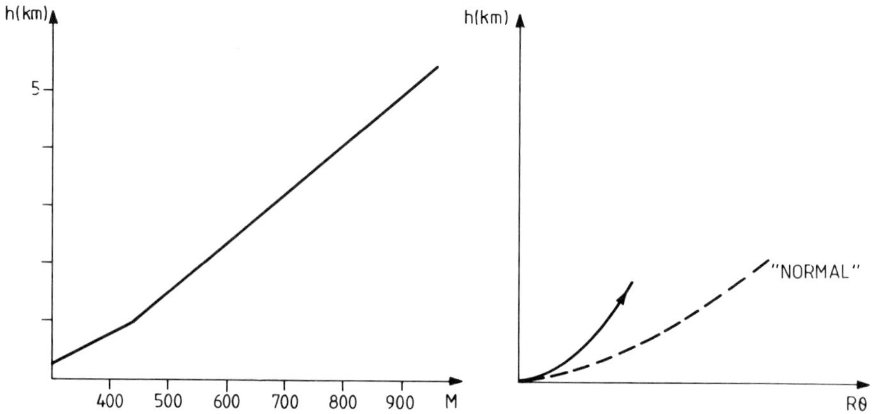

Fig. 11.14 Infrarefraction.

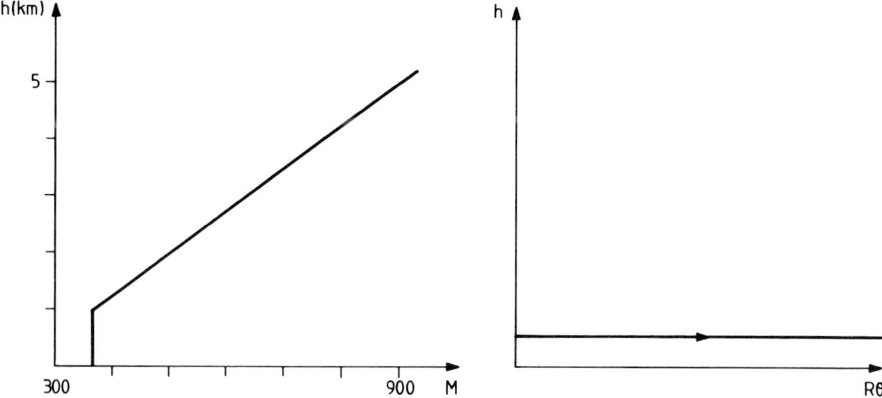

Fig. 11.15 Abnormal propagation.

Figure 11.16 represents a real case (measured over the Indian Ocean) of a superrefraction. Dry and hot winds coming from the land create a rapid increase of temperature above the sea from the sea level up to some tens of metres as well as a rapid reduction of humidity, which explains the shape of the curve. The minimum value of M is obtained for an altitude of about 25 m. If paths are created below that altitude with small values of ϕ_0, in the flat earth representation they are unable to go above a certain altitude h for which ϕ^2 would be negative.

For instance if $h_0 = 0$ $(M_0 = 390)$, and if $\phi_0^2/2$ is less than 43×10^{-6} $(M_0 - M_1 = 43)$, the light path will become horizontal below h_1, it will go down to the earth, be more or less reflected on the earth and then will be again unable to go above h_1, possibly remaining within such a 'duct' for a considerable distance,

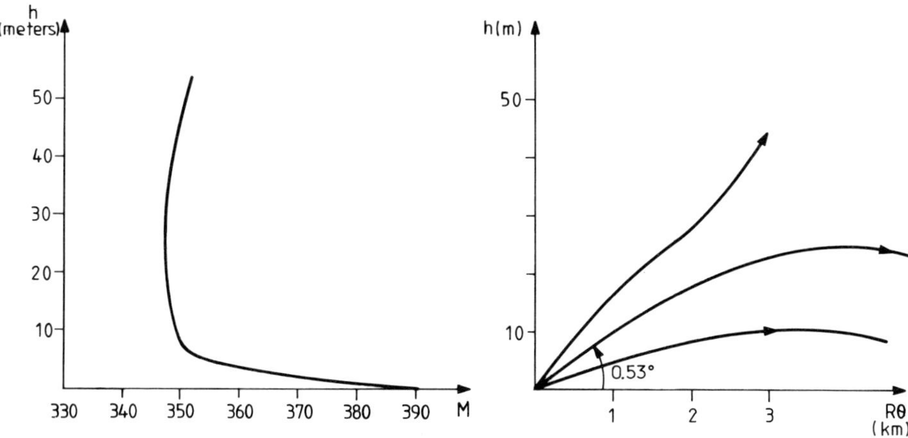

Fig. 11.16 Abnormal propagation with ducts.

ensuring long range for detection or for communication. Trajectories beginning with ϕ_0 above $(43\times10^{-6})^{1/2}$ i.e. 9.3 milliradians or 0.53 ° will not present this particularity.

In fact to be complete let us remark that the duct is somewhat similar to a waveguide and waveguides have a cut-off frequency. It is true also in our present case: wavelengths can propagate in the duct only if their wavelength is less than about $\lambda_{max} = 2.5\,h\sqrt{\Delta M \times 10^{-6}}$. In our example (Fig. 11.16) $h = 25$ m, $\Delta M = 43$ and $\lambda_{max} = 0.4$ m. If the wavelength is around 0.4 m a part of the power escapes from the duct.

Ducts also exist in temperate regions when the earth is cool and the sun is warming the higher parts of the atmosphere creating partial ducts, and local reflections from a radar to the ground, in such a way that the ground clutter is detected, produces echoes that the radar people call 'angels', which could move on the radar screen as a consequence of movements in the atmosphere.

Even if the propagation is nearly normal, there is often a duct between the ground and a few metres, with ΔM equal to a few units. Such a duct has no practical effect on decimetric waves and larger waves. In centimetric wavelengths it could have a partial effect reducing the attenuation above the horizon.

11.3.4 Attenuation in the atmosphere

Water vapour, water drops and some gases create an attenuation of microwaves in the atmosphere. This attenuation depends on the wavelength λ and on the

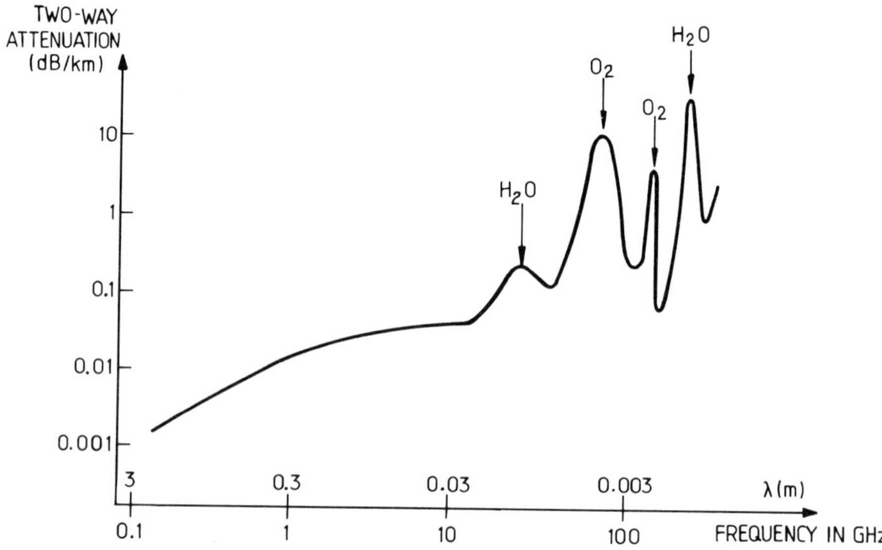

Fig. 11.17 Attenuation of microwaves in air.

composition of the atmosphere. Very low for large wavelengths, it increases on average with frequency and may forbid the utilization of some millimetric wavelengths.

On the other side, for secure communications, this attenuation can be helpful, which explains why 60 GHz (absorption by oxygen) communications are used between aircraft belonging to the same patrol.

Attenuation is given in decibels per kilometre for short trajectories, completely within the troposphere. Figure 11.17 gives a typical curve (20 ° with 60% relative humidity) of two-way (radar applications) attenuation versus frequency in conventional microwaves (of course when used for communications and/or radio relays, attenuation in dB has to be divided by 2).

For longer trajectories, the attenuation reaches a value independent of the range since the trajectory is only partially in the troposphere and the Table 11.2 gives, versus elevation and frequency, the two-way maximum attenuation (in dB) in the same conditions as for Fig. 11.17.

To that attenuation possibly has to be added attenuation from water drops (rain or fog) also given in dB/km. Curves of Fig. 11.18 indicate the relevant attenuation.

In infrared, similar phenomena exist and are represented by the graph of Fig. 11.19 giving a typical transmittance of the atmosphere in visible and infrared.

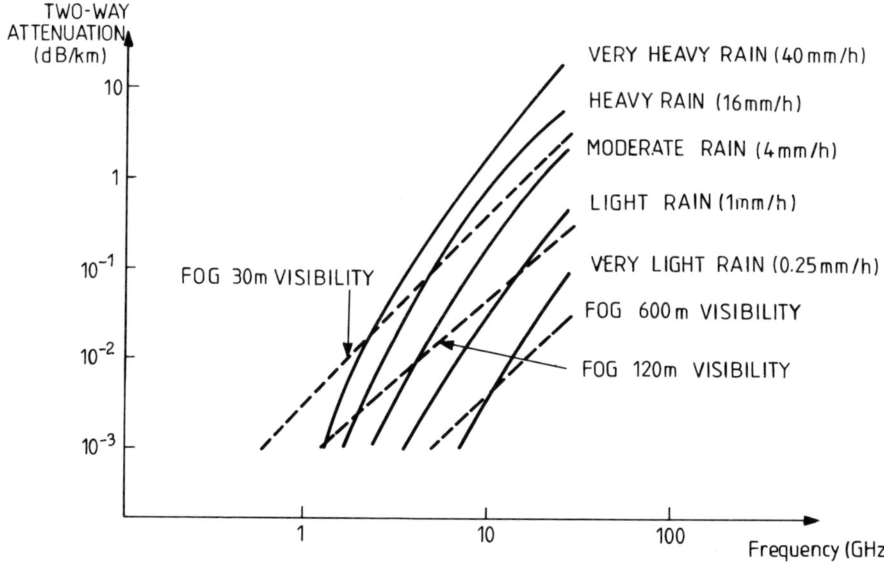

Fig. 11.18 Attenuation of microwaves in rain and fog.

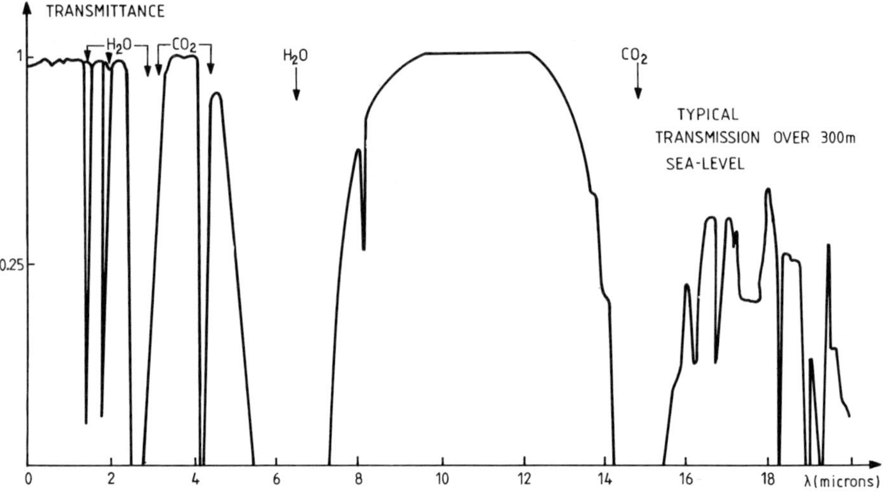

Fig. 11.19 Typical optical and infrared attenuation in the atmosphere.

Table 11.2 Two-way maximum attenuation (dB)

Elevation in degrees	*Frequency* (GHz) 0.1	0.3	1	3	10
0	0.2	1.1	2.7	3.3	4.5
1	0.2	0.7	1.6	2.0	2.6
2	0.1	0.5	1.2	1.4	1.8
5	0.1	0.3	0.6	0.7	0.9
10		0.2	0.3	0.3	0.4

11.4 TROPOSPHERIC SCATTER

When the transmitter and receiver are no longer within optical line of sight, the propagation attenuation estimated according to the former presentations increases rapidly. Improvements made to systems (more transmitted power, better receiver sensitivity) then indicated that beyond a certain distance the actual values of attenuation were smaller than the estimated losses.

The various explanations differ, and probably none of them would today be able to simulate the phenomena with enough accuracy to predict attenuation well enough, but in any case all concur to say (Fig. 11.20) that a mechanism of transmission exists, created by the heterogeneities of the refractive index situated in the volume common to both beams (transmission and reception). Those heterogeneities scatter the microwave power in all directions and in particular in the direction of the receiving antenna. Since those heterogeneities fluctuate with time, the level of the received signal also fluctuates, with long term fluctuations and rapid fluctuations.

It has been found experimentally that the average level of the received signal is mainly linked to the mean value of the vertical gradient of the refractive index in the common volume, and the following empiric law can be used, based upon the comparison of a large number of measurements in various climates:

$$A = 30 \log_{10} F + 30 \log_{10} D + 1.5 \frac{\mathrm{d}N}{\mathrm{d}h} + 102 \qquad (11.2)$$

where A is the attenuation (in dB) between antennas assumed to be isotropic, F is the frequency in Mhz, D is the distance TR in km, and h is in km.

Example: $F = 1000$ MHz, $D = 200$ km, $\frac{\mathrm{d}N}{\mathrm{d}h} = -40$ and $A = 201$ dB. If two identical antennas are used on transmission and on reception, with gain G (in dB), attenuation is reduced by $2G$.

Example: With the same parameters, $G = 20$ dB, average attenuation will be reduced down to 161 dB.

Figure 11.21 gives an idea of the actual laws A versus D at 1 GHz, showing significant differences according to the climatic region under consideration. Fluctuations are also different according to the climatic region under consideration and Fig. 11.22 gives some examples of distribution laws.

In fact within a given climatic area, seasonal fluctuations are more important for short ranges: for instance, in temperate climate (around 200 or 300 km), attenuation in winter is about 15 dB above the attenuation during summer. It is the contrary in desert conditions. For large distances the effect of the season

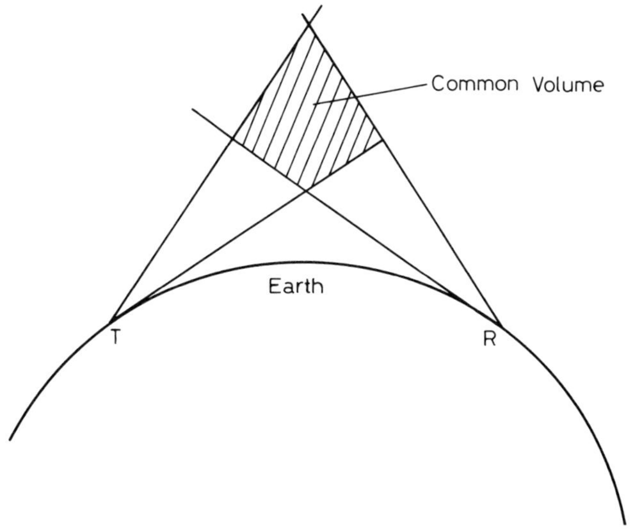

Fig. 11.20 Tropospheric scatter.

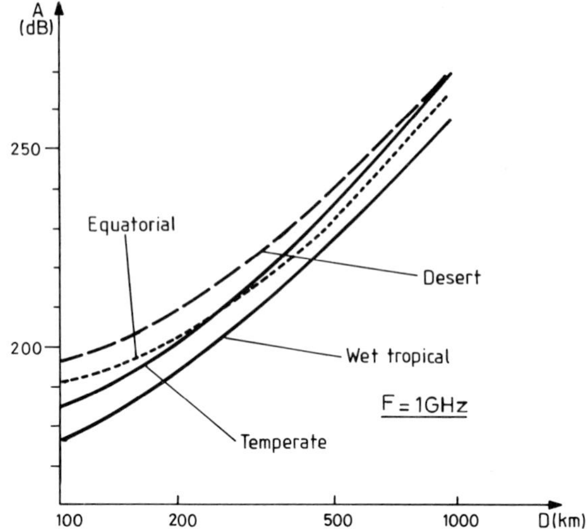

Fig. 11.21 Tropospheric attenuation under various climates.

is less important since the 'common volume' is at higher altitudes where the atmosphere is less varying.

But aside from these long term variations, it is clear that since the mechanism of tropospheric scatter comes from the addition of the re-radiation by a very large number of heterogeneities, the number and position of which change more or less rapidly versus time, thus the result is also fluctuating

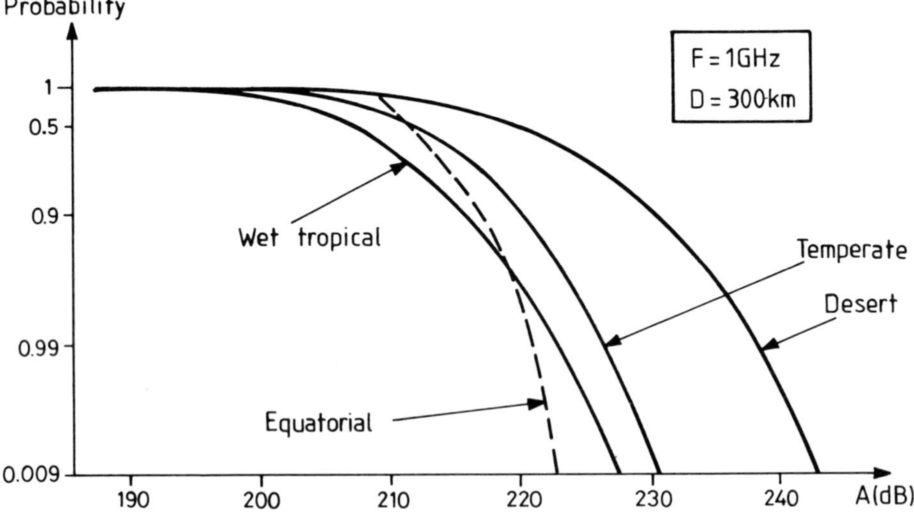

Fig. 11.22 Statistics of tropospheric attenuation.

(according to something similar to a Rayleigh's distribution). Those fast fluctuations have to be considered in addition to the former ones: one of their main effects is to limit the instantaneous bandwidth of the signal that could be communicated between antennas. Fluctuations can be reduced by using diversity reception: since the received signal fluctuates with time but obviously also versus the location of the receiver, using two receivers gives more of a chance of having one of them at a lucky position.

Another consequence of the mechanism of scattering or the fact that the signal received at the receiving antenna location fluctuates with the position is clearly that, if the receiving antenna is large, the field has not a constant phase all along its surface, thus the addition of all components (officially 'in phase' e.g at the primary feed), gives a result less than that expected: there is a loss in the effective gain of the antenna. That loss is empirically not far from what is obtained by the following formula, valid if both antennas (transmission and reception) are identical. The loss in decibels is not far from

$$(G_T + G_R)\left\{1 - \exp\left[-\frac{((G_T + G_R)/148)^4}{1 + ((G_T + G_R)/148)^4}\right]\right\} \tag{11.3}$$

approximation only valid if $G_T + G_R < 120$ dB.
Examples.

$$G_T + G_R = 40 \text{ dB, loss of } 0.2 \text{ dB}$$

$$G_T + G_R = 60 \text{ dB, loss of } 1.5 \text{ dB}$$

$$G_T + G_R = 80 \text{ dB, loss of } 6 \text{ dB}$$

$$G_T + G_R = 90 \text{ dB, loss of } 10 \text{ dB}$$

$$G_T + G_R = 100 \text{ dB, loss of } 16 \text{ dB}$$

$$G_T + G_R = 110 \text{ dB, loss of } 23 \text{ dB.}$$

Problem Let us consider a troposcatter link using for transmission and reception, two identical circular antennas uniformly illuminated with a diameter $\phi = 5$ m at a distance D of 200 km (with $dN/dh = -40$). Using formulae (11.2) and (11.3) the problem is to compute attenuation versus frequency F (in MHz) At $F = 100$ MHz ($\lambda = 3$ m)

$$G_T = \frac{4\pi}{\lambda^2} \cdot \frac{\pi\,\phi^2}{4} = \frac{\pi^2\,\phi^2}{\lambda^2} = 14.4 \text{ dB} = G_R$$

$$G_T + G_R = 28.8 \text{ dB.}$$

The loss from (11.3) is negligible. Attenuation with isotropic antennas would be $A_1 = 171$ dB. Antenna gain reduces that by 28.8 dB down to $A_2 = 171 - 28.8 = 142.2$ dB.

At $F = 200$ MHz ($\lambda = 1.5$ m)

$$A_1 = 171 + 9 = 180 \text{ dB}$$

$$(G_T + G_R)_{th} = 38.8 + 12 = 40.8 \text{ dB}$$

Because of (11.3)

$$(G_T + G_R)_{act} = 40.8 - 0.2 \text{ dB} = 40.6 \text{ dB}$$

$$A_2 = 180 - 40.6 = 139.4 \text{ dB}$$

At $F = 500$ MHz ($\lambda = 0.6$ m)

$$A_1 = 171 + 21 = 192 \text{ dB}$$

$$(G_T + G_R)_{th} = 28.8 + 28 = 56.6 \text{ dB}$$

Because of (11.3)

$$(G_T + G_R)_{act} = 56.6 - 1.2 = 55.4 \text{ dB}$$

$$A_2 = 192 - 55.4 = 136.6 \text{ dB.}$$

And so on; results are found as below:
$F = 1000$ MHz

$$A_1 = 201 \text{ dB}$$

$$(G_T + G_R)_{th} = 68.8 \text{ dB} \qquad (G_T + G_R)_{act} = 66.6 \text{ dB}$$

$$A_2 = 134.4 \text{ dB.}$$

$F = 2$ GHz

$$A_1 = 210 \text{ dB}$$

$$(G_T + G_R)_{th} = 80.8 \text{ dB} \qquad (G_T + G_R)_{act} = 74.5 \text{ dB}$$

$$A_2 = 135.5 \text{ dB.}$$

$F = 5$ GHz

$$A_1 = 222 \text{ dB}$$

$$(G_T + G_R)_{th} = 96.8 \text{ dB} \qquad (G_T + G_R)_{act} = 82.9 \text{ dB}$$

$$A_2 = 139.1 \text{ dB.}$$

Results are represented on Fig. 11.23. Optimum frequency is found around 1.5 GHz corresponding to $A_1 = 134.3$ dB. A similar computation for $\phi = 20$ m gives an optimum frequency around 350 MHz with a value of A_1 around 117 dB.

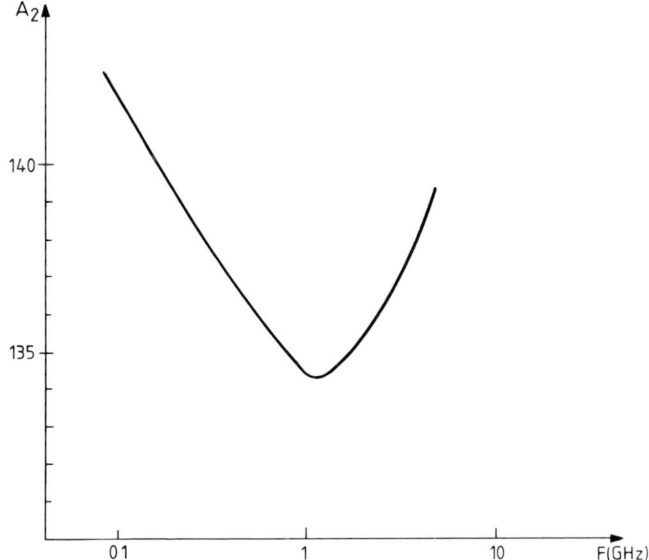

Fig. 11.23 Looking for an optimum frequency in troposcatter.

11.5 REFLECTION ON PLANE DISCONTINUITIES

Let us recall that when an electromagnetic wave meets a plane discontinuity (from Maxwell's equations):

the tangential component of the electric field has no discontinuity;
the tangential component of the magnetic field has no discontinuity;
the perpendicular component of the magnetic induction has no discontinuity;
while the perpendicular component of the electric induction is discontinuous, except if the density of superficial electric charges is zero.

11.5.1 Reflection of a plane wave on a perfect conductor

In vertical polarization (see Fig. 11.24)

$$E_i \cos\theta = E_r \cos\theta \qquad E_i = E_r$$

(If $\theta = 0$ $\boldsymbol{E}_i$ is the vector opposite to $\boldsymbol{E}_r$) $H_i = H_r$ and $\boldsymbol{H}_i$ and $\boldsymbol{H}_r$ are two opposite vectors for all.
In horizontal polarization (see Fig. 11.25)

$$H_i \cos\theta = H_r \cos\theta \qquad H_i = H_r$$

(If $\theta = 0$ $\boldsymbol{H}_i$ is the vector opposite to $\boldsymbol{H}_r$) $E_i = E_r$ and $\boldsymbol{E}_i$ and $\boldsymbol{E}_r$ are opposite for all θ.)

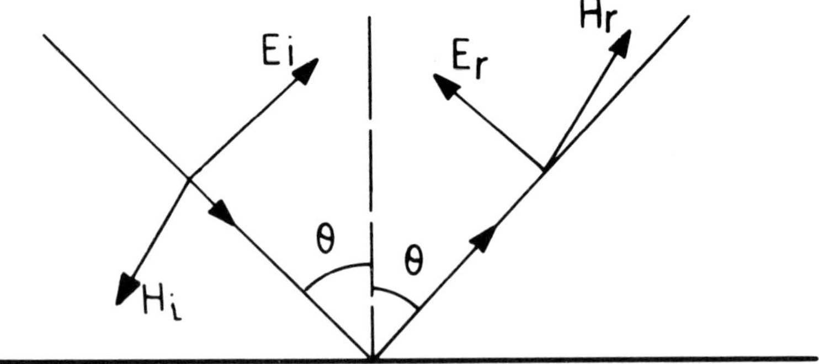

Fig. 11.24 Metallic reflection (vertical polarization).

As a result, a reflection against a metallic plane perpendicular to the direction of a radar or against a metallic sphere changes the direction of the electric field: if the polarization is circular, equivalent to an electrical field rotating regularly when propagating, if it is circular clockwise before reflection it becomes circular counter-clockwise when, after reflection, it propagates in the other direction. This is the reason why rain drops (very similar to spherical objects) when illuminated by a radar transmitting a circular polarization, reflect a wave circularly polarized in the other direction which does not in practice enter the radar antenna. In fact a plane metallic discontinuity reacts exactly as a mirror. As a consequence, if an antenna is above a ground which can be considered as metallic, it is possible to consider that the system is composed by the real antenna plus its (virtual) image (Fig. 11.26).

The field at an elevation θ which should be $Kg(\theta)$ in completely free space, will become $Kg(\theta) + Kg(-\theta) \exp(2\pi(2h/\lambda) \sin\theta)$, where $g(\theta)$ and $g(-\theta)$ are the gain (in field) respectively for θ and $-\theta$. The result is a deep modulation of the radar coverage obtained in free space as represented in Fig. 11.27, where the dotted line is the graph of the radar coverage in free space, the continuous line is the coverage for $h = 20\,\lambda$ and the cross-line the coverage for $h = \lambda$.

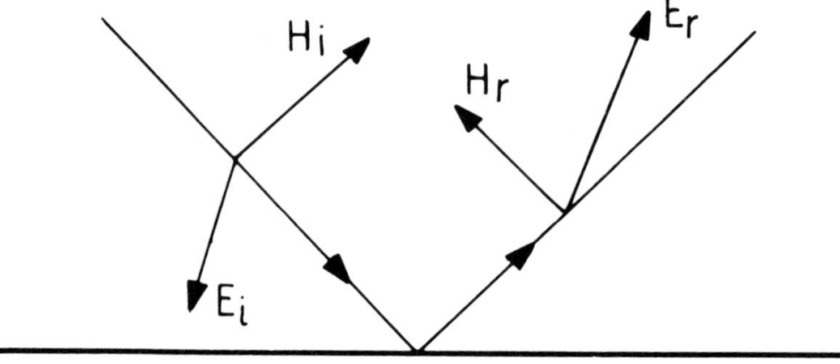

Fig. 11.25 Metallic reflection (horizontal polarization).

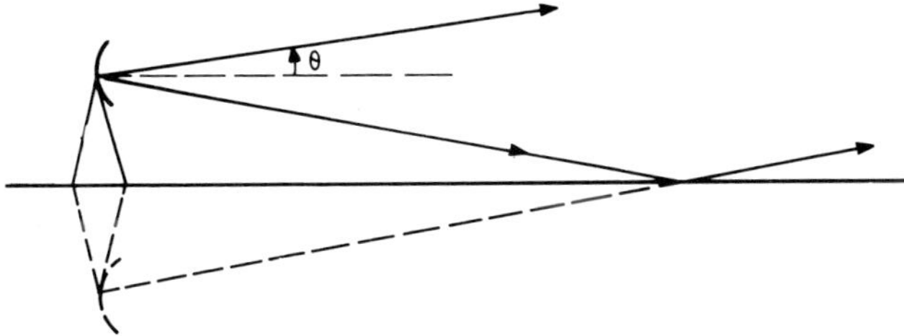

Fig. 11.26 Reflection over the earth.

Fortunately, the surface of the earth is not always a perfect plane mirror since the ground (or the sea) are not flat but rough, mainly if the wavelength is small. Even if the ground is flat enough, it is not always a metallic mirror (for $\lambda < 150$ m, a dry ground is rather a dielectric with losses (see later – section 11.5.3) and that is also true for the sea at wavelengths less than 0.3 m). Figure 11.28 gives an idea of what occurs (for $\lambda = 0.1$ m) above a quiet sea in vertical polarization (continuous line) or above ground or agitated sea both with $h = 20\,\lambda$.

11.5.2 Reflection of a plane wave on a perfect dielectric

In 'vertical' polarization (see Fig. 11.29)

(Note that the electric field is not 'vertical', but in the vertical plane). Since the tangential component of the electric field is continuous:

$$\cos\theta_1\,(E_i - E_r) = (\cos\theta_2)\,E_T. \tag{11.4}$$

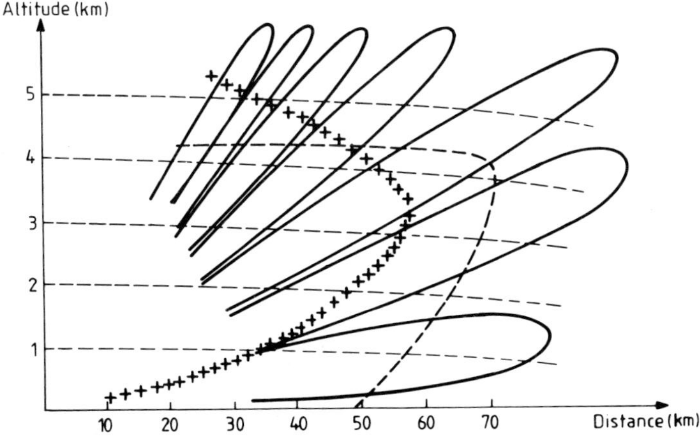

Fig. 11.27 Radar coverage modification by reflection on the ground.

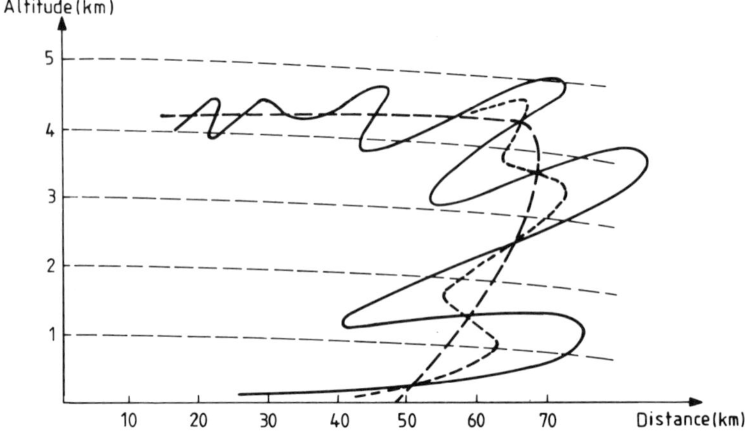

Fig. 11.28 Radar coverage modification by reflection on the ground.

Since the tangential component of the magnetic field is continuous

$$H_i + H_r = H_T \tag{11.5}$$

and since the perpendicular component of the electric induction is continuous

$$\varepsilon_1 (E_i + E_r) \sin \theta_1 = \varepsilon_2 E_T \sin \theta_2. \tag{11.6}$$

Since $H = \sqrt{(\varepsilon_0 \varepsilon_r / \mu_0 \mu_r)}\, E$ (11.5) becomes

$$\sqrt{\frac{\varepsilon_1}{\mu_1}} (E_i + E_r) = \sqrt{\frac{\varepsilon_2}{\mu_2}} E_T \tag{11.7}$$

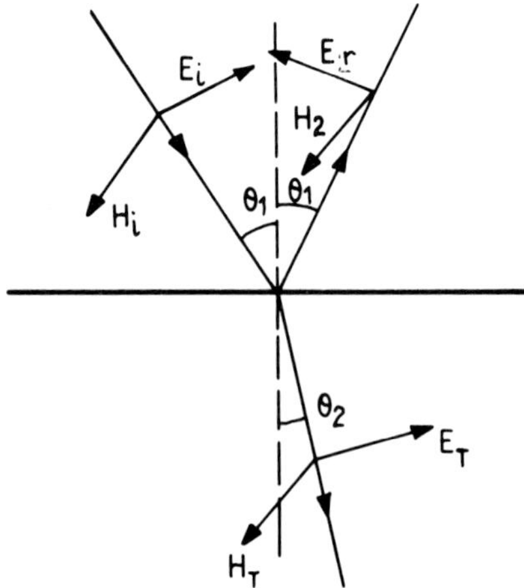

Fig. 11.29 Reflection on a perfect dielectric (vertical polarization).

Dividing (11.6) by (11.7) it becomes

$$\sqrt{\varepsilon_1 \mu_1} \sin \theta_1 = \sqrt{\varepsilon_2 \mu_2} \sin \theta_2$$

or $n_1 \sin \theta_1 = n_2 \sin \theta_2$
(which is Snell's relation).

Solution of the above relations gives

$$R_V = \frac{E_r}{E_i} = \frac{\cos \theta_1 \cdot \sqrt{(\varepsilon_2/\mu_2)} - \cos \theta_2 \sqrt{(\varepsilon_1/\mu_1)}}{\cos \theta_1 \cdot \sqrt{(\varepsilon_2/\mu_2)} + \cos \theta_2 \sqrt{(\varepsilon_1/\mu_1)}}$$

$$R'_V = \frac{H_r}{H_i} = R_V$$

$$T_V = \frac{E_T}{E_i} = \sqrt{\frac{\mu_2 \varepsilon_1}{\mu_1 \varepsilon_2}} \cdot (1 + R_V)$$

$$T'_V = \frac{H_T}{H_i} = 1 + R_V$$

$R_V = 0$ when

$$\cos \theta_1 \sqrt{\frac{\varepsilon_2}{\mu_2}} = \cos \theta_2 \sqrt{\frac{\varepsilon_1}{\mu_1}}$$

since

$$\sin \theta_1 \sqrt{\varepsilon_1 \mu_1} = \sin \theta_2 \sqrt{\varepsilon_1 \mu_1}$$

$$\varepsilon_1 \tan \theta_1 = \varepsilon_2 \tan \theta_2$$

which eventually gives

$$\tan^2 \theta_1 = \frac{\varepsilon_1 \mu_2 - \varepsilon_2 \mu_1}{\varepsilon_1 \mu_1 - \varepsilon_2 \mu_2} \cdot \frac{\varepsilon_2}{\varepsilon_1}$$

$$\tan^2 \theta_2 = \frac{\varepsilon_1 \mu_2 - \varepsilon_2 \mu_1}{\varepsilon_1 \mu_1 - \varepsilon_2 \mu_2} \cdot \frac{\varepsilon_1}{\varepsilon_2}$$

If $\mu_1 = \mu_2 = 1$

$$\tan^2 \theta_1 = \frac{\varepsilon_2}{\varepsilon_1} = \left(\frac{n_2}{n_1}\right)^2 \qquad \tan^2 \theta_2 = \frac{\varepsilon_1}{\varepsilon_2} = \left(\frac{n_1}{n_2}\right)^2$$

θ_1 is the Brewster incidence, for which the refracted trajectory is perpendicular to the reflected one (if it exists) (Fig. 11.30). θ_1 and θ_2 are also defined by

$$\sin \theta_1 = \cos \theta_2 = \sqrt{\frac{\varepsilon_2}{\varepsilon_2 + \varepsilon_1}}$$

$$\sin \theta_2 = \cos \theta_1 = \sqrt{\frac{\varepsilon_1}{\varepsilon_2 + \varepsilon_1}}$$

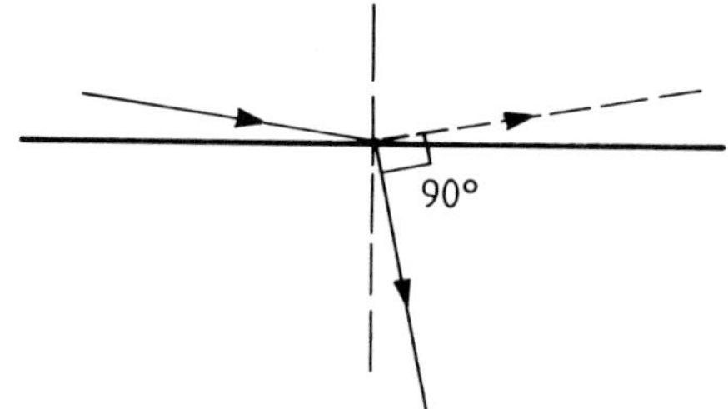

Fig. 11.30 Reflection on a perfect dielectric (Brewster's angles).

Brewster's angles if $\mu_1 = \mu_2 = 1$.
Under that assumption ($\mu_1 = \mu_2 = 1$), a frequent situation, then

$$R'_V = \frac{n_2 \cos\theta_1 - n_1 \cos\theta_2}{n_2 \cos\theta_1 + n_1 \cos\theta_2} = R'_V$$

$$T_V = \frac{n_1}{n_2}(1 + R_V)$$

$$T'_V = 1 + R_V$$

In horizontal polarization (see Fig. 11.31)

Computations made from $\cos\theta_1 (H_i - H_r) = (\cos\theta_2) H_T$, $E_i + E_r = E_T$ and $\mu_1 (H_i + H_r) \sin\theta_1 = \mu_2 H_T \sin\theta_2$ give

$$R_h = \frac{E_r}{E_i} = R'_h = \frac{H_r}{H_i} = \frac{\cos\theta_1 \sqrt{\dfrac{\varepsilon_1}{\mu_1}} - \cos\theta_2 \sqrt{\dfrac{\varepsilon_2}{\mu_2}}}{\cos\theta_1 \sqrt{\dfrac{\varepsilon_1}{\mu_1}} + \cos\theta_2 \sqrt{\dfrac{\varepsilon_2}{\mu_2}}}$$

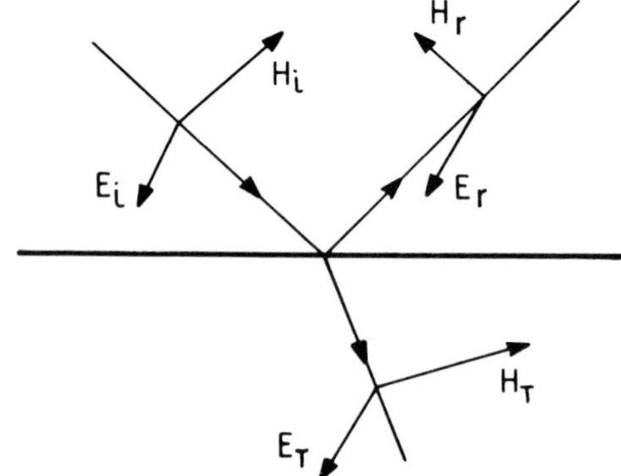

Fig. 11.31 Reflection on a perfect dielectric (horizontal polarization).

$$T_h = \frac{E_T}{E_i} = 1 + R_h$$

$$T'_h = \frac{H_T}{H_i} = \sqrt{\frac{\varepsilon_2 \mu_1}{\varepsilon_1 \mu_2}} \cdot (1 + R_h)$$

If $\mu_1 = \mu_2 = 1$

$$R_h = \frac{n_1 \cos\theta_1 - n_2 \cos\theta_2}{n_1 \cos\theta_1 + n_2 \cos\theta_2}$$

and since $n_1 \sin\theta_1 = n_2 \sin\theta_2$

$$R_h = \frac{\sin(\theta_2 - \theta_1)}{\sin(\theta_2 + \theta_1)}$$

$$T_h = 1 + R_h$$

$$T'_h = \frac{n_2}{n_1}(1 + R_h)$$

First example: $\mu_1 = \mu_2 = 1$, $2\,n_1 = n_2 = 2$; curves giving $|R_V|$ and $|R_h|$ versus θ_1, are shown in Fig. 11. 32.
Second example: $\mu_1 = \mu_2 = 1$, $n_1 = 2$, $n_2 = 1$; for $\theta_1 > 30\,°$ where $\sin\theta_1 = 0.5$ and $\sin\theta_2 = 1$ there is a total reflection as is well known. Figure 11.33 gives $|R_V|$ and $|R_h|$ versus θ_1 for $\theta_1 < 30\,°$.

11.5.3 Reflection of a plane wave on a dielectric with losses

In that case the absolute permittivity becomes complex:

$$\varepsilon_0 = \varepsilon + \frac{\sigma}{j\,\omega}$$

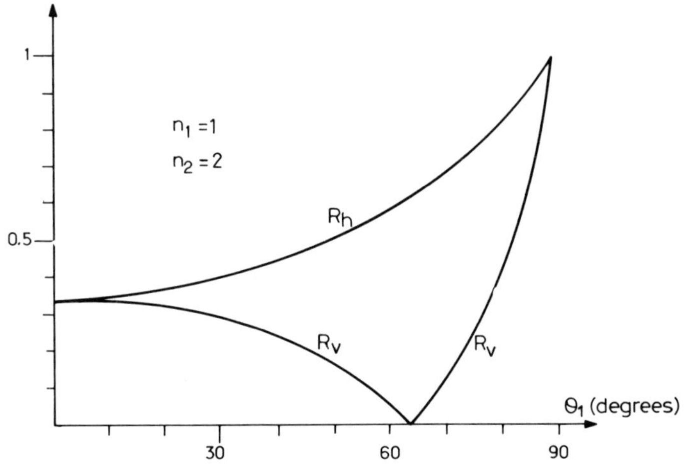

Fig. 11.32 Reflection on a perfect dielectric.

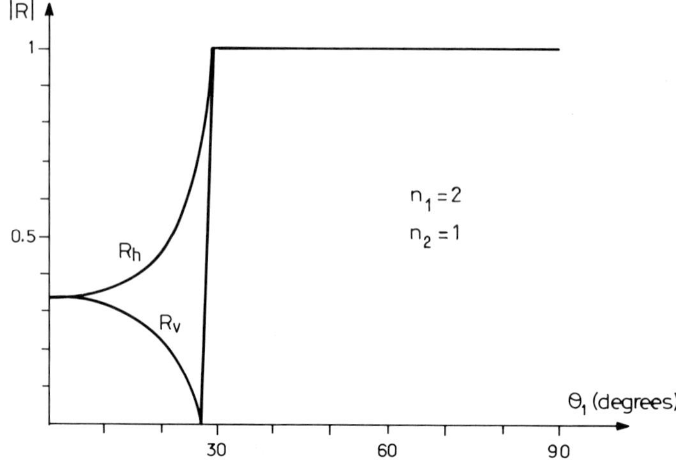

Fig. 11.33 Reflection on a perfect dielectric.

where σ is the conductivity and ω the angular frequency. Since $\omega = 2\pi F$ and $F = c/\lambda$, the relative complex permittivity becomes

$$\varepsilon + \frac{\sigma\lambda}{j\,2\,\pi c} \cdot \frac{1}{\varepsilon_0}$$

with $\varepsilon_0 = (1/36\pi) \times 10^{-9}$ F/m which for the complex relative permittivity eventually gives

$$\varepsilon - j\,60\,\sigma\lambda.$$

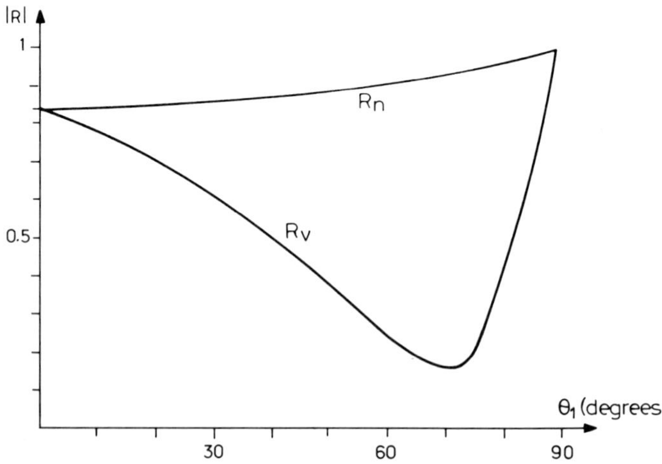

Fig. 11.34 Reflection on a dielectric with losses.

If $\varepsilon >> 60\ \sigma\lambda$ losses could be neglected, if $60\ \sigma\lambda >> \varepsilon$ the dielectric is in fact metallic.

Between those two extremes, curves giving $|R_V|$ and $|R_h|$ are similar to those represented in Fig. 11.34, with a 'pseudo-Brewster' angle. For instance, for frequencies below about 1 GHz the sea could be considered as metallic and above as a dielectric. For dry ground that 'transition frequency' is between 500 kHz and 10 MHz.

11.6 IONOSPHERIC SCATTER

Reflection on ionosphere is of general use for communication at long distance using frequencies around 10 MHz, which are definitely not microwaves. But since that is also used for detection over the horizon (OTH radars), we shall say a few words about it.

Roughly, simplifying the ionosphere and considering it as constituted by one layer, a critical frequency is associated to it (which depends on the nature of the layer) in such a way that a wavelength λ, transmitted at an elevation angle α, will be reflected by the ionosphere if (in first approximation)

$$\sin \alpha < \frac{f_c}{f}$$

f_c depending on the time and the season.
Example. We want to detect a target close to the surface of the earth at a distance of 2400 km and we assume that the ionized layer is at an altitude of 110 km, with a critical frequency of 3.5 MHz. Assuming that the radius of the earth is 6400 km, the path of the radar radiation is horizontal at the departure and α is 0.19 radian (Fig. 11.35).

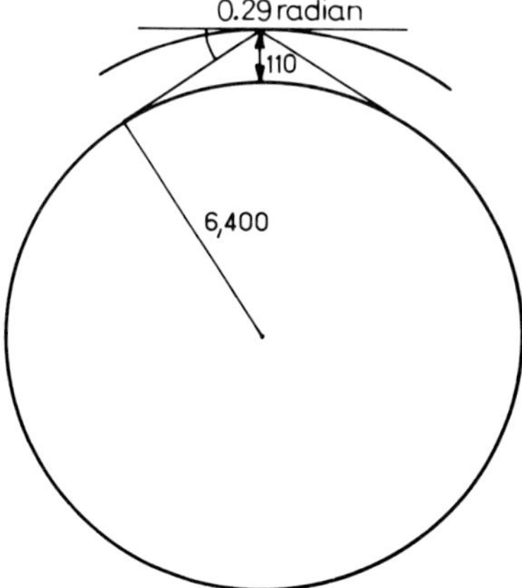

Fig. 11.35 Ionospheric scatter.

$$f < \frac{3.5}{\sin(0.19)}$$

$$f < 18.5\,\text{MHz (maximum useable frequency, MUF)}$$

The frequency has to be less than 18.5 MHz to ensure the detection at 2400 km (and the transmitted power, as well as the antenna gain, large enough).

Reflection on the ionosphere is not a completely permanent phenomenon and depends enormously on the solar activity, thus one cannot ensure a 100% detection.

12

Microwave antennas

Serge Drabowitch

12.1 FOREWORD

This chapter on microwave antennas is not a dissertation on such antennas; such dissertations are often works in themselves consisting of several volumes (Collin and Zucker, 1969, Roubine *et al.*, 1987). Nor is it a dissertation on electromagnetism (Jones, 1964) Rather it should be considered as an introduction for students and engineers who already know the basis of electromagnetism as summarized in Maxwell equations.

Microwave antennas can be considered as special optical instruments. Their study is based on the same mathematical models as those for coherent light instrument optics: physical optics, geometrical optics, while insisting, of course, on the vectorial aspect which allows representation of the polarization properties. In this field as in others, it is essential to keep in mind that, for a particular problem, the choice of a mathematical model implies that the hypotheses which it assumes be valid.

We have covered somewhat extensively the radiating aperture concept that applies to numerous antennas, if not all of them. The difficult problem of receiving antennas has not been neglected, however its presentation has been simplified, without forgetting the problem of incoherent noise reception. After the description of the main types of current microwave antennas (horns, reflectors, passive or active module array antennas), we briefly explore a very topical subject, that of signal-processing antennas. A familiar example of such antennas is that of angular tracking monopulse antennas. Today they experience widespread development in the form of adaptive antennas (antijamming systems) or synthetic aperture antennas used in imaging applications.

We end the chapter by evoking the use of spectral estimation methods at high resolution. However, we don't enter into detail since we only intend to prepare the interested reader for works on this subject.

12.2 WAVES AND RAYS

12.2.1 Waves

Antennas radiate or receive electromagnetic waves. A physical image of wave motion can be obtained by throwing a stone into a still pond. Waves propagate from the point of impact at a certain speed c, known as the phase velocity. The distance between two successive peaks is the wavelength λ. The time required for two successive peaks to pass in front of a given point is the period T and the number of peaks seen each second at a given point is the frequency f. This gives the following fundamental equations (Fig. 12.1):

$$\lambda = cT \tag{12.1}$$

$$\boxed{f = \frac{1}{T} = \frac{c}{\lambda}} \tag{12.2}$$

In a vacuum, electromagnetic waves propagate at a speed c very close to 3×10^8 m/s.

For example: $\lambda = 3$ cm, $c = 3 \times 10^8$ m/s, $f = 10\,000$ MHz, $T = 0.1$ ns.

Representation of a sinewave

The amplitude A of a sinewave propagating along an axis, Or, in direction of unit vector $\boldsymbol{u}$, at a constant velocity and with no distortion, is a sinusoidal function of a linear relationship between the abscissa, r, of the point of observation and time t:

$$A(r,\ t) = A_0 \cos 2\pi f\left(t - \frac{r}{c}\right) \tag{12.3}$$

which can be written

$$A(r,\ t) = A_0 \cos(2\pi f t - \phi) \tag{12.4}$$

where ϕ is the phase angle at point r:

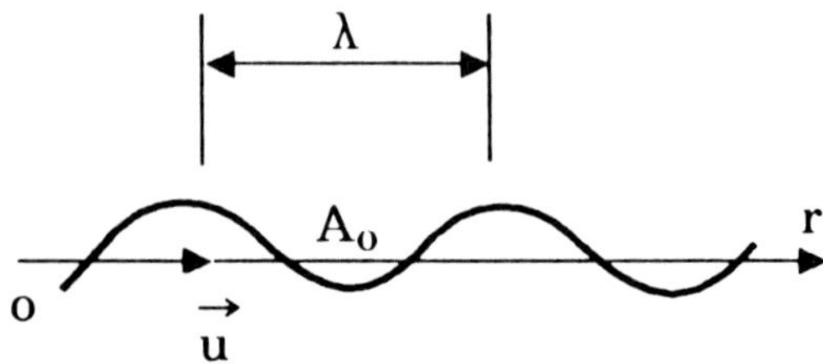

Fig. 12.1 Representation of a sine wave.

$$\phi = 2\pi f \frac{r}{c} = 2\pi \frac{r}{\lambda} \cdot \tag{12.5}$$

This formula shows that the phase rotates by 2π radians for a shift of one wavelength in the direction of propagation. Note that the constant $k = 2\pi/\lambda$ is sometimes referred to as the wave number.

Polarization

A wave on the pond surface is an oscillation transverse to the direction of propagation (in this case, vertical). The direction of oscillation is known as the wave polarization and can be represented by a unit vector $\boldsymbol{p}$ ($|\boldsymbol{p}| = 1$, $\boldsymbol{p} \cdot \boldsymbol{u} = 0$)*. The polarization of an electromagnetic wave is always that of the electric field (expressed in volts/metre) and can, therefore, be written:

$$\boldsymbol{E}(r, t) = \boldsymbol{p}\, A_0 \cos(2\pi f t - \phi). \tag{12.6}$$

Complex notation

The above expression can be considered as the real part of a complex vector $\boldsymbol{\epsilon}$:

$$\boldsymbol{E}(r, t) = R\,\{\boldsymbol{\epsilon}\,(z, t)\} \tag{12.7}$$

$\boldsymbol{\epsilon}\,(z, t)$ is defined by:

$$\boldsymbol{\epsilon}\,(r, t) = \boldsymbol{p}\, A_0\, [\exp(j\, 2\pi f t)]\, \exp(-j\phi) \tag{12.8}$$

When we consider the phenomena at a given frequency, f, all the field or current equations will contain the same exponential frequency factor between the square brackets. This can therefore be taken as a constant, and we can work directly with equations of the form $\boldsymbol{E}$

$$\boldsymbol{E}(r, t) = \boldsymbol{p}\, A_0 \exp(-j\phi). \tag{12.9}$$

This complex notation occurs consistently in the calculations.

Plane waves

The waves we generated on the surface of the pond are circular waves: the isophase points, or wavefronts, form concentric circles. When this case is generalized to three dimensions, the waveforms can become far more complex. However, they can always be decomposed into sums of plane waves or spherical waves. We shall, therefore, now consider these waves.

Let us represent space by a system of orthogonal axes Oxyz (Fig. 12.2). For a plane wave propagating along a unit vector $\boldsymbol{u}$, the field observed at any point M in a plane P $\perp$ $\boldsymbol{u}$ will have the same phase.

* The notation $\boldsymbol{p} \cdot \boldsymbol{u}$ means the scalar product of vectors $\boldsymbol{p}$ and $\boldsymbol{u}$. The vectorial product is noted by $\boldsymbol{p} \times \boldsymbol{u}$.

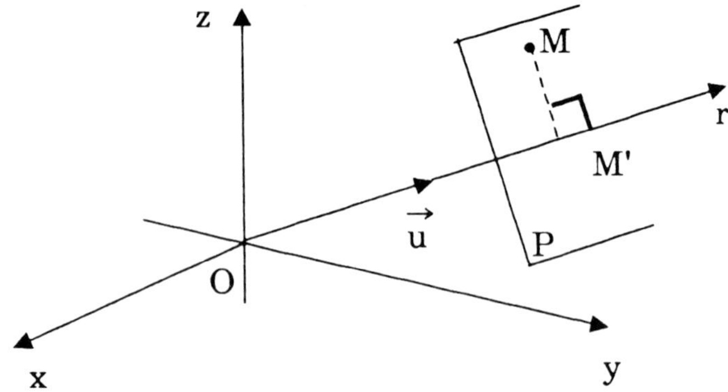

Fig. 12.2 Plane wave propagation in rectangular coordinates.

The equation for this plane can be written in the form of a condition applicable at any point M in this plane: its projection on $\boldsymbol{u}$ is a fixed point M′ regardless of the position of M. This can be written:

$$\mathbf{OM} \cdot \boldsymbol{u} = \mathrm{OM}' = r \tag{12.10}$$

Equations (12.5) and (12.9) therefore show that the equation of the electric field at M is:

$$\boxed{\boldsymbol{E}(\mathrm{M}) = \boldsymbol{p}\, A_0 \exp\left(j\frac{2\pi}{\lambda}\,\mathbf{OM} \cdot \boldsymbol{u} \right).} \tag{12.11}$$

Let α, β and γ be the direction cosines (or coordinates) of vector $\boldsymbol{u}$:

$$\alpha^2 + \beta^2 + \gamma^2 = 1. \tag{12.12}$$

Let x, y and z be the coordinates of point M. We can then write:

$$\boldsymbol{E}\,(x, y, z) = \boldsymbol{p}\, a_0 \exp\left[j2\pi\left(\frac{x}{\lambda}\alpha + \frac{y}{\lambda}\beta + \frac{z}{\lambda}\gamma \right) \right]. \tag{12.13}$$

This expression is fundamental to the study of radiation by antennas (section 12.3.3).

The electromagnetic structure of a plane wave

So far, we have only considered an electric field $\boldsymbol{E}$ (volts/metre) whose direction $\boldsymbol{p}$ defines the wave polarization (see below). We must also associate a magnetic field $\boldsymbol{H}$, measured in amperes per metre, to this electric field.

A plane wave possesses the following properties.

1. The magnetic field $\boldsymbol{H}$ is perpendicular to the electric field. Vectors $\boldsymbol{E}$, $\boldsymbol{H}$ and $\boldsymbol{u}$ form a right-angle trihedron, (Fig. 12.3).
2. The two oscillations are in phase.

3. The ratio between the $\boldsymbol{E}$ and $\boldsymbol{H}$ amplitudes is a constant η, known as the wave impedance which is a characteristic of the medium through which the wave propagates (in vacuum $\eta = 120\,\pi$ ohm). This gives:

$$\boldsymbol{E} = \eta\, \boldsymbol{H} \times \boldsymbol{u}\,. \tag{12.14}$$

Power density – Poynting vector

The Poynting vector is the following vectorial product:

$$\boldsymbol{P} = \frac{1}{2}\boldsymbol{E} \times \boldsymbol{H}\,.$$

It can be shown that the power density, in watts per square metre (W/m^2), transported by a plane wave is equal to the modulus of the Poynting vector:

$$\frac{\mathrm{d}W}{\mathrm{d}S} = |\boldsymbol{P}| = \frac{|\boldsymbol{E}|^2}{2\eta}\cdot \tag{12.15}$$

It is worth noting that a plane wave cannot physically exist on its own since it is extended to infinity in space and, therefore, carries infinite power. However, it is frequently used as an intermediate step in calculations and sometimes represents reality quite closely.

Spherical waves

The spherical wave, on the other hand, is possibly the closest to representing reality. If we consider a point source of radiation generating power W, and located at the origin, O, of the coordinates, simple symmetry tells us that the radiated wavefronts are spheres centred on O (Fig. 12.4). Let us consider a wavefront at radius r. Its surface area is $4\pi r^2$ and the power density is, therefore:

$$\frac{\mathrm{d}W}{\mathrm{d}S} = \frac{W}{4\pi r^2}\cdot \tag{12.16}$$

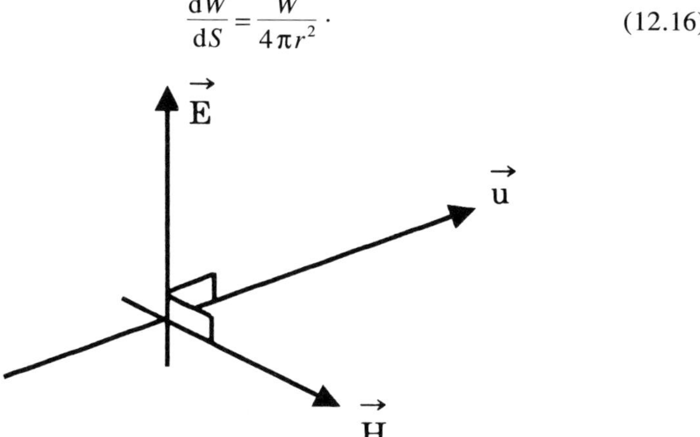

Fig. 12.3 Electromagnetic structure of a plane wave.

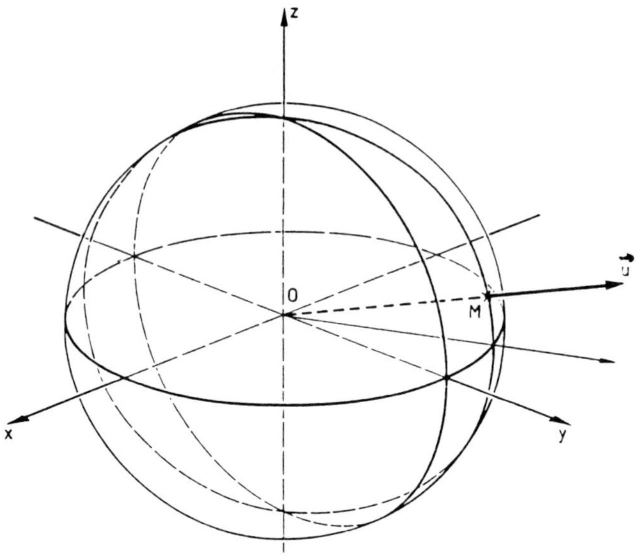

Fig. 12.4 Propagation of a spherical wave.

Let us now assume that the wave structure of the electromagnetic field is locally plane. The power density (12.16) is, therefore, also given by (12.15). This shows that the value of the electric field at a distance r from an emitter of power W is:

$$E(r) = \frac{\sqrt{60\,W}}{r}. \tag{12.17}$$

So the field amplitude of a spherical wave is inversely proportional to the distance. This is an important result.
For example, when $W = 100$ W and $r = 10\,000$ m, $E = 7.7$ mV/m
Let us now look at the phase: above, we saw that the phase varies with distance r along any direction $\boldsymbol{u}$; the phase is calculated by formula (12.5). To conclude, the field strength of a spherical wave is given by:

$$\boldsymbol{E}(r) = \frac{E_0}{r}\,\boldsymbol{p}\exp\left(-j\frac{2\pi r}{\lambda}\right). \tag{12.18}$$

Where $\boldsymbol{p}$ is the polarization unit vector.

Boundary conditions

Up to now, we have considered plane or spherical waves propagating in unlimited space completely free of conductive obstacles (either metal or dielectric). In practice, such obstacles always exist.

Boundary conditions, i.e. the conditions in the immediate vicinity of such obstacles, affect the electromagnetic field. The field in these areas can be calculated by considering it as the sum of plane or spherical waves whose propagation equations we now know. It can be seen, therefore, that these conditions determine the field structure throughout space: we shall now consider a simple example of this process applied to the reflection of a plane wave from a conductive plane.

Boundary conditions generated by a perfectly conductive surface

This is often the case of a metal surface. Even if it is of a certain thickness, the electrical current remains localized on the surface ('skin effect') (Fig. 12.5). There is a current density $\boldsymbol{J}_s$ (A/m) in the surface layer, associated with an electric field $\boldsymbol{E}_s$ normal to the surface (or zero) and a magnetic field $\boldsymbol{H}_s$ tangential to the surface; vectors $\boldsymbol{E}_s$, $\boldsymbol{H}_s$ and $\boldsymbol{J}_s$ form a right-angle trihedron. Also,

$$|\boldsymbol{H}_s| = |\boldsymbol{J}_s| \qquad \text{(amperes per metre).}$$

The magnetic field is equal to the current density. In addition, it can be shown that the electric field is proportional to the surface density of the electric charge.

Let us just reiterate, without going into details, the boundary conditions imposed by a non-ideal conductor (Roubine *et al.*, 1987): continuity of the tangential components of electric and magnetic fields and the normal component of magnetic induction.

Reflection of an incident plane wave from a plane reflector

Let us consider a conductive plane and let $\boldsymbol{n}$ be the unit vector normal to the surface at point A. Let us now consider a wave incident along $\boldsymbol{u}$ in plane P, such that it forms angle θ with $\boldsymbol{n}$ (Fig. 12.6), with P the plane normal to the

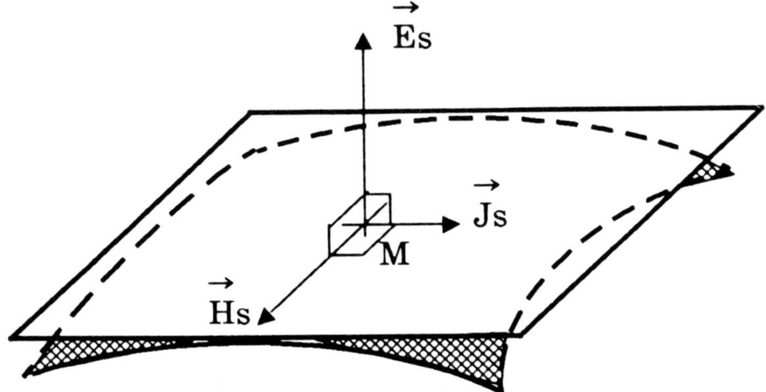

Fig. 12.5 Boundary conditions on a conducting surface.

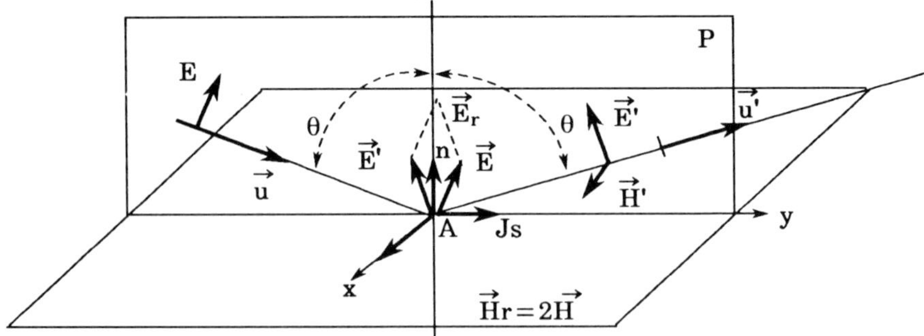

Fig. 12.6 Reflection of an incident plane wave from a plane reflector.

reflector containing $\boldsymbol{u}$ (incidence plane). It can easily be seen that the boundary conditions are satisfied if a 'reflected' wave is superimposed on this incident wave, such that the reflected wave propagates in direction $\boldsymbol{u}'$ where $\boldsymbol{u}$ and $\boldsymbol{u}'$ are symmetrical about $\boldsymbol{n}$ (this is simply the well-known Snell's optical law: the angle of reflection equals the angle of incidence).

1st case: the electric field $\boldsymbol{E}$ is in the plane of incidence P

To obtain the resultant electric field $\boldsymbol{E}_r$ normal to the surface, we simply superimpose a reflected wave polarized in the same plane P, in phase with the incident wave at A and propagating along $\boldsymbol{u}'$. We then check that the magnetic field at A is tangential. Its amplitude is double the incident magnetic field amplitude. The result is a surface current, parallel to plane P, whose density is:

$$|J_s| = 2|H| \; . \tag{12.20}$$

2nd case: the incident electric field is perpendicular to P (i.e. parallel to the reflector)

This case is left to the student as an exercise: what is the reflected wave phase at A? What is the resultant electric field amplitude at A? And the magnetic field amplitude? And the current?

Note that the interaction between the resultant magnetic field $\boldsymbol{H}_r$ and the current $\boldsymbol{J}_s$ generates a Laplace force perpendicular to the plane. This is the 'radiation pressure' phenomenon.

Important generalization: the physical optics approximation

When considering the effects of an incident wave on a curved conductive surface, the physical optics approximation (POA) is equivalent to accepting that the field structure, close to any point on the surface, is the same as that generated by a plane surface tangential to this point (Fig. 12.5). This approximation is widely used and, in fact, corresponds to assuming that the radius of curvature of the surface is not too small in relation to the wavelength. The approximation is not valid in areas around sharp edges or tips. Another approximation which can be used, but which is even more restricted than the POA is the geometrical optics approximation (GOA) (discussed below).

Polarization

We have already defined polarization as being the direction $\boldsymbol{p}$ in which the electric field oscillates. This direction is always transverse to the direction of propagation. This definition describes linear polarization, but it is possible to create elliptical polarization by combining two mutually perpendicular linear polarizations. Let us suppose that this sheet of paper is a wavefront. Then let us define a rectangular frame of reference with axes O_x, and O_y and unit vectors $\boldsymbol{u}$ and $\boldsymbol{v}$ (Fig. 12.7).

When two waves with polarization $\boldsymbol{u}$ and $\boldsymbol{v}$ respectively are superimposed, the resulting polarization's unit vector is:

$$\boldsymbol{p} = a\boldsymbol{u} + b\boldsymbol{v} \tag{12.21}$$

with the normality condition:

$$|a|^2 + |b|^2 = 1. \tag{12.22}$$

If coefficients a and b are real, the component waves are in phase and the resultant polarization is linear but at an angle α such that:

$$\tan \alpha = b/a. \tag{12.23}$$

This representation can be extended to the general case by considering complex components.

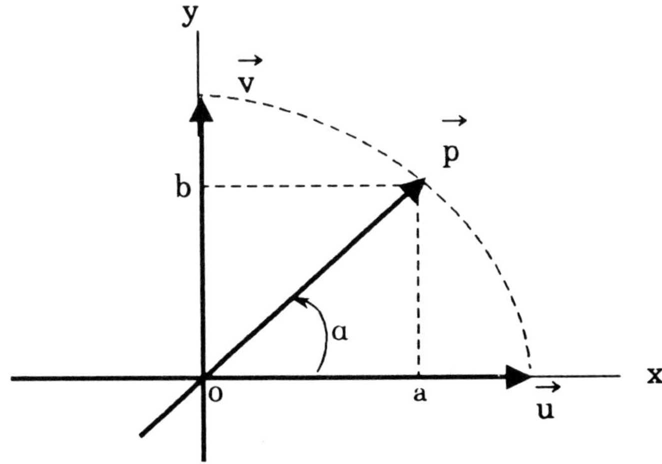

Fig. 12.7 Polarization vector: linear polarization.

Circular polarization

Let us assume that the components have the same amplitude but are 90 ° out of phase, leading or lagging. If we use the 'real' notations, we will obtain, for example:

$$x(t) = \frac{1}{\sqrt{2}} \cos 2\pi ft \qquad y(t) = \frac{\pm 1}{\sqrt{2}} \sin 2\pi ft \tag{12.24}$$

The resultant electric field is a vector which rotates at frequency *f*, clockwise or anticlockwise depending on which phase leads or lags. With the complex notations, we shall have:

$$a(t) = \mathbf{R}\left\{\frac{1}{\sqrt{2}} \exp j2\pi ft\right\} \qquad b(t) = \mathbf{R}\left\{\frac{\pm 1}{\sqrt{2}} \exp(\pm j\pi/2) \exp(j2\pi ft)\right\} \tag{12.25}$$

whence

$$a = \frac{1}{\sqrt{2}} \qquad b = \pm \frac{j}{\sqrt{2}} \cdot \tag{12.26}$$

The polarization unit vector for the two possible circular polarizations will therefore be:

$$\boldsymbol{C} = \frac{\boldsymbol{u} + j\boldsymbol{v}}{\sqrt{2}} \qquad \boldsymbol{C}' = \frac{\boldsymbol{u} - j\boldsymbol{v}}{\sqrt{2}} \cdot \tag{12.27}$$

Elliptical polarization

If the two components are mutually perpendicular but do not have the same amplitude, the resultant field will describe an ellipse whose axes are O$\boldsymbol{u}$ and O$\boldsymbol{v}$. The real notations will be:

$$x(t) = a_0 \cos 2\pi ft \qquad y(t) = \pm b_0 \sin 2\pi ft\,. \tag{12.28}$$

The result will, therefore, be an ellipse with half-axes a_0 and b_0 (Fig. 12.8). The ratio b_0/a_0 is termed the ellipticity.

We shall now go back to our complex notations and note the polarizations:

$$\boldsymbol{p} = a_0\, \boldsymbol{u} + j\, b_0\, \boldsymbol{v} \qquad \boldsymbol{q} = a_0\, \boldsymbol{u} - j\, b_0\, \boldsymbol{v} \tag{12.29}$$

Since a_0 and b_0 are real, the normality condition yields: $|\boldsymbol{p}|^2 = 1$, hence:

$$a_0^2 + b_0^2 = 1. \tag{12.30}$$

The ellipse is inscribed in a rectangle whose half-diagonal length is 1.

Exercise

We shall now look at the general case for components with any amplitude and any phase; their form will be, for example:

$$x(t) = a_0 \cos 2\pi ft \qquad y(t) = b_0 \cos(2\pi ft - \phi). \tag{12.31}$$

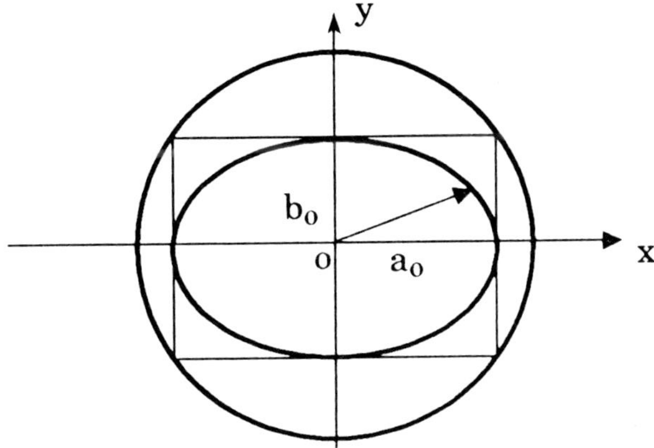

Fig. 12.8 Elliptical polarization.

It is left to the reader to determine the complex notations and the ellipse parameters, i.e. the ellipticity and the direction of the axes.

Changing the reference polarizations

Formula (12.21) expresses the polarization $\boldsymbol{p}$ based on the orthogonal linear polarizations $\boldsymbol{u}$ and $\boldsymbol{v}$. As equations (12.27), which define circular polarizations $\boldsymbol{C}$ and $\boldsymbol{C}'$, are orthogonal linear transforms, it is therefore possible to express the same polarization $\boldsymbol{p}$ based on the component polarizations $\boldsymbol{C}$ and $\boldsymbol{C}'$. This will yield:

$$\boldsymbol{p} = a'\,\boldsymbol{C} + b'\,\boldsymbol{C}' \tag{12.32}$$

where

$$a' = \frac{a + jb}{\sqrt{2}} \qquad b' = \frac{a - jb}{\sqrt{2}} \tag{12.33}$$

This satisfies the requirement for normalization: $|a'|^2 + |b|^2 = 1$.

Hence any polarization can be considered as a combination of two inverse circular polarizations (Fig. 12.9). Note that the above linear transforms frequently tend to represent the polarization as a 2-element column matrix:

$$\begin{bmatrix} p \end{bmatrix} = \begin{bmatrix} a \\ b \end{bmatrix}. \tag{12.34}$$

12.2.2 Fundamental propagation laws

The physical laws relative to radiation and propagation are studied in works on electromagnetism (Baker and Copson, 1950). For microwave antennas, whose dimensions are generally large with respect to the wavelength, a simplified approximate form can be given. This form is based on the concepts from physical optics (PO) and geometrical optics (GO) (Born and Wolf, 1964, Silver, 1949). These concepts are those of wave and wavefronts, and the Huygens–Fresnel principle.

Wavefront

If at a same point, the electric and magnetic fields oscillate with the same phase, it is possible to define the equiphase surfaces, i.e. the wavefront, which generalizes into any given surface certain properties which we have indicated with respect to spherical waves and plane waves. The case is frequently considered, where a wavefront has a plane wave structure locally; the Poynting vector allows definition of a power flux transported normally to the wavefront, expressed in watts per square metre (12.15). This is at the basis of the ray concept used in geometrical optics.

Huygens–Fresnel's principle on the propagation of waves

This principle is in fact a consequence of the Maxwell equations. Each point M of a wavefront (S) can be considered as a secondary source yielding an elementary spherical wave. The field at a point P is thereby obtained by superposition of these basic waves, given their differences in phase when they reach the point P.

Let E(M) be a given scalar component of the electric field at a point M of the wavefront (Fig. 12.10). The corresponding component of the electric field E'(P), observed at a far point P, at a distance $r = \mathrm{MP}$, is given by the expression:

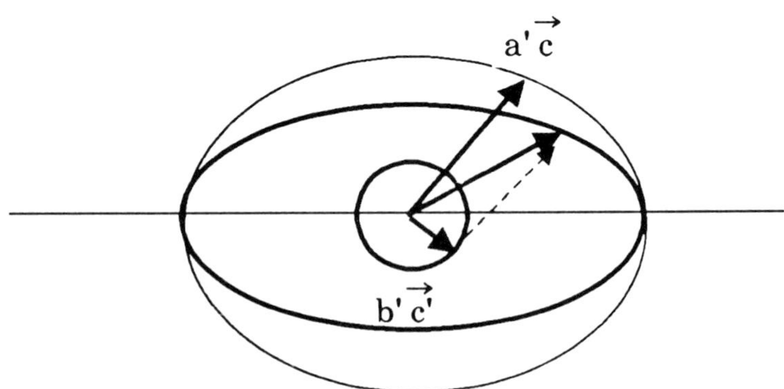

Fig. 12.9 Representation of an arbitrary polarization as a linear combination of two orthogonal circular polarizations.

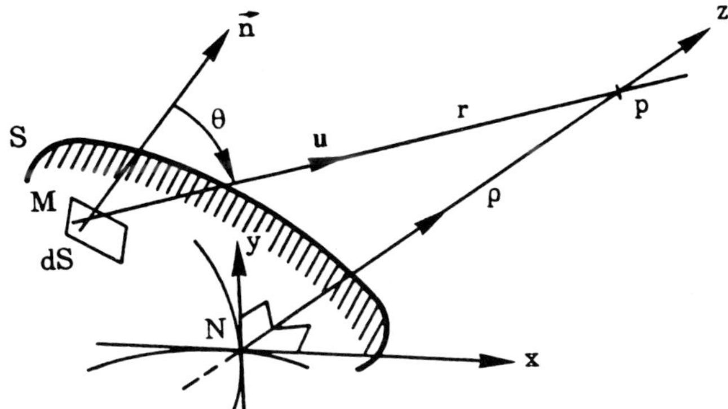

Fig. 12.10 Stationary phase principle.

$$E'(P) = \frac{j}{\lambda} \int_S \gamma E(M) \frac{\exp\left(-j\frac{2\pi r}{\lambda}\right)}{r} \mathrm{d}S \tag{12.35}$$

We recognize in this expression the mathematical form of a spherical wave (12.18). The factor γ or 'obliquity factor' has a value often close to 1.

It decreases slowly with respect to angle θ between the normal $\boldsymbol{n}$ to the surface at M and the direction $\boldsymbol{u}$ of point P. Often it can be considered that $\gamma \simeq \cos\theta$ or even $\gamma \simeq 1$ since this factor is generally not critical for microwave antennas as we will see below. Note that a relation identical to (12.35) exists between magnetic fields.

Note that if S is not an equiphase surface, the same formulae apply provided the phase variation is included in the expression of E(M), which becomes a complex scalar function.

Stationary phase principle

This principle allows transition between Huygens–Fresnel wave optics and geometrical optics (Fig. 12.10). Given PN a normal dropped from P onto a wave surface S. Assume that *this normal is unique, in other words that P is not a focal point.* NP is the z axis. Axes Ox and Oy are tangent to the main sections of this surface in N. Let NP $= p$.

We can then state, based on 12.35:

$$E'(P) = \frac{j}{\lambda} \exp\left(-j\frac{2\pi}{\lambda}p\right) \int_S \gamma E(M) \frac{\exp\left\{-j\frac{2\pi}{\lambda}(r-p)\right\}}{r} \mathrm{d}S \tag{12.36}$$

This integral can be represented as a sum of elementary vectors $\delta \boldsymbol{V}$ in the complex plane. The amplitude of each vector is:

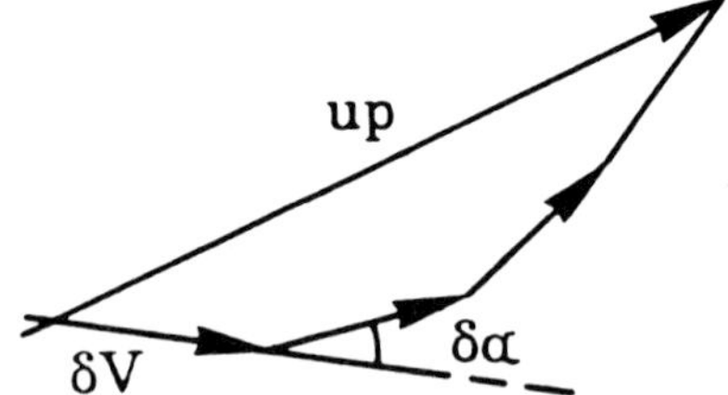

Fig. 12.11 Contribution of surface element ΔS near P.

$$\delta V = \gamma \frac{E(M)}{r} \, \mathrm{d}S.$$

Its difference in phase with the corresponding vector at the base of the normal in N is:

$$\delta\alpha = \frac{2\pi}{\lambda}(r - p).$$

Let us examine the contribution of a given surface element ΔS which does not contain the normal. If λ is large, the angles between the vectors are small. The addition vector graph has the aspect of a curve and the resultant is in general significantly different from zero (Fig. 12.11).

If λ is small, the angles $\delta\alpha$ are large and the vector graph looks like a tight spiral, wrapped around itself. In this case $E(\mathrm{P})$ will be almost null (see Fig. 12.12).

Hence in the regions where the component wave phase varies rapidly, the resultant field is somewhat cancelled out by interference. The situation is different in the neighbourhood of the normal base, N.

The phase function

$$\phi(x, y) = \frac{2\pi}{\lambda}(r - p) = \left[\sqrt{x^2 + y^2 + (p - z)^2} - p\right]\frac{2\pi}{\lambda}$$

is stationary in the neighbourhood of this point. Consequently, phase varies slowly even if λ is small. The result is that only the contribution of points in

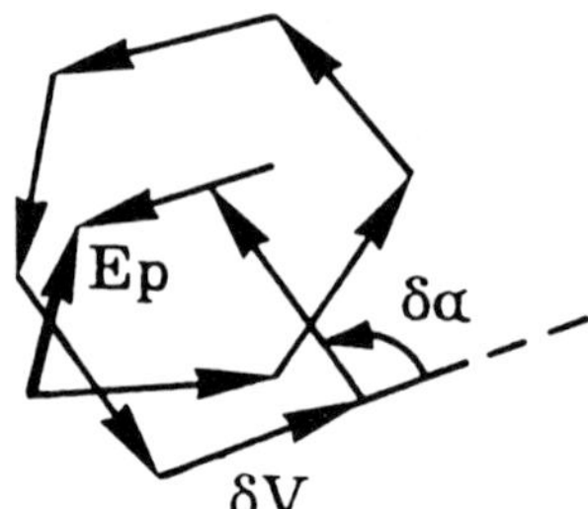

Fig. 12.12 Contribution of surface element ΔS far from P.

the neighbourhood of the normal are significant in the expression of the field at P. Note that for these points, angle θ is narrow. Therefore $\gamma \simeq 1$.

We can therefore formulate this **stationary phase principle** as follows: for short wavelengths, the electrical state at a point P only depends, as a first approximation, on the electrical states at the points of the wave surface neighbouring the normal passing through P.

We can state a simple mathematical relationship connecting the electrical states at N and at P, given the curvature radii of the wave surface at N.

The integral (12.36) reduces to:

$$E(\mathrm{P}) = \frac{j}{\lambda} E_{\mathrm{N}} \frac{\exp\left(-j\dfrac{2\pi}{\lambda}p\right)}{p} \iint \exp\{-j\,\phi\,(x,\ y)\}\,\mathrm{d}x\,\mathrm{d}y.$$

Given the equation of the surface in the neighbourhood of N and extending the integral to infinity, which is allowed by the stationary phase principle, we find:

$$E(\mathrm{P}) = E_{\mathrm{N}} \exp\left(-j\frac{2\pi}{\lambda}p\right) \sqrt{\frac{R_1 R_2}{(R_1+p)(R_2+p)}} \qquad (12.37)$$

(R_1 and R_2 are the main curvature radii at N).

Ray theory and geometrical optics

Rays

The stationary phase principle establishes a point-to-point correspondence between two successive wave surfaces (Fig. 12.13). It sets up a relation between the points of two wave surfaces which have a common normal NN′. The set of these normals form the light rays. Here, this concept has a purely mathematical meaning, it does not assume the possibility of physically isolating such rays, but it allows extension, within the limits of the validity of the stationary phase principle, of the principles and methods of geometrical optics.

For ultrashort waves, two main methods are generally used: when the wavelength is short enough and we are examining the relation between electrical states of non-focal points, the stationary phase principle allows us to use the methods of **geometrical optics** which we will review on the following pages. For focal points, diffraction has to be examined using Huygens formulae (section 12.3.3 and 12.5.3), i.e. **physical optics**.

Electromagnetic energy flux and geometrical optics

It can be shown that the rays defined above are supports for the Poynting vectors of electromagnetic waves. These rays are therefore lines of energy propagation. Let us consider two successive wave surfaces and a ray tube intercepting

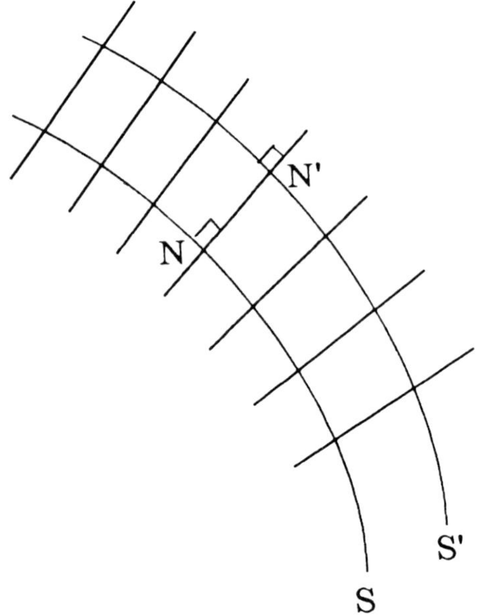

Fig. 12.13 Point-to-point correspondence between successive wavefronts.

surfaces S_1 and S_2 (Fig. 12.14), the condition of the conservation of energy implies the following relationship between Poynting vectors P_1 and P_2:

$$|\boldsymbol{P}_1|\,\mathrm{d}S_1 = |\boldsymbol{P}_2|\,\mathrm{d}S_2$$

$$\frac{1}{2\eta} E_1^2\,\mathrm{d}S_1 = \frac{1}{2\eta} E_2^2\,\mathrm{d}S_2 \qquad (12.38)$$

This relation allows establishment of the relation (12.37) which we deduced from the Huygens–Fresnel principle.

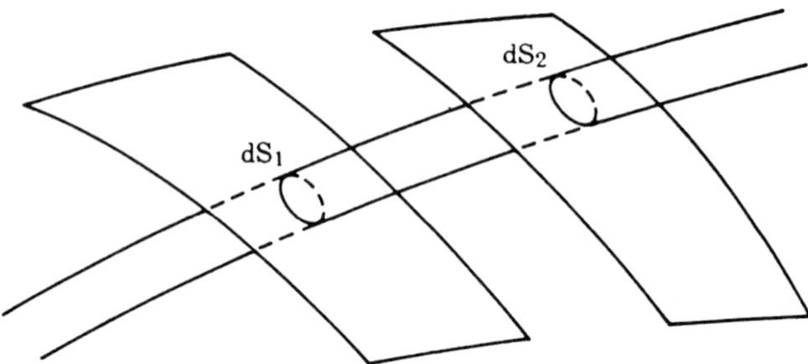

Fig. 12.14 Conservation of energy between successive wavefronts.

Optical path

A fundamental principle of geometrical optics is the Fermat's principle which is often taken as a postulate. Before formulating it we have to define the concept of the optical path. The optical path, ΔL, along a curve Γ between points P_1P_2 is defined by the curvilinear integral:

$$\Delta L = \int_\Gamma n \mathrm{d}s \tag{12.39}$$

where n is the refraction index of the medium crossed and $\mathrm{d}s$ length elements. Example: if we consider two neighbouring wave surfaces, the optical paths followed by the rays are equal because the rays are normal to these two wave surfaces.

Fermat principle

If P_1 is a light source and P_2 is an observation point, the path followed by the light between P_1 and P_2 is such that the path P_1P_2 is stationary (in general, minimum).

Consequently:

1. In a uniform medium, the rays are straight.
2. The Snell–Descartes laws are translated by the formula:

$$n_1 \sin i_1 = n_2 \sin i_2$$

   where i_1, i_2 are the angles of the rays with the normal to the interface between the two media with index n_1 and n_2 respectively.

Optical path law — applications

Let us consider the optical paths separating a wave surface associated to the incident rays (S_1) and a wave surface associated with emerging rays (S_2). Consider two rays going from one surface to the other ABC, A′B′C′ (Fig. 12.15).

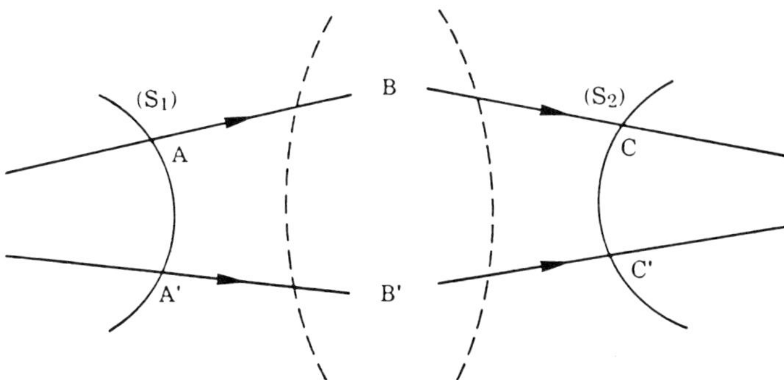

Fig. 12.15 Illustration of the optical path law.

Based on Fermat's principle, we can demonstrate the following basic relationship which expresses the optical path law.

$$\int_{\mathrm{ABC}} n\mathrm{d}s = \int_{\mathrm{A'B'C'}} n\mathrm{d}s\,. \tag{12.40}$$

When moving along a luminous light path from an incident wave surface to an emergent wave surface, the optical path and consequently the phase variation, are independent of the chosen light path.

Application of the law: parabolic reflector

The optical path law we have just stated often constitutes a simpler and more direct method to determine reflection or refraction surfaces than the Snell's laws.

For example: Let us determine a reflector transforming a spherical wave into a plane wave. Let F, the centre of the spherical wave, be the focus of the radiation. Let FX be the direction of radiation of the emerging plane wave. The reflector sought must have a circular symmetry about this axis. Take the plane of the figure as the meridian plane (Fig. 12.16). Let S be the apex of the reflector. Its 'focal distance' is defined by:

$$\mathrm{SF} = f$$

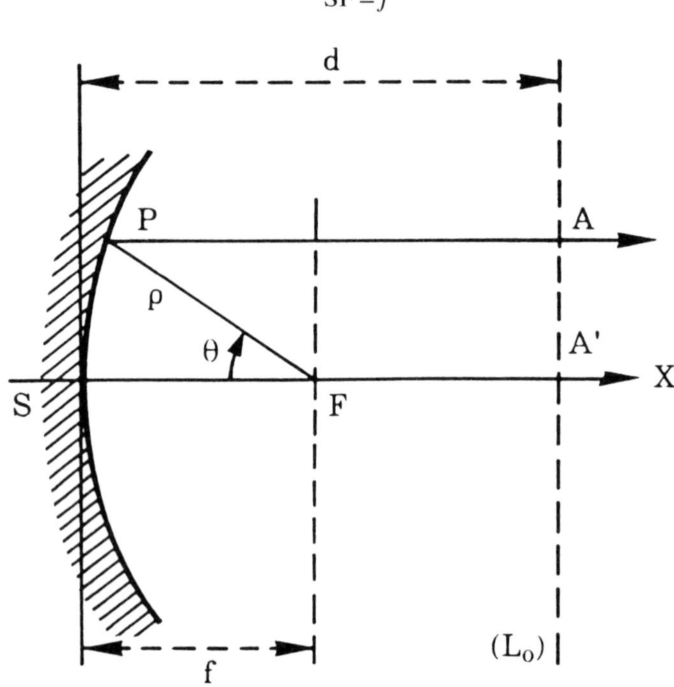

Fig. 12.16 Parabolic reflector.

Let L_0 be the intersection of the emerging plane wave with the meridian plane, at distance d from the apex S ($SA' = d$). Let the two diverging rays FP and FS (axial radius) be associated with the spherical waves and the parallel waves PA and SA′ be associated with plane waves. The equality of the optical paths FPA and FSA′ is written:

$$FP + PA = \text{constant} = FS + SA'$$

or, in polar coordinates,
$$\rho + \rho\cos\theta + d - f = f + d \tag{12.41}$$

$$\rho = 2f/1 + \cos\theta.$$

This the equation of a parabola. In contrast with the preceding calculation, the Snell's laws yield a differential equation of the surface. It is left to the reader to determine this equation.

12.3 THE ANTENNA AS A RADIATING APERTURE

12.3.1 The radiation from an antenna and the equivalent aperture method

Let us consider any antenna A and a surface S which completely surrounds the antenna, including its feeder system (Fig. 12.17). The following theorem (Baker and Capson, 1950) can be proven: If a component (electric or magnetic) of the electromagnetic field is known at any point M on this surface, we can deduce the radiated field at any point P outside the surface.

If a wavefront crosses surface S, the field at P can be deduced by applying the Huygens' principle in the form described in section 12.2.2 and equation (12.35). This result is important for the following reasons.

1. In most microwave antennas, surface S is reduced to a plane in which the electromagnetic field is negligible outside a certain area referred to as the antenna's 'equivalent aperture'. We shall give a few examples below.

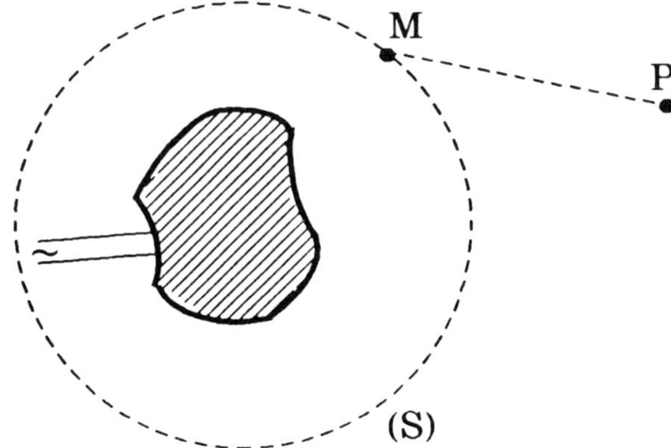

Fig. 12.17 Schematic representation of equivalent aperture method.

2. The electromagnetic field on this surface, or at least a first approximation, can frequently be calculated by geometrical optics methods.
3. The radiated field at an external point M can then be calculated by methods derived from applications of Huygens' principle.

12.3.2 Examples of microwave antennas and equivalent apertures

The electromagnetic horn

Horns will be discussed in more detail in section 12.5.2. The electromagnetic horn (Fig. 12.18) is, perhaps, the simplest microwave antenna but a large number of variants are possible. The structure is, in fact, a waveguide, whose cross-section increases progressively along the guide up to plane (S), at which point the cross-section forms the equivalent radiating aperture. The cross-section of the horn can be any shape but is usually rectangular or circular.

The fact that there are no discontinuities along the horn avoids generating higher spurious propagation modes. Consequently, the field at any point M in the aperture is defined by the propagation modes used for excitation in the feeder guide. However, we can intentionally create discontinuities (multimode horns) or change the longitudinal section of the horn to change field distribution on the aperture (illumination law) and the radiation pattern (section 12.4.1).

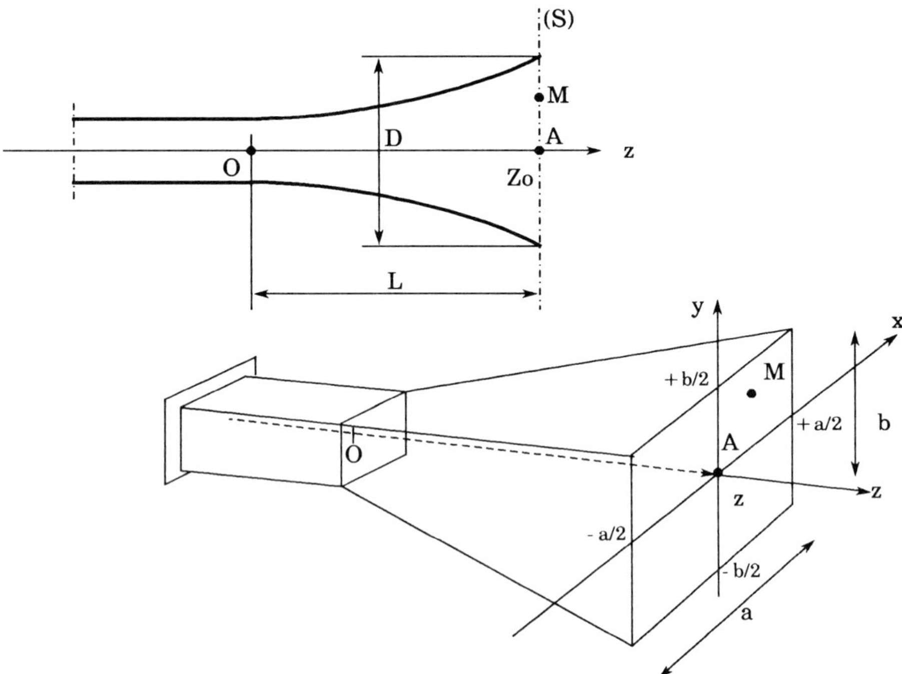

Fig. 12.18 Electromagnetic horn antennas.

The parabolic reflector

The parabolic reflector will be examined in more detail in the section 12.5.3. We saw above that a parabolic reflector will convert the spherical wave generated by a radiation source (for example, a horn) into a plane wave. If we consider a plane (P) through the focus F, the electromagnetic field at any point M can be approximately calculated using the principle of transmitted energy conservation in ray cones and tubes (Fig. 12.19). In this way, we define a radiating aperture in plane P, whose diameter is approximately equal to the reflector diameter D. If we follow a ray FM′M, we can see that the field is only attenuated in the conical section FM′ and remains constant along M′M. If we have a point source at F, the expression for the field at M takes the form:

$$E_M = E_0 \frac{1}{\mathrm{M'F}} = E_0 \frac{1}{\rho} \tag{12.42}$$

where E_0 is a constant which depends on the source.
The vector radius ρ is given by formula (12.41). If we introduce:

$$\mathrm{FM} = r = \rho \sin\theta \qquad (r \leqslant D/2). \tag{12.43}$$

We can write:

$$E_M(r) = \frac{E_0}{f} \frac{1}{1 + \left(\frac{r}{2f}\right)^2}. \tag{12.44}$$

The illumination of the equivalent aperture therefore decreases symmetrically about the axis (Fig 12.20)

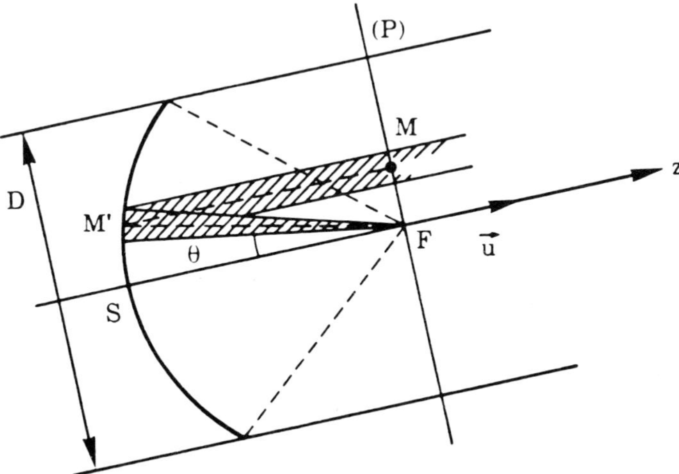

Fig. 12.19 Parabolic reflector.

Fig. 12.20 Illumination law of an equivalent aperture.

The array antenna

Array antennas will be examined in section 12.5.6. The most conventional form of an array antenna comprises a set of N elementary antennas uniformly distributed along a straight line or over a plane (P) (Fig. 12.21). This plane therefore forms the plane of the equivalent aperture, which is limited to the size of the array (D). The elementary antennas can be fed in various ways, for example by lines and couplers which may include phase-shifters, amplifiers or attenuators. These components can, therefore, be used to vary the electromagnetic illumination law $E_{(M)}$ — and consequently the directional properties — as required. This is the main advantage of array antennas over the antennas described earlier.

12.3.3 Far field radiation of an aperture

Introduction

The field existing in the plane of the aperture is not always equiphased: phased arrays (section 12.5.6) use such variations in phase to orient the maximum

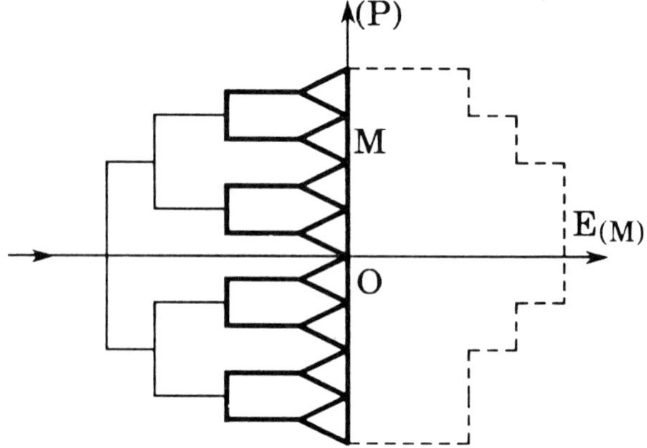

Fig. 12.21 Array antenna with corporate feed.

radiation in a given direction. However, the radiation of an aperture can be determined to a good approximation by applying Huygens–Fresnel's principle (section 12.2.2) to all points, M, of the aperture provided the field phase at M is considered. The field in an outside point P can then be considered as the sum of the basic spherical waves emitted at all points M, i.e. weighted by the amplitude, the polarization and the phase existing in M.

To calculate the radiation of an aperture, we generally assume that the field is known for all points of the aperture. In fact, we only know the field 'applied' to the aperture. The calculation of the resulting field is a diffraction problem which, strictly, requires an integral equation. We have shown that the wave radiated at a large distance is spherical and that the field's local structure is that of a plane wave, in particular, polarization $\boldsymbol{p}$ is normal to the direction of propagation $\boldsymbol{u}$.

The result is that we cannot arbitrarily set the field value in the aperture but only a transverse component (i.e. contained in the aperture plane) of one of the electric or magnetic fields. We will therefore use the following procedure:

1. we assume that the transverse component $\boldsymbol{E}_{\mathrm{T}}(\mathrm{M})$ of the electrical field in the aperture plane is known;
2. using Huygens–Fresnel's formula (section 12.2.2) we deduce the corresponding component $\boldsymbol{E}_{\mathrm{T}}(\mathrm{P})$, parallel to the aperture, of the field radiated at a far point P in direction $\boldsymbol{u}$;
3. knowing that the total field $\boldsymbol{E}(\mathrm{M})$ is normal to the direction $\boldsymbol{u}$ we can deduce it from the component $\boldsymbol{E}_{\mathrm{T}}(\mathrm{P})$ found.

Note that we can also use the same formulae to deduce the radiated magnetic field from the magnetic field on aperture, then the electric field, since the structure of the radiated field is locally that of a plane wave (section 12.3.3).

Calculation of the transverse component $E_{\mathrm{T}}(\mathrm{P})$ of the field radiated by an aperture using the transverse component $E_{\mathrm{T}}(\mathrm{M})$ of the electric field at the aperture

Let us examine (Fig. 12.22) an aperture (S) of contour (C) in a plane xOy, with axis Oz of the unit vector $\boldsymbol{n}$ normal to the aperture. Let us consider a point P at a great distance $OP = r_0$ in the direction with unit vector $\boldsymbol{u}$ and coordinates (or direction cosines) α, β, γ.

$$|\boldsymbol{u}|^2 = \alpha^2 + \beta^2 + \gamma^2 = 1.$$

The transverse field at P is obtained by applying the Huygens–Fresnel's formula to each of the components of $\boldsymbol{E}_{\mathrm{T}}$ (M) in the plane xOy (formula 12.35). Therefore:

$$\boldsymbol{E}'_{\mathrm{T}}(\mathrm{P}) = \frac{j}{\lambda} \int_{(\mathrm{S})} \gamma \boldsymbol{E}_{\mathrm{T}}(\mathrm{M}) \frac{\exp\left(-j\dfrac{2\pi r}{\lambda}\right)}{r} \mathrm{d}S. \tag{12.45}$$

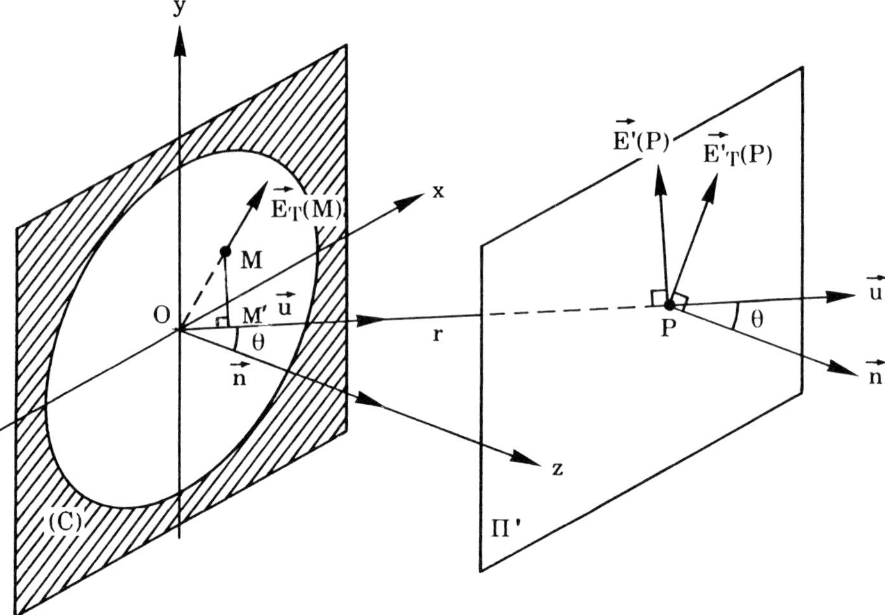

Fig. 12.22 Calculation of the transverse component of $\boldsymbol{E}$-field radiated from an aperture.

Let M′ be the normal projection of M on OP. Since point P is at infinity, we obtain:

$$r = \mathrm{MP} = \mathrm{OP} - \mathrm{OM}' = \mathrm{OP} - \mathbf{OM} \cdot \boldsymbol{u} = r_0 - \mathbf{OM} \cdot \boldsymbol{u}.$$

The attenuation factor proportional to $1/r$ is assimilable to r_0: it is therefore constant and can be removed from the integral. The same applies to factor γ which is constant since the angle $(\boldsymbol{n}, \boldsymbol{u})$ is the same at all points M of the aperture. We therefore have:

$$\boldsymbol{E}'_{\mathrm{T}}(\mathrm{P}) = j \frac{\exp\left(-j2\pi \dfrac{r_0}{\lambda}\right)}{r_0 \lambda} \gamma \int_{(\mathrm{S})} \boldsymbol{E}_{\mathrm{T}}(\mathrm{M}) \exp\left(j2\pi \frac{\mathbf{OM}}{\lambda} \cdot \boldsymbol{u}\right) \mathrm{dS}. \quad (12.46)$$

Characteristic radiation function

It is useful to use a 'reference field' to express the directional properties independently of the antenna's feed power, for example, the maximum field on aperture E_0. We can therefore define the aperture's illumination law by a dimensionless vectorial function:

$$\boldsymbol{f}_{\mathrm{T}}(\mathrm{M}) = \frac{\boldsymbol{E}_{\mathrm{T}}(\mathrm{M})}{E_0}.$$

The transverse field radiated in P is written:

$$\boldsymbol{E}'_{\mathrm{T}}(\mathrm{P}) = j \frac{\exp\left(\dfrac{-j2\pi r_0}{\lambda}\right)}{\left(\dfrac{r_0}{\lambda}\right)} E_0 \gamma \boldsymbol{F}_{\mathrm{T}}(\boldsymbol{u}) \tag{12.47}$$

$\boldsymbol{F}_{\mathrm{T}}(\boldsymbol{u})$ is the antenna's characteristic radiation function:

$$\boldsymbol{F}_{\mathrm{T}}(\boldsymbol{u}) = \frac{1}{\lambda^2} \int f_{\mathrm{T}}(\mathrm{M}) \exp\left(j2\pi \frac{\mathbf{OM}}{\lambda} \cdot \boldsymbol{u} \right) \mathrm{dS}. \tag{12.48}$$

Using the direction cosines of $\boldsymbol{u}$ and the cartesian coordinates (x, y) of M, we will obtain:

$$\boldsymbol{F}_{\mathrm{T}}(\alpha, \beta) = \frac{1}{\lambda^2} \int f_{\mathrm{T}}(x, y) \exp\left\{ j2\pi \left(\frac{x}{\lambda} \alpha + \frac{y}{\lambda} \beta \right) \right\} \mathrm{d}x\, \mathrm{d}y. \tag{12.49}$$

We can simplify this expression by using λ as a unit and by setting:

$$\frac{x}{\lambda} = \nu \qquad \frac{y}{\lambda} = \mu \tag{12.50}$$

whence

$$\boxed{\boldsymbol{F}_{\mathrm{T}}(\alpha, \beta) = \int\int_{(\mathrm{S})} f_{\mathrm{T}}(\nu, \mu) \exp\left(j2\pi (\nu\alpha + \mu\beta) \right) \mathrm{d}\nu\, \mathrm{d}\mu.} \tag{12.51}$$

Whence the basic result that *the characteristic radiation function of an aperture is the bidimensional Fourier transform of the law of illumination*. Likewise, the transverse field is expressed based on (12.47) by setting:

$$r_0/\lambda = \rho$$

$$\boldsymbol{E}_{\mathrm{T}}(\mathrm{P}) = j \frac{\exp(-j2\pi\rho)}{\rho} E_0 \gamma \boldsymbol{F}_{\mathrm{T}}(\alpha, \beta). \tag{12.52}$$

Calculation of the total field in P from its transverse component (Fig. 12.23)

The local structure of the total electric field is that of a plane wave, it is therefore normal to the direction of propagation $\boldsymbol{u}$

$$\boldsymbol{E}'(\mathrm{P}) \cdot \boldsymbol{u} = 0$$

Furthermore, it is the sum of its transverse component $E'_{\mathrm{T}}(\mathrm{P})$, which we have just calculated, and longitudinal component.

$$\boldsymbol{E}'(\mathrm{P}) = \boldsymbol{E}'_{\mathrm{T}}(\mathrm{P}) + E'_{\mathrm{Z}}(\mathrm{P})\, \boldsymbol{n}$$

It follows that:

$$\boldsymbol{E}'_T \cdot \boldsymbol{u} + \boldsymbol{n} \cdot \boldsymbol{u}\, E'_Z = 0.$$

Whence:

$$\boldsymbol{E}'(\mathrm{P}) = \boldsymbol{E}'_T(\mathrm{P}) - \frac{\boldsymbol{E}'_T \cdot \boldsymbol{u}}{\boldsymbol{n} \cdot \boldsymbol{u}}\, \boldsymbol{n} = \frac{\boldsymbol{E}_T(\boldsymbol{n} \cdot \boldsymbol{u}) - \boldsymbol{n}(\boldsymbol{E}'_T \cdot \boldsymbol{u})}{\boldsymbol{n} \cdot \boldsymbol{u}}.$$

We recognize the expression of a double vectorial product in the numerator. The denominator is the direction cosine which is approximately equal to the obliquity factor γ:

$$\gamma = \boldsymbol{n} \cdot \boldsymbol{u}.$$

Therefore:

$$\boldsymbol{E}'(\mathrm{P}) = \frac{1}{\gamma}\,(\boldsymbol{n} \times \boldsymbol{E}'_T(\mathrm{P})) \times \boldsymbol{u}. \tag{12.53}$$

The total field is obtained from the characteristic radiation function through the expression (12.52) of the transverse field:

$$\boldsymbol{E}'(\mathrm{P}) = j\,\frac{\exp(-j2\pi\rho)}{\rho}\, E_0\,[\boldsymbol{n} \times \boldsymbol{F}_T(\alpha,\ \beta)] \times \boldsymbol{u}. \tag{12.54}$$

The characteristic function $\boldsymbol{F}_T(\alpha, \beta)$ is the Fourier transform given in (12.51).

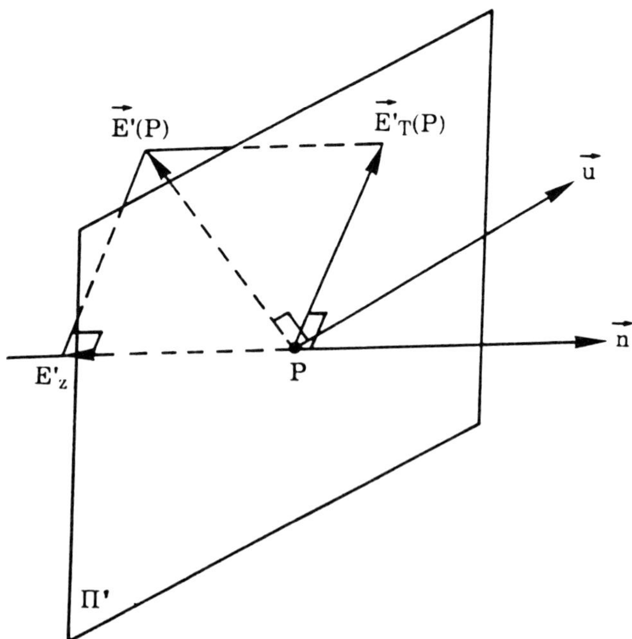

Fig. 12.23 Calculation of total field from its transverse component.

Expression from the magnetic field

We obtain an identical expression for the magnetic field $\boldsymbol{H}'(\mathrm{P})$.

$$\boldsymbol{H}'(\mathrm{P}) = j\,\frac{\exp(-j2\pi\rho)}{\rho}\,H_0\,[\boldsymbol{n}\times\boldsymbol{H}_{\mathrm{T}}(\alpha,\ \beta)]\times\boldsymbol{u}\,. \tag{12.55}$$

Where $\boldsymbol{H}_{\mathrm{T}}(\alpha,\ \beta)$ is the magnetic field law of illumination:

$$\boldsymbol{H}_{\mathrm{T}}(\alpha,\ \beta) = \frac{\boldsymbol{H}_{\mathrm{T}}(\mathrm{M})}{H_0}\,\cdot$$

Since the far field always has a plane wave structure (section 12.2.1), we can directly deduce $\boldsymbol{E}'(\mathrm{P})$ from the expression (12.55), i.e. from the transverse magnetic field on the aperture:

$$\boldsymbol{E}'(\mathrm{P}) = \eta\,\boldsymbol{u}\times\boldsymbol{H}'(\mathrm{P}), \tag{12.56}$$

η being the wave impedance.

12.3.4 Two important examples of basic antennas: dipole and slot

Radiation of a slot (Figs. 12.24 and 12.25)

Let us assume a narrow slot of length L, in a conductive plane xOy. The structure of the field in the neighbourhood of the slot is similar to that of a two-wire line or a slotted line (Vol. I, Ch. 19). In plane xOy, where the slot is located, the electric field is only 'transverse', i.e. contained in this plane, in the region bounded by the slot. Outside, the transverse component is null since the electric field is orthogonal to the conductive walls. For a half-wave slot, the distribution of the field along the slot is practically a stationary wave whose equation is:

$$\boldsymbol{E}(x) = \boldsymbol{e}_y\,E_0\cos 2\pi\frac{x}{\lambda}\,\cdot \tag{12.57}$$

The magnetic field can be deduced, but its distribution law is more complex. We shall therefore calculate the field at a very far distance using only the transverse electric component, given by the preceding equation, and using the formulae (12.50) and (12.54) which yield a radiated field of the form:

$$\boldsymbol{E}'(\boldsymbol{u}) = \boldsymbol{u}\times(\boldsymbol{n}\times\boldsymbol{e}_y)\,E_0\int_{-\lambda/4}^{+\lambda/4}\cos 2\pi\frac{x}{\lambda}\exp\left(j2\pi\frac{x}{\lambda}\cos\theta\right)\mathrm{d}x. \tag{12.58}$$

In this expression we can define $\boldsymbol{u}$ in spherical coordinates where the polar axis is e_x (θ colatitude, ψ longitude). These spherical coordinates have the unit vectors e_θ and e_ψ on a sphere (S). Calculation of the characteristic function thereby yields:

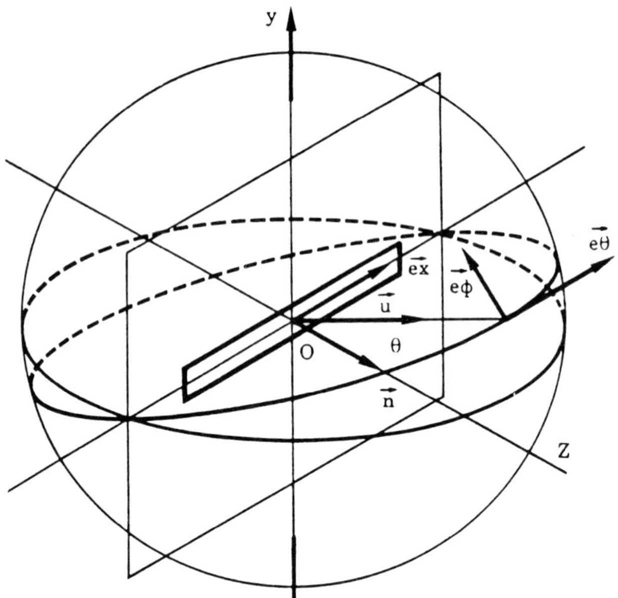

Fig. 12.24 Radiation from a slot in spherical geometry.

$$E'(u) = u \times e_x \frac{\cos\left(\frac{\pi}{2}\cos\theta\right)}{\sin^2\theta} \tag{12.59}$$

that is:

$$E'(u) = u \times e_\psi \frac{\cos\left(\frac{\pi}{2}\cos\theta\right)}{\sin\theta} \tag{12.60}$$

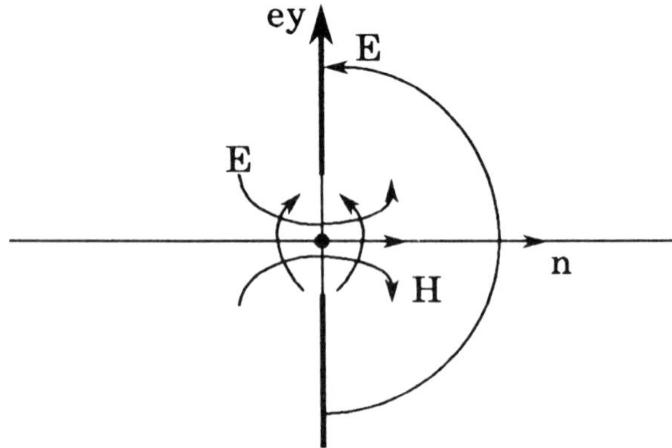

Fig. 12.25 Electromagnetic field configurations in Fig. 12.24.

We obtain a semi-toroidal characteristic radiation surface: the radiation in the axis of this slot is null. The polarization of the electric field, $\boldsymbol{e}_\psi$ is transversal with respect to the axis of the slot (Fig. 12.26).

Radiation of a dipole

The dipole is formed by a conductive bar or a metal strip cut in the middle, and fed through a line, for example a two-wire line (Fig. 12.27). A large number of dipole radiation theories have already been postulated (King, 1956). They are all based on the solution of an integral equation and are shown in electromagnetism courses. Here we will present an approximate theory based on the concept of an equivalent aperture.

Consider the complementary dipole of the preceding slot, i.e. the dipole obtained by removing the metal and metallizing the slot. This narrow dipole is assumed to be excited at its centre through a narrow cut. Let us assume that a purely longitudinal stationary current crosses it:

$$\boldsymbol{I}(x) = \boldsymbol{e}_x I_0 \cos 2\pi \frac{x}{\lambda} \cdot \tag{12.61}$$

The result from the conditions at the boundaries is that, on the dipole, the magnetic field in the plane xOy is purely transversal (contained in the plane xOy). Outside the dipole, this magnetic field does not have components in xOy (Fig. 12.28). The electric field can be deduced from it: on the dipole its

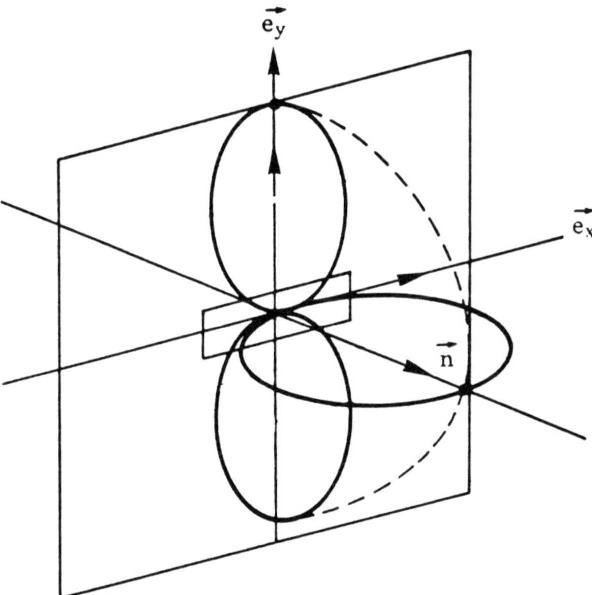

Fig. 12.26 Radiation from a slot — characteristic surface.

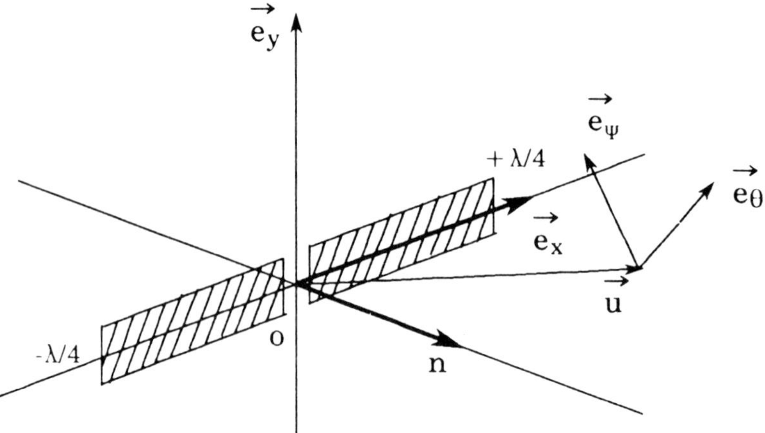

Fig. 12.27 Radiation from a dipole.

transverse component is null since it is perpendicular to the dipole, however, it has a more complex distribution. Therefore we shall deduce the field radiated at a large distance from the transverse magnetic field only:

$$\boldsymbol{H}(x) = \boldsymbol{e}_y H_0 \cos 2\pi \frac{x}{\lambda} \tag{12.62}$$

using formula (12.55), H_T being calculated from $H(x)$ by the Fourier Transform. We then obtain the electric field using (12.56):

$$\boldsymbol{E}'(\boldsymbol{u}) = \boldsymbol{u} \times \int_{-\lambda/4}^{+\lambda/4} (\boldsymbol{n} \times \boldsymbol{e}_y) \times \boldsymbol{u} H_0 \cos 2\pi \frac{x}{\lambda} \exp\left(-j2\pi \frac{x}{\lambda} \cos\theta\right) \mathrm{d}x$$

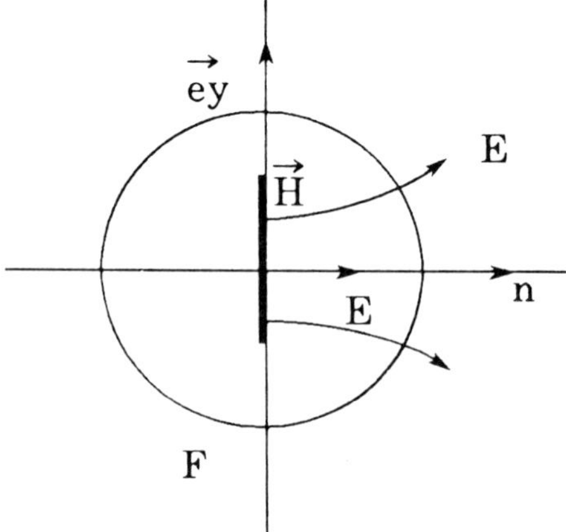

Fig. 12.28 Electromagnetic field configurations in Fig. 12.27.

$$\boldsymbol{E}'(\boldsymbol{u}) = \boldsymbol{u} \times (\boldsymbol{e}_x \times \boldsymbol{u}) H_0 \int_{-\lambda/4}^{+\lambda/4} \cos 2\pi \frac{x}{\lambda} \exp\left(j2\pi \frac{x}{\lambda} \cos\theta \right) \mathrm{d}x. \quad (12.63)$$

Which gives:

$$\boldsymbol{E}'(\boldsymbol{u}) = \boldsymbol{e}_\theta \frac{\cos\left(\frac{\pi}{2}\cos\theta\right)}{\sin\theta}. \quad (12.64)$$

Conclusion: Babinet's theorem

We can see that this expression is identical with that of the slot (equation 12.60), except that here the electric field is polarized in the plane of the dipole — which is obvious because of symmetry. The characteristic function of the dipole is therefore the same as that of the complementary slot, provided the electric field and the magnetic field are permutated. This result is general whatever the shape of the slot and the corresponding dipole: this is an important aspect of Babinet's theorem. We can also demonstrate that the product of the radiation resistances of the dipole (Z) and of the slot (Z′) is constant (Roubine *et al.*, 1987):

$$ZZ' = \left(\frac{\eta_0}{2}\right)^2 \simeq 3600\,\pi^2. \quad (12.65)$$

12.3.5 Examples of radiating apertures

Rectangular aperture

Radiation of a rectangular aperture whose law of illumination is separable (Fig. 12.29) Let D and D' be the aperture dimensions. In this case the transverse field at a point M on the aperture is the product of two functions with reduced coordinates:

$$\nu = \frac{x}{\lambda} \qquad \mu = \frac{y}{\lambda}.$$

If the polarization is fixed (for example vertically) the field is described by a simple scalar function:

$$E_T(\nu, \mu) = E_1(\nu)\, E_2(\mu). \quad (12.66)$$

Let:

$$\nu_0 = \frac{D}{2\lambda} \qquad \mu_0 = \frac{D'}{2\lambda}. \quad (12.67)$$

The integral (12.51) can then be described in the form of the product of two integrals:

$$F(\alpha,\ \beta) = F_1(\alpha)\, F_2(\beta) \tag{12.68}$$

$$F_1(\alpha) = \int_{-\nu_0}^{+\nu_0} E_1(\nu)\exp(j2\pi\alpha\nu)\,\mathrm{d}\gamma.$$

$$F_2(\beta) = \int_{-\mu_0}^{+\mu_0} E_2(\mu)\exp(j2\pi\beta\mu)\,\mathrm{d}\gamma. \tag{12.69}$$

Let us consider, for example, the first integral; if we consider ν as a frequency, function $F_1(\alpha)$ behaves as a temporal signal of which $E_1(\nu)$ is the spectrum. This spectrum is bounded by dimension D of the aperture; we can see that ν_0 behaves as the cut-off frequency. This is why variables ν and μ, defined in this way, are often called spatial frequencies. We will develop these analogies with the theory of bounded-spectrum signal in the last part (section 12.6.2).

Exercise: case of a narrow slot. In this case, for example, $\mu_0 = D'/2\lambda$ is very small. If we set $E_2(\mu) = E_0$, we obtain:

$$F_2(\beta) = 2\,\mu_0\, E_0 = \text{constant}.$$

Radiation is omnidirectional around the slot:

$$F(\alpha,\ \beta) = 2\,\mu_0\, E_0\, F_1(\alpha). \tag{12.70}$$

Let us calculate the characteristic function $F_1(\alpha)$ associated with a law of constant illumination:

$$E_1(\nu) = E_0.$$

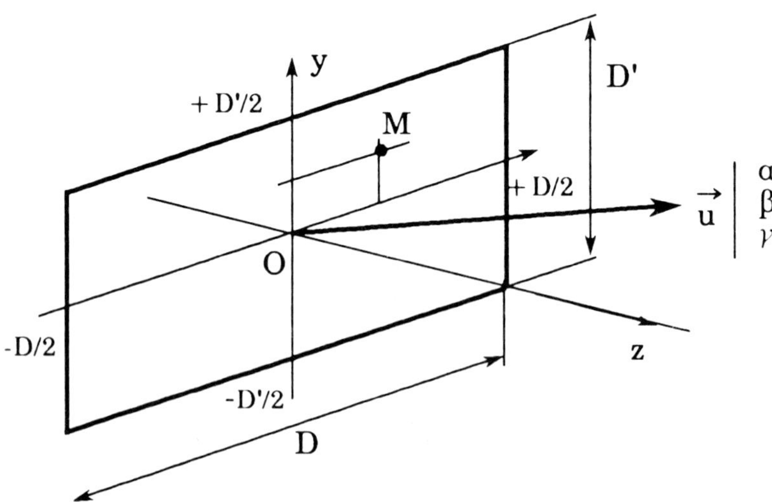

Fig. 12.29 Radiation from a rectangular aperture.

We find the typical result:

$$F_1(\alpha) = 2\gamma_0 E_0 \frac{\sin 2\pi \nu_0 \alpha}{2\pi \nu_0 \alpha} = \frac{D}{\lambda} E_0 \frac{\sin \pi \frac{D}{\lambda} \alpha}{\pi \frac{D}{\lambda} \alpha}. \tag{12.71}$$

This function is shown in Fig. 12.30. Radiation is maximum in a normal direction ($\alpha = 0$). The angular half-width between the nearest zeros is:

$$\boxed{\Delta\alpha = 1/2\nu_0 = \lambda/D} \tag{12.72}$$

This is approximately the 3 dB beamwidth: i.e. the angular width where the amplitude is reduced by 3 dB.

This result is general: the angular aperture of the radiated beam is inversely proportional to aperture D measured in wavelengths.

Circular aperture

This problem occurs in numerous practical cases: circular horns, parabolic reflectors, Cassegrain antennas, etc. Given a circular aperture of diameter D (Fig. 12.31). Consider a point M of the aperture with polar coordinates r, ϕ. Let us assume constant polarization in the aperture; a scalar function $f(r, \phi)$ is sufficient for describing the illumination law. Let $\boldsymbol{u}$ be a direction with spherical coordinates (θ, ψ). The characteristic function of radiation is the Fourier transform of illumination f. Using (12.47), it can be written as:

$$\boldsymbol{F}(\boldsymbol{u}) = \frac{1}{\lambda^2} \int_0^{D/2} \int_0^{2\pi} f(r, \phi) \exp\left(j2\pi \frac{\mathbf{OM}}{\lambda} \cdot \boldsymbol{u} \right) \mathrm{d}S. \tag{12.73}$$

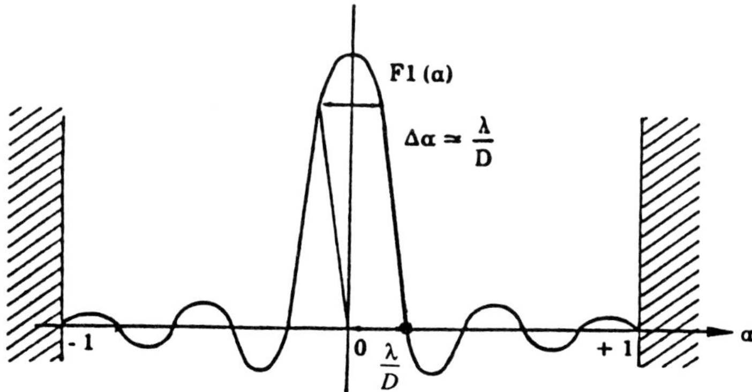

Fig. 12.30 Constant illumination of a rectangular aperture gives the characteristic function $F_1(\alpha)$.

Or as

$$F(\theta,\psi)=\frac{1}{\lambda^2}\int_0^{D/2}\int_0^{2\pi} f(r,\phi)\exp\left(j2\pi\frac{r}{\lambda}\sin\theta\cos(\phi-\psi)\right) r\,\mathrm{d}r\,\mathrm{d}\phi. \tag{12.74}$$

Case of an illumination with circular symmetry

The function $f(r, \phi)$ is independent of ϕ and takes the form $f(r)$. For a parabolic reflector, for example, this type of radiation can be obtained through a primary Huygens source which also has a circular symmetry, for example a corrugated horn (see section 12.5.2). The characteristic function only depends on θ:

$$F(\theta)=\frac{1}{\lambda^2}\int_0^{D/2} f(r)\left[\int_0^{2\pi}\exp\left(j2\pi\frac{r}{\lambda}\sin\theta\cos(\phi-\psi)\right)\mathrm{d}\phi\right] r\,\mathrm{d}r.$$

The integral between brackets is independent of ψ. It represents the contribution to the radiation of a ring in the aperture included between rays r and $r+\mathrm{d}r$. It is calculated using a zero-order Bessel function:

$$\int_0^{2\pi}\exp\left(j2\pi\frac{r}{\lambda}\sin\theta\cos(\phi-\psi)\right)\mathrm{d}\phi = 2\pi j J_0\left(\frac{2\pi r}{\lambda}\sin\theta\right).$$

Which reduces to the calculation of a simple integral:

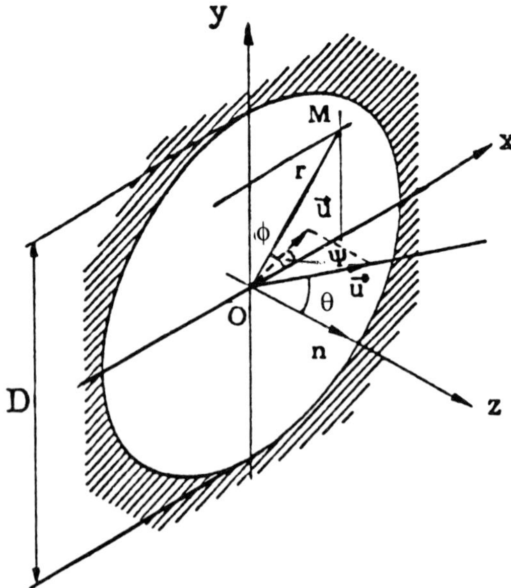

Fig. 12.31 Radiation from a circular aperture.

$$F(\theta) = j\frac{2\pi}{\lambda^2}\int_0^{D/2} f(r)\, J_0\left(\frac{2\pi r}{\lambda}\sin\theta\right) r\,dr. \tag{12.75}$$

Case of a constant illumination $f_0(r) = 1$.

We will see in section 12.4.1 that it is this illumination which produces maximum aperture gain in its axial direction. The characteristic function is written:

$$F_0(\theta) = \frac{2\pi j}{\lambda^2}\int_0^{D/2} J_0\left(\frac{2\pi r}{\lambda}\sin\theta\right) r\,dr.$$

It is expressed by a first-order Bessel function:

$$F_0(\theta) = j\frac{\pi D^2}{4\lambda^2}\,\frac{2J_1\left(\frac{\pi D}{\lambda}\sin\theta\right)}{\frac{\pi D}{\lambda}\sin\theta}. \tag{12.76}$$

We can normalize this expression by its value in the axial direction. We obtain the well known result:

$$\boxed{H_0(\theta) = \frac{F_0(\theta)}{F_0(0)} = \frac{2J_1\left(\frac{\pi D}{\lambda}\sin\theta\right)}{\frac{\pi D}{\lambda}\sin\theta}} \tag{12.77}$$

which is often written as:

$$H_0(\theta) = \Lambda_1(v); \qquad \left(v = \frac{\pi D}{\lambda}\sin\theta\right)$$

The characteristic surface of radiation has circular symmetry. This surface can be represented by a meridian cross-section, namely the aperture radiation pattern (Fig. 32, plot Λ_1).
Keep in mind the following characteristics:

1. aperture at 3 dB (radian):

$$\theta 3\ \mathrm{dB} = 1.03\frac{\lambda}{D} \tag{12.78}$$

2. level of 1st sidelobe: the first ring of sidelobes around the main lobe has a relative level of 0.132, i.e. $N_{\mathrm{dB}} = -17.6$ dB

Case of a weighted illumination

Constant illumination yields maximum axial gain but with the drawback of relatively high sidelobes. To decrease these lobes, weighted illumination is often used, i.e. decreasing towards the edges of the aperture. This type of illumination can be represented by a Bessel function, a Gaussian law (section 12.5.2) or simply by a parabolic law (Fig. 12.33):

$$f_1(r) = 1 - \left(\frac{2r}{D}\right)^2. \tag{12.79}$$

In the latter case, the characteristic function is written:

$$H_1(\theta) = 8\,\frac{J_2\left(\dfrac{\pi D}{\alpha}\sin\theta\right)}{\left(\dfrac{\pi D}{\lambda}\sin\theta\right)^2} \tag{12.80}$$

which reduces to:

$$H_1(\theta) = \Lambda_2(v).$$

With J_2 a second-order Bessel function, the primary sidelobe is lowered (– 24.6 dB) at the price of enlarging its width at 3 dB ($\theta'_{3dB} = 1.27\,\lambda/D$ radians) and reduction in gain (plot Λ_2, Fig. 12.32).

Note that we can combine constant illumination and parabolic illumination yielding a certain relative level on the edges, this is often the case for a parabolic reflector illuminated by a primary feed (section 12.5.3). In this case, the pattern results from the combinations of functions H_0 and H_1 with the adequate weighting. In general, illumination is of the form:

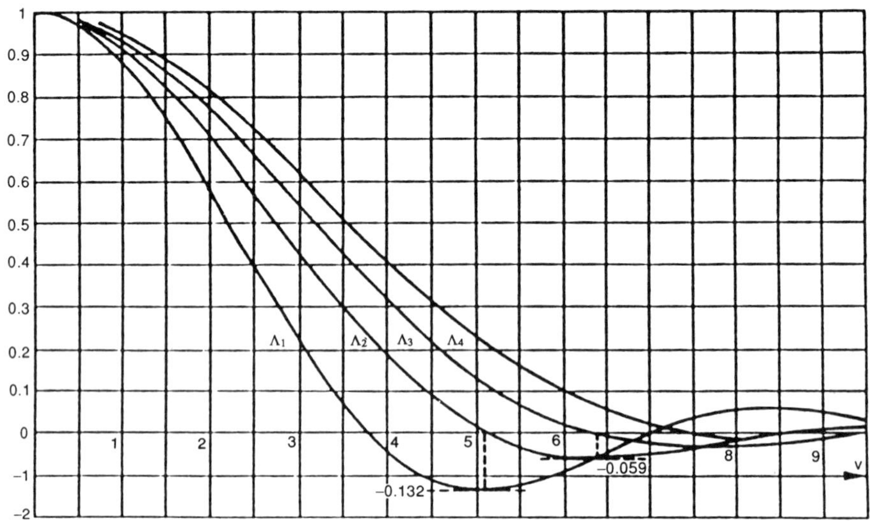

Fig. 12.32 Aperture radiation pattern.

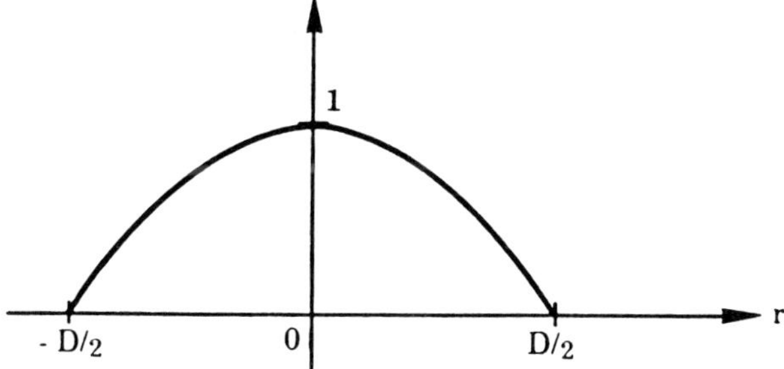

Fig. 12.33 Parabolic law for weighted illumination.

$$f_n(r) = \left[1 - \left(\frac{2r}{D}\right)^2\right]^n.$$

Yielding curves Λ_{n+1} in Fig. 12.32.

12.3.6 Case where the field on the aperture has a plane wave structure locally

Reference polarization (Ludwig, 1973)

In all cases, it is possible to express the radiated field from the transverse field which is both electric and magnetic; by taking, for example the half-sum of expressions (12.53) and (12.56) and taking into account (12.55). If we expand the Fourier transforms we obtain:

$$\boldsymbol{E}(\boldsymbol{u}) = -j \frac{\exp\left(-j2\pi \frac{R}{\lambda}\right)}{2\lambda R} \boldsymbol{u}$$

$$\times \int [\boldsymbol{n} \times \boldsymbol{E}_{\mathrm{T}} + \eta(\boldsymbol{n} \times \boldsymbol{H}_{\mathrm{T}}) \times \boldsymbol{u}] \exp\left(j2\pi \frac{\mathbf{OM}}{\lambda} \cdot \boldsymbol{u}\right) \mathrm{d}S. \qquad (12.81)$$

This expression is known as the Kottler formula. However, it should be kept in mind that we cannot specify at the same time independently and *a priori* fields $\boldsymbol{E}_{\mathrm{T}}$ and $\boldsymbol{H}_{\mathrm{T}}$, one of the two determines the other. However, this formula is often convenient, especially when these components, at least locally, have a plane wave structure. In this case, relation (12.14) is written:

$$\eta\, \boldsymbol{n} \times \boldsymbol{H}_{\mathrm{T}} = -\boldsymbol{E}_{\mathrm{T}}. \qquad (12.82)$$

And then:

$$\boldsymbol{E}(\boldsymbol{u}) = j\frac{\exp\left(-j2\pi\dfrac{R}{\lambda}\right)}{2\lambda R}\,\boldsymbol{u}\times\left[(\boldsymbol{n}+\boldsymbol{u})\times\int \boldsymbol{E}_{\mathrm{T}}(\mathrm{M})\,\exp\left(j2\pi\frac{\mathbf{OM}}{\lambda}\cdot\boldsymbol{u}\right)\mathrm{d}S\right]. \tag{12.83}$$

The field now has a *characteristic polarization which will be used as a reference for all the other cases*. Assume that the aperture is excited in horizontal polarization:

$$\boldsymbol{E}_{\mathrm{T}}(\mathrm{M}) = E_{\mathrm{T}}(\mathrm{M})\,\boldsymbol{e}_x \tag{12.84}$$

($\boldsymbol{e}_x$, $\boldsymbol{e}_y$, and $\boldsymbol{e}_z$) are unit vectors of the system of axes forming the frame of reference (Fig. 12.34). This yields:

$$\boldsymbol{E}(\boldsymbol{u}) = \boldsymbol{p}j\frac{\exp\left(-j2\pi\dfrac{R}{\lambda}\right)}{2\,\lambda R}\int E_{\mathrm{T}}(\mathrm{M})\,\exp\left(j2\pi\frac{\mathbf{OM}}{\lambda}\cdot\boldsymbol{u}\right)\mathrm{d}S \tag{12.85}$$

where $\boldsymbol{p}$ defines the polarization:

$$\boldsymbol{p} = \boldsymbol{u}\times[(\boldsymbol{n}+\boldsymbol{u})\times\boldsymbol{e}_x] \tag{12.86}$$

or, according to the double vectorial product:

$$\boldsymbol{p} = [(\boldsymbol{n}+\boldsymbol{u})\cdot\boldsymbol{u}]\boldsymbol{e}_x - (\boldsymbol{e}_x\cdot\boldsymbol{u})(\boldsymbol{n}+\boldsymbol{u}) \tag{12.87}$$

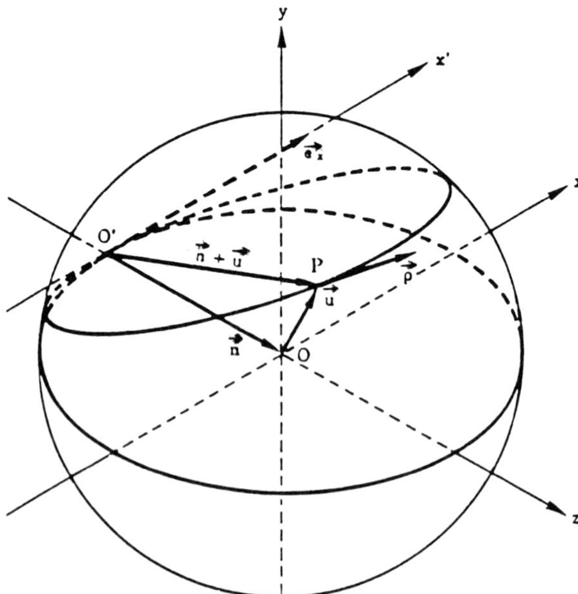

Fig. 12.34 Geometrical interpretation of reference polarization.

or by using the direction cosines of $\boldsymbol{u}$:

$$\boldsymbol{p} = \boldsymbol{e}_x (1 + \gamma) - (\boldsymbol{n} + \boldsymbol{u})\,\alpha. \tag{12.88}$$

Geometrical interpretation

Consider a sphere (S) with centre O, with a large radius we will take as the unit (Fig. 12.34). Let O′ be a point of intersection of this sphere with axis Oz. This gives us:

$$\mathbf{O'O} = \boldsymbol{n}.$$

Let O′x' be an axis parallel to Ox, hence it also carries the unit vector: $\boldsymbol{e}_x$. Let P be the point of the sphere such that:

$$\mathbf{OP} = \boldsymbol{u}$$

We can see that:

$$\mathbf{O'P} = \boldsymbol{n} + \boldsymbol{u}.$$

The above formula (12.88) can then be written:

$$\boldsymbol{p} = \boldsymbol{e}_x (1 + \gamma) - \mathbf{O'P}\,\alpha. \tag{12.89}$$

It shows that polarization $\boldsymbol{p}$ is in the plane ($\mathbf{O'P}$, $\boldsymbol{e}_x$), i.e. the plane passing through axis O′x' and point P.

So the lines of polarization of the electric field on the sphere (S) are the intersections of this sphere with the beam of planes passing through axis O′x' tangent to this sphere.

Likewise, the magnetic field lines are the circles of intersection of (S) with the beam of planes passing through axis O′y' tangent to the sphere, parallel to Oy.

Huygens sources

All antennas whose polarization possesses the above properties are called Huygens sources. Most antennas are not Huygens sources, but certain antennas approach them, this applies to bimode horns and corrugated horns (section 12.5.2). Field lines thus defined form a system of rectangular coordinates which allow definition of any antenna's polarization by its components at any point P of sphere (S). Let us define the unit vectors of these coordinates, which we will call Huygens coordinates for simplicity (Fig. 12.35).

Unit vectors of the Huygens coordinates

We can easily calculate the norm of vector $\boldsymbol{p}$, and then deduce the unit vector of polarization $\boldsymbol{e}_\alpha$:

$$\boldsymbol{e}_\alpha = \frac{1}{1+\gamma}\boldsymbol{p} = \boldsymbol{e}_x - \frac{\boldsymbol{n}+\boldsymbol{u}}{1+\gamma}\,\alpha \tag{12.90}$$

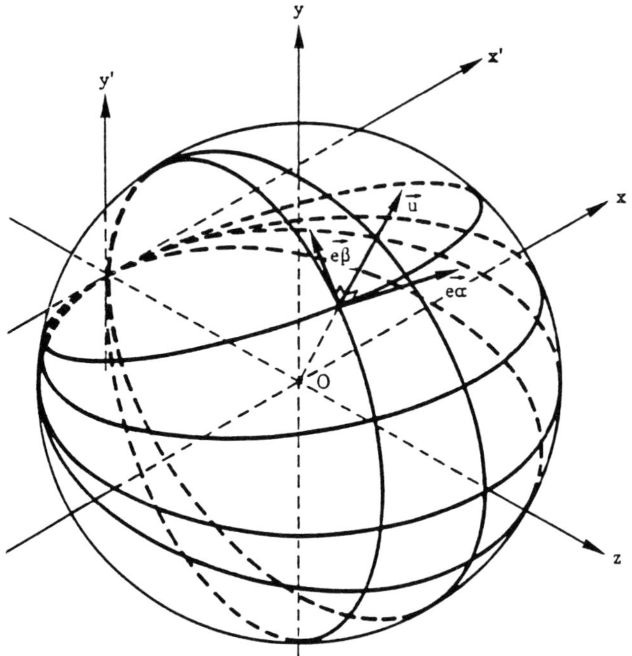

Fig. 12.35 Huygens coordinates.

$\boldsymbol{p}$ being given by (12.88). The field $\boldsymbol{E}(\boldsymbol{u})$ can be written from (12.85):

$$\boldsymbol{E}(\boldsymbol{u}) = \boldsymbol{e}_\alpha j \frac{\exp\left(-j\dfrac{2\pi R}{\lambda}\right)}{\lambda R}\left(\frac{1+\gamma}{2}\right)\int E_{\mathrm{T}}(\mathrm{M})\ \exp\left(j2\pi\frac{\mathbf{OM}}{\lambda}\cdot\boldsymbol{u}\right)\mathrm{d}S. \tag{12.91}$$

Likewise we define the unit vector $\boldsymbol{e}_\beta$ of the lines of the magnetic field from comparable formulae:

$$\boldsymbol{q} = \boldsymbol{e}_y\,(1+\gamma) - (\boldsymbol{n}+\boldsymbol{u})\,\beta$$

$$\boldsymbol{e}_\beta = \frac{\boldsymbol{q}}{1+\gamma} = \boldsymbol{e}_y - \frac{\boldsymbol{n}+\boldsymbol{u}}{1+\gamma}\,\beta. \tag{12.92}$$

12.3.7 Geometrical properties of Huygens coordinates

Properties of axial symmetry

Consider a Huygens source, the following property plays an important part in antennas which have axial symmetry. On sphere (S), the magnetic field lines can be deduced globally from the electric field lines by a rotation of $\pi/2$ radians. The lines of the electric and magnetic fields are therefore orthogonal both locally and globally.

Relations with spherical coordinates

Consider the beam of planes cutting sphere (S) along the Huygens lines (Fig. 12.36). Consider the effect of a reversal, with pole O′ and any power: it transforms a sphere (S) into plane π normal to $\boldsymbol{n}$. The beam of planes cut on π grid lines which form a system of cartesian coordinates. Any point P′ on this plane can be located either by this system of coordinates, with unit vectors $\boldsymbol{e}_x$, $\boldsymbol{e}_y$, or in a system of polar coordinates with unit vectors $\boldsymbol{e}_\alpha$, $\boldsymbol{e}_r$. We know that these two systems of unit vectors can be deduced from each other through a simple rotation of angle ϕ.

$$\boldsymbol{e}_x = \boldsymbol{e}_r \cos\phi - \boldsymbol{e}_\phi \sin\phi$$
$$\boldsymbol{e}_y = \boldsymbol{e}_r \sin\phi + \boldsymbol{e}_\phi \cos\phi. \tag{12.93}$$

If we go back to sphere S, we can see that in a similar matter we can locate point P either using the Huygens coordinates which we have defined (with unit vectors e_α, e_β) or by spherical coordinates with unit vectors e_θ, e_ϕ. These spherical coordinates can be deduced clearly by inversion of the polar coordinates of P′. We therefore find the same transformation relation between spherical coordinates and Huygens coordinates that we saw between polar and cartesian coordinates. The Huygens unit vectors are deduced from the unit vectors of spherical coordinates ($\boldsymbol{e}_\theta$, $\boldsymbol{e}_\phi$) through a simple rotation of polar angle ϕ (Fig. 12.37):

$$\boldsymbol{e}_\alpha = \boldsymbol{e}_\theta \cos\phi - \boldsymbol{e}_\phi \sin\phi$$
$$\boldsymbol{e}_\beta = \boldsymbol{e}_\theta \sin\phi + \boldsymbol{e}_\phi \cos\phi. \tag{12.94}$$

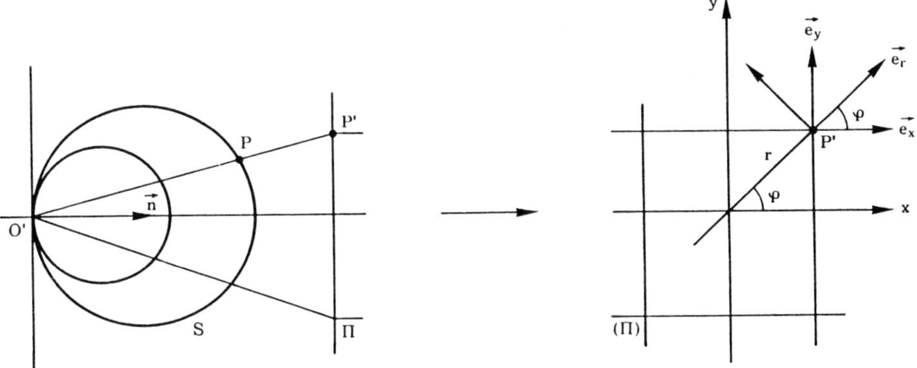

Fig. 12.36 Plane intersecting a sphere along the Huygens lines and transformation of the sphere onto a plane.

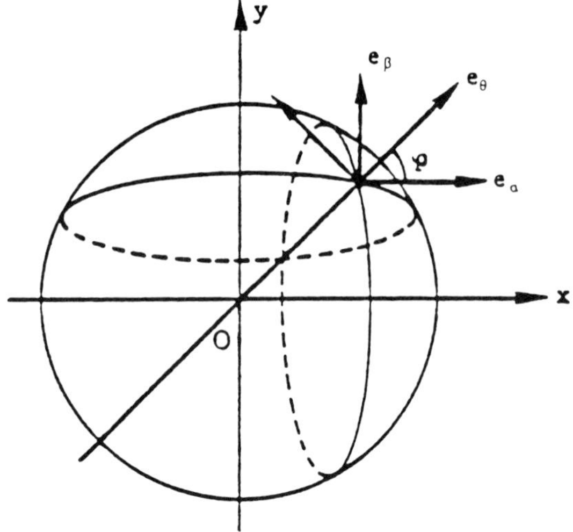

Fig. 12.37 Polarization components of a field.

Polarization components of a field

We can see that a field of polarization $\boldsymbol{p}$ can be expressed either by its spherical components or its Huygens components in the following form:

$$\boldsymbol{p} = p_\theta \boldsymbol{e}_\theta + p_\phi \boldsymbol{e}_\phi$$

$$\boldsymbol{p} = p_\alpha \boldsymbol{e}_\alpha + p_\beta \boldsymbol{e}_\beta. \tag{12.95}$$

These components can be deduced from each other using a simple rotation matrix.

$$\begin{pmatrix} p_\theta \\ p_\phi \end{pmatrix} = \begin{pmatrix} \cos\phi & \sin\phi \\ -\sin\phi & \cos\phi \end{pmatrix} \begin{pmatrix} p_\alpha \\ p_\beta \end{pmatrix}. \tag{12.96}$$

12.3.8 Near-field aperture radiation

In this section and to simplify we will limit ourselves to the scalar aspect of radiation, polarization being assumed as known. Consider a circular aperture of diameter D, source of a nearly constant equiphase illumination (Fig. 12.38). A first approximation yields three main zones of radiation of an equiphase aperture.

The far field zone

The far field zone is the Fraunhoffer region that we have just examined. We shall assume its occurrence for distances greater than:

$$Z_3 \geqslant 2\frac{D^2}{\lambda}. \tag{12.97}$$

Example: $\lambda = 10\,\text{cm}$, $D = 10\,\text{m}$, $Z_3 \geqslant 2\,\text{km}$. This condition requires installation of test towers at large distances to measure antennas under a far-field condition.

A very near field zone

A very near field zone, or Rayleigh region exists for distances less than:

$$Z_1 \leqslant \frac{D^2}{2\lambda}. \tag{12.98}$$

In the above example, this zone reaches 500 m. At this distance the energy density is comparable to that on the reflector. In this zone the field remains more or less within a radiation tube of a diameter equal to that of the aperture. This roughly is a field of application of a geometrical optics (section 12.2.2).

An intermediate zone

An intermediate zone, between the two preceding limits, is often called the Fresnel zone. This is the field 'rearrangement' zone between the illumination law on the aperture and the pattern at a large distance.

To calculate the field at a point P of this zone (Fig. 12.22), we can use the Fresnel – Huygens principle (12.35), without using the approximation of fields at a large distance. We can however do a quadratic approximation of the phase factor. Let us consider the vector **MP**. We can write:

$$\mathbf{MP} = \mathbf{OP} - \mathbf{OM}$$

$$|\mathbf{MP}|^2 = r^2 + (x^2 + y^2) - 2\,\mathbf{OM} \cdot \mathbf{OP}$$

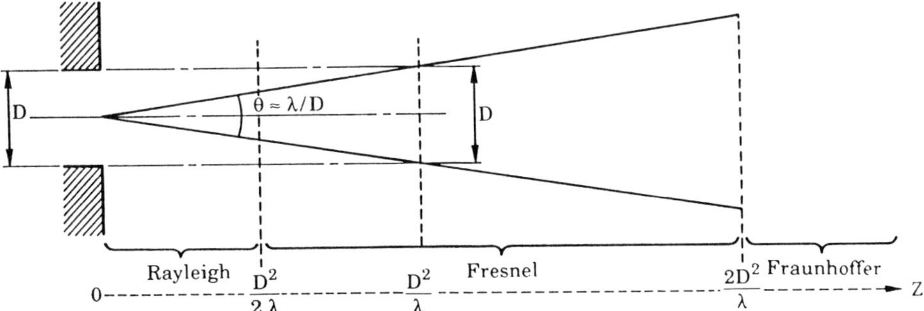

Fig. 12.38 Circular radiating aperture showing near field and far field zones.

$$|\mathbf{MP}|^2 = r^2\left[\frac{1+x^2+y^2}{r^2} - 2\,\frac{\mathbf{OM}\cdot\mathbf{OP}}{r^2}\right] \qquad \mathbf{MP} \approx r + \frac{x^2+y^2}{2r} - \mathbf{OM}\cdot\boldsymbol{u}.$$

If $f(x, y)$ is the illumination at point M of the aperture, a scalar component of the field in P is:

$$E_P(\alpha, \beta, r) = \frac{j}{\lambda}\,\frac{\exp\left(-j\,\frac{2\pi r}{\lambda}\right)}{r}$$

$$\int_{(S)} f(x, y)\exp\left\{j\,\frac{2\pi}{\lambda}\left(\frac{x^2+y^2}{2r}\right)\right\}\exp\left\{-j\,2\pi\left(\alpha\,\frac{x}{\lambda}+\beta\,\frac{y}{r}\right)\right\}\mathrm{d}x\,\mathrm{d}y. \quad (12.99)$$

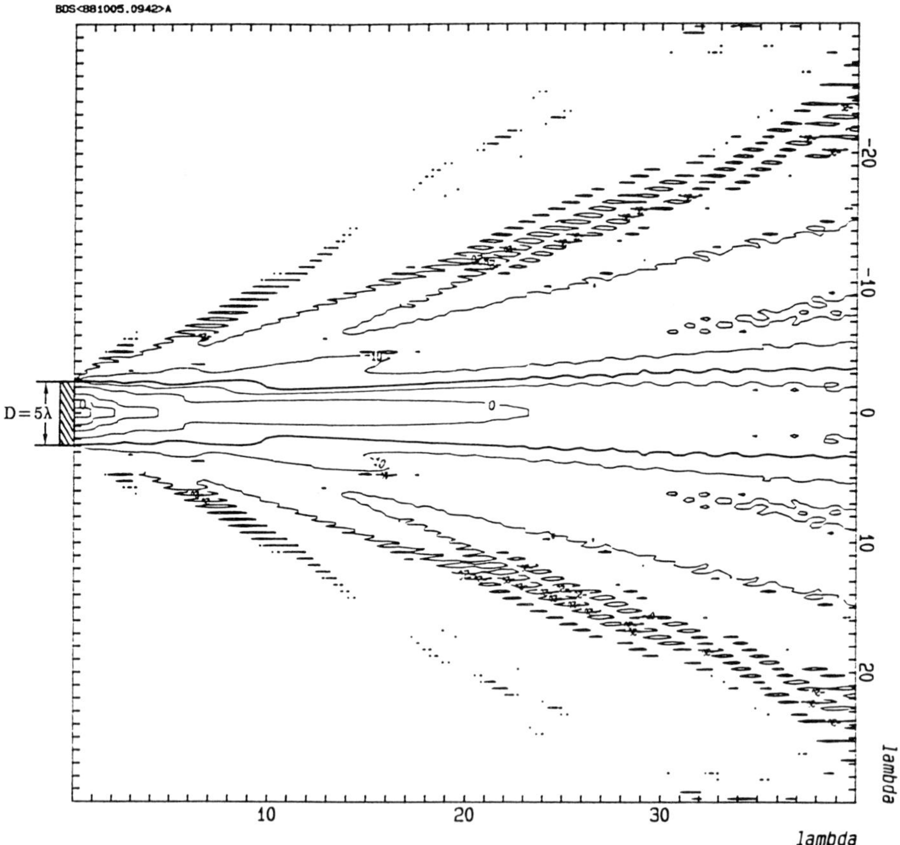

Fig. 12.39 Near field distribution for an aperture with $D = 5\,\lambda$ (lines of equal amplitudes).

Here again, the field is the Fourier transform of the illumination law, provided it is assigned a quadratic phase factor. This factor disappears if distance is increased again which will then yield the classic result of a far away field. Figure 12.39 shows the distribution of the nearby field of an aperture with the dimension $D = 5\ \lambda$. The figure shows the isolevel lines at 0 dB, – 3 dB, – 5 dB, – 10 dB, – 20 dB.

12.4 BASIC ANTENNA CONCEPTS

The examples of antenna which we have considered and the radiation properties of equivalent apertures now allow us to understand the meaning and scope of the basic concepts we are now going to introduce. These concepts generally apply both to reception and transmission antennas, but it is preferable to consider these two separately to facilitate understanding.

12.4.1 Transmitting antennas

Radiation characteristic function

We have already defined the characteristic function of a radiating aperture. In general, the antenna is identified by a system of any coordinates (Oxyz) (Fig. 12.40), and it can be shown that the field radiated at a large distance (r) in a direction with unit vector $\boldsymbol{u}$ is a spherical wave with a local plane wave electromagnetic structure.

If we consider the electric field, the expression will be:

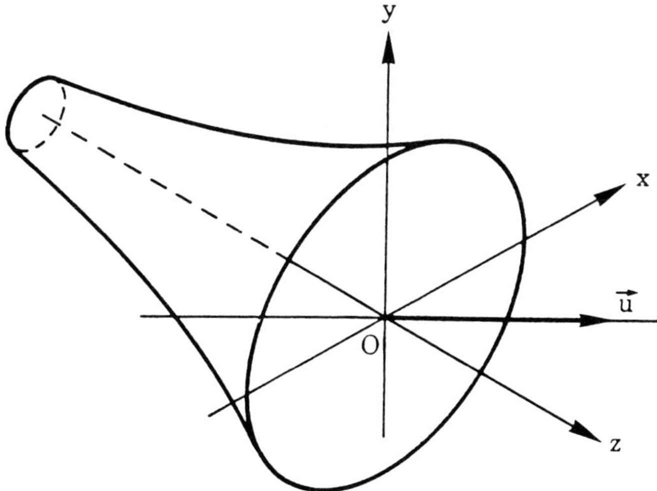

Fig. 12.40 System of coordinates for an antenna.

$$\boldsymbol{E}(\boldsymbol{u}, r) = E_0 \frac{\exp\left(-j\dfrac{2\pi r}{\lambda}\right)}{\left(\dfrac{r}{\lambda}\right)} \boldsymbol{F}(\boldsymbol{u}) . \qquad (12.100)$$

In this expression, E_0 depends only on the power fed to the antenna and $\boldsymbol{F}(\boldsymbol{u})$ depends only on the direction $\boldsymbol{u}$ of the point of observation. This is a complex vectorial function which characterizes the relative amplitude, the relative phase and the polarization of the far field in the direction $\boldsymbol{u}$.
We can therefore set:

$$\boldsymbol{F}(\boldsymbol{u}) = \boldsymbol{p}(\boldsymbol{u}) \, |\boldsymbol{F}(\boldsymbol{u})| \, \exp(j\Phi(\boldsymbol{u})) \qquad (12.101)$$

$\boldsymbol{p}$ is the unit polarization vector normal to $\boldsymbol{u}$. $|F|$ represents the amplitude and Φ the phase. Let us consider each of these last two factors separately.

The phase law $\Phi(\boldsymbol{u})$

Isophase surfaces — phase centre
If the phase is independent of $\boldsymbol{u}$, these surfaces or wavefronts are spheres whose centre is the 'antenna's phase centre'. In general, isophase surfaces are defined by the condition:

$$\Phi(\boldsymbol{u}) - 2\pi \frac{r}{\lambda} = \text{constant}. \qquad (12.102)$$

This associates a vector $r\boldsymbol{u}$ with any direction $\boldsymbol{u}$. The extremity of this vector describes an isophase surface. If the envelope of the normals to this surface is a point, this point is the phase centre. The antenna is said to be stigmatic.

The astigmatism concept. Example (Fig. 12.41)

Let us consider a wide-angled but very flat horn with its apex at C and a virtually linear aperture AB. It can be seen that the wave surface is toroidal: a section in a plane of symmetry perpendicular to AB gives a phase centre at O, i.e. the centre of the aperture, while a section in plane CAB gives a phase centre at C. Distance CO is the antenna astigmatic distance.

Amplitude law $|\boldsymbol{F}(\boldsymbol{u})|$ *— radiation characteristic surface — radiation patterns*
The characteristic radiation surface is the geometrical position of the tip Q of the vector:

$$\mathbf{OQ} = \boldsymbol{u} \, |\boldsymbol{F}(\boldsymbol{u})| . \qquad (12.103)$$

This surface is used to display the maximum or secondary radiation zones (Fig. 12.42). In general, only a few plane sections of this surface, known as radiation patterns, are represented.

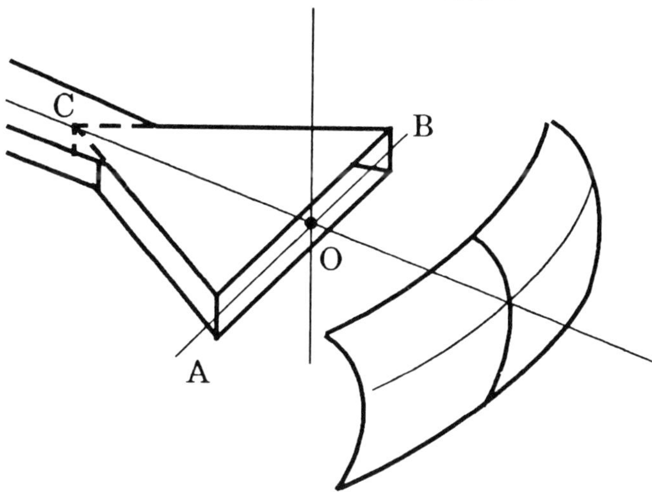

Fig. 12.41 Illustration of the astigmatism concept.

If, for example, the surface is symmetrical about an axis Oz, it is fully represented by its meridian. This type of diagram can be drawn in cartesian coordinates with an arbitrarily selected angular scale (Fig. 12.43), with an abscissa proportional to the angular parameter:

$$\tau = \sin \theta.$$

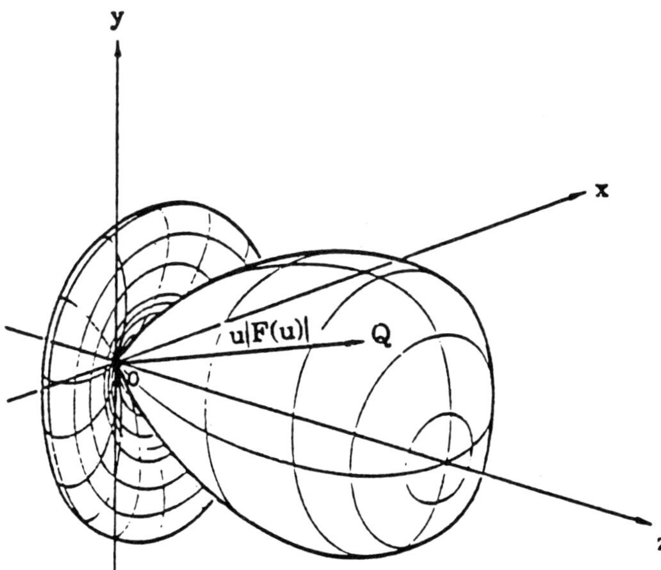

Fig. 12.42 Example of a characteristic radiation surface.

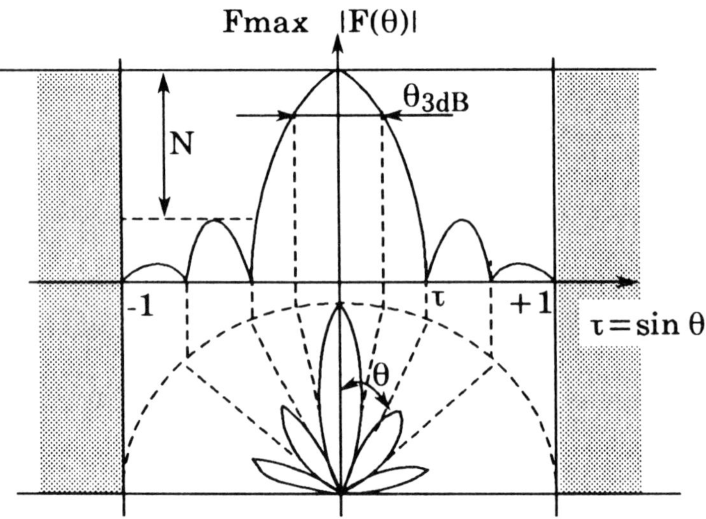

Fig. 12.43 Radiation pattern in cartesian and polar coordinates.

There is generally an angular range of maximum radiation (the main lobe). It is measured by the 3 dB beamwidth, θ_{3dB}, defined as follows:

$$|F|/|F_{max}| \geqslant 1/\sqrt{2}$$

or:

$$20 \log \frac{|F_{max}|}{|F|} \leqslant 3 \text{ dB}.$$

The level of sidelobes is an important characteristic. It is frequently expressed in decibels relative to the maximum level:

$$N_{dB} = 20 \log \left| \frac{F}{F_{max}} \right|. \tag{12.104}$$

If the pattern is not symmetrical about an axis, there are frequently two orthogonal planes of symmetry as, for example, in the case of the flat horn.

In this case, the main patterns consist of the main cross-sections of the characteristic surface by the planes of symmetry xOz and yOz (Fig. 12.44).

Finally, a section transverse to the direction of maximum radiation is sometimes used. This applies particularly to geostationary satellite antennas used for telecommunications: the transverse section, or spot, is the area of the earth's surface illuminated by the antenna. One of the antenna designer's tasks is to optimize the shape of the spot and the patterns to match the country covered.

Scattered sidelobes

The distribution of sidelobes, widely separated from the main lobe, is frequently chaotic and extremely complex. In this case, the overall level of the sidelobes

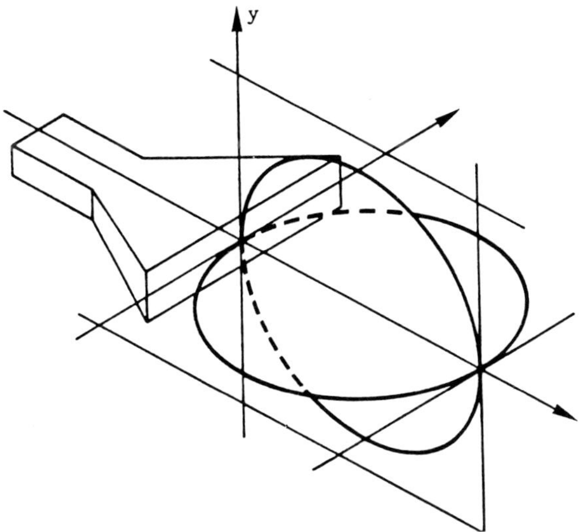

Fig. 12.44 Main cross-sections of the characteristic surface for a flat horn.

is determined statistically using a distribution function; we then define a mean relative level in different angular sectors (Fig. 12.45):

$$m = \frac{\dfrac{1}{\theta_2 - \theta_1} \displaystyle\int_{\theta_1}^{\theta_2} |F|^2 \, d\theta}{|F_{max}|^2} \qquad (12.105)$$

An example of this statistical approach will be given for array antennas in section 12.5.6, equation (12.222).

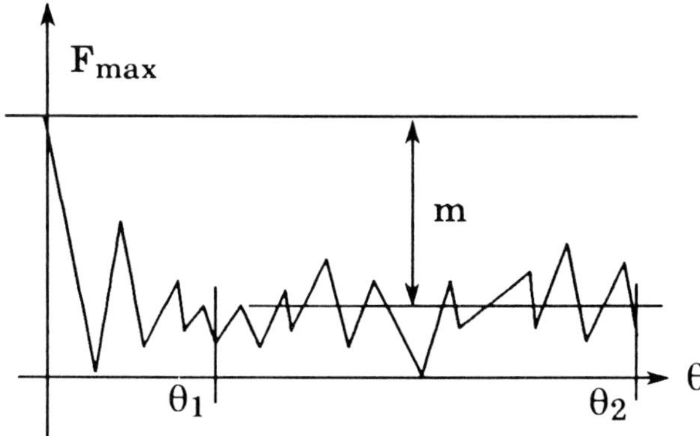

Fig. 12.45 Determination of mean relative sidelobe level.

Directivity of an antenna

The '3 dB' concept is not sufficient to represent the directional properties of the antenna since it characterizes the main lobe but not the proportion of the energy radiated in this lobe since the sidelobes can themselves represent a significant proportion of the total energy. We therefore define directivity as follows, based on the concept of the radiated power density per steradian. The steradian is the unit of solid angle; the solid angle Ω of a cone is the surface it defines on a concentric sphere of unit radius (Fig. 12.46).

The surface defined by a solid angle Ω on a concentric sphere with radius r will be given by:

$$S = \Omega\, r^2. \tag{12.106}$$

For the complete spherical surface the solid angle is:

$$\Omega = 4\pi\, r^2/r^2 = 4\pi \text{ steradian.}$$

Let us now consider the field radiated by an antenna at point P on a spherical wave front with a large radius r. It is given by the formula (12.100)

$$\boldsymbol{E}(\boldsymbol{u}, r) = E_0 \frac{\exp\left(-j2\dfrac{\pi r}{\lambda}\right)}{\left(\dfrac{r}{\lambda}\right)} \boldsymbol{F}(\boldsymbol{u}). \tag{12.104}$$

Since the local structure of the field is a plane wave, the power density in W/m^2 is given by equation (12.15).

$$\frac{\mathrm{d}W}{\mathrm{d}S} = \frac{|\boldsymbol{E}|^2}{2\eta} = \frac{E_0^2}{2\eta}\frac{\lambda^2}{r^2}\,|\boldsymbol{F}(\boldsymbol{u})|^2 \tag{12.108}$$

We eliminate the distance factor (r) by considering the power density per steradian:

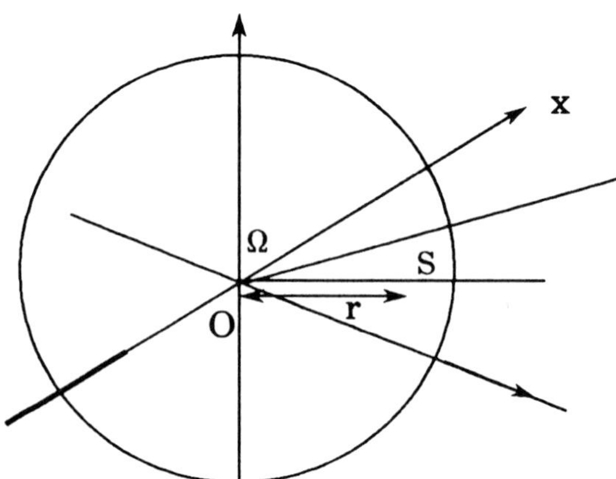

Fig. 12.46 Solid angle for defining antenna directivity.

$$\frac{\mathrm{d}W}{\mathrm{d}\Omega} = \frac{\mathrm{d}W}{\mathrm{d}S}\frac{\mathrm{d}S}{\mathrm{d}\Omega} = \frac{E_0^2\,\lambda^2}{2\,\eta}\,|\boldsymbol{F}(\boldsymbol{u})|^2. \qquad (12.109)$$

Where dΩ is the solid angle of an elementary cone subtended by the elementary area dS.

Directivity of the antenna in direction $\boldsymbol{u}$ (or gain excluding losses)

This is defined by the ratio between the power density per steradian in direction $\boldsymbol{u}$ and the mean power density radiated throughout space. If W_r is the total power radiated by the antenna, we can write:

$$\boxed{D(\boldsymbol{u}) = \frac{\mathrm{d}W/\mathrm{d}\Omega}{W_r/4\pi}\,.} \qquad (12.110)$$

Expression of directivity based on the characteristic function

W_r is, obviously, the sum of the power radiated in all directions in space, i.e. the integral of the power densities:

$$W_r = \int_{(4\pi)} \frac{\mathrm{d}W}{\mathrm{d}\Omega}\,\mathrm{d}\Omega\,. \qquad (12.111)$$

Equation (12.109) therefore gives us:

$$D(\boldsymbol{u}) = \frac{|\boldsymbol{F}(\boldsymbol{u})|^2}{\dfrac{1}{4\pi}\displaystyle\int |\boldsymbol{F}(\boldsymbol{u})|^2\,\mathrm{d}\Omega} \qquad (12.112)$$

This fundamental formula shows us that the mean of $D(\boldsymbol{u})$ for all directions in space is equal to 1:

$$\boxed{\frac{1}{4\pi}\int_{(4\pi)} D(\boldsymbol{u})\,\mathrm{d}\Omega = 1.} \qquad (12.113)$$

This is an important property which any antenna satisfies.

Gain of an antenna in direction $\boldsymbol{u}$

The concept of gain makes it possible to use the power W fed to the antenna instead of the radiated power W_r. The ratio between these two quantities is the antenna 'efficiency', which is limited by reflection and losses:

$$\rho_1 = \frac{W_r}{W} \leqslant 1. \qquad (12.114)$$

The gain is therefore defined as:

$$\boxed{G(\boldsymbol{u}) = \frac{\mathrm{d}W/\mathrm{d}\Omega}{W_r/4\pi} = \rho_1\, D(\boldsymbol{u}).} \tag{12.115}$$

Giving the requirement:

$$\boxed{\frac{1}{4\pi}\int G(\boldsymbol{u})\,\mathrm{d}\Omega = \rho_1 \leqslant 1.} \tag{12.116}$$

Application of the gain concept

What is the field radiated by an antenna fed by a power W (watts), at a distance r in the direction where the antenna gain is G? The density radiated per steradian is given (using (12.115)) by:

$$\frac{\mathrm{d}W}{\mathrm{d}\Omega} = \frac{WG}{4\pi}.$$

Moreover:

$$\frac{\mathrm{d}W}{\mathrm{d}\Omega} = r^2\frac{\mathrm{d}W}{\mathrm{d}S} = \frac{r^2}{2\,\eta}\,|E|^2.$$

We can therefore write:

$$|E|_{(\mathrm{V/m})} = \frac{\sqrt{60\,WG}}{r_{(\mathrm{m})}}. \tag{12.117}$$

A basic example: the dipole

A dipole consists of two radiating rods (Fig. 12.47). These form a straight line and are fed, 180 ° out of phase, by a feeder, for example a two-wire line.

We have seen (section 12.3.4) that in the frequently occuring case of a half-wave dipole, the characteristic function is:

$$\boldsymbol{F}(\theta) = \boldsymbol{e}_\theta \frac{\cos\left(\dfrac{\pi}{2}\cos\theta\right)}{\sin\theta}.$$

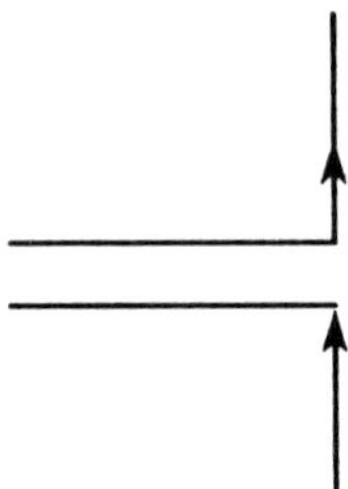

Fig. 12.47 Two wire feeder and two-rod dipole antenna.

For a short dipole (Hertzian dipole), we can use the approximate form (Roubine *et al.*, 1987):

$$\boldsymbol{F}(\theta) = \boldsymbol{e}_\theta \sin\theta\,.$$

Exercise: to calculate the dipole gain in an equatorial direction ($\theta = \pi/2$). This is an immediate application of formula (12.112). The solid angle element, expressed in spherical coordinates, can be written:

$$\mathrm{d}\Omega = \sin\theta\,\mathrm{d}\theta\,\mathrm{d}\phi.$$

This gives us:

$$G\left(\frac{\pi}{2}\right) = \frac{\left|F\left(\frac{\pi}{2}\right)\right|^2}{\frac{1}{4\pi}\int_0^{\pi}\int_0^{2\pi} |F(\theta)|^2 \sin\theta\,\mathrm{d}\theta\,\mathrm{d}\phi} = \frac{3}{2}$$

i.e., 1.76 dB. The gain of a half-wave dipole is slightly higher, approx. 2.15 dB.

The gain of a radiating aperture

Let us consider (Fig. 12.48) an aperture (S) in plane xOy. Let $\boldsymbol{n}$ be the unit vector of Oz normal to this aperture. In a direction $\boldsymbol{u}$, the directivity (or gain excluding losses G) is given by (12.110), i.e.

$$G(\boldsymbol{u}) = \frac{4\pi}{W_r}\frac{\mathrm{d}W}{\mathrm{d}\Omega}\,. \qquad (12.118)$$

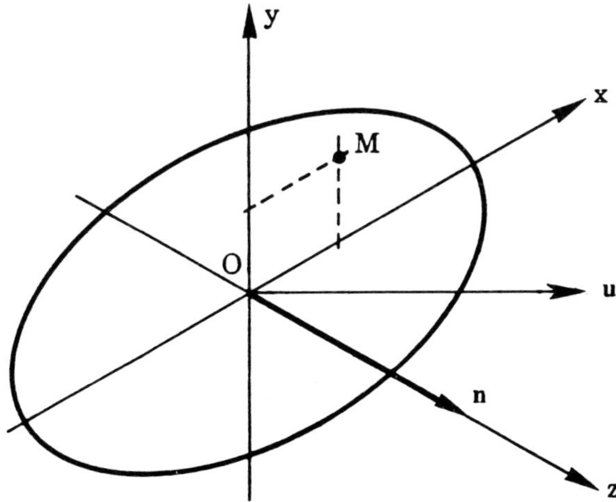

Fig. 12.48 Radiating aperture for gain calculation.

The radiated power W_r is the power which passes through the radiating aperture. If the structure of the field on the aperture is, at least locally, a plane wave, it can be calculated from formula (12.15) by the integral of the square of the electric field modulus on the aperture (12.15)

$$W_r = \frac{1}{2\eta} \int_{(S)} |\boldsymbol{E}(\mathrm{M})|^2 \, \mathrm{d}S . \tag{12.119}$$

The power density $\mathrm{d}W/\mathrm{d}\Omega$ in direction $\boldsymbol{u}$ can be calculated by applying the following equation to the field radiated in this direction:

$$\frac{\mathrm{d}W}{\mathrm{d}\Omega} = \frac{r_0^2 \, \mathrm{d}W}{\mathrm{d}S} = r_0^2 \frac{|\boldsymbol{E}(\boldsymbol{u})|^2}{2\eta} . \tag{12.120}$$

Field $\boldsymbol{E}(\boldsymbol{u})$ is calculated from the expression for a field radiated by an aperture (12.46) and (12.53). If $\boldsymbol{u}$ is negligibly different from $\boldsymbol{n}$, then, for the total field, this relation applies:

$$|\boldsymbol{E}(\boldsymbol{u})| = \frac{1}{r_0 \lambda} \left| \int_{(S)} \boldsymbol{E}(\mathrm{M}) \exp\left(j2\pi \frac{\mathbf{OM}}{\lambda} \cdot \boldsymbol{u} \right) \mathrm{d}S \right| . \tag{12.121}$$

The aperture gain in direction $\boldsymbol{u}$ is therefore given approximately by:

$$|G(\boldsymbol{u})| = \frac{4\pi}{\lambda^2} \frac{\left| \int_{(S)} \boldsymbol{E}(\mathrm{M}) \exp\left(j2\pi \frac{\mathbf{OM}}{\lambda} \cdot \boldsymbol{u} \right) \mathrm{d}S \right|^2}{\int_{(S)} |\boldsymbol{E}(\mathrm{M})|^2 \, \mathrm{d}S} . \tag{12.122}$$

Gain in the normal direction $\boldsymbol{n}$

Here the transverse field and the total field are identical, hence:

$$G(\boldsymbol{n}) = \frac{4\pi}{\lambda^2} \frac{\left| \int \boldsymbol{E} \, \mathrm{d}S \right|^2}{\int |\boldsymbol{E}|^2 \, \mathrm{d}S} . \tag{12.123}$$

Note that it may happen that only a fraction K_S of the radiated power crosses the aperture; this is the case of a parabolic reflector fed by a primary feed. Part of its pattern spills over around the reflector. In this case, the expression found in has to be multiplied by K_S (section 12.5.3).

Theoretical maximum gain

It is reasonable to ask what illumination $E(\mathrm{P})$ will give the maximum normal gain. The answer is to be found in the properties of the Schwartz inequality.

Assuming two complex functions $\boldsymbol{u}$ and $\boldsymbol{v}$ which can be integrated in the range (S), the following inequality can be demonstrated:

$$\frac{\left|\int \boldsymbol{u}\cdot\boldsymbol{v}\,\mathrm{d}S\right|^2}{\int |\boldsymbol{u}|^2\,\mathrm{d}S \int |\boldsymbol{v}|^2\,\mathrm{d}S} \leqslant 1. \tag{12.124}$$

Equality is obviously obtained when $\boldsymbol{u}=\boldsymbol{v}^*$†. In this case let us enter:

$$\boldsymbol{u}=\boldsymbol{E}_0 \text{ (constant)}, \quad \boldsymbol{v}=\boldsymbol{E}(\mathrm{M}).$$

The Schwartz inequality is then written:

$$\frac{\left|\int \boldsymbol{E}_0\cdot\boldsymbol{E}(\mathrm{M})\,\mathrm{d}S\right|^2}{\int |\boldsymbol{E}(\mathrm{M})|^2\,\mathrm{d}S \int |\boldsymbol{E}_0|^2\,\mathrm{d}S}\cdot \tag{12.125}$$

Therefore:

$$\frac{1}{S}\,\frac{\left|\int \boldsymbol{E}(\mathrm{M})\,\mathrm{d}S\right|^2}{\int |\boldsymbol{E}(\mathrm{M})|^2\,\mathrm{d}S} \leqslant 1.$$

Which gives us the following fundamental result for the theoretical maximum gain of an aperture:

$$\boxed{G(\boldsymbol{n}) \leqslant \frac{4\pi S}{\lambda^2}}\cdot \tag{12.126}$$

The maximum gain is obtained for:

$$\boldsymbol{E}(\mathrm{M})=\boldsymbol{E}_0.$$

In other words, for constant amplitude, phase and polarization illumination. In addition the basic assumption means that the illumination must be locally a plane wave.

† The notation $\boldsymbol{u}^*$ designates the complex conjucate vector $\boldsymbol{u}$.

Illumination factor of an aperture

This is the ratio of its real gain to maximum theoretical gain, i.e.:

$$K = \frac{G(\boldsymbol{n})}{\frac{4\pi S}{\lambda^2}} = \frac{1}{S} \frac{\left| \int \boldsymbol{E}(\mathrm{M})\, \mathrm{d}S \right|^2}{\int |\boldsymbol{E}(\mathrm{M})|^2\, \mathrm{d}S}. \tag{12.127}$$

Numerical example

What is the minimum diameter of a circular radiating aperture which will give a gain of 50 dB?

The surface area of an aperture with diameter D is:

$$S = \pi \frac{D^2}{4}.$$

Its theoretical maximum gain is therefore:

$$G_{\max} = \left(\frac{\pi D}{\lambda}\right)^2. \tag{12.128}$$

Here, we are looking for:

$$\left(\frac{\pi D}{\lambda}\right)^2 \geqslant 10^5.$$

We therefore have approximately: $D/\lambda \geqslant 100$. For example, with a frequency of 3000 MHz, i.e. a wavelength of 10 cm, the diameter must be at least 10 m. Exercise: the above aperture is the locus of a parabolic radial illumination of the form:

$$\boldsymbol{E}(r) = E_0 \left(1 - \left(\frac{2r}{D}\right)^2\right)$$

What is the illumination factor of this aperture?

Note that the expression for the area element in polar coordinates (r, α) is:

$$\mathrm{d}S = r\, \mathrm{d}r\, \mathrm{d}\alpha .$$

Answer: $K = 0.75$.

12.4.2 Receiving antennas

Transmitting–receiving reciprocity

Consider a transmission antenna (for example a dipole) with an input impedance Z_0 at its accesses (AB) (Fig. 12.49). Assume that this antenna is used for

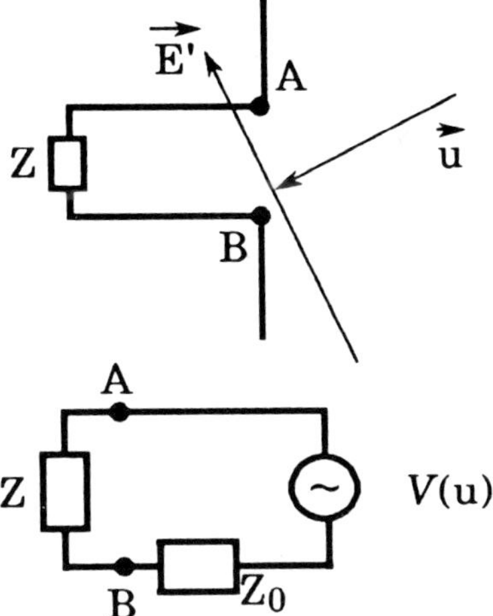

Fig. 12.49 Transmitting–receiving reciprocity.

reception. For a plane incident wave, of complex amplitude $\boldsymbol{E}'$ coming from direction $\boldsymbol{u}$, it can be demonstrated, using the reciprocity principle, that this antenna is the equivalent of a generator, with internal impedance Z_0 and electromotive force $V(\boldsymbol{u})$ proportional to the *scalar product of incident field* $\boldsymbol{E}'$ *by the characteristic function* $\boldsymbol{F}(\boldsymbol{u})$ of this antenna (defined in section 12.4.1).

$$V(\boldsymbol{u}) = A\boldsymbol{E}' \cdot \boldsymbol{F}(\boldsymbol{u}) . \tag{12.129}$$

A is a scalar coefficient. This result is sometimes translated by saying that the directive properties of a receiving antenna are the same as on transmission. The role of polarization has to be added, the scalar product of the preceding formula shows that $|V|$ is maximum if polarizations are concordant, null if they are orthogonal.

The power dissipated in load Z (representing the receiver) is maximum if the matching condition is met:

$$Z = Z_0^* .$$

This maximum value is the available power:

$$W' = \frac{|V|^2}{8\,R_0} = \frac{|A|^2}{8\,R_0}\,|\boldsymbol{E}' \cdot \boldsymbol{F}(\boldsymbol{u})|^2 \tag{12.130}$$

R_0, the real part of Z_0, is the **radiation resistance** of the the antenna.

Available power

Let us define the antenna's and the incident field's polarization unit vectors ($\boldsymbol{p}$ and $\boldsymbol{q}$ respectively) as:

$$p = \frac{\boldsymbol{F}(\boldsymbol{u})}{|\boldsymbol{F}(\boldsymbol{u})|} \qquad q = \frac{\boldsymbol{E}'}{|\boldsymbol{E}'|}. \tag{12.131}$$

The relation (12.130) can be put in the form:

$$W' = B\delta\,|\boldsymbol{E}'|^2\,G(\boldsymbol{u}) \tag{12.132}$$

where δ is the polarization factor: $\delta = |\boldsymbol{p}\cdot\boldsymbol{q}|^2 \leqslant 1.$ (12.133)

$G(\boldsymbol{u})$ is the gain of the antenna on transmission.

To set the value of coefficient B, consider the particular case of an aperture (S) which, on transmission, is the source of uniform illumination having the structure of a plane wave. For an incident wave coming from direction $\boldsymbol{n}$, normal to this aperture, having an adapted polarization, such an aperture behaves as an absorbing wall, all of the incident power W_0' is absorbed. This yields:

$$W_0' = S\frac{|\boldsymbol{E}'|^2}{2\eta}. \tag{12.134}$$

Moreover, according to (12.132) (with $\delta = 1$)

$$W_0' = B\,|\boldsymbol{E}'|^2\,G_0(\boldsymbol{n}).$$

We know that the normal gain of this aperture (12.126) is:

$$G_0(\boldsymbol{n}) = \frac{4\pi S}{\lambda^2}.$$

We can conclude that:

$$B = \frac{1}{2\eta}\frac{\lambda^2}{4\pi}. \tag{12.135}$$

Whence the expression of available power on reception with respect to antenna gain.

$$\boxed{W' = \delta\frac{|\boldsymbol{E}'|^2}{2\eta}\frac{\lambda^2}{4\pi}G(\boldsymbol{u})} \tag{12.136}$$

Absorption area of receiving antenna

The absorption area S' of any given antenna is that of a perfect aperture which in all absorbs a power equal to the power absorbed by the antenna. It therefore satisfies the following equation:

$$W' = \delta\frac{|\boldsymbol{E}'|^2}{2\eta}S'.$$

Hence according to (12.136)

$$S'(\boldsymbol{u}) = \frac{\lambda^2}{4\pi} G(\boldsymbol{u})$$

or:

$$G(\boldsymbol{u}) = \frac{4\pi\, S'(\boldsymbol{u})}{\lambda^2}. \tag{12.137}$$

Examples:

isotropic antenna: $G = 1$, $S' = \lambda^2/4\pi$;

half-wave dipole: $S' = 0.13\,\lambda^2$;

case of a radiating aperture.

The gain of a radiating aperture is given by formula (12.123). It can be used to deduce the corresponding absorption area:

$$S'(\boldsymbol{n}) = \frac{\left| \int_{(S)} \boldsymbol{E}(\mathrm{M})\, \mathrm{d}S \right|^2}{\int |\boldsymbol{E}(\mathrm{M})|^2\, \mathrm{d}S}. \tag{12.138}$$

The formula (12.126) shows that we always have:

$$\frac{S}{S'} = K \leqslant 1. \tag{12.139}$$

Coefficient K is the illumination efficiency. It may be likened to the illumination factor (12.127).

Application: radiolink equation

Given a link at distance r between a transmission antenna, with gain G (in the direction of the receiver), fed with power W, and a receiving antenna, of gain G' (in the direction of the transmitter) (Fig. 12.50), evaluate power W'_{R} delivered to the receiver.

The incident field on the receiving antenna is given by formula (12.117) of section 12.4.1:

$$E' = \frac{\sqrt{60\, WG}}{r}.$$

The received power is given by (12.136) with receiving gain G'

$$W'_{\mathrm{R}} = \delta \frac{E'^2}{2\eta} \frac{\lambda^2}{4\pi} G'.$$

From which is deduced:

$$\boxed{\frac{W'_{\mathrm{R}}}{W} = \frac{GG'\lambda^2}{(4\pi\, r)^2}\, \delta.} \tag{12.140}$$

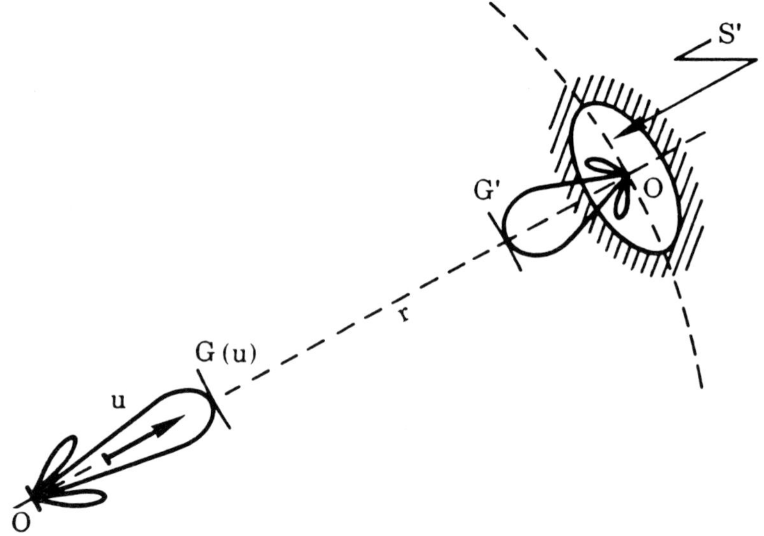

Fig. 12.50 Schematic representation of a microwave radio link.

Keep in mind that δ is the polarization factor (12.133): it emphasizes the importance of a good concordance of transmission and reception polarizations.

Radar cross-section (RCS)

This concept is introduced in radar theory (see Chapter 11) to characterize the radar-reflectivity of any type of target. This is the absorption area σ (square metres) of an antenna that is equivalent to a target which sends back absorbed energy isotropically (unit gain) (Fig. 12.51).

Radar equation

Given a transmitter with gain G, fed with power W, a target with radar cross-section σ located at a distance r, a receiver with gain G' (and radar cross-section S'), calculate the received power W'. The power density incident on the target results from the definition of gain G (formula (12.115)):

$$\frac{\mathrm{d}W}{\mathrm{d}S} = \frac{1}{r^2}\frac{\mathrm{d}W}{\mathrm{d}\Omega} = \frac{WG}{4\pi r^2}. \qquad (12.141)$$

The power scattered by the target is given (by definition of RCS) by:

$$W_{\mathrm{D}} = \sigma\frac{\mathrm{d}W}{\mathrm{d}S} = \frac{WG\sigma}{4\pi r^2}. \qquad (12.142)$$

The power density available on the receiving antenna is given by the same formula as (12.141) with a unit gain:

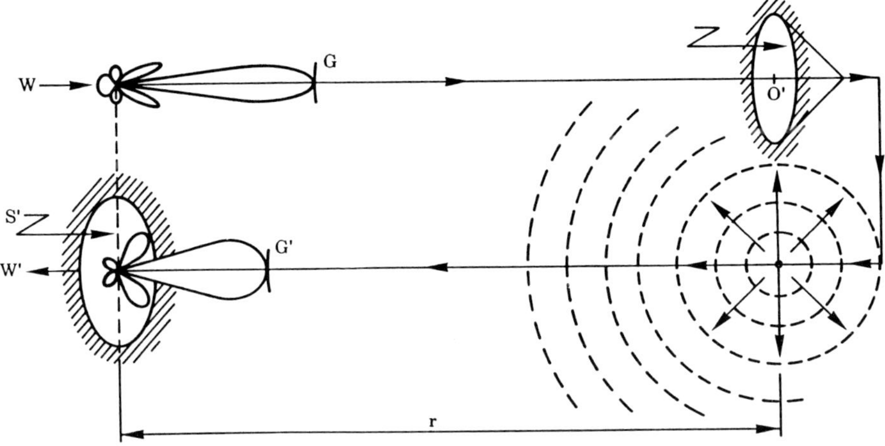

Fig. 12.51 Schematic representation of radar cross-section.

$$\frac{\mathrm{d}W'}{\mathrm{d}S} = \frac{W_D}{4\pi r^2} = \frac{WG\sigma}{(4\pi r^2)^2} .$$

The received power is therefore:

$$W' = S' \frac{\mathrm{d}W'}{\mathrm{d}S} = \frac{WGS'\sigma}{(4\pi r^2)^2} .$$

The absorption area S' is expressed with respect to reception gain by (12.137). Finally we obtain the radar equation:

$$\boxed{\frac{W'}{W} = \frac{GG'\lambda^2\sigma}{(4\pi)^3 r^4} .} \tag{12.143}$$

Exercise Calculate the above ratio for a radar fitted with a parabolic reflector transmitting–receiving antenna of diameter $D = 3$ m, operating at a frequency $f = 3000$ MHz. Reflector efficiency is $K = 1$. The target is an aircraft with a radar cross-section $\sigma = 1\ \mathrm{m}^2$ at distance $r = 100$ km.

Note that the preceding formula assumes propagation in free space. In fact the ground and the atmosphere play an important part [see chapters on propagation (Volume 2, Chapter 11) and radar (Volume 3)].

Reception of incoherent noise

Review: Planck's law

A body that perfectly absorbs all electromagnetic radiation is said to be a black body. Assume that there is thermodynamic equilibrium at an absolute temperature T with the surrounding medium. An element $\mathrm{d}S$ on its surface (Fig. 12.52) radiates in an elementary solid angle $\mathrm{d}\Omega$ about a direction $\boldsymbol{u}$ at an angle θ with

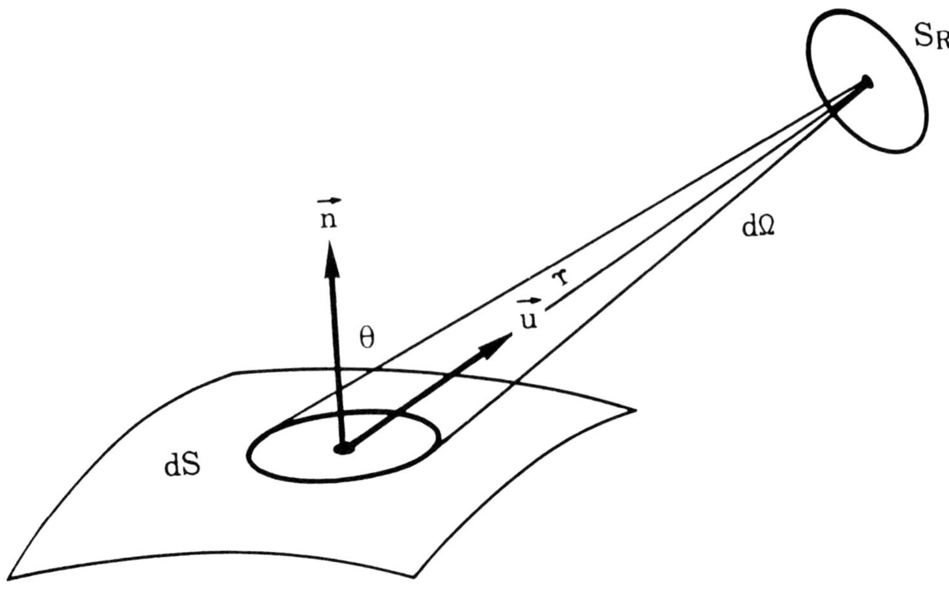

Fig. 12.52 Reception of outside noise.

the normal in the elementary band $(f,\ f+\mathrm{d}f)$ a power given by:

$$\mathrm{d}W = E(f,\ T)\,\mathrm{d}S \cos\theta\,\mathrm{d}\Omega\,\mathrm{d}f \tag{12.144}$$

where E, the spectral density of the emitting body is given by:

$$E = \frac{2hf^3}{c^2}\,\frac{1}{\exp(hf/kT)-1} \tag{12.145}$$

where h is Planck's constant $(6.55\times 10^{-34}$ J.s), and k is Boltzmann's constant $(1.38\times 10^{-23}$ J/K), i.e. $h/k = 4.75\times 10^{-11}$. At the radiowave limit ($f=3\times 10^{11}$ Hz), $hf/k \simeq 15$ K. If the temperature is not too low (a few tens of degrees) we can assume:

$$E \simeq \frac{2hf^3}{c^2}\,\frac{kT}{hf} = \frac{2kT}{\lambda^2}. \tag{12.146}$$

Reception of outside noise

An area S_R normal to direction $\boldsymbol{u}$, at a distance r will receive a power of:

$$\mathrm{d}W = E\,\mathrm{d}S\cos\theta\,\frac{S_R}{r^2}\,\mathrm{d}f = ES_R\,\mathrm{d}\Omega\,\mathrm{d}f \tag{12.147}$$

where $\mathrm{d}\Omega$ is now the solid angle under which dS is seen from point S_R (Fig. 12.52).

If S_R is the area of absorption of an antenna that only extracts from the disordered polarization of thermal radiation the one it is designed for:

$$\mathrm{d}W = \frac{1}{2} E \frac{\lambda^2}{4\pi} G \mathrm{d}\Omega \, \mathrm{d}f \tag{12.148}$$

or

$$\mathrm{d}W = kT \, \mathrm{d}f \, \frac{G\mathrm{d}\Omega}{4\pi} . \tag{12.149}$$

Random distribution of incoherent noise sources

If the noise sources in two distinct directions $\boldsymbol{u}$ and $\boldsymbol{u}'$ are incoherent, the fields that they radiate will not interfere with each other; their powers will add to each other (see section 12.6.5). Under these conditions we can consider an angular distribution of noise temperature $T(\boldsymbol{u})$. If an antenna aimed in direction $\boldsymbol{u}$ has a gain $G(\boldsymbol{u}, \boldsymbol{u}')$ in direction $\boldsymbol{u}'$, an obvious extension of (12.149) gives the total noise received by the antenna:

$$W = k \, \Delta f \frac{1}{4\pi} \int_{(4\pi)} T(\boldsymbol{u}') \, G(\boldsymbol{u}', \boldsymbol{u}) \mathrm{d}\Omega$$

$$W = k \Delta f T_{\mathrm{A}}(\boldsymbol{u}). \tag{12.150}$$

T_{A} is here the antenna's noise temperature:

$$\boxed{T_{\mathrm{A}}(\boldsymbol{u}) = \frac{1}{4\pi} \int T(\boldsymbol{u}') \, G(\boldsymbol{u}', \boldsymbol{u}) \mathrm{d}\Omega .} \tag{12.151}$$

This is the 'spatial' mean of the product of the gain by the angular distribution of the noise temperatures.

Low-noise antenna for space links

Space links which are used in various applications (spatial telecommunications, planetary probes, etc.) involve very large distances, much greater than those for earth links (altitude of a geostationary satellite: 36 000 km, mean Earth-to-Moon distance: 360 000 km, mean Earth-to-Mars distance: 150×10^6 km).

The result is that the fields received on the ground (or power densities in $\mathrm{W/m^2/Hz}$) are very weak with respect to the various sources of parasitic noise. Therefore to obtain an adequate signal-to-noise ratio, earth stations are designed based on a twofold objective. That is to maximize the signal received by using large surface antennas which have very high gain (diameters of several tens of metres), and to minimize parasitic noise of external and internal origins by choosing low noise temperature antennas and receivers (hence the name: cold antennas).

1. The problem of gain. The cost of large antennas is obviously high, therefore the surface of the antenna has to be used in the best way possible by acting on:

(a) the antenna's illumination law — technique of conformal cassegrain antennas (section 12.5.4);

(b) primary feed — its patterns affect in part the illumination law of the antenna as well as losses by spillover (section 12.5.2, electromagnetic horns);

(c) precision of antenna's construction — this item is essential for large antennas because it affects considerably the cost of the antenna. It can be shown that a standard deviation σ on the reflector results in an increase of the sidelobe level and a drop in gain:

$$\Delta G(\mathrm{d}B) \simeq 700\left(\frac{\sigma}{\lambda}\right)^2 .$$

2. Noise temperature problem. Parasitic noises in reception are of both internal and external origin.

   External noises (earth's thermal radiation, cosmic and atmospheric noises) can, *a priori* be picked up by the antenna's radiating lobes, and are expressed by overall antenna temperature T_A. We have seen (formula (12.151)) that the latter is the mean of the angular distribution $T(\boldsymbol{u})$ of external noises, a mean weighted by the gain function $G(\boldsymbol{u})$ of the antenna.

   Therefore, the problem for a cold antenna involves minimizing function G outside of the main lobe, especially in directions where $T(\boldsymbol{u})$ is large (ground directions especially). This problem is essentially solved by the choice of the type of antenna and certain precautions taken for its fabrication.

   Concerning internal noises, the largest contribution is due to the noise temperature of the receiver T_R. Current practice is to use amplifiers with field effect transistors of GaAs (T_R ranging from 50 to 75 K in C-band).

   Another cause of internal noise is due to various ohmic losses which can occur before amplification (7 K for 0.1 dB of losses). An estimate of these various losses must therefore always be set up with care.
3. Station's factor of merit. The overall signal-to-noise ratio of a station is proportional to the factor of merit defined as the ratio between the maximum gain of the antenna (given the various losses) and the total noise temperature of this system (antenna temperature, receiver temperature, temperature due to losses): $M = G/T$.

   An antenna project should be carried out with a view to finding the minimum total cost which yields a given factor of merit.

12.5 MAIN TYPES OF MICROWAVE ANTENNAS

12.5.1 Properties of antennas with axial symmetry

Antennas with axial symmetry are in wide use; tracking antennas, parabolic reflector antennas, conic horn antennas, cassegrain antennas, array antennas with circular cross-section, etc. They possess certain properties of great practical importance.

General expressions of fields

Consider a given structure (metallic, dielectric, etc.) formed by the rotationally symmetric surfaces about an axis Oz. For example, this structure can be a monomode, multimode or corrugated horn (section 12.5.2), a centred system with reflectors (sections 12.5.3–4) and lenses, etc.

It is assumed to be excited through a waveguide on the same axis Oz, transmitting uniquely its fundamental propagation mode in a polarization of any given unit vector $\boldsymbol{p}$. Note that higher modes can exist in other parts of the structure.

Consider a rotationally symmetric surface (S) with the same axis Oz. This surface can be for example the radiating aperture plane, a surface bearing on the border of the aperture, or a sphere (Γ) of a large radius R where the radiated patterns are observed.

We will show that on such a surface the electromagnetic field necessarily has a very characteristic mathematical form. We will deduce a simple criterion of the absence of radiated cross-polarization. Other consequences will be drawn with respect to 'hybrid' modes (section 12.5.1).

Any point M of (S) is characterized by polar coordinates (r, ψ) where r designates the curvilinear abscissa along a meridian. Consider for example the case of excitation of the structure in horizontal polarization:

$$\boldsymbol{p} = \boldsymbol{e}_x \tag{12.152}$$

The results over any surface (S) in a distribution of the electromagnetic field characterized by its radial and tangential components are:

$$E(\mathrm{M})\begin{cases} E_r(r,\ \psi) \\ E_\psi(r,\ \psi) \end{cases} \qquad H(\mathrm{M})\begin{cases} H_r(r,\ \psi) \\ H_\psi(r,\ \psi) \end{cases} \tag{12.153}$$

Observe that this distribution has the same symmetries as excitation $\boldsymbol{p}$. This necessarily yields:

$$\begin{aligned} &E_r\left(r,\ \frac{\pi}{2}\right) = 0 \qquad &&H_r(r,\ 0) = 0 \\ &E_\psi(r,\ 0) = 0 \qquad &&H_\psi\left(r,\ \frac{\pi}{2}\right) = 0 \end{aligned} \tag{12.154}$$

Excitation $\boldsymbol{p} = \boldsymbol{e}_x$ can be decomposed into two partial excitations $\boldsymbol{p}'$ and $\boldsymbol{p}''$, at right angles to each other, with orientations ψ and $\psi - \pi/2$ (See Fig. 12.53):

$$\boldsymbol{p} = \boldsymbol{e}_x = \boldsymbol{p}' \cos\psi + \boldsymbol{p}'' \sin\psi \tag{12.155}$$

Excitations $\boldsymbol{p}'$ and $\boldsymbol{p}''$ give rise, by symmetry, to the field distributions deduced respectively from (12.153) by rotations ψ and $\psi - \pi/2$. Their expressions are

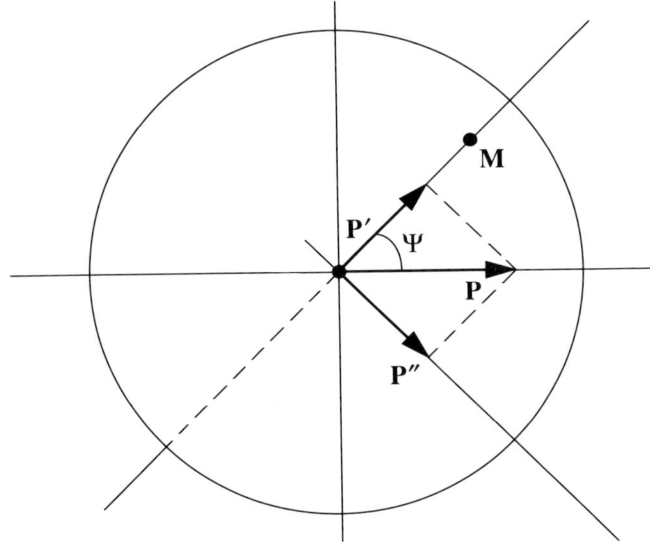

Fig. 12.53 Superposition of partial excitations to obtain the initial field distribution.

obtained by writing the conservation of the radial and tangential components in these rotations. Since the excitation is monomode, their superposition yields the initial distribution.

Therefore, for the electric field we obtain:

$$E_r(r,\ \psi) = E_r(r,\ 0)\cos\psi + E_r(r, \pi/2)\sin\psi$$
$$E_\psi(r,\ \psi) = E_\psi(r,\ 0)\cos\psi + E_\psi(r, \pi/2)\sin\psi \tag{12.156}$$

and similar expressions for the magnetic field.
Given the symmetry conditions (12.154), we obtain relations of the form:

$$\begin{cases} E_r = a\cos\psi \\ E_\psi = -\,b\sin\psi \end{cases} \qquad \begin{cases} H_{r+} = c\sin\psi \\ H_{\psi+} = d\cos\psi \end{cases} \tag{12.157}$$

Or in cartesian components:

$$\begin{cases} E_x = A(r) + B(r)\cos 2\psi \\ E_y = B(r)\sin 2\psi \end{cases} \qquad \begin{cases} H_{x+} = C(r)\sin 2\psi \\ H_{y+} = D(r) - C(r)\cos 2\psi \end{cases} \tag{12.158}$$

with:

$$H_+ = \eta_0 H$$

η being the wave impedance (see equation 12.14).

Properties of radiated patterns – general form

If surface (S) is a sphere Γ of a large radius R (polar coordinates $\theta = r/R,\ \phi$) the fields are the radiated diagrams in classical terms. Their form is therefore:

$$E_\theta(\theta,\ \phi) = E_\theta(\theta,\ 0)\cos\phi = E_\alpha(\theta,\ 0)\cos\phi$$

$$E_\phi(\theta,\ \phi) = E_\phi\left(\theta,\ \frac{\pi}{2}\right)\sin\phi = E_\alpha\left(\theta,\ \frac{\pi}{2}\right)\sin\phi \qquad (12.159)$$

where:

$$E_\alpha(\theta,\ 0) \qquad \text{and} \qquad E_\alpha(\theta,\ \pi/2)$$

are the patterns taken from the main E (electric) and H (magnetic) planes. (The magnetic fields can be deduced from the fact that the wave is plane locally.)

Taking into account the properties of Huygens coordinates (section 12.3.6), we therefore deduce the expression of the radiated patterns in co-polarization and crossed polarization E_α, E_β:

$$E_\alpha(\theta,\ \phi) = \frac{E_\alpha(\theta,\ 0) + E_\alpha(\theta,\ \pi/2)}{2} + \frac{E_\alpha(\theta,\ 0) - E_\alpha(\theta,\ \pi/2)}{2}\cos 2\phi$$

$$E_\beta(\theta,\ \phi) = \frac{E_\alpha(\theta,\ 0) - E_\alpha(\theta,\ \pi/2)}{2}\sin 2\phi \qquad (12.160)$$

Hence cross-polarization is maximum in the diagonal planes ($\phi = \pm\pi/4$) Fig. 12.54).

In these planes the pattern observed is the half-difference of the 'E plane' and 'H plane' patterns.

1. A criterion of null cross-polarization is the identity of 'E plane' and 'H plane' patterns (in amplitude and phase).
2. In co-polarization, the pattern observed in a diagonal plane is the mean of the 'E plane' and 'H plane' patterns.

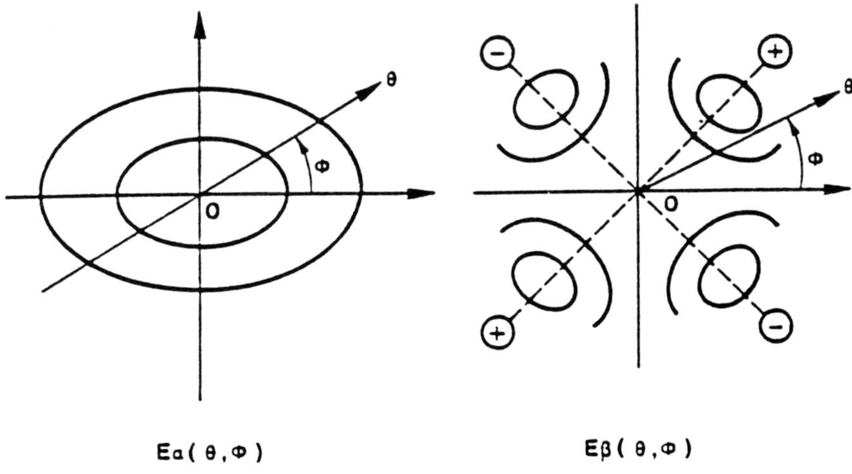

Fig. 12.54 Radiation patterns in the copolar (left) and cross-polar (right) polarizations.

3. If the second condition is satisfied, the pattern observed in co-polarization is independent from ϕ (characteristic rotationally symmetric surface).

Condition of polarization purity. Hybrid waves

We have seen from formulae (12.160) that the criterion of absence of cross-polarization is equivalent to the identity of the patterns observed in the plane of polarization $\boldsymbol{p}$ and in the orthogonal plane. Under these conditions and by definition (section 12.3.6) the antenna is a Huygens source; its magnetic field lines are deduced from the electric field lines by a rotation of $\pi/2$ about the axis (section 12.3.7). This property can be verified whatever the surface with axial symmetry where the field is observed; in particular in the radiating aperture plane, the same property of symmetry is verified (Fig. 12.55). Therefore at all points $M(r, \psi)$ of such an aperture we obtain:

$$\begin{aligned} E_r(r, \psi) &= a(r)\cos\psi \qquad & H_r^+ &= -a(r)\sin\psi \\ E_\psi(r, \psi) &= -b(r)\sin\psi \qquad & H_\psi^+ &= b(r)\cos\psi \end{aligned} \tag{12.161}$$

This symmetry property is also true for the longitudinal components. This is why these waves, which do not enter into classical TE, TM categories are said to be 'hybrid'.

Going back to cartesian coordinates, by transposing formula (12.96) by substituting p_α and p_β by E_x and E_y and p_θ and p_ϕ by E_r, E_ψ and likewise for

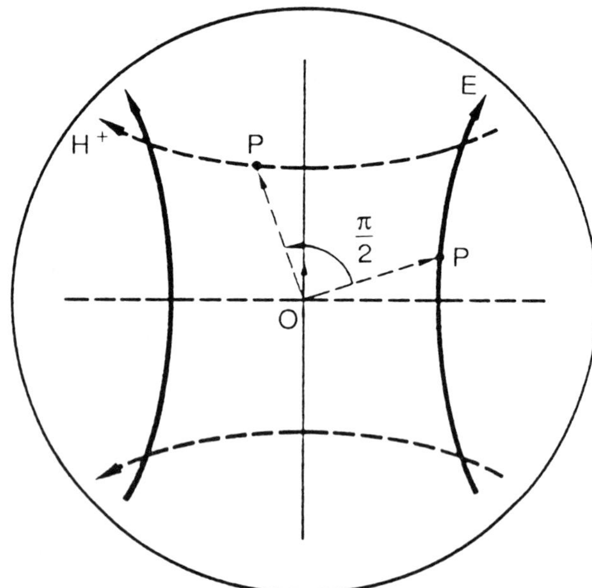

Fig. 12.55 A Huygens source exhibits electric and magetic fields superposition under rotation through 90 °.

the magnetic field, we obtain the following equations for the general structure of the field on the aperture of a Huygens source:

$$\begin{aligned} E_x &= A + B\cos 2\psi \qquad & H_x^+ &= B\sin 2\psi \\ E_y &= B\sin 2\psi \qquad & H_y^+ &= A - B\cos 2\psi \end{aligned} \tag{12.162}$$

(with $H^+ = \eta H$).

12.5.2 Electromagnetic horns

General properties

We have already seen that a horn is formed by the section of a waveguide with an increasing cross-section ended by the radiating aperture. The other end (connected to the feed waveguide) is the throat (Fig. 12.18). This type of antenna is widely used as a primary feed and offers mechanical and radioelectrical advantages. It is easy to protect it from weather by covering it with a radome which is transparent to the radiation. Internal pressurization prevents water condensation and improves power capability. Moreover, several different radiating characteristics can be obtained according to the type of horn used as well as excitation conditions: polarization, propagation mode. There are several different categories of horns: monomode, multimode, corrugated. In addition, in each category, horns differ by their geometry: aperture angle, form of transverse and longitudinal sections. All of these possibilities allow finding a horn that solves a given problem. To evaluate the radiating characteristics, we can often assume that the horn is equivalent to a radiating aperture whose illumination is the result of the excited propagation mode(s). Sometimes higher modes of various origins have to be taken into account: discontinuity formed by the aperture itself, parasitic reflections, etc. However, if the horn is powered by a guide transporting its fundamental mode, the longitudinal section of the horn ensures progressive transition up to the aperture and prevents the appearance of higher interference modes. The illumination obtained is therefore a true image of the field distribution of the mode being used (monomode horn). Conversely, if the horn has discontinuities, higher modes are generated which modify the illumination law and consequently, the pattern (multimode horn). The same horn can be fed by several waveguides; this yields several patterns.

It is often instructive to consider the operation of a horn on reception. If for example it is a primary source, the horn is immersed in an incident field (E', H') whose structure is the diffraction pattern of the focusing system. For example in the case of a monomode horn, the diffraction pattern produces a set of modes of various orders in the neighbourhood of the aperture which are likely to propagate towards the throat of the guide up to the planes whose cross-section corresponds to the cut-off of the modes involved; finally, only the working mode remains: the horn acts as a mode filter.

Horns with a narrow angle at the apex and open guides

An open guide is the limiting case of a horn whose angular aperture is null. Examination of its radiation can give a good approximation of that of a horn with a finite angular aperture but which is sufficiently narrow so that the wave emerging from the horn is very close to a plane wave (Fig. 12.56). For example in the case of a cone of revolution, with the angle at apex 2α, of length L and whose radiating aperture is a circle (C) of diameter $AA' = 2a$, the maximum longitudinal difference between the spheric wave with centre O′ and radius $L = OO'$ and the aperture plane is given by approximate relations (Fig. 12.56):

$$\Delta z = L(1 - \cos \alpha) \simeq L \frac{\alpha^2}{2} = \frac{a^2}{2L} \cdot \tag{12.163}$$

If the maximum phase error criterion imposes a difference that is less than a fraction of the wavelength: $\Delta z \leqslant \lambda/4$ for example, we obtain:

$$L \geqslant 2\, a^2/\lambda. \tag{12.164}$$

Under these conditions, we can see that a horn whose aperture, $2a$, is greater than a few wavelengths, has to be very long. However, such horns are used as a primary feeds for illuminating systems with a long focal distance or afocal systems (Cassegrain). It is possible to completely cancel out this phase error by adding a reflector horn or an auxiliary lens.

Flared horns

When the angular aperture of a horn is not negligible, variations in the phase of illumination have to be taken into account (Silver, 1949, King, 1956). A

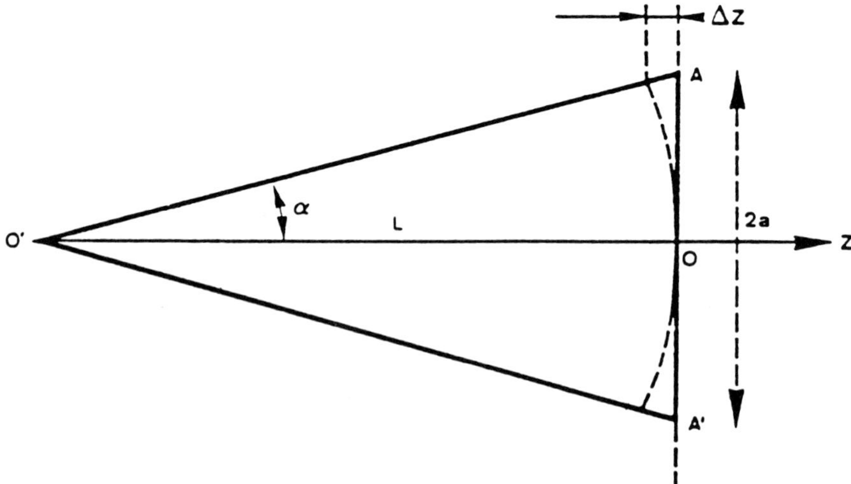

Fig. 12.56 Horn with a narrow angle.

quadratic approximation of these variations leads to classical results that we will summarize. We will find them later on in the discussion of corrugated horns.

1. For a given aperture a/λ, the main lobe of the pattern remains close to that of the corresponding equiphase illumination as long as the phase error is less than $\pi/2$. The phase centre is near the centre of the aperture. The width of the beam varies as λ/a. It is therefore sensitive to frequency,
2. If the length of the horn decreases, the 'nulls' between sidelobes fill, then the main lobe spreads out and becomes deformed. The phase law becomes complex. The phase centre moves away from the aperture and approaches the apex of the horn,
3. For a given length, L, of the horn, if the aperture $2a$ increases, the beamwidth first decreases, passes through a minimum, then increases again and tends toward the geometrical angle 2α of the horn.

Multimode horns (Drabowitch, 1969)

A multimode horn allows synthesis of the illumination law of its aperture – and therefore its pattern – by superposition of several modes of propagation at suitable doses. A multimode horn can have several independent feed points associated with independent patterns. This is the case for monopulse tracking processes which gave birth to this technique. The multimode horns can form suitable primary feeds which contribute to the antenna's gain factor and noise temperature qualities when the frequency band covered is not too large. The possibility of synthesizing the law of illumination of an open guide by superposition of modes is based on the general property of orthogonality of the transverse components of the fields associated with these modes: the latter form an orthogonal base which allows linear expression of any law of illumination with the coefficients obtained by projecting onto this base the illumination to be synthesized. The necessary modes are generated by a 'moder'.

Examples

1. Multimode horn, *H* plane. Figure 12.57 shows the diagram of an *H*-plane moder. If inputs A and B are excited in phase opposition through a magic T, only mode TE_{20} propagates up to the throat and produces a difference pattern that can be used by a monopulse tracking antenna. If inputs A and B are excited in phase, the 'even' modes TE_{10} and TE_{30} are excited. Their superposition yields, as shown in Fig. 12.57, an illumination S that closely resembles the diffraction spot of a focusing system and is likely to yield an excellent axial gain factor of the system.
2. Bimodal circular horn (Potter, 1963). Figure 12.58 shows how superposition of modes TE_{11} and TM_{11} on the waveguide's circular aperture yields an illumination with almost perfect circular symmetry and hence almost

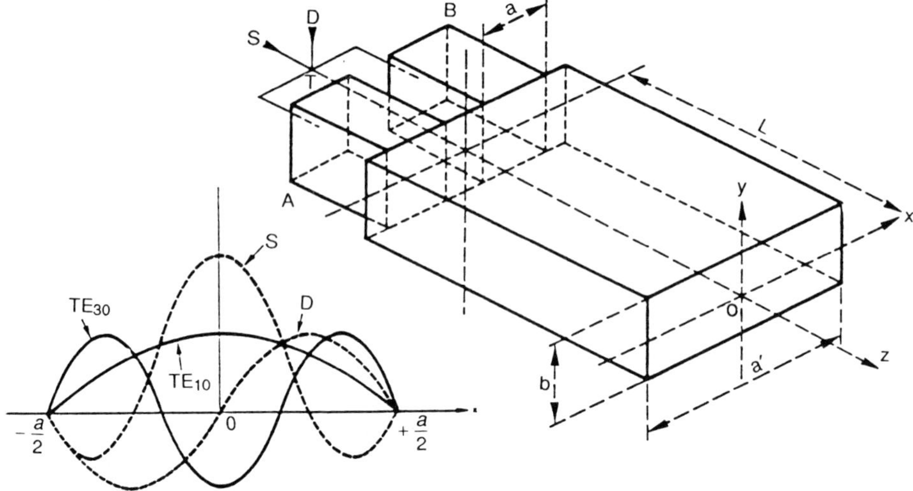

Fig. 12.57 Diagram of an H-plane moder; left: modes superposition.

completely free of cross-polarization. The generation of mode TM_{11} can be obtained by a simple discontinuity of the guide's diameter as shown in Fig. 12.59.

Hybrid modes and corrugated horns (Clarricoats and Saha, 1971)

Cross-polar-free circular aperture

We have established the general expressions (12.157) of the field on the aperture of any given structure of revolution under the sole condition that the excitation guide only transmits its fundamental mode. The expression (12.159) of radiated fields enabled us to formulate the condition of polarization purity, which entails more restrictive forms (12.161) or (12.162). The latter express the fact that the lines of the magnetic field are deduced from the electric field lines by $\pi/2$ rotation about axis Oz. Is it possible to imagine a waveguide transmitting a mode possessing these properties?

Fig. 12.58 Superposition of TE_{11} and TM_{11} modes.

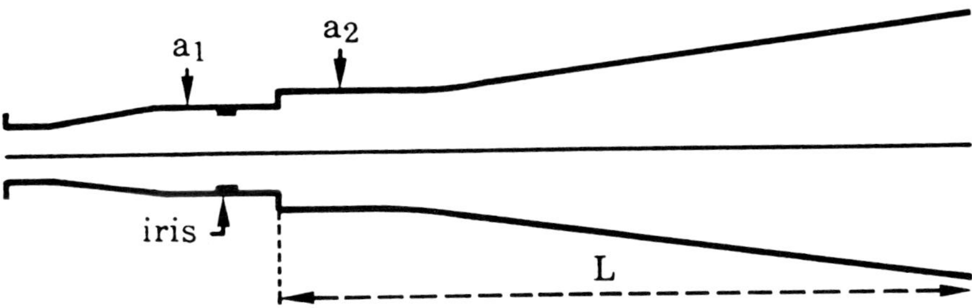

Fig. 12.59 Circular waveguide horn for generating TM_{11} mode.

Corrugated guides

Constructing a guide with anisotropic superficial conductivity

Consider a circular guide of radius *a* with smooth conducting walls (Fig. 12.60). At any point M of the surface, conductivity is assumed to be infinite, both longitudinally (Mz) as well as transversely (Mψ). We know that boundary conditions impose an electric field normal to the wall ($E\psi = 0$) and a magnetic field tangent to the wall ($H_r = 0$) whose amplitude is related to the longitudinal current I_Z. These conditions are therefore different for E and H; the corresponding field lines in any given right angle section therefore are also of different shapes; they cannot satisfy the condition of axial symmetry. To meet this condition, it is necessary to imagine a circular guide whose walls are such that the conditions relative to the magnetic field are the same as for the electric field. Therefore $H_\psi = 0$. We have seen that the transverse magnetic field is related to the longitudinal current I_Z (Fig. 12.61). To cancel it $I_Z = 0$, the wall has to have a null longitudinal conductivity ($Y_Z = 0$) while maintaining an infinite transverse conductivity ($Y_\psi = \infty$).

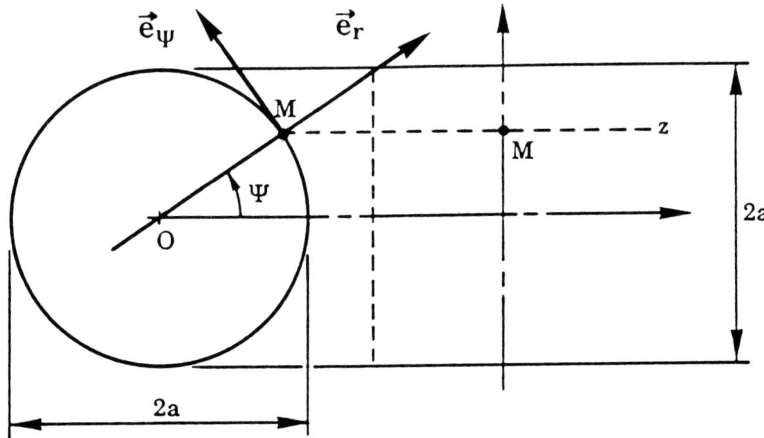

Fig. 12.60 Circular guide with smooth reflecting walls.

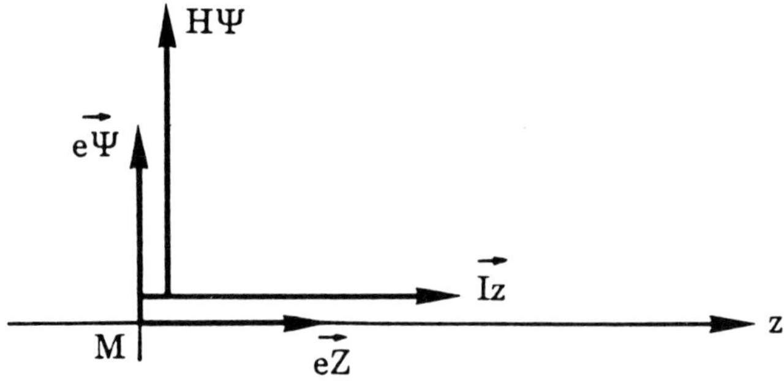

Fig. 12.61 Transverse magnetic field and longitudinal current.

In practice this condition can be met as follows: we know that a quarter-wave trap (Fig. 12.62) can be made by a short-circuited line. We obtain an infinite input impedance, i.e. a null admittance: $Y_{AB} = 0$ across input terminals A, B. The current across the line is therefore null. This property can be maintained for the full length of the line by stacking a large number of these traps. The resultant effect is a corrugated guide (Fig. 12.63). In practice, three to four corrugations per wavelength are sufficient to block out longitudinal currents and obtain the desired properties.

The traps are carved into a circular guide, with the result that the depth, p, corresponding to condition $Y_Z = 0$, is in general slightly greater than a quarter wavelength. As a general rule, if the guide is to transmit a certain frequency band, depth p should be chosen so that conductance is always capacitive or null, but never inductive, so as to prevent the appearance of generally unwanted surface waves. The conditions that give zero longitudinal admittance are called 'hybrid balance conditions'.

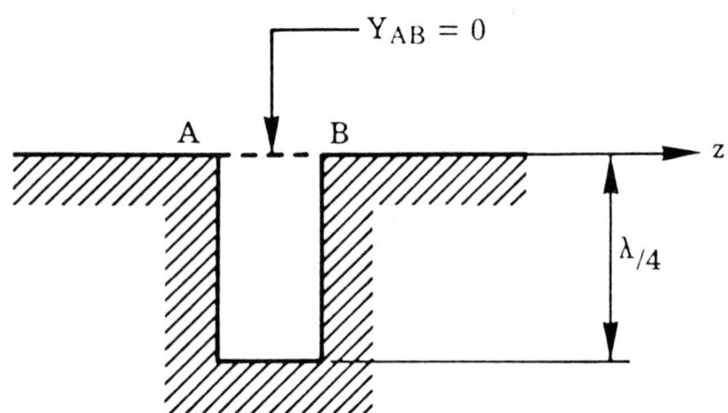

Fig. 12.62 Quarter-wave trap.

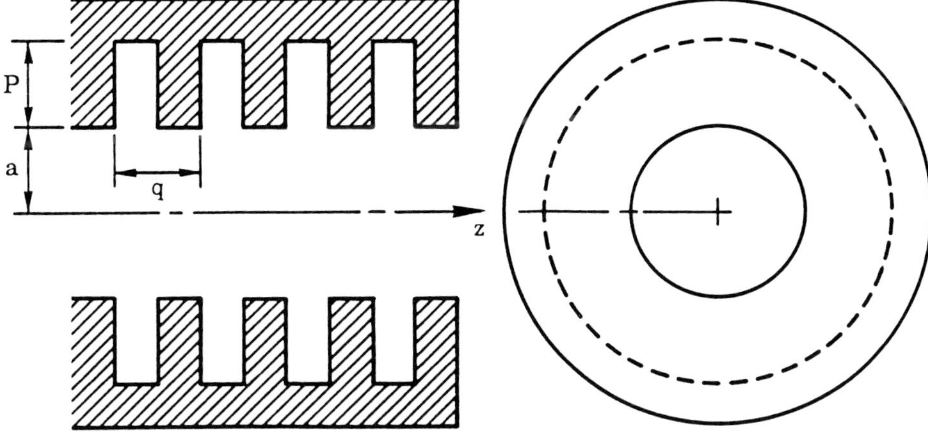

Fig. 12.63 Corrugated circular waveguide.

Equation of field lines in a transverse section (Clarricoats and Saha, 1971)

Thorough analysis of modes which can be carried by this type of structure has been made in numerous papers. The most generally used hybrid mode is denoted HE_{11}. The electric and magnetic field equations are of the form (12.161) with:

$$a = J_0\left(\frac{2\pi r}{\lambda}\sin\theta_0\right) + \tan^2\frac{\theta_0}{2}\,J_2\left(\frac{2\pi r}{\lambda}\sin\theta_0\right)$$

$$b = J_0\left(\frac{2\pi r}{\lambda}\sin\theta_0\right) + \tan^2\frac{\theta_0}{2}\,J_2\left(\frac{2\pi r}{\lambda}\sin\theta_0\right) \qquad (12.165)$$

J_0 and J_2 being order zero and order two Bessel functions.

Angle θ_0 is related to the cut-off wavelength λ and a and to the guided wavelength λ_g:

$$\sin\theta_0 = \frac{\lambda}{\lambda_c} \qquad \cos\theta_0 = \frac{\lambda}{\lambda_g}\,.$$

The latter is the same for a smooth circular waveguide of the same radius:

$$\lambda_c = 3.41\,a. \qquad (12.166)$$

Very far from the cut-off ($\lambda << \lambda_0$), θ_0 is small, the terms in J_2 can be ignored and the equations can be written as:

$$E_r = E_0\,J_0\left(\frac{2\pi r}{\lambda_c}\right)\cos\phi \quad \eta H_r = E_0\,J_0\left(\frac{2\pi r}{\lambda_c}\right)\sin\phi$$

$$E_\psi = E_0 J_0\left(\frac{2\pi r}{\lambda_c}\right)\sin\phi \qquad \eta H_\psi = E_0 J_0\left(\frac{2\pi r}{\lambda_c}\right)\cos\phi \tag{12.167}$$

i.e. a field with constant polarization in cartesian coordinates:

$$\begin{aligned} E_x &= E_0 J_0\left(1.841\frac{r}{a}\right) & H_x &= 0 \\ E_y &= 0 & H_y &= \frac{E_0}{\eta} J_0\left(1.841\frac{r}{a}\right) \end{aligned} \tag{12.168}$$

Radiation patterns

Radiation from an open-ended corrugated waveguide
The patterns are given by formula (12.75) from expressions (12.168) of the field on the aperture. The integrals are calculated explicitly using Lommel's formulae. The pattern obtained is independent of ϕ as expected and free of cross-polarization. If the aperture is large enough, illumination follows, as we have seen, an 'apodized' law decreasing towards the periphery of the aperture. This is favourable to a very low level of sidelobes as verified by experimental calculation.

Radiation from a corrugated horn (Aubry and Bitter, 1975, Kildal, 1988)

1. *Introduction* Radiation from corrugated horns excited in HE_{11} mode presents the same remarkable properties of symmetry as open-ended corrugated guides; in the neighbourhood of hybrid balance conditions, illumination of the aperture satisfies the condition for zero cross-polarization. Its expression can be given with classical horn approximations: same shape as in a guide of the same diameter transmitting the same mode with a quadratic phase characteristic corresponding to a pseudo-spherical wave dome of radius L (the length of the generator of the cone defining the horn). In addition, since the horn's aperture is assumed to be large enough, we can see that the field has a locally plane wave structure. Therefore it is solely determined by the electric field. A point M on the aperture of radius a (polar coordinates (ρ, ϕ), cartesian coordinates (ε, η)) is therefore the source of an illumination defined by the scalar expression:

$$E_M = J_0\left(u_{11}\frac{\rho}{\alpha}\right)\exp\left(-j\frac{2\pi}{\lambda}\frac{\rho^2}{2L}\right). \tag{12.169}$$

Radiation at a given distance can be obtained using Fresnel's approximation formulae (12.99). We therefore obtain a simple 'radial' integral. Field polarization is that of a Huygens source. A particularly simple and accurate calculation method consists of decomposing the illumination into a series of Laguerre – Gauss functions (Roubine, 1987). The particular form of function J_0 means that, in a first approximation; only the first term need be considered:

$$J_0\left(u_{11}\frac{\rho}{\alpha}\right) \simeq \exp\left(-\frac{\rho^2}{w^2}\right). \tag{12.170}$$

The scatter w can be chosen by expressing under these two forms the conservation of the energy encircled in the aperture. We then obtain $w \cdot a/\sqrt{2}$. This approximation yields fairly rigorous, simple and complete results which are at least qualitative values for other types of horns (bimodal circular horn for example).

2. *Radiation in the Fresnel region* Radiation in a near zone has to be known when this type of horn is to be used to illuminate an auxiliary mirror (in the case of the Cassegrain antenna, for example). At any point M (coordinates x, y, z) (Fig. 12.64), the field is given by Fresnel formulae which here are completely integrated. Since the Gaussian function is the eigenfunction of the Fourier transform, the field takes on a Gaussian form:

$$E_p(r,\ z) = A\ \exp\left(-j2\pi\frac{r^2}{2\lambda R}\right)\exp\left(-\frac{r^2}{w'^2}\right). \tag{12.171}$$

This result is backed up by experimental results which give very low sidelobes (Fig. 12.65). Parameters w' and R characterize respectively the scatter and the position of the phase centre.

3. *Scatter*

$$w'^2 = \frac{2}{n^2}\frac{z^2\lambda^2}{a^2} + \left(1 + \frac{z}{L}\right)^2\frac{a^2}{2}. \tag{12.172}$$

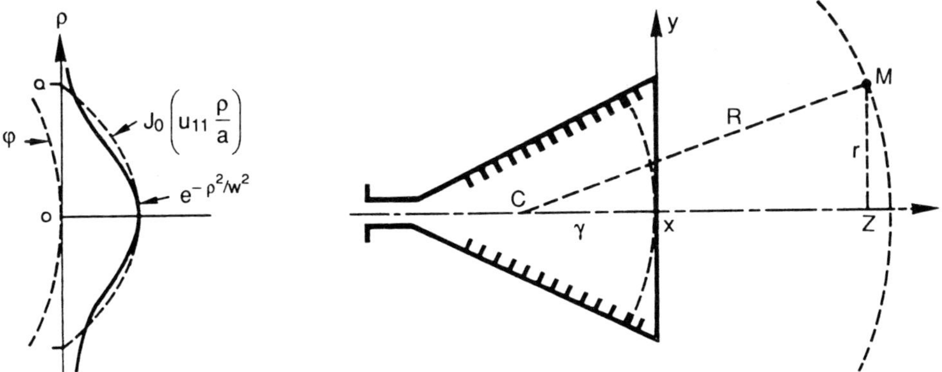

Fig. 12.64 J_0 Bessel function approximation (left) and coordinates of radiation in the Fresnel zone.

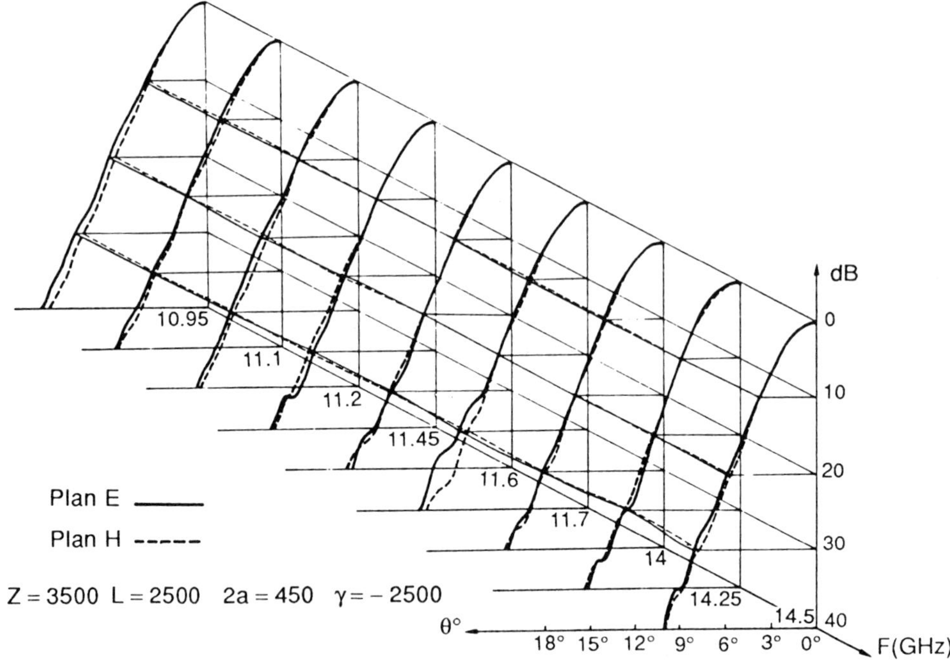

Fig. 12.65 Experimental results in the Fresnel zone.

(a) In the neighbourhood of the aperture: $z \cong 0,\ w'^2 = \dfrac{a^2}{2} = w^2$. Scatter is the same as that of illumination.

(b) At far distances:

$$\frac{w'^2}{z^2} \cong \frac{2}{\pi^2}\frac{\lambda^2}{a^2} + \frac{a^2}{2L^2}\cdot \tag{12.173}$$

(i) For $L = \infty$ (equiphase aperture), the beamwidth is defined, in radians, by:

$$\theta_0 = \frac{w'}{z} = \frac{\sqrt{2}\,\lambda}{\pi a} \approx \frac{\lambda}{2a}\cdot \tag{12.174}$$

This is the classical formula.

(ii) For a small L (fairly open horn):

$$\theta_0 = \frac{w'}{z} \approx \frac{a}{\sqrt{2}\,L} \tag{12.175}$$

The beamwidth is given by the geometry of the horn. It is in principle independent of frequency, as can be expected according to the stationary phase conditions. This property is particularly well proven in practice for corrugated horns. It is one of their advantages (Fig. 12.65).

4. *Phase centre* The curvature of the wave surfaces is defined by the abscissa $\gamma = (z - R)$ of the phase centre:

$$\gamma = -\frac{1 + \frac{z}{L}}{\frac{1}{L} + z\left(\frac{1}{L^2} + \frac{4\lambda^2}{\pi^2 a^4}\right)} \tag{12.176}$$

   (a) Near zone: small z: $\gamma = -L \cdot C$ near the apex S of the cone (at infinity if the aperture is equiphase).
   (b) Far zone: $z >> L$.
      (i) L large (narrow horn): $\gamma \simeq 0 \cdot C$ near 0, centre of the aperture.
      (ii) L small (wide horn): $\gamma \simeq -L \cdot C$ near the apex S of the cone.

   We are under stationary conditions.
5. *Optimum horn with minimum scatter* The problem is the following: given a horn L of a limited length for size reasons. This horn is to be used to illuminate a zone in a plane normal to the axis, located at distance z. What horn aperture a yields the maximum concentration of the field (or the least scatter) on this plane? This problem occurs, for example, in the case of a project for a Cassegrain antenna. The existence of a minimum is obvious from the expression of scatter (12.172). It is obtained by deriving w'^2/z^2 with respect to a^2:

$$a_{\text{opt.}} = \sqrt{\frac{2}{\pi}}\sqrt{\frac{\lambda L}{1 + \frac{L}{z}}}. \tag{12.177}$$

   At a far distance ($z = \infty$), we obtain the minimum beamwidth and the optimum aperture of the horn:

$$\theta_{0\text{opt.}} = \sqrt{\frac{2}{\pi}}\sqrt{\frac{\lambda}{L}} \quad \text{and} \quad a_{\text{opt.}} = \sqrt{\frac{2\lambda L}{\pi}} \tag{12.178}$$

12.5.3 Parabolic reflector

Introduction

We have referred to the parabolic reflector several times. It is frequently used in all the fields covered by radar, telecommunications and radioastronomy. It is a simple way to obtain high gains and directivity. Many of its properties are a reference for other types of symmetrical structures such as a Cassegrain antenna (section 12.5.4).

A parabolic reflector has a primary feed whose phase centre (section 12.4.1) should coincide with the focal point. A primary feed with axial symmetry is often used which yields a characteristic surface which has the same symmetry:

it is a Huygens source (section 12.3.6). A good example of such a feed is a corrugated horn (section 12.5.2). The reflector assembly of such a feed has the same axial symmetry; it therefore forms a Huygens source itself for which the properties are known, in particular, polarization (section 12.3.6). If the primary feed is not a Huygens source, we know that the level of cross-polarization results from the difference of the patterns observed in the two orthogonal main planes of symmetry (section 12.5.1).

Equivalent aperture (Fig. 12.66)

We can see that the reflector determines in a plane perpendicular to the axis near the reflector — for example plane π passing through F, a circular domain limited to diameter D of the reflector which functions as a radiating aperture.

We have seen (section 12.3.2) the expression of illumination in the case of a spot source. In the general case, this illumination law (electrical field in M) is deduced from the pattern of the primary feed. In fact, since the energy conserves itself in the ray tubes, the field in M, except for the phase, is the same as the field in M′ resulting from the radiation of the primary feed. Let W be the power radiated by the latter. Therefore (see Basic antenna concepts, formula 12.117).

$$E(\mathrm{M}) = \frac{1}{\rho}\sqrt{60\,Wg(\boldsymbol{u})} \qquad (12.179)$$

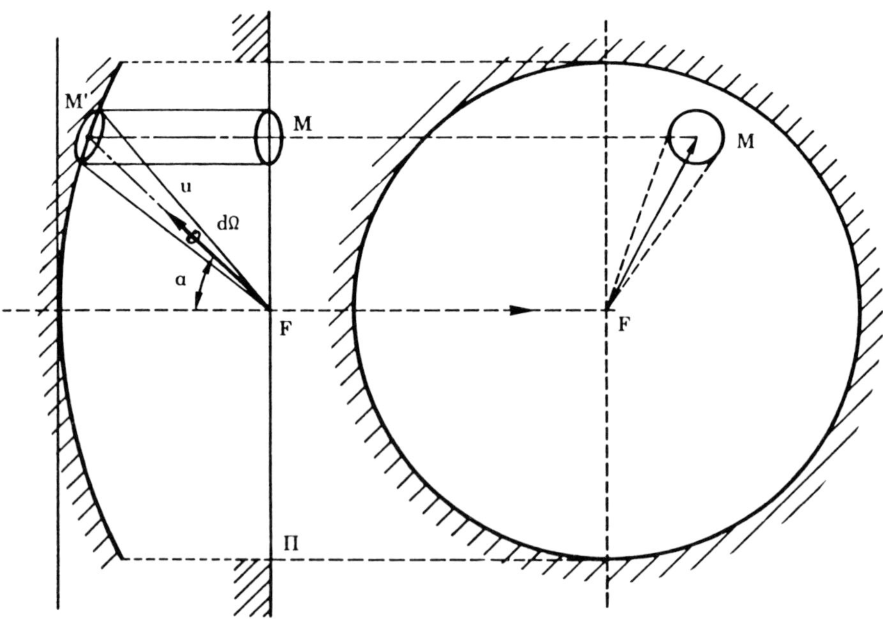

Fig. 12.66 Equivalent aperture of a parabolic reflector.

with $\boldsymbol{u}$ the unit vector of direction FM′, $g(\boldsymbol{u})$ gain of the primary feed, and $\rho = \mathrm{FM}'$. The phase of E, $\phi(\mathrm{M})$, is the exact duplicate of phase $\psi(\boldsymbol{u})$ of the primary feed.

Knowing the illumination law of the equivalent aperture, we can deduce all of the antenna's properties: gain and pattern.

Paraboloid gain

Let W be the transmitter power. Sent into an omnidirectional feed (of unit gain), it creates at distance r_0 a field (12.117):

$$E_1 = \frac{\sqrt{60\,W}}{r_0}\,.$$

Sent into the antenna, it creates a given field (in modulus) based on the formula (12.46):

$$E_1' = \frac{1}{\lambda r_0} \int E\,\mathrm{d}S \qquad (12.180)$$

where E is the field on the reflector given by (12.179).
The gain of the paraboloid with respect to the isotropic antenna is therefore:

$$G_1 = \left|\frac{E_1'}{E_1}\right|^2 = \frac{1}{\lambda^2}\left|\int \frac{\sqrt{g(u)}}{\rho}\,\mathrm{d}S\right|^2. \qquad (12.181)$$

Example 1

A reflector with a long focal distance (Fig. 12.67) is seen from the focal point as a solid angle:

$$\Omega = \frac{S}{F^2}\,.$$

The primary pattern – assumed ideal – is a cone with a solid angle Ω. Its gain is therefore:

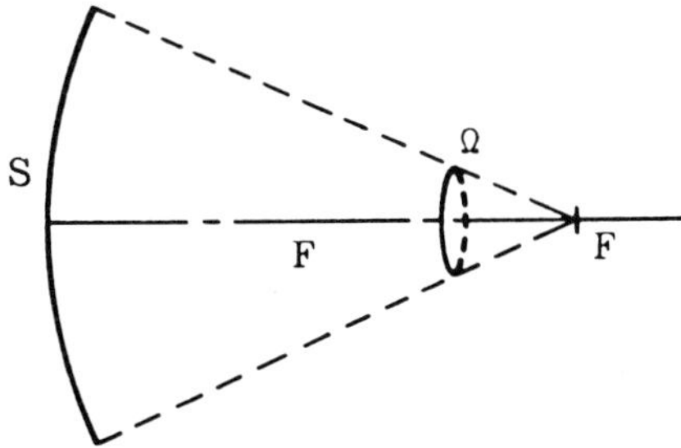

Fig. 12.67 Reflector with a long focal distance.

$$g = \frac{4\pi}{\Omega}.$$

In this case, formula (12.181) yields for the paraboloid:

$$G_1 = \frac{4\pi S}{\lambda^2}.$$

We can verify that this is the theoretical maximum gain of an aperture of surface area S.

Example 2

Reflector illuminated by an axially symmetric primary pattern ($g(\boldsymbol{u}) = g(\alpha)$). Calculation yields for the 'gain factor'; the ratio of the actual gain to the theoretical maximum gain:

$$K = \frac{1}{\tan\frac{\alpha_0}{2}} \left| \int_0^{\alpha_0} \sqrt{g(\alpha)} \tan\frac{\alpha}{2}\, \mathrm{d}\alpha \right|^2 \qquad (12.182)$$

where α_0 is the apparent half-diameter of the reflector seen from focal point F.
In the case where:

$$g(\alpha) = g_n \cos^n \alpha,$$

the gain factor can be calculated. By integrating $g(\alpha)$ to check out the normality condition (12.116), we find:

$$g_n = 2(n+1).$$

Under these conditions, it can be shown (Fig. 12.68) that, for any n, the curves of K all passed through a maximum near 0.82 when the primary directivity is such that:

$$N = 10 \log \frac{g(\alpha_0)}{g(0)} \simeq -10 \text{ dB}.$$

Whence the practical rule: level N of the primary pattern observed on the edge of the reflector should be approximately – 10 dB lower than its value in the axis.
Actually, due to side and scattered radiation and losses, the primary feed always has a gain g_n less than the theoretical value found above and the maximum gain factor of the paraboloid is therefore given by:

$$K_{\max} \simeq 0.82 \cdot \frac{g_n}{2(n+1)}.$$

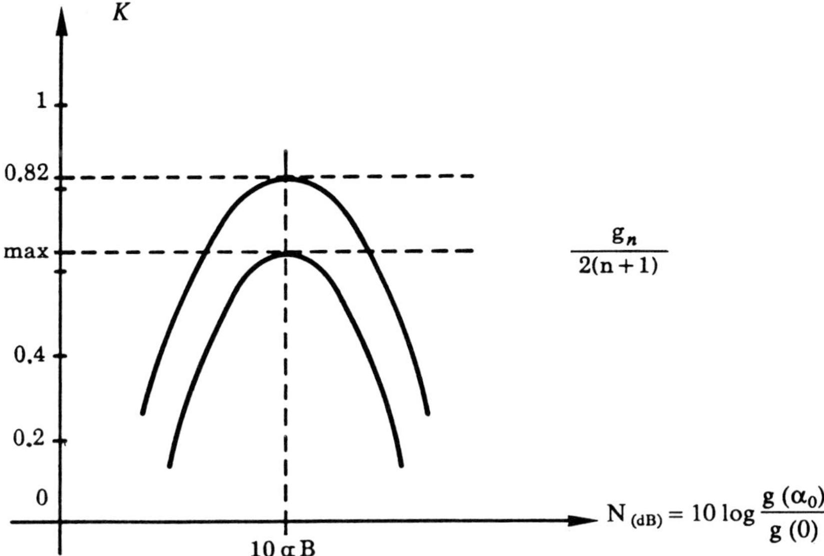

Fig. 12.68 Gain factor of a parabola with respect to the primary feed directivity: theoretical (upper) and measured (lower).

For example:

$$\frac{g_n}{2(n+1)} = 0.80$$

whence:

$$K_{max} \simeq 65\%.$$

The actual gain factor is often less due to other interference effects: blocking by the primary feed and its supports, manufacturing tolerances, etc. (Silver, 1949).

Radiation pattern

If the primary feed is a Huygens source, polarization of illumination on the aperture is constant. The illumination law can be described by a scalar function f(M). We have seen that the characteristic function is the result of the two-dimensional Fourier transform of this illumination. The problem has been treated in section 12.3.5 (circular aperture).

Operation on reception

A radiating source at infinity on the axis produces a plane incident wave which, after reflection, yields a spherical wave converging towards the focal point. The resulting field distribution in the focal plane is therefore the image of the source. This image is not a spot: we will briefly analyse its structure.

Problem posed

Consider a cap of a spherical wave Σ converging towards a point O, the focal point. Its apparent half-diameter seen from O is ϕ_0. We are looking for the resultant distribution in the focal plane xOy perpendicular to the focal axis SO (Fig. 12.69). Length F = SO is the focal distance. This distribution is called the diffraction pattern (it is also called the diffraction spot).

Let U_Σ (M) be a scalar component of the field at a point M of the wave surface Σ (we assume that on this surface the field has a plane wave structure at least locally). According to the Huygens–Fresnel formula (12.35) the corresponding component U(P) at point P of the focal plane is given by:

$$U(\mathrm{P}) = \frac{j}{\lambda} \int\int_\Sigma \gamma\, U_\Sigma(\mathrm{M}) \frac{\exp\left(-j\dfrac{2\pi R}{\lambda}\right)}{R}\, \mathrm{d}S \tag{12.183}$$

(R = MP, F = MO). We will transform this expression and show that the diffraction pattern U(P), is the two-variable Fourier transform of the U_Σ(M) distribution, provided certain approximations are accepted.

Note that the calculation that follows is purely scalar. There is a vectorial calculation in the bibliography which yields magnetic and electric field lines

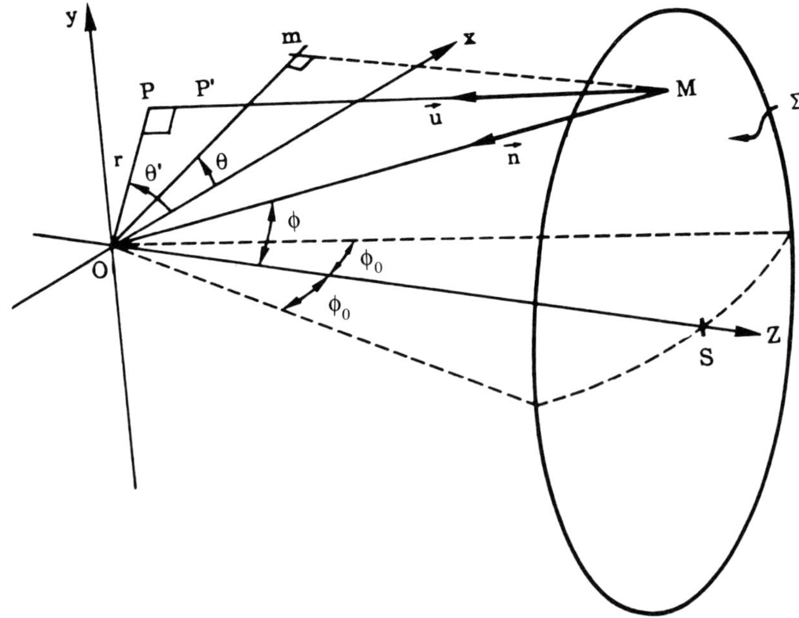

Fig. 12.69 Geometry for computing the diffraction pattern.

in a diffraction spot (Roubine *et al.*, 1987, Minett *et al.* 1968). In fact these lines necessarily have the property of axial symmetry mentioned in section 12.3.7 on hybrid waves. Effectively, an axial plane wave has this symmetry. This symmetry is conserved after reflection on a circularly symmetrical mirror. In the diffraction spot, the magnetic field lines are globally deduced from the electric field lines by a $\pi/2$ rotation about the axis. Their equations are therefore of the form (12.161) or (12.162), (for a long focal distance we can say that $B \simeq 0$).

Transformation of a field expression

Since P is in the neighbourhood of the focus we can make the following approximations:

1. $\gamma \simeq 1$. In effect, angle $(\boldsymbol{n}, \boldsymbol{u}) = \mathrm{OMP}$ is narrow.
2. $1/R \simeq 1/F$ (F focal distance), this factor can be extracted from the integral. Therefore:

$$U(\mathbf{OP}) = \frac{j}{\lambda F} \int\int_{\Sigma} U_{\Sigma}(\mathbf{OM}) \exp\left(-j\frac{2\pi R}{\lambda}\right) \mathrm{d}S \tag{12.184}$$

3. Evaluate the exponential term

$$R^2 = |\mathbf{PM}|^2 \qquad \mathbf{PM} = \mathbf{OM} - \mathbf{OP}$$

$$R^2 = |\mathbf{OP}|^2 + |\mathbf{OM}|^2 - 2\,\mathbf{OP}\cdot\mathbf{OM} = r^2 + F^2 - 2\,\mathbf{OP}\cdot\mathbf{Om}$$

Om is the projection of **OM** on $x\mathrm{O}y$, and $|\mathbf{OP}| = r$.
Approximately:

$$R \simeq F\left(1 - \frac{\mathbf{OP}\cdot\mathbf{Om}}{F^2} + \frac{r^2}{2F^2}\right) = F - \frac{\mathbf{OP}\cdot\mathbf{Om}}{F} + \frac{r^2}{2F}.$$

Which yields:

$$U(\mathbf{OP}) = \frac{j}{\lambda}\,\frac{\exp\left(-j2\pi\dfrac{F}{\lambda}\right)}{F}\exp\left(-j\frac{2\pi}{\lambda}\frac{r^2}{2F^2}\right)$$

$$\int\int_{\Sigma} U_{\Sigma}(\mathbf{Om}) \exp\left(-j\frac{2\pi}{\lambda}\frac{\mathbf{OP}\cdot\mathbf{Om}}{\mathbf{F}}\right)\mathrm{d}S. \tag{12.185}$$

Given:

$$\frac{\mathbf{OP}}{\lambda} = \boldsymbol{\nu} \qquad \frac{\mathbf{Om}}{\mathbf{F}} = \boldsymbol{\mu} \qquad \frac{\mathrm{d}S}{\lambda^2} = \mathrm{d}\Sigma \qquad \frac{F}{\lambda} = \Phi$$

$$U(\mathbf{OP}) = U(\boldsymbol{\nu}) = j\,\frac{\exp(-j2\pi\Phi)}{\Phi}\,\exp\left(-j\,\frac{2\pi}{\lambda}\,\frac{r^2}{2F}\right)$$

$$\iint_{\Sigma} U_{\Sigma}(\boldsymbol{\mu})\,\exp(-j2\pi\,\boldsymbol{\mu}\cdot\boldsymbol{\nu})\,\mathrm{d}\Sigma. \qquad (12.186)$$

We can recognize the Fourier transform of $U_{\Sigma}(\mu)$ in the integral.

This result is fundamental. It is the basis of a large number of properties for diffraction patterns.

Note that we have noticed a quadratic phase variation in r. This expresses the fact that the focal surface is not rigorously plane; this term disappears if the diffraction pattern is observed at P′ located on a sphere with centre S, vertex of the spherical dome Σ, and with radius SF, instead of observation at P of xOy. This gives:

$$U(\mathrm{P}') = j\,\exp(-j2\pi\Phi)\,\frac{1}{\Phi}\iint_{\Sigma} U_{\Sigma}(\boldsymbol{\mu})\,\exp(-j2\pi\,\boldsymbol{\mu}\cdot\boldsymbol{\nu})\,\mathrm{d}\Sigma. \qquad (12.187)$$

Expression of the diffraction pattern by a simple integral

In a frequent particular case, where the illumination law U_{Σ} is axially symmetric, the diffraction pattern can be expressed in the form of a simple integral. In effect, by using spherical coordinates, the element of area can be written:

$$\mathrm{d}\Sigma = \frac{F^2}{\lambda^2}\,\sin\Phi\,\mathrm{d}\Phi\,\mathrm{d}\theta$$

and:

$$\boldsymbol{\nu}\cdot\boldsymbol{\mu} = \frac{r}{\lambda}\,\sin\Phi\,\cos(\theta-\theta').$$

We obtain:

$$U(P') = j\,\exp(-j2\pi\,\Phi)\,\frac{F^2}{\Phi\,\lambda^2}\,V(P') = j\Phi\,\exp(-j2\pi\,\Phi)\,V(P')$$

with

$$V(\mathrm{P}') = \int_0^{\phi_0}\int_0^{2\pi} U_{\Sigma}(\phi)\,\exp[\,j2\pi\,\nu\,\sin\phi\,\cos(\theta-\theta')]\,\sin\phi\,\mathrm{d}\phi\,\mathrm{d}\theta. \qquad (12.188)$$

At this point, the rest of the calculation is perfectly comparable to that of the radiation of a circular aperture illuminated by an illumination law with the same symmetry (section 12.3.5).

Particular cases: quasi uniform illumination law
The calculation can be completely carried out in the case where:

$$U_\Sigma(\phi) = U_0 \cos\phi.$$

This represents an illumination law of an incident wave slightly decreasing with respect to ϕ: for example, for $\phi = 30°$, $U_\Sigma = 0.866$. We can then use $\sin\phi$ as a differential element. A first integration yields:

$$V(P') = 2\pi j U_0 \int_0^{2\pi} J_0\,(2\pi\gamma \sin\phi) \sin\phi \,\mathrm{d}(\sin\phi).$$

Let

$$Z = 2\pi\nu \sin\phi.$$

We can demonstrate the following relationship:

$$\int_0^{Z_0} J_0\,(Z)\, Z \,\mathrm{d}Z = Z_0\, J_1(Z_0)$$

where J_1 is the first-order Bessel function.
It then follows that:

$$U(\nu) = -\exp(-j2\pi\Phi)\; U_0 \pi\Phi \sin^2\phi_0 \left[2\,\frac{J_1(2\pi\,\nu \sin\phi_0)}{2\pi\,\nu \sin\phi_0}\right].$$

Which can be written as:

$$U(\nu) = -\exp(-j2\pi\Phi)\; \frac{U_0}{2}\left(\frac{\pi D}{\lambda}\right)\sin^2\phi_0 \left[2\,\frac{J_1(2\pi\,\nu \sin\phi_0)}{2\pi\,\nu \sin\phi_0}\right]. \qquad (12.189)$$

where D is the pupil diameter, U_0 is the field at the centre of the spherical wave cap, and ϕ_0 is the apparent half-diameter of the pupil seen from the focal point.

Examination of the diffraction pattern on the focal surface

The variable here is equal to r/λ, with r the distance to the focal axis. The shape of the diffraction pattern is characterized by the function between brackets:

$$\boxed{\psi\,(\nu) = 2\,\frac{J_1\,(2\pi\nu \sin\phi_0)}{2\pi\nu \sin\phi_0}\,.} \qquad (12.190)$$

It is represented by a surface of revolution having a central lobe (central diffraction spot) whose maximum is normalized to 1 and a series of annular lobes (bright rings) separated by null-field circular voids (dark rings) (Fig. 12.70).

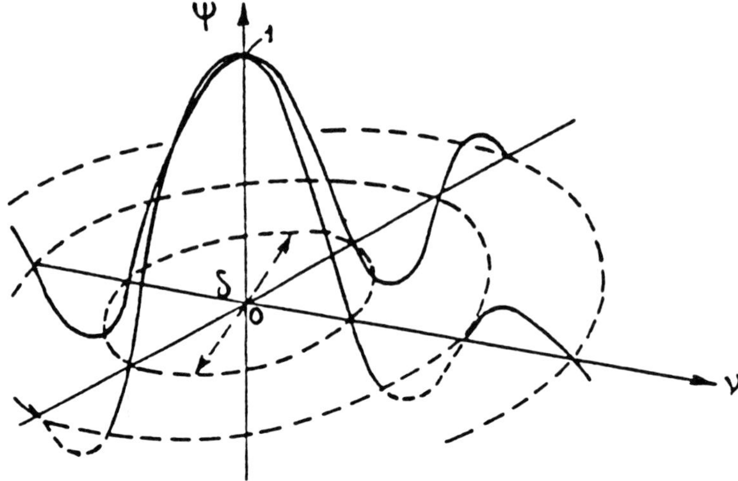

Fig. 12.70 Diffraction pattern on the focal surface.

Diameter of the central diffraction spot

An important characteristic is the diameter δ of the central lobe (or the first dark ring). It is given by the first zero of function J_1:

$$2\pi \frac{\delta}{2\lambda} \sin \phi_0 = 3.83.$$

We can deduce the classical formula used in physical optics:

$$\boxed{\delta = \frac{1.22\,\lambda}{\sin \phi_0}.} \qquad (12.191)$$

Level of the first secondary ring

Compared to the maximum of the main lobe, this level is 0.132, or around − 17.5 dB.

Energy distribution in the diffraction spot

It is interesting to know the percentage of energy contained in the centre lobe and more generally in a circle with radius r_1. This is given by the integral of $|\psi(\nu)|^2$

$$p(\nu_1) = \frac{\int_0^{\nu_1} |\psi|^2 \nu \, d\nu}{\int_0^{\infty} |\psi|^2 \nu \, d\nu} = 1 - J_0^2 (2\pi\nu_1 \sin \phi_0) - J_1^2 (2\pi\nu_1 \sin \phi_0).$$

If we consider the successive dark rings $J_1 = 0$, then:

$$p(\nu_1) = 1 - J_0^2\,(2\pi\,\nu_1 \sin\phi_0).$$

The results are shown in Fig. 12.71.

Applications

Optimum dimensions of a primary horn.

Consider a parabolic reflector operating for example as a receiver. Outside the focal plane, according to the stationary phase principle, the field distribution is approximately described by the theory of rays.

At the focal point we have obtained the diffraction pattern. To collect electromagnetic energy concentrated there, an electromagnetic horn connected to a waveguide is often used. For the incident field to enter the horn effectively, the distribution has to be close to the distribution corresponding to the propagation mode in the guide.

If the horn is too small, this condition is met approximately, but energy is lost around it. If the horn is large, this condition is generally not met. The optimum is generally obtained when the dimensions of the horn are close to those of the central diffraction spot.

Note that we can show that the optimization condition of a horn on reception corresponds to the condition known at transmission whereby the pattern of the horn intersects the edges of the pupil at a level approximately 10 dB lower than the radiation maximum (section 12.5.3).

The proportion of energy intercepted by the reflector picked up by the horn represents the gain factor of the antenna, K. With a simple horn we obtain, $K_{max} \leqslant 0.6$, approx.

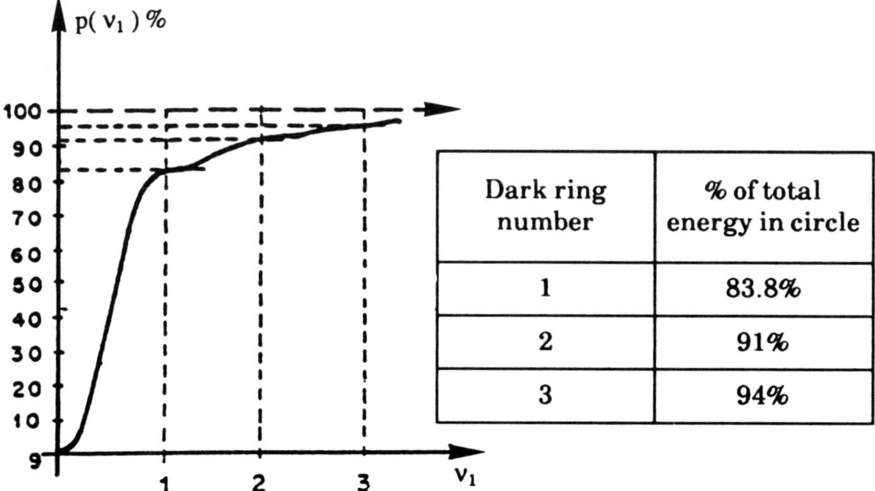

Dark ring number	% of total energy in circle
1	83.8%
2	91%
3	94%

Fig. 12.71 Energy distribution in the diffraction spot.

Gain factor and convolution

More rigorously, if the focal length is long enough (ϕ_0 small) we can demonstrate the following property:

If $\psi(P)$ is the diffraction pattern (electric field), $\mathbf{f}(P)$ the field distribution corresponds to the propagation mode in the horn (Fig. 12.72, 73).

The 'gain factor' (section 12.5.3) can be considered as the result of the product of convolution between ψ and f:

$$K = \frac{\left| \iint \psi \cdot f \, \mathrm{d}S \right|^2}{\iint |\psi|^2 \, \mathrm{d}S \iint |f|^2 \, \mathrm{d}S} \tag{12.192}$$

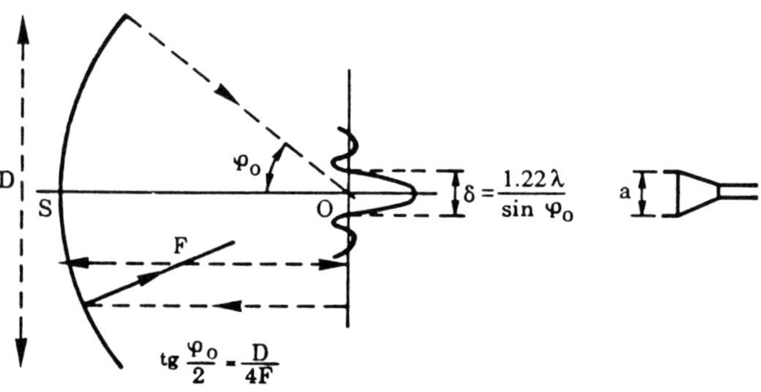

Fig. 12.72 Diffraction spot of a parabolic reflector.

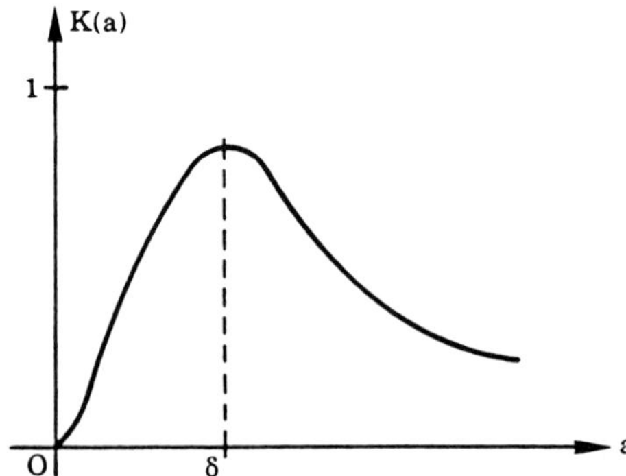

Fig. 12.73 Gain factor variation resulting from the interaction between diffraction spot and primary aperture *a*.

We can say that it is the 'cosine' of the two distributions ψ and f. According to the Schwarz inequality, $\eta \leqslant 1$

$$K = 1 \qquad \text{if} \qquad f = \psi^*.$$

The gain factor is therefore maximum if the illumination law of the primary feed is the duplicate of that of the diffraction spot, with an opposite phase. Here we find a generalization of the concept of a matched filter. The problem can be solved approximately by using multimode horns or corrugated horns (section 12.5.2).

12.5.4 Cassegrain antenna

Introduction

Guillaume Cassegrain was a sculptor and a founder at Chartres. In 1672, he proposed a two-mirror telescope: the small mirror is convex and sends the image to the centre of the main mirror which has a hole drilled in it to enable observation.

In geometrical optics, the meridians of the reflectors have to be calculated so that the image of a point at infinity on the axis is 'stigmatic' i.e. is also a point. A classical rigorous solution to this problem has been defined by Newton: if an intermediate image, also stigmatic, is imposed, application of the optical path law shows that the main mirror has to be parabolic and the secondary mirror has to be hyperbolic.

For large antennas operating with wavelengths in the centimetre range, the adoption of Cassegrain structures was motivated at the start by the same advantages offered by the optical telescopes: the presence of an image near the centre of the main mirror allows placement of the primary feed in an easily accessible location. Other advantages showed up afterwards.

1. The first involves the antenna's noise temperature. When a classical parabolic reflector is turned towards the sky, its primary feed is orientated towards earth; it is therefore likely to pick up noise from the ground. Conversely, the feed in a Cassegrain antenna is turned towards the sky as well, it only receives signals which come from the sky. It is much less sensitive to noises from the ground.
2. Another advantage lies in the possibility of obtaining an excellent gain factor. In effect, the presence of two reflectors allows choosing, to a certain extent, the form of illumination of the equivalent aperture. If it is constant, it leads to maximum gain, this is the principle of so-called 'conformal' Cassegrain antennas, the meridians no longer have a simple equation but the manufacturing cost does not increase (Williams, 1965). This is why this technique is widespread for satellite telecommunications earth stations (section 12.4.2).

Geometry (classical solution) (Roubine et al., 1987)

The Cassegrain antenna has two axially symmetric reflectors. The primary feed located at the vertex of the main reflector illuminates the latter via an auxiliary reflector located in front of it. The two surfaces are calculated so that after a double reflection, the primary wave becomes plane. The classical solution consists in choosing a hyperboloid and paraboloid with the same focal point. The primary feed phase centre C′ coincides with one of the foci F' of the hyperboloid (Fig. 12.74).

We shall place ourselves in a plane passing through the axis of revolution SS′. We will use the following notation:

1. effective diameters of the reflectors: D, D′;
2. path of a ray from the primary feed: C′MM′P;
3. vertex of the main and auxiliary reflectors: S, S′;
4. focus of the paraboloid, coincident with one of the foci of the hyperboloid: C;
5. second focus of the hyperboloid and phase centre of the primary feed: C′;
6. hyperboloid magnification:

$$m = \frac{\mathrm{C'S'}}{\mathrm{S'C}}$$

(see below).
The following approximate relationship can be demonstrated

$$D/D' \simeq m.$$

The surfaces of the reflectors are defined, for example, by the following equations:

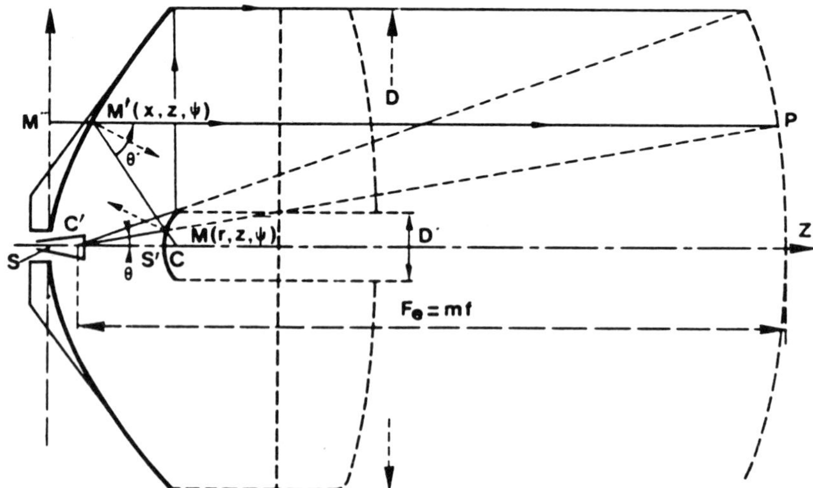

Fig. 12.74 Geometry of a Cassegrain antenna.

paraboloid:

$$\frac{X}{2F} = \tan\frac{\theta'}{2} \tag{12.193}$$

and hyperboloid:

$$m = \frac{\tan\theta'/2}{\tan\theta/2} \tag{12.194}$$

Principal surface (Fig. 12.74)

The principal surface of the system is by definition, the locus of the points of intersection 12.of rays C′M from the primary feed with the emerging rays M′P parallel to the axis. In the classical case of a paraboloid-hyperboloid pair, it can be found by combining equations (12.193) and (12.194) for the two reflectors, that this locus is a paraboloid whose focal distance (called equivalent focal length) is: $F_e = mF$ (with F the focal distance of the paraboloid).

The radioelectrical characteristics of the Cassegrain antenna, as for any axially symmetric system, are defined by its principal surface. In the geometrical optics approximation, analysis of the Cassegrain system can be reduced to that of a paraboloid (section 12.5.3).

Blocking by the auxiliary reflector (R′)

Blocking effect

The illumination of the system spreads over a domain of diameter D. It has a centre hole of diameter D' due to the shadow formed by the auxiliary reflector. This illumination can be considered as the sum of the non-disturbed illumination and of a central parasitic interference illumination, of an amplitude equal to the maximum amplitude of the former and in phase opposition with it. The global radiation pattern results from the superposition of these illuminations (Fig. 12.75). In particular, the parasitic illumination produces a pattern m times wider than the main lobe in the form of a side radiation which diminishes the axial gain.

Let A be the effective area of the antenna ($A = KS$, with K the gain factor and S the geometrical projected area of the main reflector) and S', the projected area of the auxiliary reflector. The central parasitic illumination is fed with the fraction S'/A of the total power and radiates with a gain proportional also to fraction S'/A. The relative power level of parasitic radiation is therefore:

$$N^2 = \left(\frac{S'}{A}\right)^2 = \frac{1}{K^2}\left(\frac{D'}{D}\right)^4 = \frac{1}{K^2 m^4} \tag{12.195}$$

Example: for $K \simeq 1$ and $D'/D = 1/10$ we have: $N^2 = 10^{-4}$, i.e. -40 dB, this typical value is often acceptable.

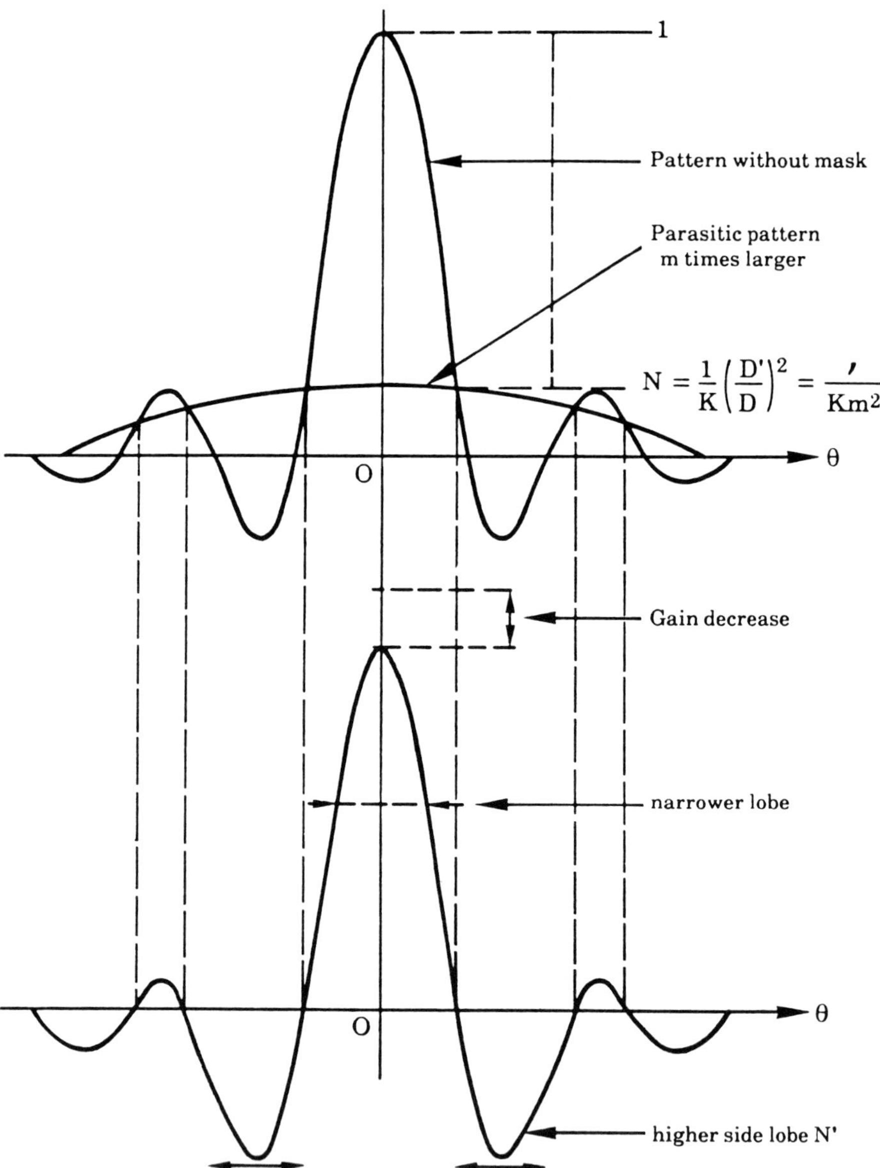

Fig. 12.75 Pattern resulting from the mask effect.

The gain decrease is measured by the relative decrease in the equivalent area, that is:

$$\frac{G'}{G} = \frac{A - S'}{A} = 1 - \frac{1}{K}\frac{S'}{S} = 1 - N. \qquad (12.196)$$

The preceding example yields a loss of: – 0.04 dB.

Note that the theoretical pattern (in the absence of blocking) has a sidelobe level N_0 (for example – 17.6 dB for a theoretical uniform illumination). The overall level N' is a result of the combination of N_0 and N. In the worst phase combination case, we find that:

$$N' = \frac{N_0 + N}{1 - N}. \tag{12.197}$$

Note that telescopes are often Cassegrain telescopes. Bright stars produce images with halos and radiating peaks which are due to the auxiliary reflector and its supports.

Elimination of the blocking effect by rotating the polarization plane

For large antennas, the above effect is often negligible (radiotelescopes, space telecommunications). Conversely, for small antennas it is a problem.

1. **Principle**. This system is used in applications where only rectilinear polarization is used. The principle of the process used is shown in Fig. 12.76. The primary feed radiates a horizontal polarization wave. The auxiliary reflector consists of horizontal metal wires which reflect this wave towards the main reflector. This reflector surface is covered with a layer which rotates the polarization plane by 90 °. The structure of this layer is described below. Polarization of the plane wave reflected by the paraboloid being vertical, this wave crosses, without reflection, the horizontal wires forming the auxiliary reflector. The blocking effect is therefore eliminated.

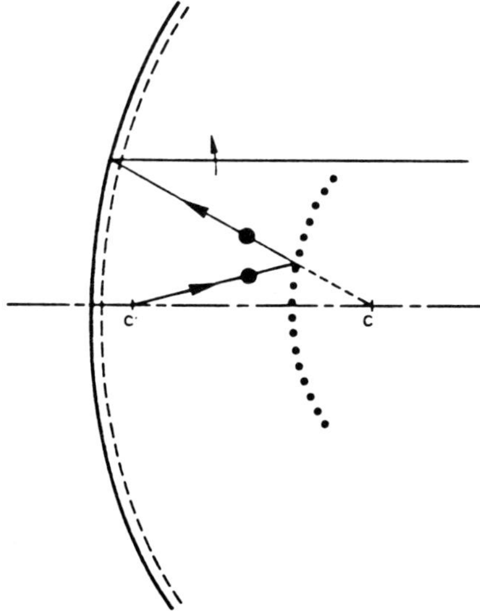

Fig. 12.76 Elimination of the mask effect by rotating the polarization plane.

2. **Structure of the polarization rotating layer**. At its simplest form, this system is a layer of parallel wires inclined 45 ° with respect to the vertical, located on a surface parallel to that of the reflector at a distance of a quarter wavelength from it.

   If we imagine (Fig. 12.77) an incident wave with horizontal polarization P_1, this wave can be considered as the superposition of two equiphase component waves, whose polarizations are inclined at 45 °: one is parallel to the wires (P_1) the other perpendicular (P_1''). The first component is reflected by the wires (P_2'), the second crosses them and is reflected by the solid reflector. It travels an extra path of a half-wavelength and after reflection is out of phase by π with respect to the first (P_2'').

   The combination of these two reflected components therefore creates a wave with vertical polarization (P_2). This operating mode assumes that the array of inclined wires is perfectly reflecting for the component whose polarization is parallel to the wires and perfectly transparent (without phase shift) for the orthogonal component. This condition becomes better approximated as the wires become thinner and more tightly packed. The residual phase shifts can often be corrected by changing the distance separating the wire layer and the reflector.

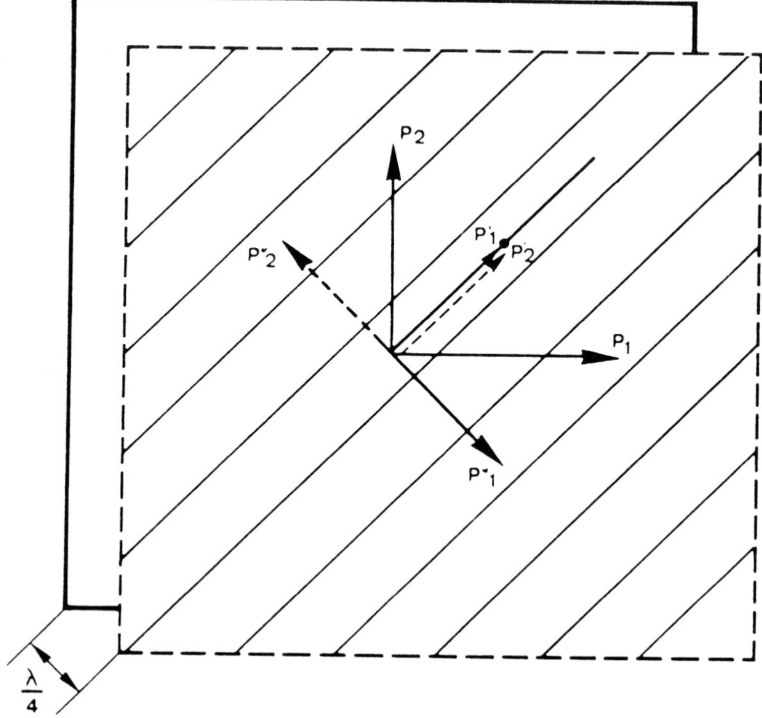

Fig. 12.77 Structure of the polarization rotating layer of parallel wires.

In principle, this system only operates at a single frequency. Practically, a relative frequency band of a few percent can be accepted. There is a system with a wider band where the wires are more widely spaced and in which a compensation phenomenon between the selectivity of reflection and transmission characteristics of the wire layer and that of the optical path separating the wires from the reflector can be worked on. The metal wires are generally held in place by a transparent plastic support.

12.5.5 Tracking systems

Introduction

Tracking systems are used to maintain a narrow beam of radiation in the direction of a mobile remote object (radar target, telecommunications satellite, etc.). The antenna is generally a focusing system set on a turret to provide angular mobility most often about a vertical axis (bearing or azimuth axis) and horizontal axis (elevation axis). The turret is driven by servomechanisms. The elevation or bearing error signals which control them are acquired by the signal(s) picked up by the antenna through different methods of interpolation. The first uses a conical scanning of the narrow beam: the error signal is the result of the time-dependent modulation of the signal received. Another method (monopulse), compares the signals received by several adjacent narrow beams. In fact, we can show that these two methods are basically a way to reduce the differences in phase of the incident wave on the antenna aperture: tracking antennas have a tendency to orientate their aperture parallel to the surface of the incident wave.

Monopulse antennas

Analysis of the diffraction pattern (Fig. 12.78)
Consider an antenna with a centered focusing system (for example a reflector) and approximately pointed in the direction of a radiating target. This target produces an image near the focus in the antenna's focal plane. This image is not a perfect point image, it is a diffraction pattern whose diameter δ of the centre spot is the result of the relative aperture of the antenna θ_0 and the wavelength in accordance with the classical formula (section 12.5.3). It is shown on Fig. 12.78.

The problem of determining the target's position with respect to the axis of symmetry of the antenna is determining the exact location of the centre of the diffraction pattern with respect to the focus. To solve this problem, compare the level of the diffraction spot at various points distributed over a circumference centred on the focus. If these levels are equal, this means that the centre of the spot coincides with the focus. We can say that the principle of a conically-scanning antenna consists of successive explorations of these points

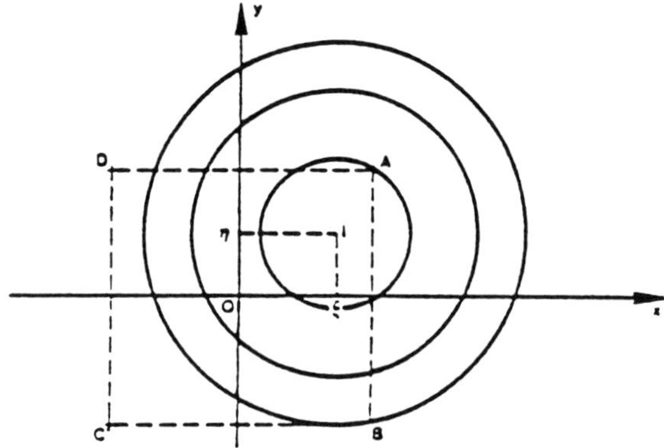

Fig. 12.78 Analysis of the diffraction pattern (isolevel lines).

through a unique primary aperture set in circular motion and in the analysis of the modulation of the signal received.

Conversely, the monopulse antenna, simultaneously compares the amplitude received through several fixed primary apertures (generally four) symmetrically located around the focus. The angular differences between the target's position and the axis of the antenna are measured by the different levels received by the apertures. The differences are compared to the sums so as to obtain information that is independent of the overall level of the received signal.

It should be noted that this method of measurement assumes that the diffraction spot has a symmetrical structure. This is a case for a point target. Conversely, for a complex target, the diffraction pattern can be distorted and the measurements inaccurate. This results in angle noise ('glint').

Sum and difference patterns

We have seen that we have to add and subtract the amplitudes received in the primary horns located around the focus. The variations of sum and difference amplitudes with respect to the angular deviations of the target form the sum and difference patterns. Let us assume that the system is operating on a plane space: the target is characterized by a single angular coordinate a (Fig. 12.79). The focal plane is reduced to a straight line where the abscissas are indicated by a coordinate x. The image of the target is a diffraction spot centred on a point of the abscissa: $\xi = fa$ (f is the focal distance) and whose amplitude law is characterized by an even function: $\psi(x - \xi)$. Let us consider two primary horns whose phase centres are placed symmetrically at a distance $a/2$ from focus F. At these points, the diffraction function amplitudes are:

$$A(\xi) = \psi\left(\frac{a}{2} - \xi\right)$$

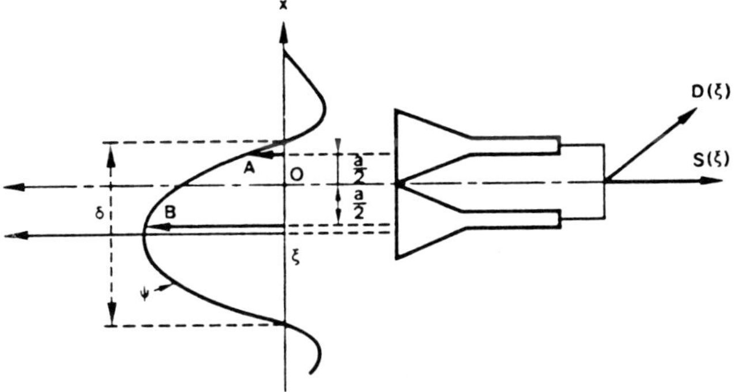

Fig. 12.79 Two-horn receiver for generating sum and difference patterns.

$$B(\xi) = \psi\left(-\frac{a}{2} - \xi\right) = \psi\left(\frac{a}{2} + \xi\right). \tag{12.198}$$

Assume, as the first approximation, that the levels received by each horn are proportional to A and B. These signals are sent to a 'magic Tee' coupler which produces the sum and difference signals. The corresponding patterns are expressed by the functions:

$$S(\xi) = A + B = \psi\left(\frac{a}{2} - \xi\right) + \psi\left(\frac{a}{2} + \xi\right)$$

$$D(\xi) = A - B = \psi\left(\frac{a}{2} - \xi\right) - \psi\left(\frac{a}{2} + \xi\right). \tag{12.199}$$

We can verify that the first function is an even function and the second function is an odd function. These functions are plotted (Fig. 12.80). Near the axis, we can replace the patterns by their principal parts:

$$S = S(0) \qquad D = D'_a(0)\, a \tag{12.200}$$

where $D'_a(0)$ denotes the derivative of $D(a)$ at $a = 0$.

To obtain angle data which is independent of the intensity of the signal received, we calculate the quotient or tracking signal:

$$X(a) = \frac{D}{S} \simeq pa \tag{12.201}$$

Coefficient p, called the tracking slope, is generally close to the reciprocal of the 3 dB-beamwidth of the 'sum' pattern:

$$\boxed{p \simeq \frac{1}{\theta_{3\text{ dB}}}.} \tag{12.202}$$

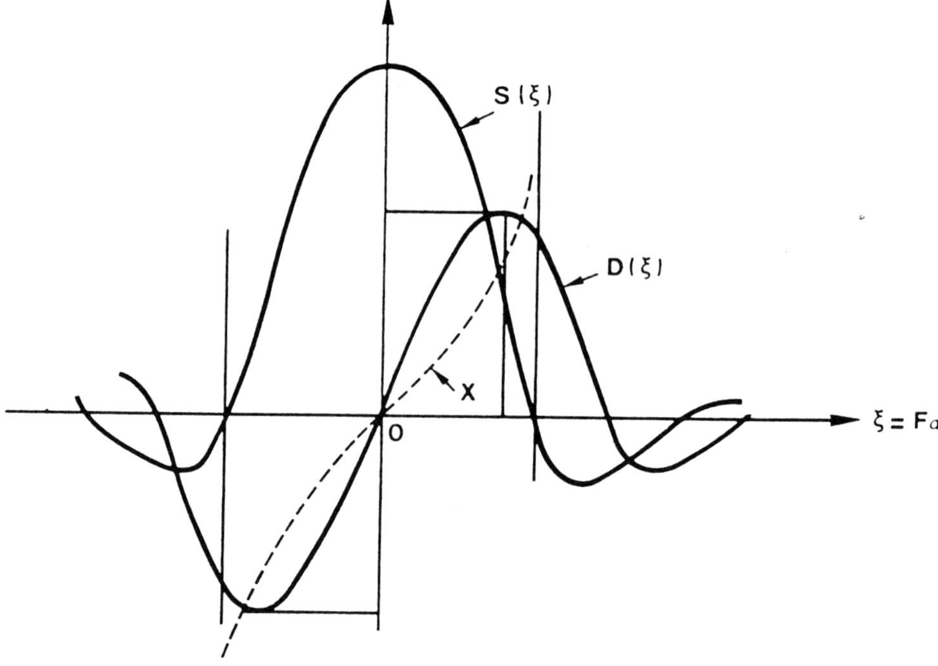

Fig. 12.80 Plot of the sum and difference patterns.

Thermal noise effect – tracking accuracy

This effect is especially critical on the difference channel: assume that a signal D (assumed demodulated) is combined with a noise represented by the stationary random function $n(t)$ with average power:

$$W_B = \overline{|n(t)|^2}.$$

The tracking signal is affected by an error:

$$\Delta X = \frac{D + n(t)}{S} - \frac{D}{S} = \frac{n(t)}{S}$$

which is translated into an angular error:

$$\Delta a = \frac{1}{p}\Delta X = \frac{1}{p}\frac{n(t)}{S}$$

whose standard deviation is the mean:

$$\sigma = \sqrt{\overline{|\Delta a|^2}} = \frac{1}{p}\frac{\sqrt{\overline{|n(t)|^2}}}{\sqrt{\overline{|S|^2}}}.$$

Let us designate the signal power on the sum channel by:

$$|\bar{S}|^2 = W_S$$

This yields:

$$\sigma = \frac{\frac{1}{p}}{\sqrt{\frac{W_S}{W_B}}} = \frac{\left(\frac{1}{p}\right)}{\sqrt{R}} \tag{12.203}$$

where R designates the signal-to-noise ratio. According to (12.202) we can write:

$$\boxed{\sigma = \frac{\theta_{3\,\text{dB}}}{\sqrt{R}}\ .} \tag{12.204}$$

Example: consider a tracking radar using a monopulse antenna of diameter $D = 3$ m radiating at a wavelength $\lambda = 5$ cm. This pulse radar has a repetition frequency $f_r = 1$ kHz. The signal-to-noise ratio per pulse is 10 dB. The integration constant for the servomechanisms is $T = 1$ s. What is the integrated signal-to-noise ratio? What is the 3 dB beamwidth of the pattern? What is the standard deviation of the tracking error?
Answer: $R = 40$ dB (i.e.10^4), $\theta_{3\,\text{dB}} = 1/50$ (or around 1 degree), $\sigma = 2 \times 10^{-4}$, i.e. 0.2 mrad.

This precision is obtained if the mechanical qualities of the antenna and the turret are adequate.

12.5.6 Phased array

Introduction

Due to their inertia, radar antenna drive turrets have time constants which can hardly be less than a second. This results in a loss of precious time when one wants to aim at several targets in succession. Conversely, a system that allows orientation of the radiation beam without inertia would allow full use of the radar's possibilities.

A way to obtain an electronically steered beam has been known for a long time, i.e. phased array antennas. The method consists of feeding an array of radiating sources through the use of phase shifters so that the variations in phase along the array follow an arithmetic progression law. The array produces a plane wave whose orientation is dependent on the phase shifts. In the 1970s the appearance of sufficiently accurate and reliable phase shifters opened the way for the development of this type of antenna.

Simultaneously, the computer, which is becoming widely used, is an ideal control and programming tool for this type of phase shifter. In effect, phase shifters used today are almost always digital. Moreover, the quantized character of these phase shifters imposes certain constraints which affect the performance of arrays (section 12.6.5).

The most elaborate form of an array antenna consists of active arrays where each element is associated with an amplifier which improves efficiency and antijamming and adaptive capabilities.

We shall begin by a classical analysis of passive arrays.

General structure of a phased array antenna (example)

General structure (Fig. 12.81)

An array antenna can have several forms but it always contains the same components. From the point of view of transmission, we find in succession:

1. a power splitter which distributes the energy among the array's radiating elements through phase shifters according to a certain amplitude law. This amplitude law does not vary during the sweep except for adaptive arrays (section 12.6.5);
2. the phase shifter array;
3. a computer to calculate the phases based on the direction of aim (called a phase shifter driver) and associated with a phase shifter control system (to supply the control voltages or currents to the diodes or ferrites in the phase

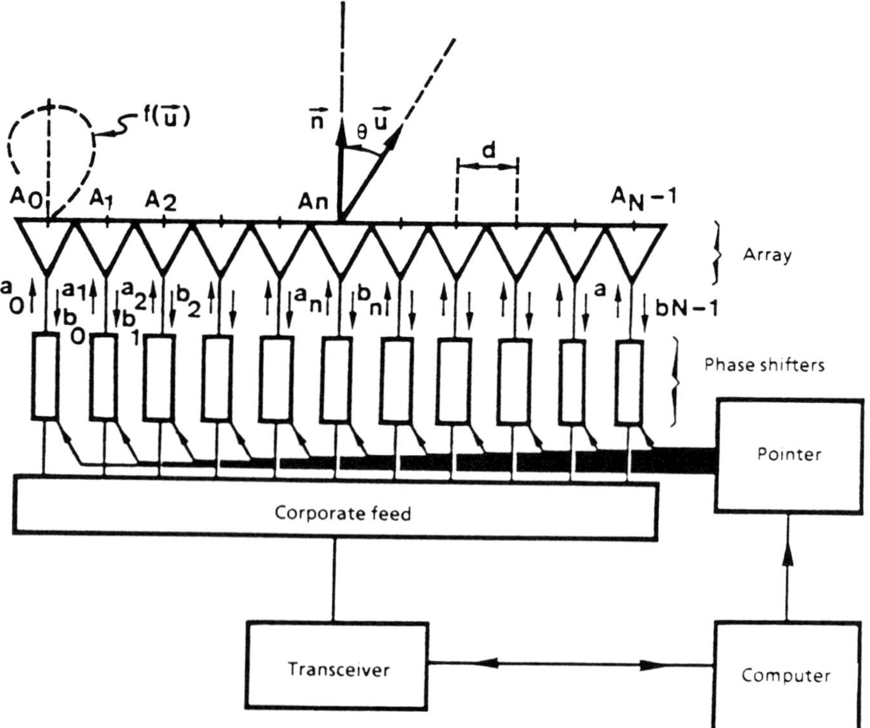

Fig. 12.81 General structure of a phased array.

shifters). The phases are given by the formula (12.207) below. We can see that they are sensitive to wavelength, and a wide bandwidth array requires delay lines in place of phase shifters;

4. the radiating elements, which can be any type of small antenna (dipoles, slots, patches, horns, helical antennas, etc.);
5. and if needed, an additional focusing system (reflector or lens).

Remark on the power splitter

As a consequence of inter-element coupling, the radiating elements present a reflection coefficient that varies with the angle of aim of the beam ('active' reflection coefficient). Therefore the elements reflect a set of waves to the power splitter outputs. If the power splitter outputs are not individually matched and decoupled, these waves are reflected again towards the radiating elements with uncontrolled phases and amplitudes and show up as an interference radiation pattern which is superimposed on the useful pattern. The couplers are therefore generally directive couplers with all their inputs matched (magic Tee, etc.).

Phase shifter structure

Phase shifters are discussed in chapters 1 and 2. Here we shall only briefly review the working principle of a typical diode phase shifter using rings.

A one-bit phase shifter has two possible states: ϕ_0, $\phi_0 + \Delta\phi$. If an input I and an output O are connected by two possible paths I and II whose lengths differ by l, the path followed by the wave can be controlled with three diodes a_1, a_1', a_2 placed in parallel to a quarter-wavelength of tees I and O (Fig. 12.82). The difference in length l introduces a difference in phase:

$$\Delta\phi = 2\pi l/\lambda. \tag{12.205}$$

If p loops of this type are placed one behind the other with individual phase shifts of the form:

$$\Delta\phi_1 = \pi \qquad \Delta\phi_2 = \frac{\pi}{2} \cdots \qquad \Delta\phi_p = \frac{2\pi}{2p} \tag{12.206}$$

we can see that the combination of the $2p$ possible states produces all the phases included between 0 and 2π with a phase quantum of $\Delta\phi_p$. These phase shifts can be implemented by a stripline circuit on a ceramic substrate but there are a large number of other manufacturing processes. The diode control voltages are obtained through transistor circuits which are themselves controlled by the phase shifter driver.

Phase shifter performance involves:

1. its precision (number of bits, precision of each phase quantum);
2. its stability in frequency (temperature has little effect on this type of phase shifter);

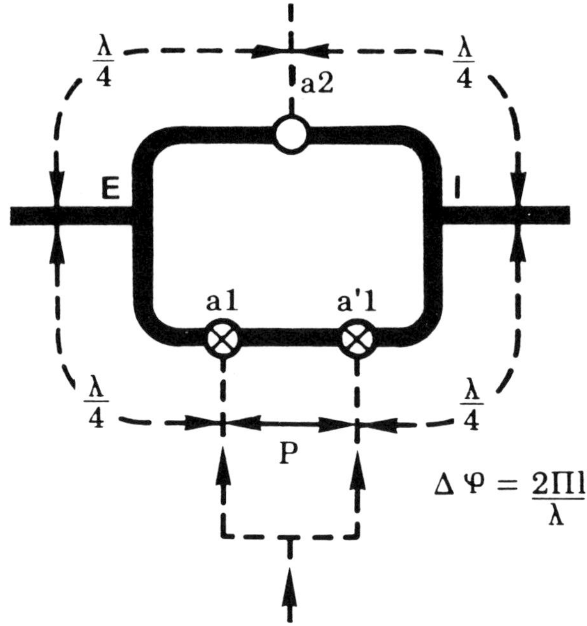

Fig. 12.82 Basic diode phase shifter loop.

3. its insertion losses;
4. its matching, its reproducibility, its price.

Currently used phase shifters generally are 2 to 5 bits. Their precision is on the order of a half bit and insertion losses are from 1 to 2 dB.

Example of array structures

Corporate parallel feed ('candelabra' feed)

This structure consists of a set of consecutive dividers. The dividers are either directive couplers or magic tees whose outputs that are not used in transmission are connected to loads. This precaution is necessary to satisfy the matching and decoupling condition mentioned above.

The *advantages* are:

plane, symmetrical, wide-bandwidth structure.

The *drawbacks* are:

1. complex structure;
2. losses in the guides;
3. difficulty in obtaining an optimal amplitude weighting law, especially if a sum channel and difference channel are wanted simultaneously (see monopulse antenna).

For small arrays, such structures can easily be built with striplines (printed circuit) and integration of phase shifters and if needed, the radiating feeds themselves (printed circuit dipoles or 'patches').

Series feed with waveguide and directive couplers (Fig. 12.83)

Energy is distributed by a centre waveguide whose energy is picked up by stepped directive couplers.

The *advantages* are:

> as opposed to the preceding feed which distributed on a surface, this one has a quasi-linear structure and because of this can take up less room (this is especially of interest for long wavelengths).

The *drawbacks* are:

1. the array is fed by a travelling wave, this results in significant frequency sensitivity of the phase law;
2. all the directive couplers are different;
3. its non-symmetrical structure is detrimental to a precise tracking antenna.

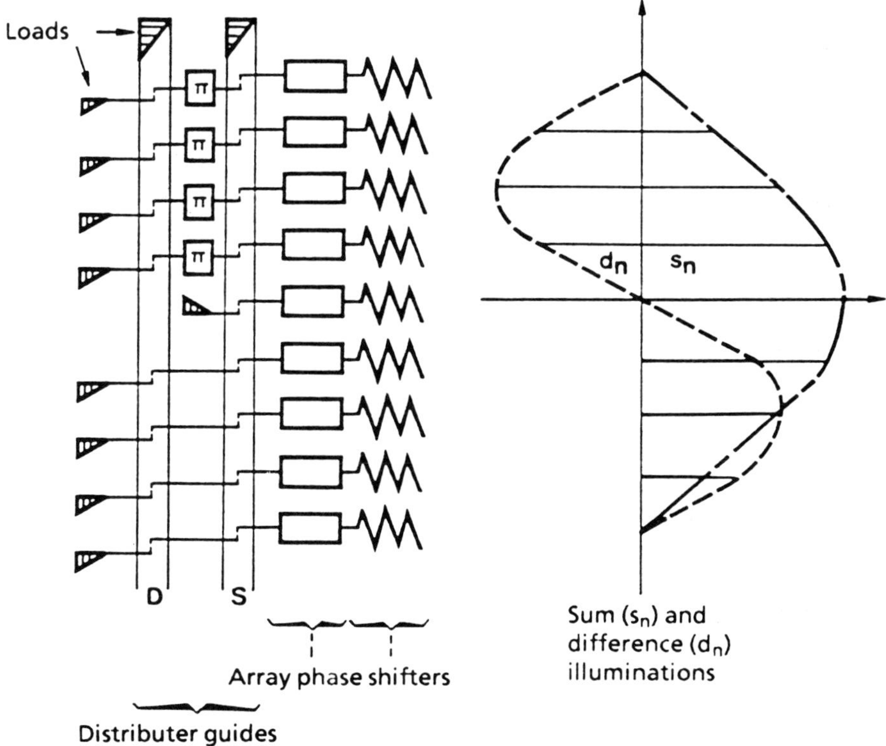

Fig. 12.83 Phased array feed with waveguides and directive couplers.

If a difference channel is wanted, this structure becomes more complicated since it needs a second feed guide and a second set of directive couplers to yield the desired odd distribution. The orthogonality of the sum and difference distributions entails decoupling of the two guides; however, interactions can occur and adjustment has to be done with great care.

Disk feed (Fig. 12.84)

Consider two parallel planes bounded by a circle of diameter D and with a centre O. A central primary feed is likely to emit a radial wave which propagates between these planes. This energy is picked up by couplers (preferably directive) located on concentric circles. The energy is then phased to:

1. compensate the phase shift that occurs during propagation of the radial wave (phase shift proportional to the radius of the circle considered);
2. provide the array with the phase gradient producing the wanted orientation of the beam.

It is possible, with a multimode source, to produce the illuminations corresponding to the various sum and difference patterns of a monopulse antenna; a circular guide carrying radial mode TM_{01} will yield wave TEM with an electric field perpendicular to the walls, which is suitable for a sum illumination. The same guide carrying mode TE_{11} in vertical or horizontal polarization will produce illuminations with odd symmetry in these same planes and will produce 'elevation difference' and 'bearing difference' patterns.

The *advantages* are:

1. plane structure, short longitudinal extent;
2. its symmetry is suitable for tracking functions.

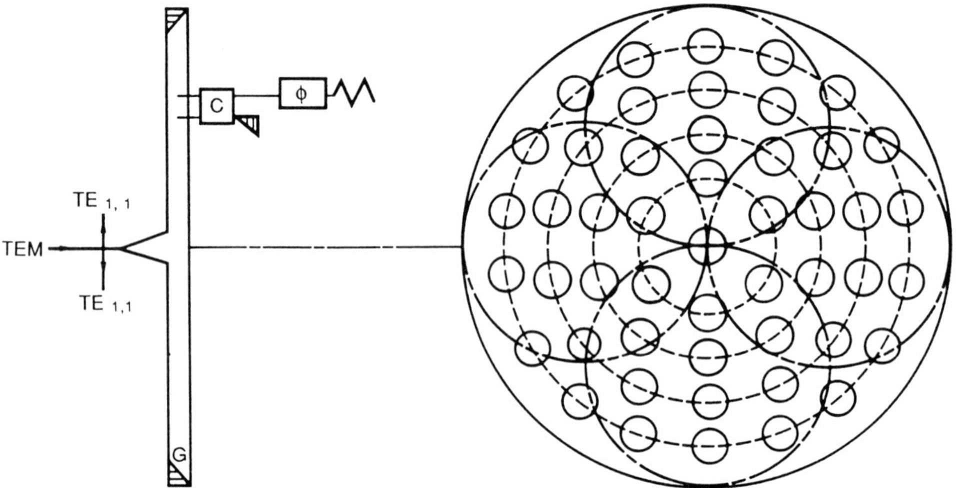

Fig. 12.84 Phased array, disk feed.

The *drawbacks* are:

since the array is fed by a travelling wave, its phase law depends critically on the frequency. The pointer has to take the frequency into account to display the correct phase.

Linear array theory

Basic equations — array factor

The properties of linear arrays can be expressed rather simply and can be easily extrapolated to the case of planar arrays (especially in the case of 'separable' structures and illuminations).

Consider therefore, a set of N identical radiating elements whose phase centres $A_0, A_1 \ldots, A_{N-1}$ are uniformly aligned on an axis (Fig. 12.81). The distance d separating two neighbouring phase centres, i.e. the array pitch, plays an important role in the properties of the array. The elements are assumed to be fed by a matched distributor according to an illumination law defined in amplitude and phase by a set of complex numbers:

$$a_0,\ a_1, \ldots, a_n, a_{N-1}.$$

If the element in A_0 is chosen as a reference, the element in A_n has, in direction of unit vector $\boldsymbol{u}$, a phase lead measured by the projection of vector $\mathbf{A_0 A_n}$ on $\boldsymbol{u}$, i.e. by the scalar product: $\mathbf{A_0 A_n} \cdot \boldsymbol{u}$.

Assume that each element in the presence of the others, connected to suitable loads, has the same pattern $\boldsymbol{f}(\boldsymbol{u})$. As a consequence of the principle of superimposition, the pattern for the full array $\boldsymbol{F}(\boldsymbol{u})$ is found by summing the patterns of the basic sources given their excitation amplitudes and their locations with respect to the chosen reference point:

$$\boldsymbol{F}(\boldsymbol{u}) = \sum_{n=0}^{N-1} a_n \boldsymbol{f}(\boldsymbol{u}) \exp\left(j \frac{2\pi}{\lambda} \mathbf{A}_0 \mathbf{A}_n \cdot \boldsymbol{u}\right) \tag{12.207}$$

$$\boldsymbol{F}(\boldsymbol{u}) = \boldsymbol{f}(\boldsymbol{u})\, R(\boldsymbol{u}) \tag{12.208}$$

$R(\boldsymbol{u})$ is the array factor that only depends on the array pitch and its excitation law. If θ is the angle between the direction $\boldsymbol{u}$ and the normal $\boldsymbol{n}$ to the array, we can write:

$$\mathbf{A}_0 \mathbf{A}_n \cdot \boldsymbol{u} = nd \sin\theta. \tag{12.209}$$

Let:

$$\tau = \sin\theta \qquad \frac{d}{\lambda} = \Delta\nu. \tag{12.210}$$

Then:

$$R(\boldsymbol{u}) = \sum_{0}^{N-1} a_n \exp(j\, 2\pi n\, \Delta\nu\tau). \tag{12.211}$$

Case of a uniform illumination and a constant phase gradient ϕ
We can then write:

$$a_n = \exp(-jn\phi) \tag{12.212}$$

The array factor is therefore of the form:

$$R(\boldsymbol{u}) = \sum_0^{N-1} \exp jn(2\pi\tau\,\Delta\nu - \phi). \tag{12.213}$$

Let:

$$x = \exp j(2\pi\tau\,\Delta\nu - \phi) \tag{12.214}$$

then:

$$R(\boldsymbol{u}) = \sum_0^{N-1} x^n = \frac{x^N - 1}{x - 1} = \left(x^{\frac{N-1}{2}}\right) \frac{x^{\left(\frac{N}{2}\right)} - x^{\left(-\frac{N}{2}\right)}}{x^{\left(\frac{1}{2}\right)} - x^{\left(-\frac{1}{2}\right)}}. \tag{12.215}$$

If abstraction is made of the constant amplitude factor $Nx^{(N-1)/2}$, then:

$$R(\boldsymbol{u}) = \frac{\sin N\left(\pi\,\Delta\nu\tau - \dfrac{\phi}{2}\right)}{N\sin\left(\pi\,\Delta\nu\tau - \dfrac{\phi}{2}\right)}. \tag{12.216}$$

The array factor R is then a periodic function represented in Fig. 12.85. Note that R is defined for all values of τ, including $|\tau| > 1$. This can be interpreted as imaginary or invisible directions. It passes through an extremum ± 1 each time that the denominator and the numerator cancel out at the same time, i.e. for:

$$\pi\,\Delta\nu\tau - \frac{\phi}{2} = k\pi \tag{12.217}$$

(k is an integer). One of these lobes is obtained for the direction τ_0 such that:

$$\pi\,\Delta\nu\tau_0 - \frac{\phi}{2} = 0. \tag{12.218}$$

A pointing direction τ_0 is therefore obtained with the phase gradient:

$$\boxed{\phi = 2\pi\,\Delta\nu\tau_0 = 2\pi\frac{d}{\lambda}\sin\theta_0.} \tag{12.219}$$

Two consecutive extremums are separated by an interval $\Delta\tau$ such that, based on (12.217):

$$\Delta\nu\,\Delta\tau = 1 \tag{12.220}$$

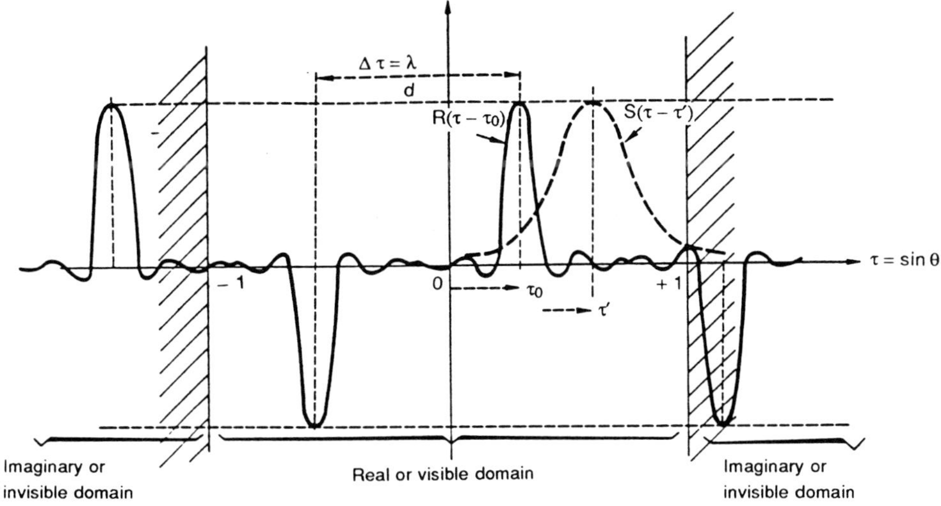

Fig. 12.85 Array factor, even illumination.

or

$$\Delta\tau = \frac{\lambda}{d} \cdot \tag{12.221}$$

These results show an important property of arrays: the existence of several grating lobes distributed periodically with a spacing inversely proportional to the array pitch. These lobes generally present a problem since they are sources of ambiguity on the direction of a radiating source (they are sometimes called grating lobes in analogy with time signals). To eliminate them or provide enough spacing, a sufficiently small pitch d has to be taken. Example for $d = \lambda/2$, $\Delta\tau = 2$, the grating lobes are outside of the $(-1, +1)$ domain of the real directions. The relative phase of the grating lobes depends on the parity of N. To show up the pointing direction τ_0 (12.218), the array factor is written:

$$R_{\tau_0}(\tau) = R(\tau - \tau_0) = \frac{\sin \pi N \Delta\nu (\tau - \tau_0)}{N \sin \pi \Delta\nu (\tau - \tau_0)} \cdot \tag{12.222}$$

The phase shift ϕ produces a translation of the array factor in the τ space. This result agrees with the one known in the domain of time-dependent signals, according to which a linear phase shift with respect to frequency does not change the form of the signal but produces an advance or a delay.

3 dB beamwidth

If we take angle θ as a variable instead of its sine τ, the array factor undergoes a deformation with respect to the pointing angle θ_0. In particular, the 3 dB beamwidth of the main lobe increases. This width is defined by the condition:

$$R = \frac{\sin [\pi N \Delta\nu (\sin\theta - \sin\theta_0)]}{N \sin [\pi \Delta\nu (\sin\theta - \sin\theta_0)]} = \frac{1}{\sqrt{2}} \qquad (12.223)$$

with:

$$\theta = \theta_0 + \frac{\theta_{3\,\mathrm{dB}}}{2}\,. \qquad (12.224)$$

This is obtained approximately for:

$$\pi N \Delta\nu \left[\sin\left(\theta_0 + \frac{\theta_{3\,\mathrm{dB}}}{2}\right) - \sin\theta_0\right] = \frac{\pi}{2} \qquad (12.225)$$

or:

$$\theta_{3\,\mathrm{dB}} \simeq \frac{1}{N \Delta\nu \cos\theta_0} = \frac{\lambda}{N d \cos\theta_0} = \frac{\lambda/L}{\cos\theta_0} \qquad (12.226)$$

where L is the length of the array. Therefore the beam widens in the $\cos\theta_0$ ratio in the sweep plane. Note that for θ_0 near $\pi/2$ (array with longitudinal radiation), another approximation has to be made which yields:

$$\theta_{3\,\mathrm{dB}} \simeq 2\sqrt{\lambda/L} \qquad (12.227)$$

Condition of grating lobe absence in the scanned zone (Fig. 12.85)
Given an array for sweeping in the $\pm\tau_M$ ($\tau_M = \sin\theta_M$) domain. The absence of an grating lobe implies that at the end of the sweep ($\tau = \tau_M$), the nearest grating lobe is in the 'imaginary' domain, i.e. in a direction $\tau < -1$, the spacing $\Delta\tau = \lambda/\Delta\nu$ between the grating lobes should be such that:

$$\frac{1}{\Delta\nu} > 1 + \tau_M \qquad (12.228)$$

hence the pitch condition:

$$\boxed{\frac{d}{\lambda} < \frac{1}{1 + \sin\theta_M}} \qquad (12.229)$$

For example: $-\ \theta_M = 30\,°$, $\sin\theta_M = \frac{1}{2}$, $\frac{d}{\lambda} < \frac{2}{3}$.

This condition plays an important role in antenna design: decreasing the pitch presents various drawbacks: increase in intersource coupling, increase in the number of total sources (and hence phase shifters), for a given antenna size (it is the size which sets the gain) hence an increase in cost. Therefore, a trade-off has to be found on this basis with respect to the objectives pursued.

The role of weighting an array illumination law
The illuminations considered up to the present are equi-amplitude levels. The corresponding patterns have near sidelobes which are rather strong: of the order of – 13 dB.

To reduce their strength, the amplitude has to be weighted to decrease the relative level of excitation of the sources at the edges of the array. This is equivalent to multiplying the illuminations described above by a continuous function; i.e. $s(v)$ with spectrum $S(\tau)$. From the Plancherel's theorem, the pattern obtained is the convolution of the preceding pattern by the pattern $S(\tau)$ (Fig. 12.86).

The result is that each array lobe resembles pattern $S(\tau)$ to a certain extent with a greater or lesser reduction of the near sidelobes. The pattern is defined by the function:

$$R_s(\tau - \tau_0) = \int_{-\infty}^{+\infty} R(\tau' - \tau_0)\, S(\tau' - \tau)\, \mathrm{d}\, \tau' . \tag{12.230}$$

Application — difference pattern For tracking antennas, use is often made of an 'S-shaped' pattern with a zero in the direction pointed to and two main lobes of opposite phase on each side (section 12.5.5 'Monopulse antennas', Fig. 12.80). This pattern is obtained using an odd weighting law for the array's illumination. Let $d(v) = -d(-v)$ be such a law. Its spectrum $D(\tau)$ is also odd. The convolution of the latter with pattern $R(\tau - \tau_0)$ produces a periodic pattern $R_D(\tau - \tau_0)$ formed by a set of anti-symmetrical lobe pairs resembling pattern $D(\tau)$ (Fig. 12.86). The particular shape of the resulting pattern, its slope near the zeros, its near sidelobes, are controlled by the shape of the weighting function $d(v)$.

Role of radiating elements directivity

The diagram of an array F is as we saw in (12.208), the product of the pattern $f(\boldsymbol{u})$ of an element in the presence of other sources by array factor $R(\boldsymbol{u})$; we

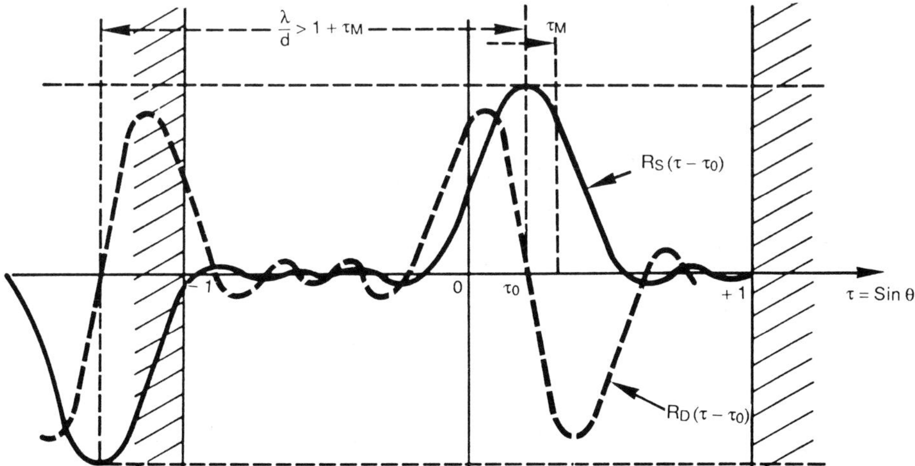

Fig. 12.86 Array factor, odd illumination.

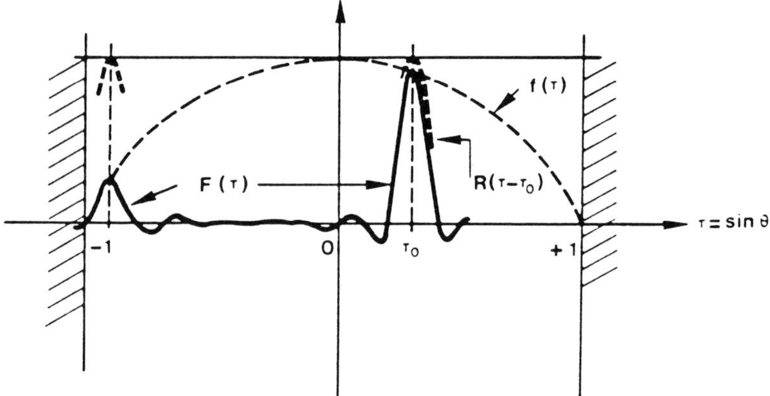

Fig. 12.87 Effect of element pattern on grating lobe levels.

can see that the first effect of element directivity is weighting of the level of any 'grating lobes' (Fig. 12.87). Here, we have $F(\tau) = f(\tau)\,R(\tau - \tau_0)$.

Another important effect is the control of variations in gain of the main lobe during sweep. $f(\boldsymbol{u})$ should vary as little as possible in the effective sweep zone, and be at a minimum outside of this zone to reduce the grating lobes. If the pitch is large enough, we know that these lobes can exist; moreover, it is possible to change the individual elements pattern by selecting the radiating element and by introducing coupling with adjacent radiating elements if needed. If this pitch is small, the choice of the element hardly affects the effective element pattern which is essentially controlled by the natural coupling between the elements.

Active module arrays

Active transmission arrays

When a phased array antenna is fed through a compact transmitter, the link between the transmitter and the radiating elements usually causes significant losses: the feeder, the rotary joint if it exists, the phase shifters all together often lose more than half of the transmitter's power.

The development of solid-state transmitters (SST) now allows isolation of the transmission function in modules that directly feed the radiating elements so as to avoid any transmission loss. Another advantage is that these transmitter modules are generally transistors which do not require a highly stabilized high voltage power supply as is the case for compact tube transmitters.

In addition, if one of the modules fails, the station can continue operation with only slightly degraded performance; operating reliability is therefore improved. Finally, the phase shifters associated with each radiating element can be low power components if they are located on the input of the modular amplifier. Low-power phase shifters can be more accurate, faster and more reliable than the high-power phase shifters used in classical techniques.

All of these advantages justify the wide development of transmission active modular array antennas.

Receiver active arrays

In a passive phased array antenna operating as a receiver, the same losses as described above for transmission are still present, which decreases the signal-to-noise ratio of the signals received. However, for certain radars whose operating range is limited by external jammers, we can see that the external signal-to-noise ratio is not changed by this attenuation.

In this case, the main interest in using active modules for reception is the ability to control the weighting in amplitude and phase of the signals received by each antenna in the array. We know that phased array antennas only allow control of the phase of such signals; variable attenuation would increase the losses to an unacceptable point. Conversely, with active modules, the amplitude can be controlled without degrading the internal signal-to-noise ratio provided a controlled attenuator is located after amplification.

Amplitude control allows modification of the receiver patterns at will, especially the level and location of the sidelobes. This capability is systematically used in adaptive antennas where weighting is slaved to certain characteristic signals from the signals received by the antenna, this is the domain of signal processing antennas (section 12.6).

Active transceiver arrays

Of course, the two functions, transmission and reception, can be combined into the same module which then becomes relatively complex. The development of active arrays is limited by the cost of equipment that can contain several hundreds or several thousands of modules. Great strides in the development of monolithic microwave integrated circuits (MMIC) using GaAs have provided the basis for operational equipment using active module array antennas.

In figure 12.88 weighting (amplitude and phase) is the same for transmission and for reception. We can, of course, put different attenuators and phase shifters in the transmission and receiving channels, but at a price of greater complexity.

Array limitations

Sweep range limits

As we have seen the 3 dB beamwidth increases with respect to sweep angle θ (12.226). This results in a drop in gain in $\cos\theta$. In fact, the effects of coupling between array elements causes an active reflection coefficient $\rho(\theta)$ which in turn causes an additional decrease in gain according to the law in $1 - |\rho|^2$. The general form of the gain versus sweep angle is therefore:

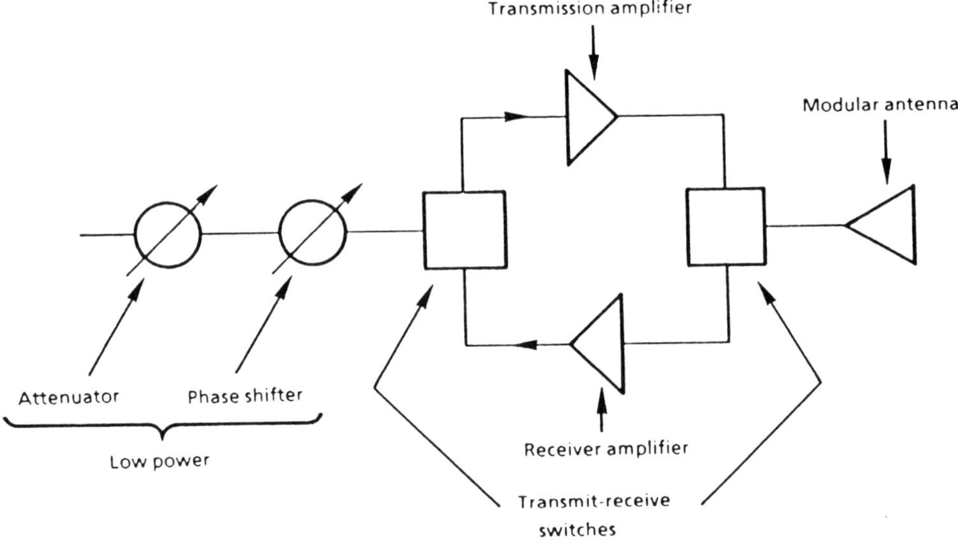

Fig. 12.88 Example of an active transceiver array.

$$G(\theta) = G_0 \cos\theta\,[1 - |\rho(\theta)|^2]. \tag{12.231}$$

The graph of function $\rho(\theta)$ depends on the particular nature of the array. It can happen that the active reflection coefficient reaches the unit in the sweep range; the gain cancels out and a blinding phenomenon occurs. To prevent this, we can change the structure of the array by adding elements such as: dielectric blades, periodic obstacles, etc. In all cases, we cannot expect to significantly exceed the sweep range of ± 45 to $\pm 60\,°$. To sweep a greater angle, we can associate several arrays shifted angularly, use a lens or a complementary reflector. This is a difficult problem (Amitay and Galindov, 1972).

Sidelobe quality limits (Amitay and Galindov, 1972)

The quality of the radiated pattern is essentially characterized by the low level of the sidelobes. Use of quantized (p-bit) phase shifters causes phase errors and increases the level of the sidelobes. We know that the theoretical phase law is linear, we can replace it approximately with a staircase function (the height of the step is equal to a phase bit). In this case, the phase error varies periodically along the array and we can observe a localized interference lobe in direction θ' such that:

$$\sin\theta' = 2^{p} \sin\theta \tag{12.232}$$

with a relative level

$$1/(2p - 1). \tag{12.233}$$

We can see that 3-bit phase shifters cause a level of -17 dB, which is often not acceptable.

We can break the periodic occurrence of the error by giving each phase shifter a known 'origin phase', that is variable from one phase shifter to another, for example according to a pseudo-random law.

The quantizing sidelobes are therefore distributed over the full range of sweep; their level approximately follows a Rayleigh's law of probability with a mean relative level expressed in decibels:

$$\boxed{L_{\mathrm{dB}} \approx 10 \log N + 6(p-1).} \tag{12.234}$$

We can see that as the number N of array sources increase, the number of phase shifter bits can be decreased.

12.6 SIGNAL PROCESSING ANTENNAS

12.6.1 Introduction: the antennas and signal theory

Signal theory saw its greatest development after the second world war (Shannon, Wiener, 1950). The development of telecommunications and radar opened unlimited fields of application. This theory allows quantitative evaluation of the capacity of a system to transmit information. At about the same time, workers in optical physics had the idea of assimilating an optical image to a two-dimensional spatial message and evaluating the quantity of information that it contained. In 1947, Prof. Duffieux developed a harmonic analysis of images both theoretically and experimentally: the performance of an optical instrument can be evaluated by its capacity to reproduce periodic patterns with shorter and shorter spatial periods (Duffieux, 1947).

This time also saw the growth of radioastronomy. Radioastronomers were often engineers faced with signal transmission problems and celestial body imaging problems; therefore they were particularly well placed to develop antennas from the point of view of harmonic analysis, which had begun to take hold in optics but which was already extensively used for circuits.

This point of view allowed setting up a relationship between antenna performance and processing of the signals that they received. In fact, in conventional systems, the antenna and the receiver ensure completely separate functions, the directive properties are set by the antenna. The spectral and temporal properties are tied to the processing circuits.

The extension of signal theory to antennas made it possible to design systems whose directive properties resulted from the association of the antenna with the signals received. A simple classical example is that of automatic monopulse tracking antennas where location of a far radiating source is done through the use of a special processing algorithm (section 12.3.5) with an angular precision that depends on both the 3 dB beamwidth of the antenna and the signal-to-noise ratio (formula (12.72)).

Today, signal processing antennas are being developed in the fields of sonar and radar. The algorithms used often imply complex calculations, but the increasing capacity of digital processing now allows their operation in real time.

An additional step can be made in this field with adaptive antennas: instead of being content with processing the antenna signals in an optimum manner, it is the antenna itself whose properties can be changed using these received signals. In this case, the antenna can form an adaptive spatial filter, controlled through one or several optimization criteria: maximum signal (tracking antennas) minimum noise (CSLC) antennas (coherent side lobes cancellation, section 12.6.5)), maximum signal-to-noise ratio, etc. There is a wealth of algorithms and diversely organized systems available to engineers. We will limit ourselves here to a few basic concepts and a few examples of application.

12.6.2 Spatial frequency concept — properties of characteristic radiation functions

The basis for the application of signal theory to antennas is the relationship that we established in (section 12.3.3) between the illumination law of a finite aperture and its characteristic radiation function, i.e. a Fourier transform. For simplification, let us consider a linear aperture of centre 0 and length L where all the points are located by their abscissas x respective to wavelength λ, i.e. by the parameter $\nu = x/\lambda$.

This aperture is assumed to be the origin of a law of illumination (in amplitude and phase) defined by a complex function $f(\nu)$ (null outside the domain defined by L). Let $\nu_0 = L/2\lambda$. The result is a radiation with characteristic function $F(\tau)$, (with $\tau = \sin\theta$, θ being the angle of the direction considered with the normal to the aperture (section 12.3.3)):

$$F(\tau) = \int_{-\infty}^{+\infty} f(\nu)\ \exp(i\, 2\pi\nu\tau)\, d\nu \qquad (12.235)$$

$F(\tau)$ is therefore the Fourier transform of a function with a bounded support in the domain $(-\nu_0, +\nu_0)$. This transform can in principle be inverted: it is the basis of methods for synthesis of illuminations radiating a given pattern (section 12.6.3).

We can see that $F(\tau)$ has the same properties as a time-domain modulated signal whose spectrum is bounded by the domain $(-\nu_0, +\nu_0)$ around the carrier frequency. Variable ν is called the spatial frequency. We will now find the properties of such signals in the field of antennas.

1. The radiation function $F(\tau)$ does not contain spectral components of spatial frequencies greater than ν_0 which is called the cut-off frequency. The antenna is the equivalent of a low-pass filter.
2. The sampling theorem applies: the function $F(\tau)$ is known for all directions if it is known in the set of directions sampled: $\tau_n = n/2\,\nu_0$. This gives:

$$F(\tau) = \sum_{-\infty}^{+\infty} F\left(n \frac{1}{2\,\nu_0}\right) \operatorname{sinc}(2\pi_0 \tau - n\pi). \tag{12.236}$$

The $F(\tau)$ pattern can be represented by a combination of linear 'cardinal sine' functions ($\operatorname{sinc} x = \sin x/x$). This functions form an 'orthogonal base'. The maximum of one of these functions coincides with the zeros of the other.

3. The Bernstein theorem applies. The relative slope of a pattern cannot exceed a limit value set by the spatial cut-off frequency ν_0. If F_1 is an upper limit of the modulus of $F(\tau)$ then:

$$\frac{1}{F_1}\left|\frac{\mathrm{d}F}{\mathrm{d}\tau}\right| \leqslant 2\pi\nu_0 = \pi\frac{L}{\lambda}. \tag{12.237}$$

The maximum slope plays an important role in the problem of pattern synthesis and in particular for automatic tracking antennas where the slope of the difference pattern determines the angular accuracy of tracking.

12.6.3 Synthesis of an aperture radiating a given pattern

In several practical applications, the problem posed to the engineer is to develop an antenna whose pattern is as close as possible to the ideal pattern for the antenna's application. This is the case for surveillance antennas, space telecommunications antennas, etc. In all cases, one starts with a more or less arbitrary approximation criterion that leads to an optimal result given the various requirements for the antenna.

Problem posed

Here we will start with a linear aperture model. This aperture is limited, at dimension D equivalent to a finite-passband filter $(-\nu_0, +\nu_0)$. Our goal is to obtain an ideal pattern $F_0(\tau)$. Is it possible to define an illumination $f_0(\nu)$ on the aperture that yields pattern F_0? If not, what is the law $f_1(\nu)$ that is physically possible giving the pattern, $F_1(\nu)$, that is as close as possible to F_0? If f_0 solves the problem, it is necessarily the Fourier transform of F_0. For it to be physically possible, the support of f_0 has to be bounded by a domain contained in $(-\nu_0, +\nu_0)$.

In a particularly important case, where the desired pattern F_0 is itself a bounded support (null pattern outside of a certain interval), the result is that f_0 is an unbounded support and therefore not physically possible. To find the best possible function f_1, a difference criterion has to be defined, a distance, between F_0 and F_1. The choice of the criterion of a minimum mean quadratic difference leads to a relatively simple solution. The following integral must be minimized:

$$I = \int_{-\infty}^{+\infty} |F_0 - F|^2 \, d\tau \tag{12.238}$$

i.e., according to Parseval's theorem:

$$I = \int_{-\infty}^{+\infty} |f_0 - f|^2 \, d\nu \tag{12.239}$$

since f is null outside of $(-\nu_0, +\nu_0)$, the result is obtained if $f = f_1 = f_0$ in this interval. The best approximation is therefore obtained with an illumination f_1 identical to the Fourier transform of the ideal pattern in the antenna aperture. The corresponding pattern F_1 is the Fourier transform of this illumination:

$$F_1 = \int_{-\nu_0}^{+\nu_0} f_1 \exp(j\, 2\pi\nu\tau) \, d\nu = \int_{-\infty}^{+\infty} f_0 \operatorname{Rect}\frac{\nu}{\nu_0} \exp(j\, 2\pi\nu\tau) \, d\nu \tag{12.240}$$

where Rect $x = 1$ for $|x| \leqslant 1$ Rect $x = 0$ for $|x| > 1$.

It is therefore the convolution product:

$$F_1 = \int_{-\infty}^{+\infty} F_0(\tau') \frac{\sin 2\pi\nu_0 (\tau - \tau')}{2\pi\nu_0 (\tau - \tau')} \, d\tau'. \tag{12.241}$$

For example, F_0 is a rectangular pattern, $F_0 = \operatorname{Rect}\dfrac{\tau}{\tau_0}$ · This gives:

$$f_0 = \frac{\sin 2\pi\nu\tau_0}{2\pi\nu_0} \qquad f_1 = f_0 \operatorname{Rect}\frac{\nu}{\nu_0}$$

$$F_1 = \int_{-\tau_0}^{+\tau_0} F_0(\tau') \frac{\sin 2\pi\nu_0 (\tau - \tau')}{2\pi\nu_0 (\tau - \tau')} \, d\tau' = Si\,[2\pi\nu_0 (\tau - \tau_0)] - Si\,[2\pi\nu_0 (\tau + \tau_0)]$$

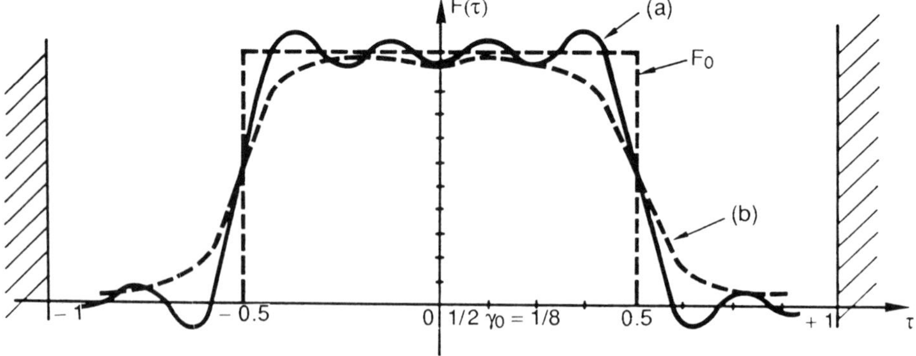

Fig. 12.89 Example of aperture synthesis.

where $Si(x)$ is defined by the integral: $Si(x) = \int_0^x \frac{\sin u}{u}\,du$.

Figure 12.89 shows F_0 and F_1 (curve a) for $\nu_0 = 4,\ \tau_0 = \frac{1}{2}$.

Shortcomings of the method — Gibbs' phenomenon This method can yield very significant differences locally between F_0 and F_1 especially in the neighbourhood of discontinuities. For the preceding example, it can be shown that if ν_0 tends to infinity, $F_1(\tau_0)/F_0(\tau_0)$ tends towards 1.09.

Generalization of the approximation method

We can see that the approximation F_1 of F_0 is none other than the result of filtering F_0 through a constant transfer function filter in band $(-\nu_0, +\nu_0)$ (function Rect ν/ν_0). We can replace this function by a different filtering function $T(\nu)$ in the same $2\nu_0$ passband. If $\psi(\tau)$ is its Fourier transform form, approximation F_T will be expressed by the convolution:

$$F_T(\tau) = \int_{-\infty}^{+\infty} F_0(\tau')\ \psi(\tau - \tau')\,d\tau'. \tag{12.242}$$

We can see that $T(\nu)$ and $\psi(\tau)$ respectively play the role of the transfer function and impulse response of a filter. The illumination law to construct on the aperture is therefore:

$$f_T(\tau) = f_0(\nu)\ T(\nu). \tag{12.243}$$

To prevent overshoot and interference lobes, we are intuitively led to find a function ψ with weak sidelobes (apodized function) which is obtained, as you know, through a bell-shaped function $T(\nu)$. The choice of a triangular function leads to a Fejer approximation:

$$T(\nu) = 1 - \frac{|\nu|}{\nu_0} \qquad \psi(\tau) = \left(\frac{\sin \pi\nu_0\tau}{\pi\nu_0\tau}\right)^2 \tag{12.244}$$

Numerical example Let us again look at the preceding numerical example. We notice that the sides of F_T are not as steep as previously. Conversely, the ripples and sidelobes are much weaker (Fig. 12.89, curve(b)).

12.6.4 Imaging problems with incoherent sources

Introduction

The imaging problem can be stated as follows: consider a spatial domain that is not well known. This domain can contain either spontaneous radiating sources (as in radioastronomy, radiometry, seismography), sources that respond to excitation or illumination (for example, radar, sonar, tomography).

We have a certain number N of sensors judiciously located (for example in a regular or irregular linear array) that allows taking N measurements. The latter are not of an infinite precision, there is always a measurement noise: noise internal to the receptors, position and stability of the sensors and stability of illumination. The problem is to find the most likely configuration of the sources given the information we have available.

Incoherence conditions (Fig. 12.90)

In numerous domains of application, we are dealing with an unknown environment of sources radiating independently of each other. This is obviously the case of radioastronomy and radiometry. It is also often the case, at least in the first approximation, in the fields of radar and sonar, especially if undesirable independent jammers are at work.

Two variable sources of complex amplitudes $z(\tau', t)$, $z(\tau'', t)$ are incoherent if the temporal mean of the hermitian product is null:

$$\overline{z(\tau', t)\; z^*(\tau'', t)} = 0 \qquad (\tau' \neq \tau'') \tag{12.245}$$

For $\tau' = \tau'' = \tau$, we define an intensity that is sometimes assimilated to noise temperature.

$$\overline{|z(\tau, t)|^2} = T(\tau). \tag{12.246}$$

Multiplicative arrays (Fig. 12.91) (Ksienski, 1965)

Multiplicative arrays form a first approach to the imaging problem. It uses the incoherent character of the sources analysed to improve the angular resolution

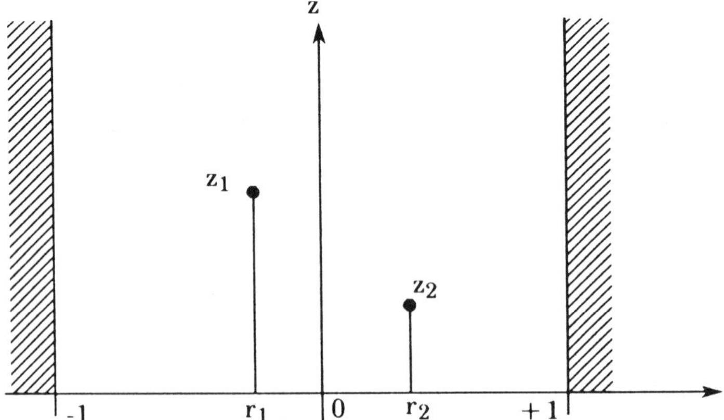

Fig. 12.90 Angular distribution of two sources.

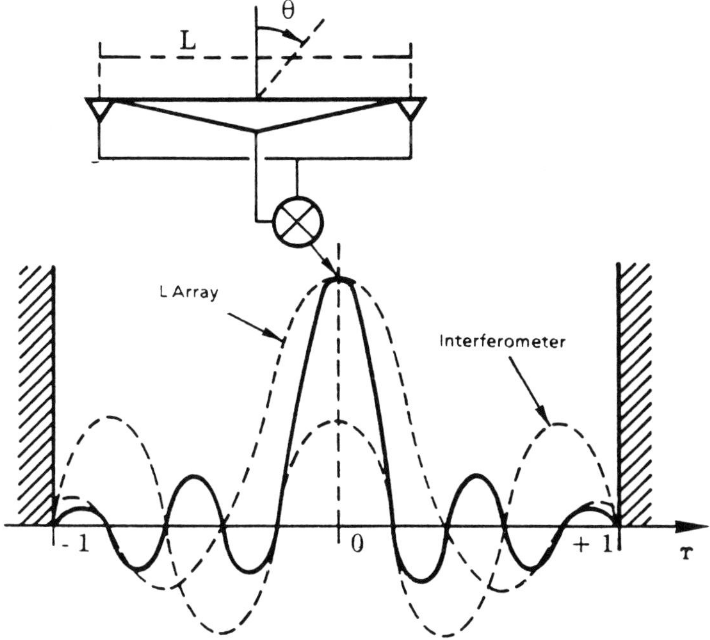

Fig. 12.91 Multiplicative array.

of the system of antennas. Given a linear array of length L, that is uniformly illuminated, its pattern has the shape of a 'cardinal sine':

$$R(\tau) = \text{sinc } 2\pi\nu_0\,\tau = \frac{\sin 2\pi\nu_0\tau}{2\pi\nu_0\tau} \tag{12.247}$$

with

$$\tau = \sin\theta \qquad \nu_0 = L/2\lambda. \tag{12.248}$$

Let us add two auxiliary antennas to each end of this array, whose receive signals are summed in phase. We have built an interferometer whose pattern is in the form:

$$I(\tau) = \cos 2\pi\nu_0\tau. \tag{12.249}$$

Let us use a coherent demodulator to obtain the mean product of the signals received:

$$S(\tau) = RI = \sin\text{c } 4\pi\nu_0\tau. \tag{12.250}$$

We obtain the pattern of an array with aperture $2L$, two times larger than the initial array. For such a system to merit the name of antenna, its response has to be linear with respect to an angular distribution of sources.

Given an angular distribution of sources with a complex amplitude $z(\tau, t)$, their incoherence is translated by the condition (12.245), (12.246), which can be written:

$$\overline{z(\tau', t)\ z^*(\tau'', t)} = T(\tau')\ \delta(\tau' - \tau'') \tag{12.251}$$

where T is the intensity of the angular distribution and δ the Dirac function ($\delta(\tau' - \tau'') = 0$ for $\tau' \neq \tau''$).

The signals picked up respectively by the array and the interferometer are of the form:

$$R(\tau) = \int z(\tau', t)\ \text{sinc}\ 2\pi\nu_0(\tau' - \tau)\ d\tau' \tag{12.252}$$

$$I(\tau, t) = \int z(\tau'', t)\ \cos 2\pi\nu_0(\tau'' - \tau)\ d\tau'' \tag{12.253}$$

And the averaged product:

$$S(\tau) = RI^* = \int\int \overline{z(\tau', t)\ z^*(\tau'', t)}\ \text{sinc}\ 2\pi\nu_0(\tau' - \tau)\ \cos 2\pi\nu_0(\tau'' - \tau)\ d\tau'\ d\tau''$$

i.e., with the incoherence property (12.251):

$$S(\tau) = \int T(\tau')\ \text{sinc}\ 4\pi\nu_0(\tau' - \tau)\ d\tau'. \tag{12.254}$$

This is a *linear* response with respect to the distribution of intensity of the sources. The multiplicative array does act as a filter.

Mills' cross (Fig. 12.92)

A system frequently used in radioastronomy is the Mills' cross which multiplies signals from orthogonal linear arrays. A direction $\boldsymbol{u}$ being identified by its direction cosines (α, β, ν), $(\alpha^2 + \beta^2 + \nu^2 = 1)$, the shapes of their patterns are given by:

$$R_1 = \text{sinc}\ 2\pi\nu_0\alpha \qquad R_2 = \text{sinc}\ 2\pi\mu_0\beta \tag{12.255}$$

The product pattern is:

$$F(\alpha, \beta) = R_1 R_2. \tag{12.256}$$

This product pattern can be considered as radiated by the array resulting from the convolution of linear arrays (co-array). Its illumination function is the inverse Fourier transform of product $R_1 R_2$:

$$F(\mu, \nu) = \text{Rect}\frac{\nu}{\nu_0}\ \text{Rect}\frac{\mu}{\mu_0} \tag{12.257}$$

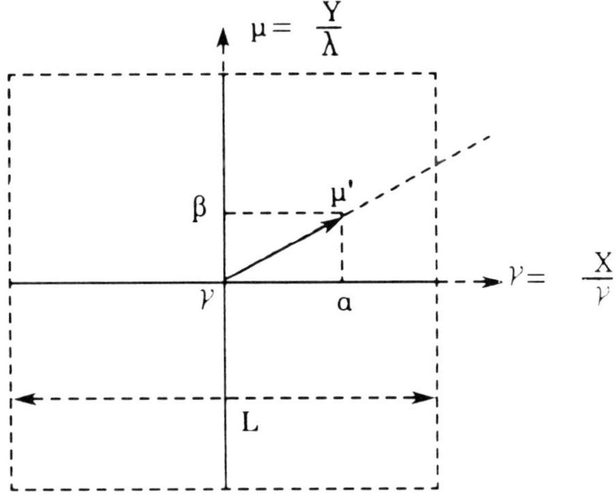

Fig. 12.92 Mills' cross.

This is a two-dimensional rectangular array: if each linear array has N sensors, with only $2N$ sensors the equivalent of a N^2 sensor array is obtained. The savings yielded by this type of antenna are significant.

12.6.5 Imaging through spectral estimations (PIEEE Special Issue, 70(9), Sept. 1982).

The most elaborate methods of incoherent source imaging are based on a fundamental property known as the Van Cittert–Zernicke theorem, which is the equivalent in the spatial domain of the Wiener–Kitchine theorem in the area of time-domain series. We know that the latter has established a Fourier transform relation between the spectral density of a stationary random time-domain series and its self-correlation function.

The Van Cittert–Zernicke theorem establishes the same relation between the angular distribution of incoherent sources and the spatial coherence function seen on a sensor array.

Van Cittert–Zernicke theorem (Born and Wolf, 1964)

Given a variable angular distribution of incoherent sources: $Z(\tau, t)$, the E-field received from the sources is assumed to be observed in various points M with abscissas x on axis $x'x$ (Fig. 12.7), we will apply the usual notation.

$$\nu = \frac{x}{\lambda} \qquad \tau = \sin\theta. \tag{12.258}$$

The field $E(\nu, t)$ results from the superposition of plane waves from the sources and weighted by their complex amplitudes $z(\tau, t)$.

$$E(\nu, t) = \int_{-\infty}^{+\infty} Z(\tau, t)\, \exp(j\,2\pi\nu\tau)\, \mathrm{d}\tau. \tag{12.259}$$

The coherence function is defined by the average:

$$C(\nu, \nu') = \overline{E(\nu, t)\, E^*(\nu', t)}. \tag{12.260}$$

The incoherence condition (12.245) yields:

$$C(\nu, \nu') = \int_{-\infty}^{+\infty} T(\tau) \exp(j\, 2\pi(\nu - \nu')\, \tau)\, \mathrm{d}\tau. \tag{12.261}$$

Let

$$\mu = \nu - \nu'.$$

We obtain the following fundamental result:

$$\boxed{C(\mu) = \int_{-\infty}^{+\infty} T(\tau) \exp(j\, 2\pi\mu\tau)\, \mathrm{d}\tau.} \tag{12.262}$$

We can conclude:

1. the coherence function is stationary along axis $x'x$: it only depends on the 'distance' μ between observation points;
2. this is the Fourier transform of the angular distribution of intensities, $T(\tau)$;
3. the relation (12.262) can obviously be inverted which yields the angular distribution of sources based on the spatial coherence function, $C(\mu)$.

Significant applications of this property will be discussed in the chapter on radioastronomy in Volume 3.

Coherence function sampling

The integral (12.262) is formally extended to infinity. In fact, the real directions are limited to the $(-1, +1)$ domain. The Shannon theorem shows that the coherence function can be sampled at a step a, such that:

$$\Delta\nu = \frac{a}{\lambda} = \frac{1}{2} \cdot \tag{12.263}$$

Inversely, the angular distribution $T(\tau)$ can be represented in the $(-1, +1)$ domain in the form of a Fourier series. We therefore have the following pair of relationships (12.264) and (12.265):

$$C(n, n') = C(n - n') = \int_{-1}^{+1} T(\tau) \exp(j\, \pi(n - n')\, \tau)\, \mathrm{d}\tau. \tag{12.264}$$

Let

$$m = n - n'.$$

We define

$$C_m = C(n - n')$$

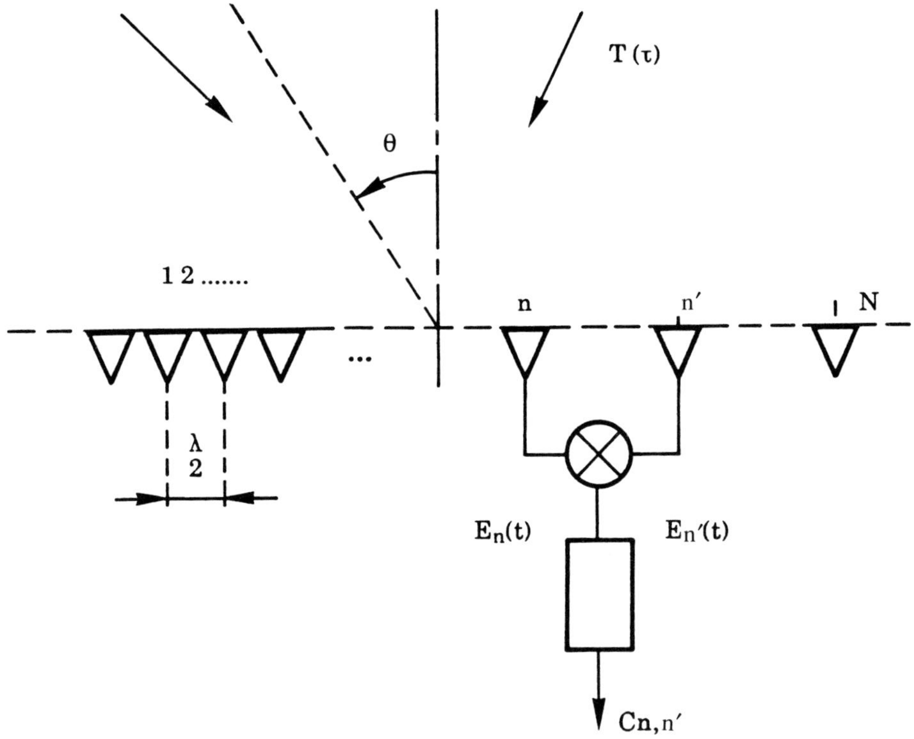

Fig. 12.93 Estimation of covariance C_m using sensors n, m.

$$\boxed{T(\tau) = \sum_{-\infty}^{+\infty} C_m \exp(-j\pi m \tau).} \tag{12.265}$$

In practice we will evaluate a covariance $C_{n,\,n'}$ by measuring the time-domain mean of the products of the components in phase and in quadrature of signals $E_n(t)$ and $E_{n'}(t)$ received by sensors n and n' (Fig. 12.93)

Spectral estimation methods

Let $T(\tau)$ be an unknown spatial distribution of incoherent source intensities, for example point radio sources (Fig. 12.94). We have seen that $T(\tau)$ is expressed from samples with covariance C_m by a Fourier series (12.265). The number of sensors available being necessarily finite, we can only have N samples with covariances from C_0 to C_{N-1}. We also know that $C_{-m} = C_m^*$. The problem is to find an estimate $T(\tau)$ of $T(\tau)$ with only the N samples being known.

The simplest method consists in considering the non-measured covariances as null ($C_N = 0$, $C_{N+1} = 0 \ldots$) and using a truncated Fourier series; this is the 'correlogram' method. In fact this assumption is far from being justified: Burg

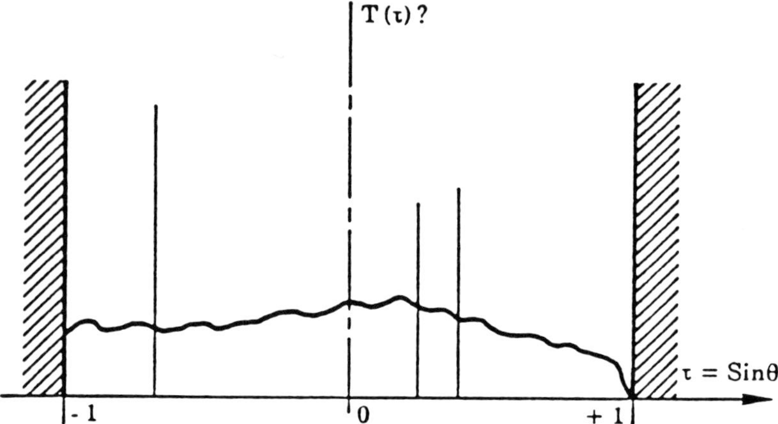

Fig. 12.94 Unknown spatial distribution of incoherent sources.

was the first, in 1967, to point out that it introduced an arbitrary hypothesis which is not contained in the measurement of other covariances: this is the basis of his method of maximum entropy which consists in extrapolating the values of the non-measured samples from the measurements made (Fig. 12.95) (Burg, 1975).

A discussion of estimation methods is not in the scope of this work, especially since this field is still under development. Let us only mention a few of the names of the most well known methods: correlogram method, AR methods (auto-regressive), MA (moving average), maximum entropy method (Burg 1967), Pissarenko's method. For more details, the reader must refer to specialized papers and works.

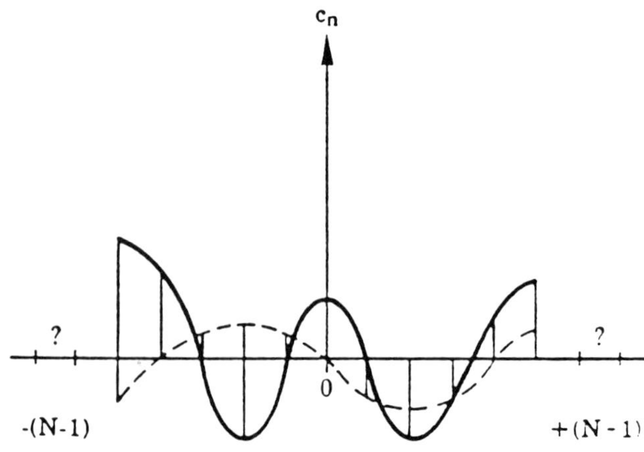

Fig. 12.95 Samples of measured covariance.

12.6.6 Adaptive antennas (Applebaum, 1976)

Adaptive arrays are another step further in the field of signal processing antennas. In imaging techniques, we have seen that an attempt is made to make the best use of the data delivered by the antenna and even, in the Burg method, guess what it does not deliver (non-measured covariances).

For adaptive antennas, we use this data to change the parameters of the antenna to better adapt it to the changes in the environment, for example, to reduce the level of interference noise from the outside and therefore favour detection of the useful signal. This is what we will see in the example of CSLC (coherent side lobe cancellation) antennas.

Adaptive antennas: CSLC method (Fig. 12.96)

Consider a receiving antenna, that we will designate as the 'main' antenna, placed in an environment formed of a limited number (N) of jammers that are incoherent with respect to each other. We will associate auxiliary antennas,

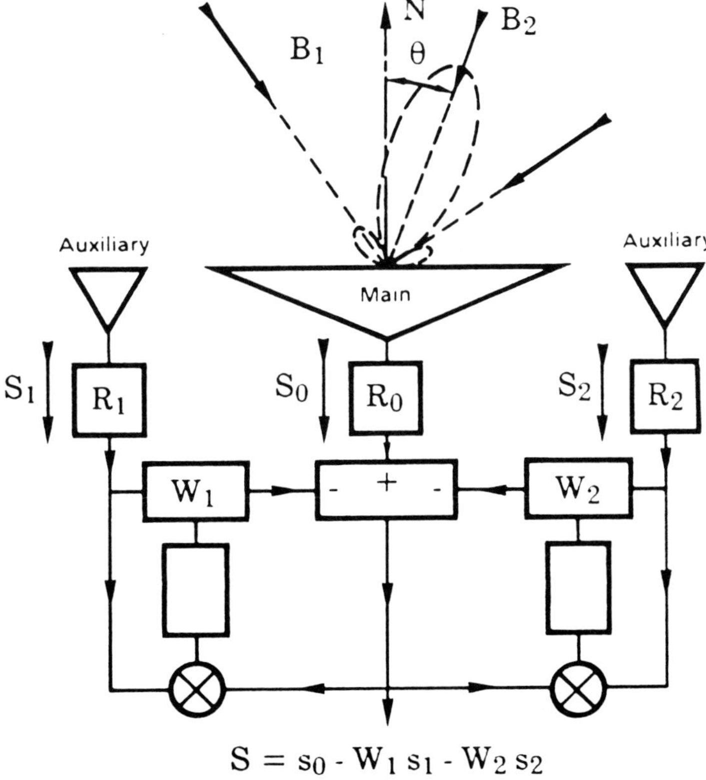

Fig. 12.96 Adaptive array: CSLC method.

whose number is greater than or equal to the number of jammers, with this antenna: Fig. 12.96 shows the case of two jammers. Signals S_1, S_2 picked up by the auxiliary antennas are subtracted from signal S_0 of the main antenna with weighting coefficients W_1 and W_2 leading to a minimum noise level. We obtain a signal:

$$S = S_0 - W_1 S_1 - W_2 S_2. \tag{12.266}$$

In practice, we measure the correlation between the signal S and auxiliary signals S_1, S_2 (in presence of noise only). Weightings W are chosen such that these correlations are minimal. It can be shown that this condition is equivalent to that of minimum noise. The signals from the demodulators are used to control weightings W. The conditions:

$$\overline{SS_1^*} = 0 \qquad \overline{SS_2^*} = 0 \tag{12.267}$$

yield the following linear system:

$$\begin{bmatrix} \overline{|S_1|^2} & \overline{S_1 S_2^*} \\ \overline{S_2 S_1^*} & \overline{|S_2|^2} \end{bmatrix} \begin{bmatrix} W_1 \\ W_2 \end{bmatrix} = \begin{bmatrix} \overline{S_0 S_1^*} \\ \overline{S_0 S_2^*} \end{bmatrix} \tag{12.268}$$

which determine the optimum weightings. Figure 12.97 shows the result obtained, nulls appear in the directions of jammers B_1 and B_2.

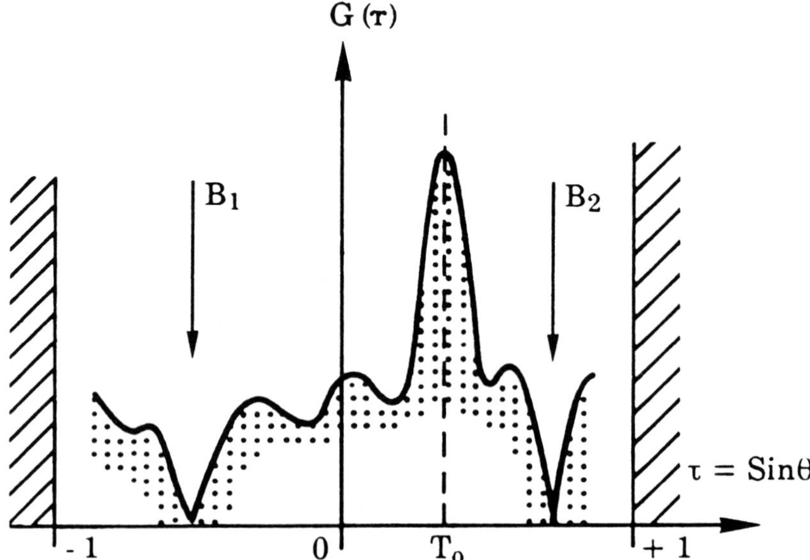

Fig. 12.97 Plot of the radiation pattern of Fig. 12.96 after a convergence of the weightings W.

BIBLIOGRAPHY

Amitay N. and Galindov (1972) *Theory and analysis of phased array antennas*, John Wiley and Sons Inc., London.

Applebaum S. P. (1976) Adaptative Arrays *IEEE Trans. on Antennas and Propagation*, **AP 24**, 5.

Aubry, C. (1989) Polarimetrie Radar – Bases Théoriques. *Revue Technique Thomson-CSF*, **20–1** (4).

Aubry C. and Bitter D. (1975) Sum and Difference radiation patterns of a corrugated conical horn by means of Laguerre-Gaussian functions *Conference proceedings, 9th European Microwave Conference Hamburg.*

Baker B. B. and Copson E. T. (1950) *The Mathematical Theory of Huygen's Principle* Oxford Clarendon Press 7.

Born M. and Wolf E. (1964) *Principles of Optics*, Pergamon, London.

Burg J. P. (1975) Maximum Entropy Spectral Analysis, Stanford CA, Stanford University.

Clarricoats P. J. B. and Saha P. K. (1971) Propagation and radiation behaviour of corrugated feeds. Part 1 and 2 – *PIEE 118* **9**.

Collin R. E. and Zucker (1969) *Antenna Theory* McGaw Hill, New York.

Drabowitch S. (1966) Multimode antennas, *Microwave Journal.*

Duffieux P. M. (1947) L'intégrale de Fourier et ses applications à l'optique, Faculté des Sciences de Besançon.

Jones (1964) The Theory of Electromagnetism, Pergamon, London.

Kildal P. S. (1988) Gaussian beam model for aperture-controlled and flareangle-controlled corrugated horn antennas, *IEE proceedings*, **135**, Pt. H, N°4.

King (1956) *The theory of linear antennas*, Harvard University Press.

Ksienski A. A. (1965) Multiplicative processing antennas for radar applications. *Radio and Electron. Eng.*, **29**.

Ludwig A. C. (1973) The definition of cross-polarisation *IEEE Trans.*, **AP–21**, 116–119.

Minett H. C., Mac, B. and Thomas, A. (1968) Field in the image space of symmetrical focus reflectors, *P.I.E.E. 15*, **10**.

Potter, P. D. (1963) A new horn antenna with suppressed side lobes and equal beamwidths. *Microwave journal.*

Roubine E., Bollomey J. C., Drabowitch S. and Ancona C. (1987) *Antennas* North Oxford Academic Publishers Ltd London.

Silver S. (1949) *Microwave Antenna Theory and Design* McGraw Hill, New York.

Wiener, N. (1950) *Extrapolation, interpolation and smoothing of stationary time series* J. Wiley and Sons, New York.

Williams, W.F. (1965) High-efficiency antenna reflectors. *Microwave Journal*, July.

(1982) Spectral estimation. *P.I.E.E Special Issue*, **70**, 9.

13

Microwave measurements

J.P. Humbert

There are five basic microwave measurements: transmission and impedance, noise, frequency and power (Fig. 13.1)

13.1.1 Transmission measurements (attenuation, insertion loss)

Attenuation and insertion loss are transmission characteristics of microwave devices. Attenuation is defined as the decrease in power level at the load caused by inserting a device between a Z_0 source and load. For a mismatch device, the attenuation factor has two terms, a reflective term caused by the mismatch, and a dissipative term.

In the scattering parameter domain, forward insertion loss is represented by the S_{21}, while reverse transmission is S_{12}. The transmission parameter may also be a positive (a gain) as well as negative (a loss) quantity. As with impedance, transmission parameters may be either scalar (amplitude only) or vector (phase and amplitude) quantities.

For attenuation measurements, the overall accuracy results from two factors, instrumentation and mismatch uncertainties. The ideal and most accurate method of measuring attenuation is to use an automatic vector analyser. Such measurements can approach the ideal because they premeasure and store source and detector imperfections and compute them out of the final result. But in practice, most measurements are scalar, and we can only attempt to

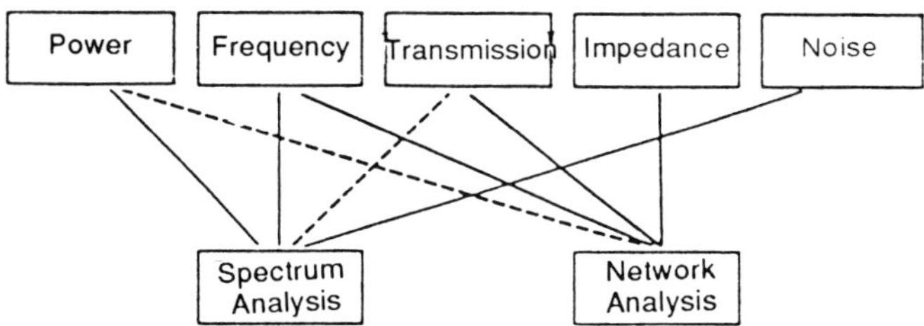

Fig. 13.1 Types of microwave measurements and appropriate measurement methods.

Table 13.1 Comparison of equivalent output and input SWR for various coaxial isolation alternatives

HP models	*Source isolation alternates*										*Detector alternates*					
	Bridges		*Coaxial pads*		*Coaxial couplers*				*Power splitter*		LBSD 8470B PDB 8473D		*Scalar analyser detectors* (See Note) 11664A/E 85025A/B		*Power sensor* 8481A 8485A	
Frequency Range	85027A 85027B		8492A 8493C		786D 787C	788C 789C	779D	11691D	11667A 11667B							
	Spec.	Typ.	Spec.	Typ.	Spec.	Spec.	Spec.	Spec.	Spec.	Typ.	Spec.	Typ.	Spec.	Typ.	Spec.	Typ.
10 MHz–1 GHz	1.15	1.1	1.15	1.08	—	—	—	—	1.1	1.05	1.15	1.08	1.25	1.16	1.1	1.02
1–2	1.15	1.1	1.15	1.08	1.13	—	—	—	1.1	1.05	1.15	1.08	1.25	1.16	1.1	1.02
2–4	1.15	1.1	1.15	1.08	1.16	—	1.2	1.2	1.1	1.05	1.15	1.08	1.25	1.16	1.18	1.05
4–6	1.15	1.1	1.15	1.08	—	1.25	1.2	1.2	1.2	1.1	1.3	1.15	1.38	1.2	1.18	1.1
6–8	1.15	1.1	1.15	1.08	—	1.25	1.2	1.2	1.2	1.1	1.3	1.15	1.38	1.25	1.18	1.1
8–10	1.15	1.1	1.25	1.1	—	1.25	1.2	1.2	1.33	1.2	1.3	1.15	1.92	1.3	1.18	1.1
10–12	1.25	1.2	1.25	1.15	—	1.25	1.2	1.2	1.33	1.2	1.3	1.2	1.92	1.43	1.18	1.1
12–14	1.25	1.2	1.35	1.15	—	—	—	1.2	1.33	1.2	1.3	1.2	1.92	1.5	1.28	1.15
14–16	1.43	1.35	1.35	1.2	—	—	—	1.2	1.33	1.2	1.4	1.2	1.92	1.6	1.28	1.15
16–18	1.43	1.35	1.35	1.2	—	—	—	1.2	1.33	1.2	1.4	1.25	1.92	1.6	1.28	1.2
18–26.5	1.75	1.5	1.25	1.7	—	—	—	—	1.29	1.2	1.36	1.25	2.2	2.0	1.25	1.18
26.5–33	—	—	—	—	—	—	—	—	—	—	2.96	2.3	—	—	—	—
26.5–40	Contact HP															
33–50	Contact HP															

Note: Some scalar analyser detectors are biased, some are not. Bias is built-in.

minimize the mismatch ambiguities, since no vector corrections are possible. Therefore a considerable amount of attention should focus on mismatch causes when making equipment selections.

Mismatch considerations and source isolation methods

The most critical consideration for all of the measuring methods is the need to maintain the source and detector impedances near Z_0 at the DUT (device under test) insertion point. The main isolation method is the levelling loop using directional couplers or bridges. Directional couplers sampling the forward power with the reference detector mounted off the main line provide the best method of maintaining the source impedance near Z_0. By using the coupler in an external power-levelling loop, the effective output match becomes primarily just the inverse of the directivity. Table 13.1 compares effective SWR (standingwave ratio) specifications for several source isolation methods.

Transmission detector mismatch

Maintaining the transmission detector near Z_0 is accomplished by using well-matched detectors, or by putting a well-matched broad-band attenuator pad in front of the detector. Video LBSD (low-barrier-Schottky-diode) detectors have good match and are better than older point-contact diodes. Schottky detectors used with scalar network analysers have moderate SWR, but wider dynamic range (60 dB+, due to biasing and shaping of the detection curve to the critically-required square-law.

Thermocouple power sensors used in the power-meter method of attenuation measurement have the best inherent SWR, and are recommended for achieving the lowest measuring uncertainty in coaxial.

Selecting the specific source isolation method and detector is straightforward, and it depends on the uncertainty and speed required. The decision should be made carefully, because many times the calculations tend to look only at instrumentation error such as range-to-range specifications and fail to account fully for the much more significant mismatch effect in the external coaxial circuit.

Scalar analyser instrumentation

A scalar system consists of four parts: a microwave source, signal separation devices, detectors to convert the microwave energy to a d.c. or low frequency signal, and an analyser which receives the detector output and displays the measurement results as a function of frequency. Figure 13.2 shows the typical network measurement configuration.

Scalar analysers depend primarily on the square-law detection function of the transmission detector component. The scalar instrument itself complements and compensates for any square-law deviation of the companion detectors to extend dynamic amplitude range over 76 dB. The compensation model

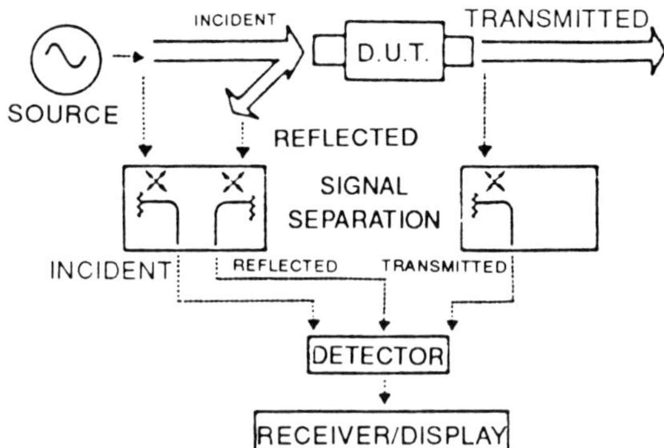

Fig. 13.2 Typical network measurement configuration using scalar analysis instrumentation.

depends on the assumption that the driving signals from the sweeper have no modulation or harmonics. Within assumptions, the technique works quite well, and deviation from the pre-compensated curves does not add much error.

When a scalar analyser is not available, the square-law method is popular because available equipment can easily be pressed into service. With any signal source and an isolating pad at the output, you simply connect a detector to the output, modulate the source, and set a reference. You then disconnect the detector, insert the DUT (device under test) between, and read the attenuated power on the meter.

The dual-channel power meter technique also features square-law detection. Since power sensors are true rms power-sensing, harmonics on the driving

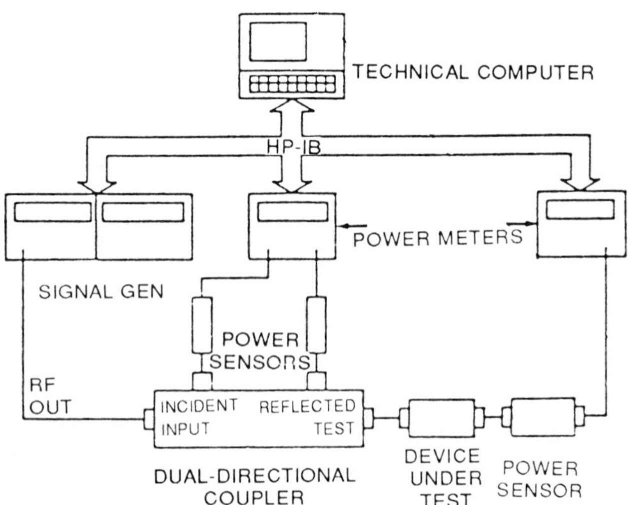

Fig. 13.3 Typical equipment set-up for dual channel power meter technique.

signal do not have an impact on the square-law characteristic. Figure 13.3 shows a typical equipment set-up. Low measuring uncertainty results because the coupler effective output match is low and the transmission sensor SWR is so low. Dynamic measuring range is 60 dB plus.

Expanding the square-law range

An interesting variation of the power sensor method which almost doubles the attenuation dynamic range depends on the use of a signal source with a programmable output attenuator. By first using the 45 to 50 dB dynamic range of the transmission sensor and its associated power meter, and then using the programmable attenuator inside the signal generator, an additional 20 or 30 dB of dynamic range can be added. The accuracy of the step attenuator, however, is not critical since the forward sensor is used to re-establish a mid-range reference.

RF-substitution method

The RF-substitution method depends on comparing an unknown inserted DUT attenuation with another accurately specified variable attenuator in the same signal path. In the simplest case, a signal generator with an accurate output attenuator drives power through the DUT into a transmission detector and indicator. After recording the reference power and removing the DUT, the generator attenation is cut in until the reference power is re-established. The difference of the signal generator output power readings is the desired attenuation value.

The RF-substitution method provides several advantages. First, the detection system no longer depends on precise square-law detection characteristics. Instead the accuracy derives mostly from the accuracy of the comparison attenuators.

IF-substition method

Two methods of IF-substitution measurement are possible for making relative power measurements.

One, the single-channel method, depends on a super-heterodyne test receiver with a highly precise and calibrated IF attenuator. After setting the reference and then inserting the RF test DUT between the signal source at the receiver input, the signal at the receiver drops. The attenuation substitution then takes place in the receiver's IF.

The second method, parallel IF substitution, depends on the IF reference attenuator in a second parallel channel and not in series with the DUT. In that way, the signal detector need not operate over a very wide dynamic range and becomes more of a detecting comparison of two extremely low levels.

Both methods of IF substitution depend critically on fully linear mixing in the down-conversion stage.

Most network analysers, being heterodyne dual-channel receivers, also measure attenuation by means of IF substitution. The biggest advantage of the vector analysers is that the system first looks back at each test port to determine its own output match, frequency by frequency. By storing the vector reflection coefficients and taking each into account, the input and output mismatch uncertainities are virtually eliminated.

13.1.2 Impedance measurement (SWR, reflection coefficient, return loss)

The impedance of a microwave component is a measure of its input or output match to a specified transmission line or path. This characteristic line impedence is usually 50 Ω. On a microwave transmission line, mismatches of impedance along the line cause reflected or reverse-travelling waves which interact with the incident wave to form a standing wave pattern. The standing wave ratio (SWR) is the ratio of standing wave maximum to minimum along the line and is directly related to the impedance mismatch. SWR is measured directly with the slotted line and sliding probe which detects the magnitude of the standing pattern along the line and reads on an indicator such as SWR meter. By common usage, it has been chosen to refer to voltage standing wave ratio as SWR. While there might be slight confusion with the term power standing wave ratio, PSWR measurements are not common.

The reflection coefficient of a device or component is another way of expressing impedance. The following relationship of ρ and SWR is frequently used in impedance work:

$$\rho = \frac{|E_r|}{|E_i|} = \frac{\text{SWR} - 1}{\text{SWR} + 1} \qquad \text{SWR} = \frac{E_{max}}{E_{min}} = \frac{1+\rho}{1-\rho}$$

E_r and E_i are the reflected and incident wave voltage. Reflection coefficient is a linear quantity varying between 0 to 1 and, being a ratio, has no units. The logarithmic expression of ρ is known as return loss and defined as $\text{dB} = -20 \log_{10}[\rho]$.

Scalar impedance measurements

In a reflection measurement, the wave of interest is the reflected wave. There are a number of methods used to determine the reflection parameter. All methods depend on some means of separating the forward travelling wave from the reflected wave.

The traditional directional coupler separates the forward and reverse waves on a transmission line. The most critical characteristic for this measurement is the directivity specification. It is the primary cause of measurement uncertainty in the scaler case. Table 13.2 shows comparison of impedance measuring techniques.

Table 13.2 Comparison of impedance measuring techniques

Measurement method	*Frequency range*	*Key HP models*	*Typical* [(1)] *uncertainty*	*Typical measurement range*			*Remarks/Cost/Accuracy/Speed*
				HP 415E SWR Meter	*Scalar/vector analyser*	*Power meter*	
COAXIAL							
Manual Slotted	2–18 GHz	809C/816B	1.02–1.06	30 dB	—	—	Lowest cost, high accuracy, slow.
Scalar analyser	0.01–18 GHz	85027A/C/8757A	34–40 dB	—	40 dB (1.01)	—	Moderate cost, fast, comprehensive.
directional	0.01–26.5 GHz	85027B/8757A	36–40 dB	—	40 dB	—	Bridges require 6 dB higher
bridge	300 kHz–3 GHz	85044A/11664A/8757A	40 dB	—	40 dB	—	power into UUT than couplers.
	0.01–4.3 GHz	85020A/8757A	40 dB @ 3 GHz	—	40 dB	—	
	0.01–2.4 GHz	85020B/8757A (75 Ω)	40 dB	—	40 dB	—	
Scalar analyser	100–2000 MHz	7780/11664A	32–34 dB	—	40 dB	—	Moderate cost, fast, comprehensive.
directional	100–4000 MHz	774–777D/11664A	40 dB	—	40 dB	—	
coupler	2–18 GHz	11692D/11664A	26–30 dB	—	40 dB	—	Couplers allow device bias.
	18–26.5 GHz	K752C/K281A/11664E	30 dB	—	40 dB	—	Lower signal into UUT device.
	26–40 GHz	R752C/R422A/R281A/B	30 dB	—	40 dB	—	Use HP 11664C with LBSD
	26–40 GHz	R752C/85026A/R281A/B	36 dB	—	40 dB	—	HP K/R/Q422A into
	33–50 GHz	Q752C/85026A/Q281A/B	33 dB	—	40 dB	—	HP 8750–scalar analysers. 2.4 mm coax for 26–50 GHz.
Power meter/	100–4000 MHz	774–777D/8484A	40 dB	—	—	40 dB	High accuracy, computes
computer	100–2000 MHz	778D/8484A	32–36 dB	—	—	40 dB	own measuring uncertainty
method	2–18 GHz	11692D/8484A	26–30 dB	—	—	40 dB	from directivity data and
	18–26.5 GHz	K752C/K281A/8485A	30 dB	—	—	40 dB	other corrections.
	26.5–40 GHz	R752C/R281A/R8486A	30 dB	—	—	40 dB	
	33–50 GHz	Q752C/Q281A/Q8486A	30 dB	—	—	40 dB	
Vector network	45 MHz–26.5 GHz	8510A/8340/41B	1.005–1.02	—	50 dB	—	Very highest accuracy, vector
analyser system	300 kHz–3 GHz	8512/13/14/15 test sets 8753A/85044/85046 test sets	1.005–1.02		50 dB		error correction techniques to remove system directivity and mismatches.

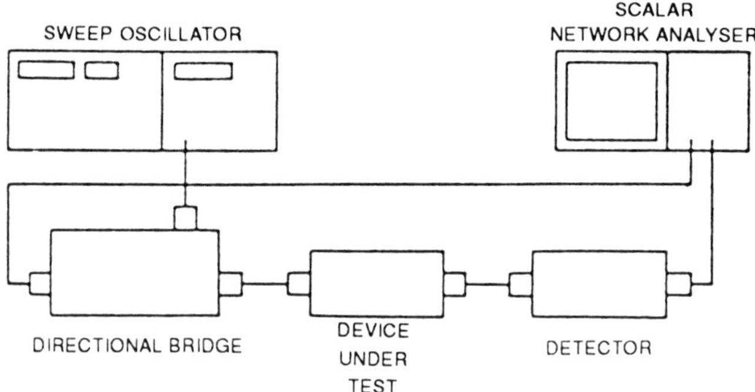

Fig. 13.4 Typical equipment set-up for scalar analyser instrumentation.

Scalar analyser instrumentation

Scalar analysers feature large amounts of internal computional power to speed the set-up, calibration, and measurements as well as to provide powerful data correction routines. In addition, the internal micro-processors control the test sweeper for driving the unit under test. Figure 13.4 shows a typical equipment set-up.

Other scalar instrumentation

There are several other scalar instrumentation alternatives which may be attractive, depending on user-available equipment. An attractive equipment alternative for measuring ratio is the dual-channel power meter, operated under desktop computer control. When the DUT product already requires power meter measurements for output power verification, the addition of a dual coupler and two sensors at the input can measure input power and input reflection, as well as output power and computed power gain. One advantage of the power meter method is that power sensors have particularly low SWR and thereby provide about the lowest uncertainty ambiguities available.

Slotted lines

Standing wave ratio may be measured directly with a slotted line as shown in Figure 13.5. While slotted lines have been losing popularity because they are manual, slow and tedious, they still serve for some measurements such as characterizing SWR of components under high power conditions. Standard couplers and detectors must remain below about 0 dBm, to maintain square-law.

The slotted line should match the device under test. As the carriage moves along the line, the probe senses the standing wave pattern, and the meter of the SWR meter reads SWR directly at the minima after referencing on the maxima.

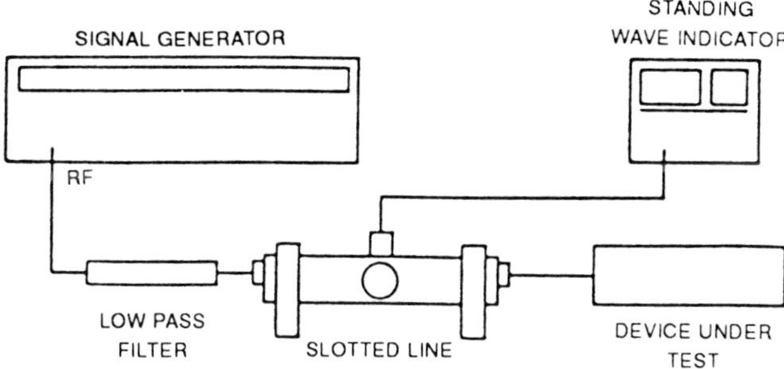

Fig. 13.5 Typical equipment set-up for slotted line SWR measurement.

With the exception of complex and expensive vector network analysers, the slotted line may still yield the lowest measurement uncertainty, especially at the higher frequencies and in wave-guide because of its inherent low residual SWR (typically 1.02).

Impedance accuracy considerations

A primary contributor to measurement uncertainty is the directivity parameter of the reverse directional coupler or the bridge. A directivity value of 40 dB yields an uncertainty of 0.01 in reflection coefficient (SWR uncertainty of 1.02). Most test systems using square-law detection are calibrated with 100% reflection. Since modern scalar analysers contain extensive memory, they easily store and correct for amplitude tracking of detectors, source mismatch, and other non-flatness of the baseline and full scale calibrations (i.e., open/short).

Square-law detection versus RF-substitution

The most popular impedance methods depend on the square-law detection characteristics of low-barrier Schottky detectors. Below – 20 dBm to the noise level, such detectors are inherently square-law. Above – 20 dBm the quasi-linear range may be 'shaped' by appropriate detector loading or by shaping circuits in the input amplifiers of the ratio meters. This can often extend the dynamic range above 60 dB. On the other hand, the basic coupler directivity specification of 40 dB effectively limits any need for square-law dynamic range over 40 dB.

For the utmost accuracy, RF-substitution techniques can remove the dependence on the square-law, by pre-inserting a calibrated amount of return-loss in the calibration cycle. For example, if the expected amount of reflection is 34 dB (SWR = 1.04), calibration values of 30 dB and 35 dB would be pre-inserted. This method is much easier to achieve in waveguide systems where high-accuracy, flat response rotary-vane attenuators are available.

Modulated versus unmodulated detection

While earlier slotted-line and SWR meter modulation operated at 1 kHz, modern scalar analysers depend on approximately 27.8 kHz. The a.c. amplification provides drift-free performance and is preferred when measurements must be made in the presence of undesired signals such as broadband noise or electromagnetic interference. The d.c. detection mode should be used when accurate power levels are to be measured, such as when sweeping the power output to determine the compression point of an amplifier or when measuring modulation-sensitive devices such as amplifiers with gain control circuits.

Measurement of directivity

The directivity parameter is the limiting value of the effective rejection a directional coupler or bridge offers the unwanted signal. Thus, when calibrating and verifying performance of the test coupler, directivity is measured by terminating the coupler with a 'perfect' zero-reflection load. Lacking the perfect load, a sliding load is used instead. Under these conditions, the only signal sneaking up into the reverse detector is directivity. By physically sliding the load element along the line to phase its small signal against the directivity signal, constructive and destructive results cause maximum and minimum deviations on the detected display. A close approximation is made for directivity by using the centre of the peak deviations.

13.1.3 Complex S-parameters measurements

S-parameters definition

The function of a network analysis is to completely characterize or describe a network so we'll know how it will behave when stimulated by some signal. For a two-port device, such as transistors, we can completely describe or characterize it by establishing a set of equations that relate the voltages and currents at the two ports. (Fig. 13.6).

In low frequencies, one such set of equations relates total voltages across the ports and total current into or out of these ports in terms of H-parameters. For example, $h_{11} = V_1/I_1$ when $V_2 = 0$.

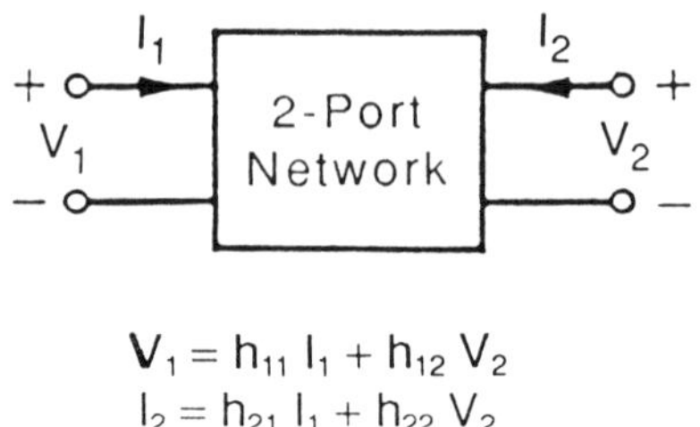

Fig. 13.6 2-port network characteristics.

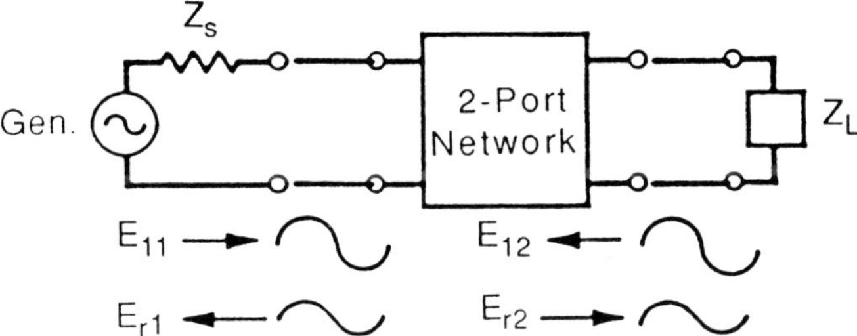

Fig. 13.7 Characterization of a 2-port network at high frequecncies – schematic diagram.

These parameters are obtained under either open or short circuit conditions. At higher frequencies, especially in the microwave domain, these operating conditions present a problem since a short circuit looks like an inductor and an open circuit has some leakage capacitance. Often, if the network is an active device such a transistor, it will oscillate when terminated with a reactive load. It is imperative that some new method for characterizing these devices at high frequencies be used.

If we embed our two-port device into a transmission line (see Fig. 13.7), and terminate the transmission line with its characteristic impedance, we can think of the stimulus signal as a travelling wave incident on the device, and the response signal as a wave reflecting from the device or being transmitted through the device. We can then establish this new set of equations relating these incident and scattered waves:

$$E_{1r} = S_{11}E_{1i} + S_{12}E_{2i} \qquad b_1 = S_{11}a_1 + S_{12}a_2$$
$$E_{2r} = S_{21}E_{1i} + S_{22}E_{2i} \qquad b_2 = S_{21}a_1 + S_{22}a_2.$$

Z_0 is the characteristic impedance of the transmission line. S_{11} is then equal to b_1/a_1 with $a_2 = 0$ or no incident wave on port 2. This is accomplished by terminating the output of the two-port in an impedance equal to Z_0. S_{11} is the input reflection coefficient with the output matched, S_{21} is the forward transmission coefficient with the output matched, S_{22} is the output reflection coefficient with the input matched, and S_{12} is the reverse transmission coefficient with the input matched.

S-parameters measurement

The system used to measure these parameters consists of:

1. an RF source;
2. a transducer or signal conditionner that splits the incoming RF signal into two channels and provides a calibrated line stretcher for equalizing the electrical lengths between the channels;

3. a harmonic frequency converter and mainframe that down converts the RF signal to a constant intermediate frequency;
4. a display unit that displays the ratio of the signals in the two channels.

S-parameters are complex quantities and all computer design programmes need both magnitude and phase information for complete characterization. Vector network analysers are tuned receivers and provide the additional capability to measure the phase response as well as the magnitude response of a device under test.

The first requirement of the measurement system is a sine wave signal source to stimulate the device under test. Since transfer and impedance functions are ratios of various voltages and currents, a means of separating the appropriate signal from the measurement port of the device under test is required. Finally the network analyser itself must detect these separated signals from the desired signal ratios, and display the results.

Careful attention to the magnitude and phase response of a communication system is essential for distortion-free transmission. By characterizing the magnitude and phase response of the errors of the test system itself, such as directivity, they can be removed from the measurement, resulting in a dramatic increase in measurement accuracy.

The vector network analyser is basically a multiple-channel tracking receiver that tunes to the test signal. The signal channels accommodate reference and test signals from the network under test, down-converts them to narrowband, constant IF signals, followed by processing, detection and display. The IF signals preserve the amplitude and phase relationships of the RF test and reference signals. Narrowband detection permits high sensitivity, wide dynamic range, and brings immunity from harmonics that might exist in the test signal source.

Currently available vector network analysers provide complete measurement capability for characterization of linear networks from 5 Hz to 100 GHz. Automatic network analysers such as the HP 8510, provide control for a complete measurement system including programming of the RF source and test set, signal processing and display of data in a variety of formats, built-in accuracy enhancement capability, and control of data output peripherals.

13.2 NOISE FIGURE MEASUREMENTS

13.2.1 Noise characteristics and measurement methods

Sensitivity and noise figure are popular system parameters that characterize the ability to process low level signals. Of these parameters, noise figure is unique in that it is suitable not only for characterizing the entire system but also the system components such as the preamplifier, mixer and IF amplifier that make up the system. By controlling the noise figure and gain of system components, the designer directly controls the noise figure of the overall system. Once the noise figure is known, system sensitivity can be easily estimated from system bandwidth.

The reason for measuring noise properties of networks is to minimize the problem of noise generated in receiving systems. Noise measurements are the key to ensuring that the added noise is minimal.

Kinds of noise

The noise characterized by noise measurements consists of spontaneous fluctuations caused by ordinary phenomena in the electrical equipment. Two principal types of such noise are thermal noise and shot noise. Thermal noise arises from vibrations and conduction electrons and holes due to their finite temperature. Some of the vibrations have a spectral content within the frequency band of interest and contribute noise to the signals. Shot noise arises from the quantized nature of current flow.

Thermal noise refers to the kinetic energy of a set of particles as a result of its finite temperature. If some particles are charged, vibrational kinetic energy may be coupled electrically to another device if a suitable transmission is provided. The power available is kTB where k is Boltzmann's constant, T is the absolute temperature, and B is the bandwith of the transmission path. That the power available depends directly on temperature is obvious. The more energy that is present in the form of higher temperature or larger vibrations, the more energy that is possible to remove per second. It should be emphased that kTB is the power available from the device. This power can only be coupled into an optimum load, i.e., a complex-conjugate impedance that is at absolute zero so that it does not send any energy back. The thermal noise power available, kTB, although dependent on bandwidth, is independent of frequency.

Shot noise is caused by the quantized and random nature of current flow. Current is not continuous but quantized, being limited by the smallest unit of charge ($e = 1.6 \times 10^{-19}$ C). Particles of charge, furthermore, also flow with random spacing. When d.c. current I_0 flows, the average current is I_0, but that does not indicate what the variation in the current is or what frequencies are involved in the random variations of current. The noisy current flowing through a load resistance forms the power variations known as shot noise.

Other random phenomena occur that are quantized in nature and can be statistically analysed in the manner of shot noise. Examples are the generation and recombination of hole/electron pairs in semiconductors and the division of emitter current between the base and collector in transistors.

A very important source of noise occurs in avalanche diodes because such devices are used as reference sources for measurement. Here a carrier achieves enough energy so that, upon collision with the crystal lattice, it is able to generate another hole/electron pair. Some mobile carriers generate two pairs, etc. The multiplication factor for the free-charge generation varies randomly and is also quantized. The noise power associated with the avalanche diode tends to be inversely proportional to current and somewhat dependent on frequency. The noise power versus frequency relation depends on the current being conducted.

Thus there are many causes of random noise in electrical devices. Noise characterization usually refers to the combined effect from all the causes in a component. The combined effect is often referred to as if it were all caused by thermal noise. Referring to a device as having a certain noise temperature does not mean that the component is that physical temperature, but merely that its noise power is equivalent to a thermal source of that temperature.

Concept of noise figure

Before showing the importance of accurate noise measurement, it is necessary to define the most popular quality factor for noise performance, noise figure. Harold Friis defined the noise figure F of a network to be the ratio of the signal-to-noise ratio at the input to the signal-to-noise ratio at the output. Thus the noise figure of a network is the decrease or degradation in the signal-to-noise ratio as the signal goes through the network. A perfect amplifier would amplify the noise at its input along with the signal. A realistic amplifier, however, also adds some extra noise from its own components and degrades the signal-to-noise ratio. A low noise figure means that very little noise is added by the network.

Noise figure is not a quality factor of network with one port; it is not a quality factor of terminations or of oscillators. Oscillators indeed generate noise and have their own quality factors like 'carrier to noise ratio' and 'single-sideband phase noise in a one hertz bandwith, X hertz from the carrier'. Noise figure also has nothing to do with modulation. It is independent of the modulation format and of the fidelity of modulators and demodulators.

Noise figure should be thought of as separate from gain. Once noise is added to the signal, subsequent gain amplifies signal and noise together and does not change the signal-to noise ratio.

The degradation in a network's signal-to-noise ratio is dependent on the temperature of the source that excites the network. To see why that is true, the ratio of the signal-to noise ratios, i.e. the noise figure is:

$$F = \frac{S_i/N_i}{S_o/N_o}$$

$$= \frac{S_i/N_i}{G_a S_i/(N_a + G_a N_i)}$$

$$= \frac{N_a + G_a N_i}{G_a N_i}$$

where S_i and N_i represent the signal and noise levels available at the input to the device under test, S_o and N_o represent the signal and noise levels available at the output, N_a is the noise added by the DUT, and G_a is the available gain of the DUT. The input noise level is usually thermal noise from the source and is referred to by kTB. Friis suggested a reference source temperature of 290 K denoted by T_o. Then equation becomes:

$$F = \frac{N_a + kT_o BG_a}{kT_o BG_a}$$

Noise figure is generally a function of frequency but it is usually independent of bandwith.

The noise figure of a DUT is independent of the signal level so long as the DUT is linear. Because of the need for linearity, any AGC circuitry must be deactivated for noise measurements. The IEEE standard definition of noise figure states that noise figure is the ratio of the total noise power output to that portion of the noise power output due to noise at the input when the input source temperature is 290 K. It is obviously related to the Friis definition by the above arguments.

Measurement of noise figure

Noise figure and effective noise temperature are usually determined by measuring two points along the straight-line noise characteristic. Figure 13.8 shows those two points corresponding to two noise source temperatures at the input; one hot (T_h) and and one cold (T_c). Although many different types of

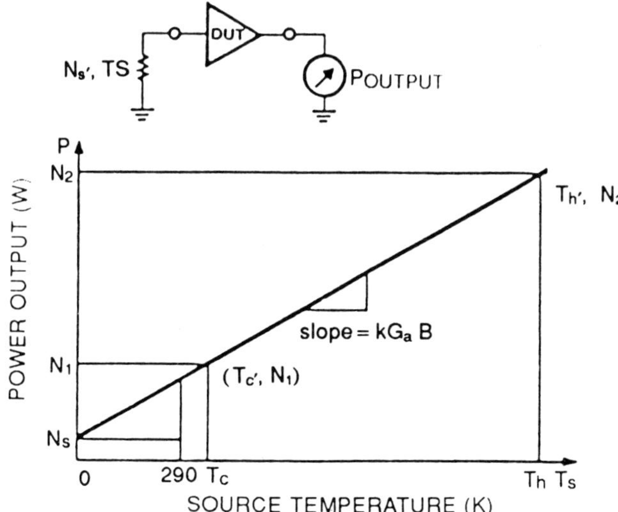

Fig. 13.8 Determination of the noise figure and effective noise temperature by measuring two points of the noise characteristic.

noise sources are used, most measurements use an avalanching semiconductor diode for T_h with an effective temperature of 10 000 K. For T_c the bias is removed from the diode so that avalanching ceases and the effective temperature corresponds to the physical temperature of the source. Manufacturers of noise sources usually specify the excess noise ratio (ENR) instead of T_h. The relationship between the two terms is:

$$\mathrm{ENR} = 10 \log \frac{T_h - T_o}{T_o}.$$

Sometimes T_{on} and T_{off} are used instead of T_h and T_c. The noise powers at the output of the DUT are often labelled N_2 (corresponding to T_h) and N_1 (corresponding to T_c). The ratio N_2/N_1, is called the Y factor.

Once the two points along the straight line are determined, ordinary algebra can be used to find noise figure F, effective input noise temperature T_e, and the gain-bandwidth product. The new noise figure meter performs the algebra with a microprocessor. The results are:

$$T_e = \frac{T_h - YT_c}{Y - 1}$$

$$F = 10 \log \frac{T_e + T_o}{T_o}$$

$$kGB = \frac{N_2 - N_1}{T_h - T_c}.$$

In summary, the noise quality factors of a linear two port are easy to interpret from the straight-line noise characteristic. In most pratical situations, the bandwidth and the frequency of measurement are determined by the noise measurement equipment.

Measurement methods

Measuring noise figure depends a great deal on a source that generates two accurately known noise powers at the frequency of interest. Both T_h and T_c must be known accurately. There are three basic ways to supply noise sources at these temperatures: avalanche diodes, hot/cold loads, gas discharge tubes. The avalanche diode is the most widely used because it is broadband and can conveniently switch between T_c and T_h under electronic control.

Measuring noise figure really boils down to measuring the ratio of N_2 to N_1 and, knowing T_h and T_c, calculating T_e and NF. This can be done with a simple system consisting of a filter tuned to the frequency of interest and a detector that has a true RMS response (Fig. 13.9). As a practical matter this system has some significant limitations, but in concept it illustrates how we can measure a component such as an amplifier or mixer or an entire receiver front-end.

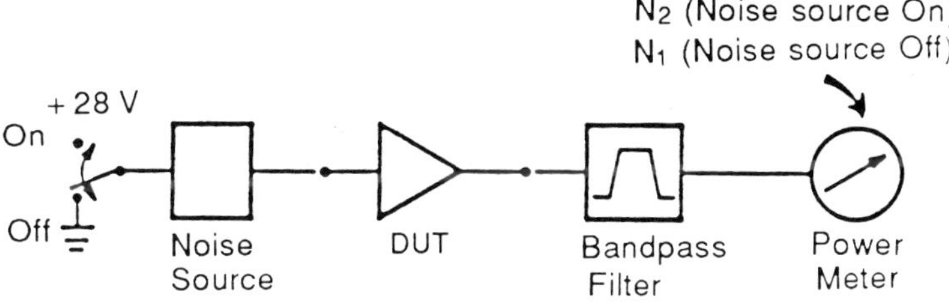

$$Y = \frac{N_2}{N_1}, \quad T_e = \frac{T_h - YT_c}{Y - 1}, \quad F = \frac{T_e}{290} + 1$$

Fig. 13.9 Simple scheme for measuring noise figure.

One limitation of the simple conceptual system, is that it only works at a single frequency determined by the bandpass filter. An obvious enhancement is to use a heterodyne front-end to allow measurements at any frequency over a band determined by the local oscillator.

Measurement at microwave frequencies

Measurements at microwave frequencies are different because the higher frequencies must be translated to a frequency accepted by the noise figure meter. A measurement system with down conversion is shown in fig. 13.10. A mixer and a local oscillator are required.

This measurement is called double sideband measurement (DSB). The mixer's noise figure is going to be a large determining factor in the noise figure. Mixer selection is more than just noise figure and match. Still another major consideration is the isolation between LO and IF ports. This is needed to reduce LO sidebands that get converted to the IF and raise the measurement

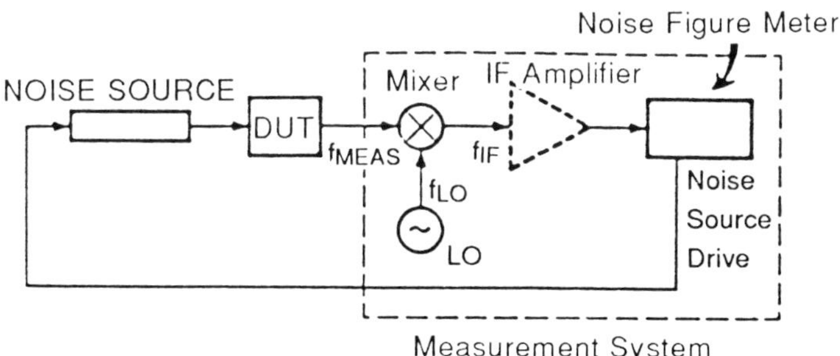

$$f_{MEAS} = f_{LO} \pm f_{IF}$$

Fig. 13.10 Noise figure measurement at microwave frequencies using a down converter.

system noise figure. The LO should have the right frequency range and sufficient power to drive the mixer. Its most important additional characteristic is its spectral purity at offset frequencies corresponding to f_{if} of the measurement system.

In general, it is easier to make DSB measurements, but some DUTs have such rapid gain or noise figure changes with frequency that SSB (single sideband) is necessary.

13.2.2 Accuracy considerations

Measurement uncertainty

There are many potential causes of noise figure measurement uncertainty. Factors affecting measurement accuracy are: extraneous signals, uncertainty due to mismatch, calibration uncertainty, system noise figure, gain and noise figure of the device, system response characteristics (DSB, SSB, harmonics).

Reducing the system noise figure is very important, particularly when measuring low gain devices. This can be done by using a low noise preamplifier. Since we are making power measurements, it is critical that the magnitude of all mismatches be as low as possible to reduce the effect of multiple mismatch uncertainty both during calibration and measurement. This is ensured by using well matched isolators or pads.

What can you expect?

There are many variables. For lowest uncertainty, we have to reduce the noise figure of the measurement system and the magnitudes of all mismatches. The uncertainty that can be expected can be as low as 0.2 dB RMS for the measurement. We have to evaluate the test system carefully and to eliminate any sharp resonances which could make the system noise figure and gain very frequency sensitive. The best results are obtained by calibrating the system at every frequency and by using low drift sources.

13.3 TIME DOMAIN REFLECTOMETRY

Time domain measurements show how a network responds to a specific waveform. By looking at the reflections of a signal off the network, the component reflections which make up the reflection coefficient measured in the frequency domain can be identified. Similarly, by looking at the signal transmitted by the device, the signal paths inside the network which make up the transmission coefficient measured in the frequency domain can be identified. The nature of a network analyser is to measure in the frequency domain, but it can also make time domain measurements by calculating time domain results from frequency domain data.

What insight does the time domain response of a device provide?
Reflection responses show the location and impedance of different elements inside the network. Transmission responses show main signal paths, isolation problems, and internal reflections.

Some examples are: identifying and analysing circuit elements, locating faults within a well known network, and identifying and removing unwanted responses.

13.3.1 Time domain measurements

A time domain measurement can be made directly by sending the known waveform out to the test network and measuring the waveform returned as a function of time. The delay of the returned waveform shows the distance to the network, and the shape of the returned wave shows the impedance of the network. Traditionally, the measurements have been made with a time domain reflectometer TDR.

An indirect time domain measurement can also be made. This is accomplished by measuring the response of the device in the frequency domain with a vector network analyser, and then using that information to predict the response of the network to a specific waveform. The key to making the time domain prediction is the inverse Fourier transform.

13.3.2 Time domain reflectometer (TDR)

A direct time domain measurement system consists of a signal generator, a signal separation device to split out the response, and an oscilloscope to display the results. A traditional block diagram is shown on Fig. 13.11.

This system can be either a step or an impulse system, depending on the source used. A step generator provides a train of very fast steps, and an impulse generator provides a train of very short impulses. A repetitive stimulus is used so that the oscilloscope can trigger.

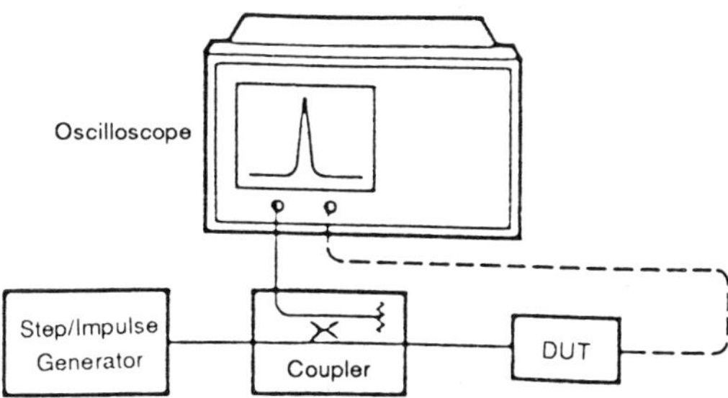

Fig. 13.11 Block diagram of a time domain reflectometer (TDR).

13.3.3 Network analyser with time domain option

The method uses a network analyser to make time domain measurements. The network analyser measures the frequency response of the device, and with the time option it calculates either the step, impulse, or band-pass response of the device, as selected when the transform is activated.

A network analyser is a very high performance time domain system when compared with a direct measurement system. A network analyser offers more bandwidth than a TDR, and the bandwidth used can be selected by the operator. The network analyser, since it has a tuned receiver, offers much better noise performance than a broadband direct measuring system. Lastly, a network analyser calculates its time axis, making it very stable and accurate. A direct system depends on the oscilloscope trigger, which leads to trace jitter.

A direct system can show non-linear system characteristics in the time domain, such as slew rate between the power supply rails. A network analyser, alternatively, has error correction to improve system specifications. And while a TDR is often a dedicated system, a network analyser is easier to use and more flexible.

13.4 PHASE NOISE MEASUREMENTS

As the performance of microwave radar and communication systems advances, certain system parameters take on increased importance. One of these parameters that must be measured is the spectral purity of microwave signal sources.

13.4.1 Phase noise and its effects on microwave systems

What is phase noise?

Frequency stability can be defined as the degree to which an oscillating source produces the same frequency throughout a specified period of time. Every RF and microwave source exhibits some amount of frequency instability. This stability can be broken down into two components: long term and short term stability.

Mathematically, an ideal sine-wave can be described by:

$$V(t) = V_0 \sin(2\pi f_0 t).$$

But an actual signal is better modelled by:

$$V(t) = [V_0 + \varepsilon(t)] \sin[2\pi f_0 t + \varphi(t)]$$

where $\varepsilon(t)$ = amplitude fluctuations and $\varphi(t)$ = randomly fluctuating phase term or phase noise.

The randomly fluctuating phase term could be observed on an ideal spectrum analyser. There are two types of fluctuating phase terms. The first, deterministic, are discrete signals appearing as distinct components in the spectral density plot. These signals, commonly called spurious, can be related to known phenomena in the signal source such as power line frequency, vibration frequencies, or mixer products.

The second type of phase instability is random in nature, and is commonly called phase noise. The sources of random sideband noise in an oscillator include thermal noise, shot noise, and flicker noise.

Many terms exist to quantify the characteristic randomness of phase noise. Essentially, all methods measure the frequency or phase deviations of the source under test in either the frequency or time domain. One fundamental description of phase instability or phase noise is the spectral density of phase fluctuations on a per-hertz basis. The term spectral density describes the energy distribution as a continuous function, expressed in units of phase variance per unit bandwidth rad^2/Hz:

$$S_\varphi(f_m) = \frac{\Delta\varphi^2_{rms}(f_m)}{\text{BW used to measure } \Delta\varphi_{rms}}.$$

Another useful measure of the noise energy is $\mathcal{L}(f_m)$, which is then directly related to $S_\varphi(f_m)$ by a simple approximation which has generally negligible error if the modulation sidebands are such that the total phase deviations are much less than 1 radian.

$$\mathcal{L}(f_m) \simeq \frac{1}{2} S_\varphi(f_m).$$

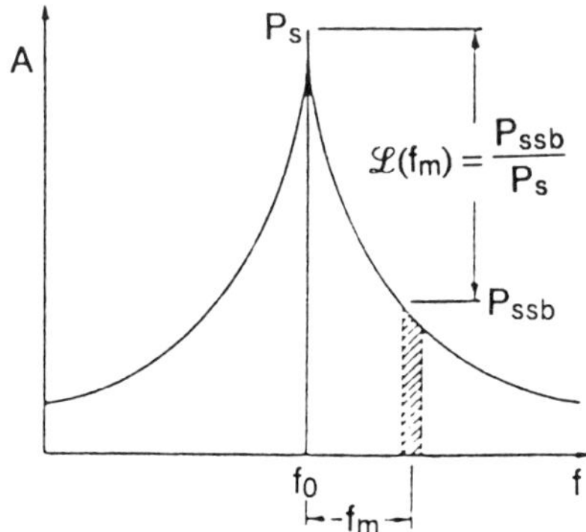

Fig. 13.12 Ratio of power density in one phase modulation sideband to total signal power.

$\mathcal{L}(f_m)$ is an indirect measure of noise energy easily related to the RF power spectrum observed on a spectrum analyser. Figure 13.12 shows how the US National Bureau of Standard defines $\mathcal{L}(f_m)$ as the ratio of the power in one phase modulation sideband to the total signal power (at an offset f_m hertz away from the carrier). The phase modulation sideband is based on a per hertz of bandwidth spectral density and f_m equals the Fourier frequency or offset frequency.

$$\mathcal{L}(f_m) = \frac{\text{power density (in one phase modulation sideband)}}{\text{total signal power}} = \frac{P_{ssb}}{P_s}$$

$$= \text{single sideband (SSB) phase noise to carrier ratio per Hz.}$$

$\mathcal{L}(f_m)$ is usually presented logarithmically as a spectral density of the phase modulation sidebands in the plot of the frequency domain, expressed in dB relative to the carrier per Hz (dBc/Hz), as shown in Fig. 13.13.

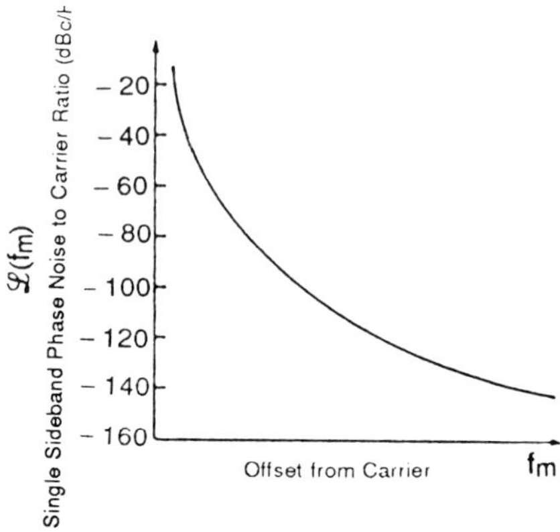

Fig. 13.13 (fm) vs fm in dBc/Hz.

Another common term for quantifying short term frequency instability (phase noise) is $S_{\Delta f}(f_m)$, the spectral density of frequency fluctuations. Again the term spectral density describes the energy distribution as a continuous function, expressed in units of frequency variance per unit bandwidth Hz^2/Hz.

$$S_{\Delta f}(f_m) = \frac{\Delta f_{rms}^2(f_m)}{\text{BW used to measure } \Delta f_{rms}}$$

Because frequency is the time rate of change of phase, the three common terms $S_\varphi(f_m)$, $\mathcal{L}(f_m)$, $S(f_m)$ can be related as shown:

$$S_\varphi(f_m) = \frac{S_{\Delta f}(f_m)}{f_m^2} \qquad \mathcal{L}(f_m) = \frac{S_{\Delta f}(f_m)}{2f_m^2}.$$

Why is phase noise important?

There are two different types of phase noise commonly specified: two-port phase noise and absolute phase noise. Two-port phase noise refers to the noise of devices. Amplifiers, mixers, and multipliers have two-port phase noise. Two-port noise results from the noise contributed by a device, regardless of the noise of the driving source. Absolute phase noise refers to the total phase noise present at the output of a source or a system. It is a function of both the device two-port phase noise and the oscillator noise.

In general, the absolute phase noise of a source is most important in the final system application. However, two-port noise of devices or synthesized sources might also be measured prior to system integration.

Phase noise on signal sources is a concern in frequency conversion applications where signal levels span a wide dynamic range. The frequency offset of concern and the tolerable level of noise at this offset vary greatly for different microwave systems. Sideband phase noise can convert into the information passband and limit the overall system sensitivity. The general case is illustrated in Fig. 13.14.

The phase noise of the local oscillator will be directly translated onto the mixer products.

Note that though the system's IF filtering may be sufficient to resolve the larger signal's mixing product ($f_1 - f_{LO}$), the smaller signal's mixing product ($f_2 - f_{LO}$) is no longer recoverable due to the translated local oscillator noise. The noise on the local oscillator thus degrades the system's sensitivity as well as its selectivity.

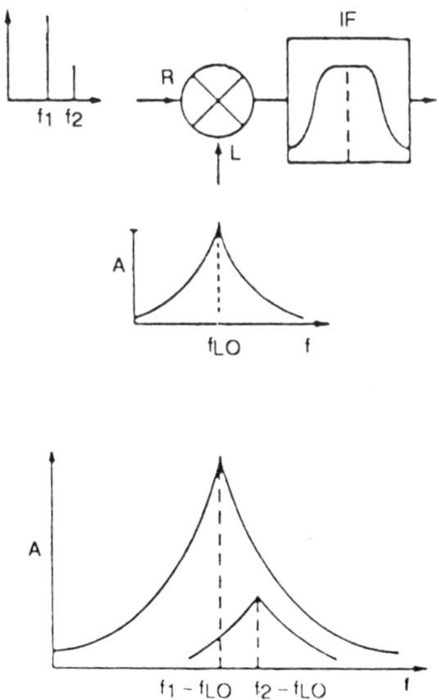

Fig. 13.14 Phase noise measurement schematic with signals in (top) and signals out (bottom).

13.4.2 Measurement of phase noise

Heterodyne frequency measurement technique

This time domain method down-converts the signal under test to an intermediate frequency. The down-converting signal must be of greater stability than the signal to be measured. Then a high resolution frequency counter repeatedly counts the IF signal frequency, with the time period between each measurement held constant. This allows several calculations of the fractional frequency difference, y, over the time period used. From these values for y, the Allan variance, $\sigma_y(\tau)$ can be computed. $\sigma_y(\tau)$ in the time domain corresponds to $S_y(f_m)$ in the frequency domain.

This method gives particularly useful results for short-term frequency instabilities occuring over periods of time greater than 10 ms (less than 100 Hz offset from the carrier in the frequency domain), where the phase noise is increasing rapidly. The heterodyne/counter method is ideal for close-in measurements on frequency standards.

Phase noise measurement with spectrum analyser

The most straightforward method of phase noise measurement inputs the test signal into a spectrum analyser, directly measuring the power density of the oscillator. However, this method may be significantly limited by the spectrum analyser's dynamic range, resolution, and its own LO phase noise.

Though this direct measurement is not useful for measurements close-in to a drifting carrier, it provides a convenient method for qualitive, quick evaluation on sources with relatively high noise. The following conditions make the measurement valid:

1. the spectrum analyser SSB phase noise at the offset of interest must be lower than the noise of the DUT;
2. since the spectrum analyser will measure total noise power, the amplitude noise of the DUT must be significantly below its own phase noise.

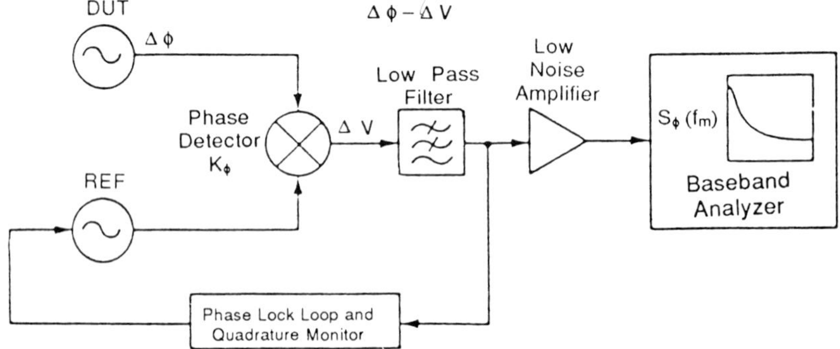

Fig. 13.15 Typical equipment set-up for phase noise measurement with two sources and phase detector.

Phase noise measurement with two sources and phase detector

This method (Fig. 13.15) uses a double-balanced mixer to convert phase fluctuations into baseband voltage fluctuations. Two signals at the same frequency f_0 are input into the mixer. The sum frequency $2f_0$ is filtered off with a low-pass filter leaving the difference frequency. If the two signals are 90° out of phase the difference frequency will be 0 Hz with an average output voltage of 0 V. Riding on this d.c. signal are a.c. voltage fluctuations that are linearly proportional to phase noise of both sources. Usually the quadrature condition is maintained by phase locking the two signals.

Phase noise measurement with a frequency discriminator

This method does not require a second reference source phase locked to the source under test (Fig. 13.16).

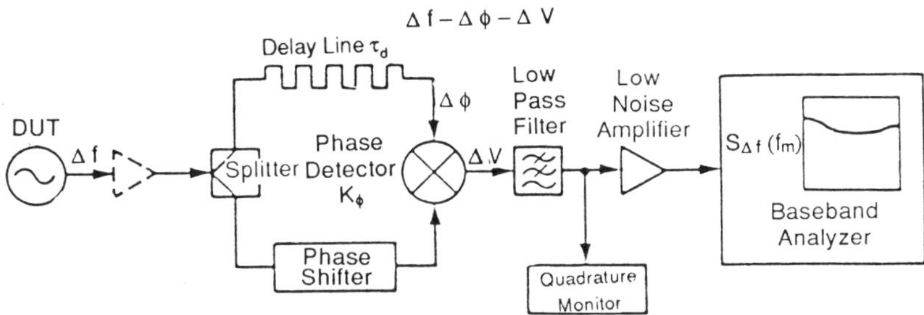

Fig. 13.16 Typical equipment set-up for phase noise measurement with frequency discriminator.

The frequency discriminator method is used for measuring sources that are difficult to phase lock. It can also be used to measure sources with high-level, low-rate phase noise, or high close-in spurious sidebands. Frequency discriminators can be implemented in several common ways including cavity resonators, RF bridges, and a delay-line/mixer.

13.5 SPECTRUM ANALYSER

Before describing the details of a spectrum analyser, let us see what a spectrum is, and why it is useful to analyse it. Usually the reference frame in signal analysis is time. For electrical events, an oscilloscope is used to show the instantaneous value as a function of time, i.e. in the **time domain**. Sometimes a sine wave is expected but the oscilloscope shows a signal which is not a pure sinusoid but has a harmonic on it, and it is not possible to measure its level.

The transformation **time domain → frequency domain** is obtained by means of the Fourier transformation: any time-domain electrical phenomenon is made up of one or more sine waves of appropriate frequency, amplitude and phase. In other words, with proper filtering it is possible to separate a time plot into separate sine waves that we can then evaluate independently. If the signal to be analysed is periodic, by Fourier analysis it is known that the constituent sine waves are separated in the frequency domain by $1/T$, where T is the period of the signal. If the time signal occurs only once, then T is infinite, and the frequency representation is a continuum of sine waves.

It should be noted that to make the transformation from the time to the frequency domain properly, the signal must be evaluated over all time, i.e. since and up to infinity. However a practical point of view is to assume that signal behaviour over several seconds or minutes is indicative of its overall characteristics.

Obviously, time domain measurements such as rise and fall times, pulse widths, repetition rates, delays, etc, can not be replaced by only frequency domain measurements.

One main area that has an interest in spectrum measurements is communications. For example, cellular radio systems must be checked for harmonics of the carrier signal that might interfere with other systems operating at the same frequencies as the harmonics. Some other interesting features of the spectrum analysis are the distortion of the message modulated onto a carrier, the spectral occupancy, the electromagnetic interference, etc.

13.5.1 The superheterodyne spectrum analysis

Figure 13.17 shows a simplified block diagram of a superheterodyne spectrum analyser. Heterodyne means to mix, i.e. translate frequency, and **super** refers

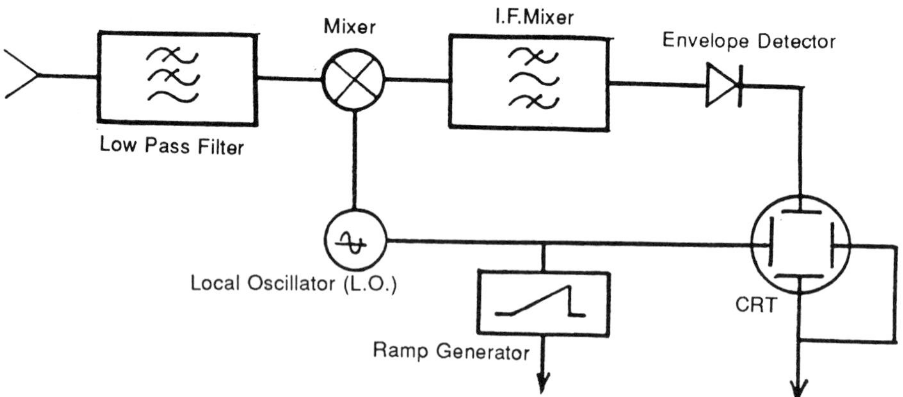

Fig. 13.17 Simplified block diagram of a superheterodyne spectrum analyser.

to the frequencies above the audio range or super audio-frequencies. We can see that the input signal passes through a low-pass filter to a mixer where it mixes with a signal from the local oscillator (LO). Because the mixer is a non-linear device its output includes not only the two original signals, but also harmonics and sums and differences. If any of the signals at the output of the mixer fall within the passband of the intermediate frequency (IF) filter, it is further processed (amplified and perhaps logged), essentially rectified by the envelope detector, and applied to the vertical plates of the cathode ray tube (CRT) to produce a vertical deflection on the CRT screen. A ramp generator simultaneously deflects the CRT beam horizontally across the screen from left to right and tunes the LO higher in frequency. This system is similar to a superheterodyne radio, but the output of a spectrum analyser is the screen of a CRT instead of a speaker, and the local oscillator is tuned electronically rather than merely by a front-panel knob.

Tuning equation — bandwidth

To what frequency is the spectrum analyser tuned? It depends; it is a function of the centre frequency of the IF filter, the frequency range of the local oscillator, and the range of frequencies allowed to reach the mixer from the outside world, i.e. allowed to pass through the low-pass filter. Of all the products emerging from the mixer, two with the greatest amplitude and therefore the most desirable are those created from the sum of the LO and the input signal, and the difference between the LO and the input signal. So if we can arrange things so that the signal we wish to examine is either above or below the LO frequency by the IF, one of the desired mixing products will fall in the pass-band of the IF filter and create a vertical deflection on the CRT.

How should one choose the LO and IF frequencies to create an analyser with the desired frequency range? Let us assume that we wish to have a tuning range from 0 to 2.9 GHz. Suppose we choose an IF of 1 GHz. This frequency is within the desired tuning range, so we could have an input signal at 1 GHz. Since the output of the mixer includes the original input signals, an input signal at 1 GHz would give us a permanent output from the mixer at the IF. The 1 GHz signal would thus pass through the system and give us a constant deflection on the CRT regardless of the tuning of the LO. The result would be a hole in the frequency range at which we could not examine signals because the CRT deflection would be independent of the LO.

This is why we shall choose an IF above the highest frequency to which we wish to tune. In some spectrum analysers of this frequency range the IF chosen is 3.6 GHz. Now, if we wish to tune from very low frequencies near 0 Hz to 2.9 GHz, over what range must the LO tune? If we start the LO at the IF (frequency of the LO – frequency of the IF = 0), and tune it upward from there to 2.9 GHz above the IF, we can cover the tuning range with the LO minus IF mixing product. With this information, then, we can generate a tuning equation:

$$f_{\mathrm{sig}} = f_{\mathrm{LO}} - f_{\mathrm{IF}}$$

where f_{sig} is the signal frequency, f_{LO} is the local oscillator frequency (LO), and f_{IF} is the intermediate frequency (IF). Since the ramp generator controls both the spot on the CRT and the LO frequency, we can now calibrate the horizontal axis of the CRT in terms of input signal frequency.

What happens if the frequency of the input signal is 8.2 GHz, for instance? As the LO tunes through its 3.6 to 6.5 GHz range, it reaches a frequency (4.6 GHz) at which it is the IF away from the 8.2 GHz signal, and once again we have a mixing product equal to the IF, creating a deflection on the CRT. In other words, the tuning equation could just as easily have been:

$$f_{\mathrm{sig}} = f_{\mathrm{LO}} + f_{\mathrm{IF}}.$$

This equation says that the architecture of Fig. 13.17 could also result in a tuning range from 7.2 to 10.1 GHz. But only if we allow signals in that range to reach the mixer. The job of the low-pass filter in Fig. 13.17 is to prevent these higher frequencies from getting to the mixer. We also want to keep signals at the IF from reaching the mixer, as described above, so the low-pass filter must do a good job of attenuating signals at 3.6 GHz as well as in the range from 7.2 to 10.1 GHz.

So, in summary, we can say that for a single band RF spectrum analyser, we would choose an IF above the highest frequency of the tuning range, have an LO tuneable from the IF up to the IF plus the upper limit of the tuning range, and include a low-pass filter in front of the mixer that cuts off below the IF.

Resolution

Frequency resolution is the ability of a spectrum analyser to separate two input sinusoids into distinct responses. But why should resolution even be a problem when Fourier tells us that a signal (read sinewave in this case) has energy at only one frequency? It would seem that two signals, no matter how close in frequency, would appear as two lines on the CRT. But a closer look at our superheterodyne receiver shows why signals appear as spread responses. The output of a mixer includes the sum and difference products plus the two originals (input and LO signals). The intermediate frequency is determined by a bandpass filter, and this filter selects the desired mixing product and rejects all other signals. Because the input signal is fixed and the LO is swept, the products from the mixer are also swept. If a mixing product happens to sweep past the IF, the characteristics of the bandpass filter are traced on the CRT. The narrowest filter in the chain determines the overall bandwidth, and in the architecture of Fig. 13.18 this filter is in the 3 MHz IF.

So unless two signals are separated enough, the traces they make fall on top of each other and look like only one response (see Fig. 13.19). Spectrum

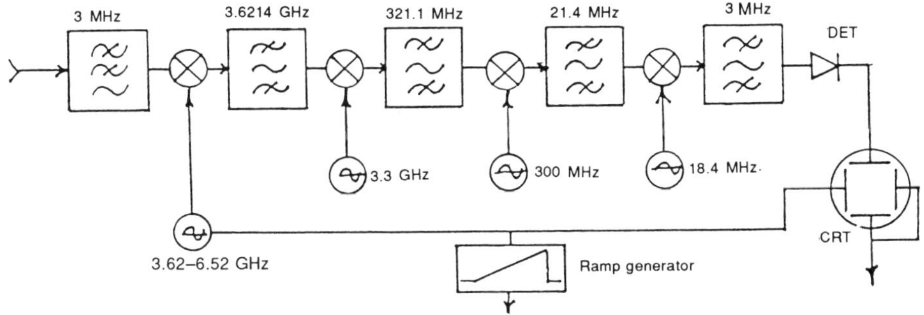

Fig. 13.18 Block diagram of a selectable resolution spectrum analyser.

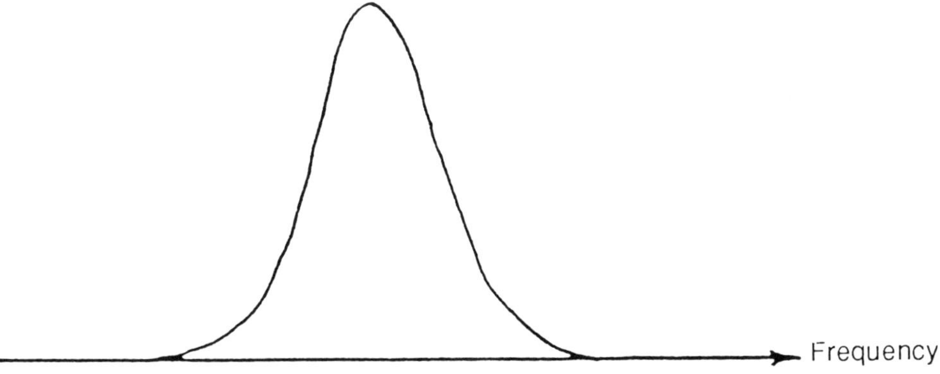

Fig. 13.19 The response characteristics of the IF filter traced on the CRT.

analysers have selectable resolution (IF) filters, so it is usually possible to select one narrow enough to resolve closely spaced signals.

There is another factor that affects the resolution of a spectrum analyser: the stability of the LOs in the analyser, particularly the first LO. YIG-tuned oscillators of the type used in most analysers (they typically tune somewhere in the 2 to 6 GHz range) exhibit residual *f*m of 1 kHz or more. This instability is transferred to any mixing products resulting from the LO and incoming signals. And it is not possible to determine whether the signal or LO is the source of the instability.

So, the minimum resolution bandwidth typically found in a spectrum analyser is determined at least in part by the LO stability. In economy analysers, where no steps are taken to improve upon the inherent residual *f*m of the YIG oscillators used in them, the minimum bandwidth is usually 1 kHz. The first LO is stabilized in mid-performance analysers, and they include 100 Hz filters.

High-performance analysers have more elaborate synthesis schemes to generate all of their LOs and so have bandwidths down to 10 Hz or less. So with the possible exception of economy analysers, any *f*m-type instability that we see on a spectrum analyser is due to the incoming signal.

Sensitivity

One of the primary uses of a spectrum analyser is searching for and the measurement of very low level signals. The ultimate limitation in these measurements is the random noise generated by the spectrum analyser itself. This noise is generated by the random electron motion through the various circuit elements of an analyser, is amplified by the various gain stages in the analyser, and ultimately appears on the CRT as a noise signal below which we cannot make measurements. A good bet for the starting point for noise seen on the CRT is the first stage of gain within an analyser. This amplifier boosts the noise generated by its input termination plus adds some of its own. So as this noise signal passes on through the system, it is typically high enough in amplitude so that the noise generated in subsequent gain stages does not materially add to the noise power. It is true that the input attenuator and one or more mixers may be between the input connector of a spectrum analyser and the first stage of gain, and all of these elements generate noise. However, the noise that they generate is at, or near, the absolute minimum (kTB), the same as the input termination of the first gain stage, so they do not affect the noise level input to, and are not amplified by, the first gain stage.

For best sensitivity in a spectrum analyser we would like to have our input signal go straight to the input of the first gain stage because this is a point of minimum effective noise. However, our input signal cannot go directly to the first gain stage because it must first be translated in frequency by one or more mixers. In addition, it must pass through some value of RF input attenuation. Any loss that is incurred represents a degradation of signal-to-noise or, looking at it in another way, an increase in the effective noise at the input to the analyser.

We cannot do anything about the conversion loss of the mixer(s), but we do have control over the RF input attenuator; and setting it, the effective noise level at the input of an analyser is modified. There are several means to achieve this. Some analysers change the IF gain in a manner opposite to changes in the RF input attenuation. Thus true input signals will remain stationary on the CRT as the input attenuator is changed while the displayed noise moves up and down on the CRT. In this case, the reference level remains unchanged. In this case, best signal-to-noise is achieved with minimum input attenuation.

Another factor affecting signal-to-noise ratio is resolution bandwidth. And because the resolution, or IF, bandwidth filters come after the first gain stage, the bandwidth selected determines the level of noise that reaches the envelope detector and, ultimately, the CRT. And because the noise is truly random and flat relative to the bandwidths encountered in the analyser, the displayed or effective noise level changes as $10 \log(\mathrm{bw}_1/\mathrm{bw}_2)$. That is, if we change the resolution bandwidth by a factor of 10, the displayed noise level changes by 10 dB. So we get best signal-to-noise ratio or best sensitivity, using the minimum resolution bandwidth available in our spectrum analyser.

Dynamic range

Dynamic range is generally thought of in terms of an analyser's ability to measure harmonically related signals, i.e. how well it can make second or third harmonic distortion measurements or third-order intermodulation measurements. When dealing with such measurements, remember that the input mixer of a spectrum analyser is a non-linear device and so always generates distortion of its own. It is not that the mixer is non-linear by accident. It must be non-linear to translate an input signal to the desired IF. But the unwanted distortion products generated in the mixer fall at the exact same frequencies as the distortion products we wish to measure on the input signal. So we might define dynamic range in this way: the ratio, expressed in dB, of the largest to the smallest signals present at the analyser's input at the same time that allows measurement of the smaller signal to a given degree of uncertainty.

To determine dynamic range versus distortion, we must first determine just how our input mixer behaves. Most analyser mixers adhere pretty well to the mathematical model, and when we look at the squared and cubed terms of the series expansion, we find the following terms when the input signal is a pure sinusoid:

$$-aE^2 \cos 2\omega t - bE^3 \sin 3\omega t$$

where E is the signal voltage at the mixer, ω is the angular frequency of the input signal and t is the time; and for two input sinusoids when we are trying to measure the third-order intermodulation:

$$cE_1^2E_2 \sin(2\omega_1 - \omega_2)t + cE_1E_2^2 \sin(2\omega_2 - \omega_1)t,$$

where E_1, E_2 are the input signal voltages and ω_1, ω_2 are the angular frequencies of the input sinusoids.

The interesting thing about these terms is that they tell us that the dynamic range due to internal distortion is a function of the signal level at the input mixer. The dynamic range may be defined as the difference in dB between the fundamental tone or tones and the internally generated distortion.

The argument of the cosine in the first term includes 2ω, so it represents the second harmonic of the input signal. The level of this second harmonic is a function of the square of the voltage of the fundamental. This fact tells us that for every dB we drop the level of the fundamental at the input mixer, the internally generated second harmonic drops by 2 dB. The second term includes 3ω, the third harmonic, and the cube of the input signal voltage. So a dB change in the fundamental at the input mixer changes the internally generated

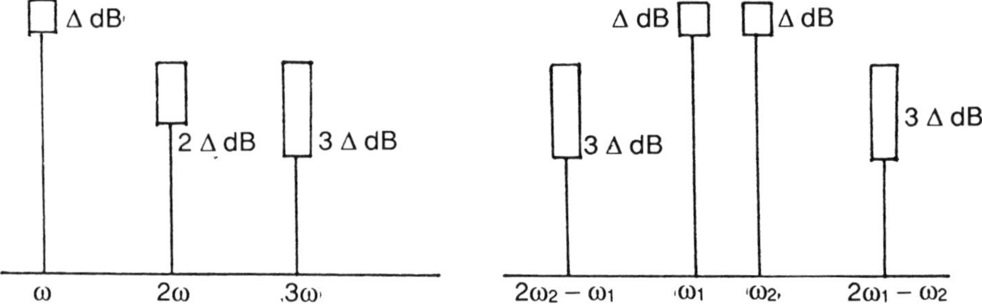

Fig. 13.20 Variation of harmonic levels for a 1 dB drop in the fundamental level(s) for second and third harmonics (left) and third order intermodulation products (right).

third harmonic by 3 dB (Fig. 13.20). So, dynamic range is variable. It depends upon the signal level at the mixer. Many analyser data sheets now include graphs to tell us how dynamic range varies, to know what level we need at the mixer for a particular measurement. However, if no graph is provided, we can draw it.

For instance, we could construct a line for third-order distortion. For example, a data sheet might say third-order distortion is 70 dB down for a level of − 30 dBm at this mixer. This is our starting point, and we would plot the point in Fig. 13.21. If we now drop the level at the mixer to − 40 dBm, referring to Fig. 13.20, we see that both third-harmonic distortion and third-order intermodulation distortion fall by 3 dB for every dB that the fundamental tone or tones fall. It is the differential that is important, so if the level at the

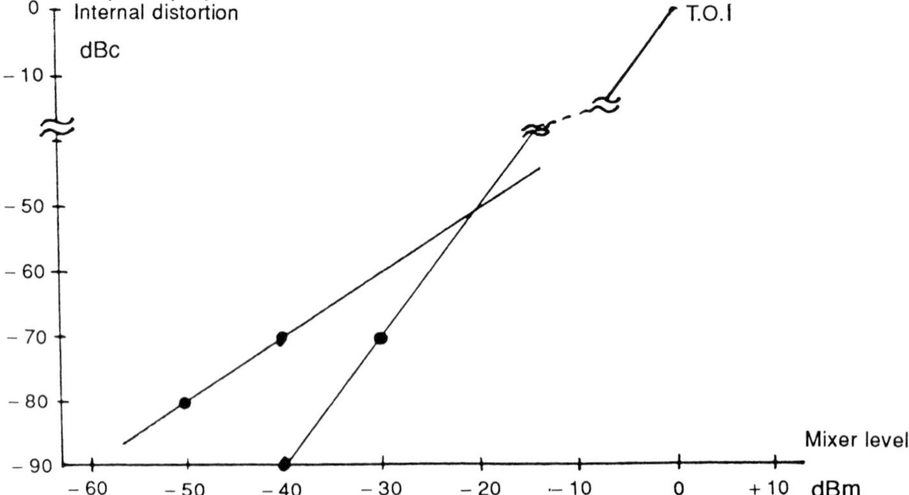

Fig. 13.21 Third harmonic distortion and third order intermodulation (T.O.I) distortion plotted as a function of level at the mixer.

mixer changes from – 30 to – 40 dBm, the differential between fundamental tone or tones and internally generated distortion changes by 20 dB. So the internal distortion is 90 dB below the fundamental(s). These two points fall on a line having a slope of 2, giving us the third order performance for any level at the mixer.

Sometimes third order performance is given as **TOI** (**third order intercept**). This is the mixer level at which the third-order distortion would be equal to the fundamental(s). This situation cannot be realized in practice because the mixer would be well into saturation. However, from a mathematical standpoint it is a perfectly good data point because we know the slope of the line. So even with TOI as a starting point, we can still determine the degree of internally generated distortion at a given mixer level.

Harmonic mixing

Up to now we described a spectrum analyser that tunes to 2.9 GHz. The most economical way of achieving an extended frequency range to tune higher in frequency is to employ **harmonic mixing**.

In developing our tuning equation above we found that we needed the low-pass filter in Fig. 13.17 to prevent higher frequency signals from reaching the mixer. The result was a uniquely-responding single-band analyser that tuned to 2.9 GHz. Now we wish to observe and measure higher frequency signals, so we must remove the low-pass filter.

Another factor that we explored while developing the tuning equation was the choice of the LO and IF frequencies. We decided that we should not have the IF within the band of interest because we ended up with a hole in our tuning range in which we could not make measurements. So we chose 3.6 GHz to move it above the highest tuning range of interest (2.9 GHz). Our new tuning range of interest is above 2.9 GHz, so it would seem logical to move the IF to a frequency below 2.9 GHz. A typical first IF for these higher frequency ranges in some spectrum analysers is 321.4 MHz. We shall use that frequency in our examples here. So for the low band, up to 2.9 GHz, our first IF is 3.6 GHz, and we must switch to a first IF of 321.4 MHz for the upper frequency bands. As shown in Fig. 13.22 the second IF is 321.4 MHz, so all we need to do is bypass the first IF when we wish to tune to the higher ranges.

We used a mathematical approach to conclude that we needed a low-pass filter. Things become a little more complex in the situation here; so we shall use a graphical approach as perhaps an easier method to see what is happening. We shall start with the simpler case, the low-band. In all of our graphs, we shall plot the LO (along the horizontal axis) and the signal frequency (along the vertical axis) as in Fig. 13.23.

Since we know that we get a mixing product equal to the IF (and therefore a response on the CRT) whenever the input signal differs from the LO by the IF, we can determine the frequency to which the analyser is tuned simply by adding the IF to or subtracting it from the LO frequency.

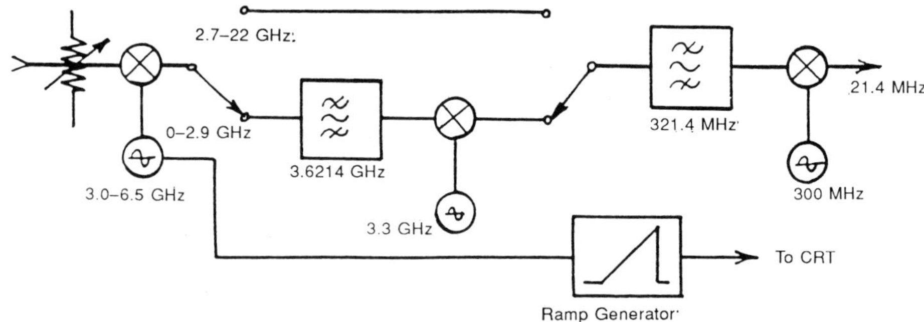

Fig. 13.22 Schematic diagram of a spectrum analyser with harmonic mixing for extended useful frequency range.

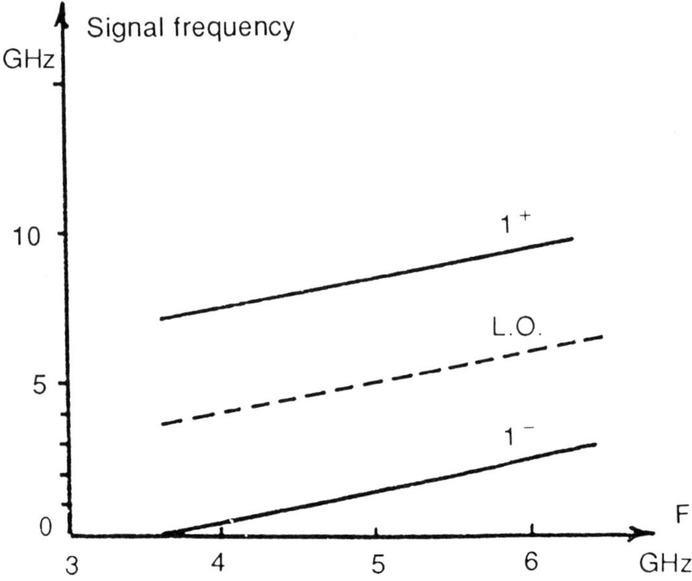

Fig. 13.23 Tuning range of a spectrum analyser with harmonic mixing versus L.O. frequency, in low band.

To determine our tuning range, then, we start by plotting the LO frequency against the signal frequency axis as shown in the dashed line in Fig. 13.23. Subtracting the IF from the dashed line gives us a tuning range of 0 to 2.9 GHz, the range that we developed in the first chapter. This line in Fig. 13.23 is labelled 1^- to indicate fundamental mixing and the use of the **minus sign** in the tuning equation. We can use the graph to determine what LO frequency is required to receive a particular signal (to display 1^- GHz signal, the LO must be tuned to 4.6 GHz) or to what signal frequency the analyser is tuned for a given LO frequency (for a LO frequency of 6 GHz the spectrum analyser is tuned to receive a signal frequency of 2.4 GHz).

Now let's add the other fundamental mixing band by adding the IF to the LO line in Fig. 13.23. This gives us the upper solid line, labelled 1^+, that indicates a tuning range from 7.2 to 10.1 GHz. Note that for a given LO frequency, the two frequencies to which the analyser is tuned are separated by twice the IF. So it would seem valid to argue that while measuring signals in the low frequency band it is unlikely that one would be bothered by signals in the 1^+ frequency range.

Let's now examine the high band, where the situation is very different. As before, we shall start by plotting the LO fundamental against the signal frequency axis and then add and subtract the IF, producing the results shown in Fig. 13.24. Note that the 1^- and 1^+ tuning ranges are much closer together, and in fact overlap, because the IF is a much lower frequency, 321.4 MHz, in this case.

The close spacing of the tuning ranges sometimes complicates the measurement process. Our system can be calibrated for only one tuning range at a time. In this case, we would choose the 1^- tuning to give us a low end frequency of 2.7 GHz so we have some overlap with the 2.9 GHz upper end of our low-band tuning range. So what are we likely to see on the CRT? If we enter the graph at an LO frequency of 5 GHz, we find that there are two possible signal frequencies that would give us responses at the same point on the CRT: 4.7 and 5.3 GHz (rounding the numbers, 5.0 + 0.3 GHz, and 5.0 − 0.3 GHz). On the

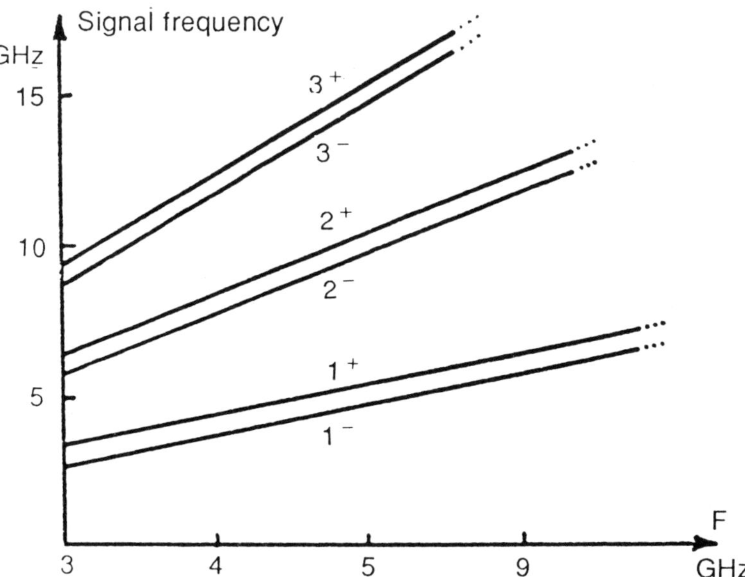

Fig. 13.24 Tuning range of a spectrum analyser with harmonic mixing versus L.O. frequency, in high band.

other hand if we enter the signal frequency axis at 5.3 GHz, we find that in addition to the 1^+ response at an LO frequency of 5 GHz, we could also get a 1^- response if we allowed our LO to sweep as high as 5.6 GHz, twice the IF above 5 GHz. Also, if we entered the signal frequency graph at 4.7 GHz, we would find in addition to the 1^- response at an LO frequency of 5 GHz, a 1^- response at an LO frequency of about 4.4 GHz (twice the IF below 5 GHz).

To differentiate between responses generated on the 1^- tuning curve, for which our analyser is calibrated, and those produced on the 1^+ tuning curve, there are several means described in detail in the manufacturer's handbooks, as the identification methods 'image' and 'shift' of Hewlett Packard. But many users work in controlled environments in which they deal with only one or two signals at a time, and in such an environment most analysers function very well. And if image signals exist, they can be filtered away with simple band-pass filters and multiple responses will not bother us if we limit our frequency span to something less than 600 MHz (twice the IF). And typically knowing the signal frequency, we can tune to the signal directly knowing that the analyser will select the appropriate mixing mode (1^-, 1^+, 2^-, ...) for which it is calibrated.

13.6 VECTOR MODULATION (I/Q) MEASUREMENTS

Digital microwave radio (DMR) are distinguished from traditional microwave links in that they use phase modulation, such as QPSK (quadrature phase shift keying), or QAM (quadrature amplitude modulation). These modulation schemes are very different from their analogue predecessors in the way in which they accomplish communication and their tolerance to various real-world impairments.

Regardless of specific type of QAM or PSK modulation chosen for a DMR, most radios use I/Q or 'vector modulation and demodulation' to put the information on and take it off the carrier. Since the I and Q signals are derived from the phase and magnitude of the radio signals, they are also valuable analysis tools. These signals combined with the appropriate measurement instrumentation can be valuable in analysing a transmission path, characterizing an amplifier's performance, or trouble-shooting a radio.

13.6.1 I/Q modulation

DMR measurements require accurate sources as well as analysis instruments. There are many situations that require the accurate generation of precisely modulated test signals of calibrated levels and stability. The most natural tool for doing this is one already used in most radios: the I/Q modulator (see Fig. 13.25). It generates signals in terms of the I and Q components inherent in most DMRs, so it seems to be an ideal choice for generation of signals to

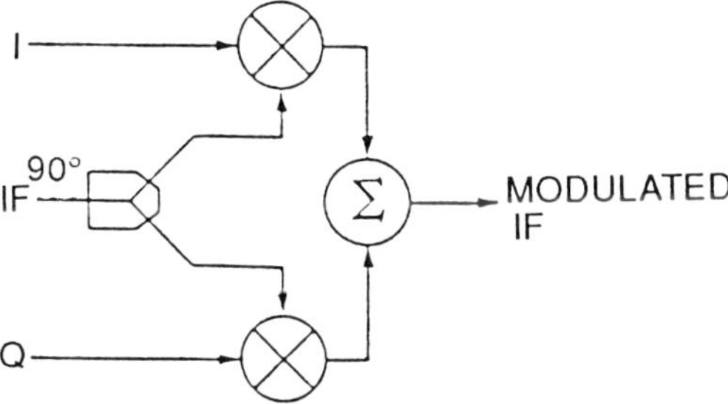

Fig. 13.25 A simplified vector modulator.

characterize a digital radio. The modulators built for radios are built to perform within a certain range of tolerances to ensure the correct operation of the radio, and these tolerances may or may not be tight enough for specific measurement requirements. In situations like simulating degradations, adding noise, or varying signal levels, a more versatile modulator is required.

The most important requirement of an instrumentation quality modulator is extremely accurate quadrature and amplitude balance. After this, it is important that the modulator has a high degree of linearity to provide precise I/Q modulations for receiver testing and demodulator calibration. The modulator's output should be very clean and free of spurious signals in the modulation bandwith so that sensitive measurements of receiver noise won't be masked by the modulator's noise.

13.6.2 I/Q demodulation

The most fundamental difference between DMR measurements and traditional communications measurements is the use of the I/Q demodulator as an analysis tool (see Fig. 13.26). The vector modulator is an I/Q demodulator capable of complex modulation analysis. Although the demodulators in many receivers provide a significant analysis capability by themselves, in some instances a more accurate or separate demodulator will be required.

Basically the I/Q demodulator consists of two double-balanced mixers and a 90° hybrid connected as shown in the figure. For analysis, the characteristics which are important in an I/Q demodulator are precise quadrature, amplitude balance and linearity. Any imperfections in these measures of performance will result in erroneous measurements of I, Q, amplitude and phase.

13.6.3 Vector modulation analysis

Although I/Q demodulation has long been used in DMR as well as other receivers, the meaning of these signals has often been hidden by the displays

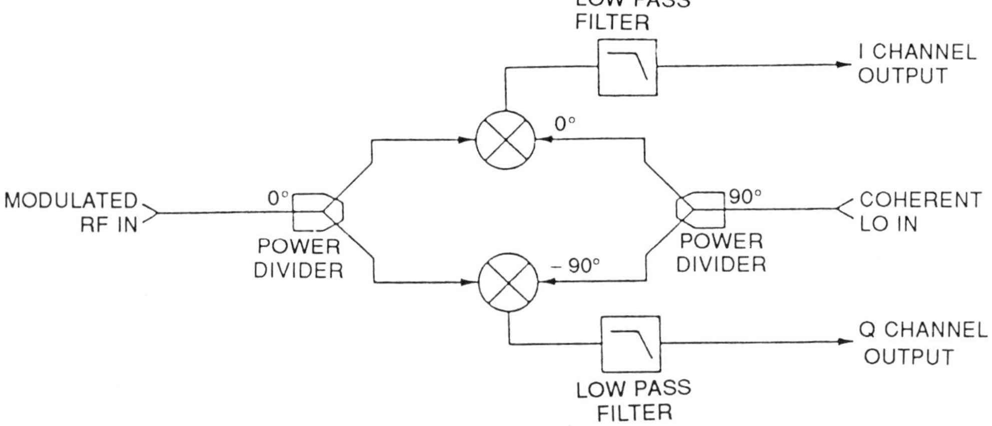

Fig. 13.26 A simplified vector demodulator.

used to show them. Usually I and Q channels have been viewed on normal dual display oscilloscopes, with one signal above the other and time along the horizontal axis. This is fine for observing eye diagrams, but the relationship between I and Q is lost in this type of representation. To understand digital modulations, and to see the relationship of I, Q and phase, the I and Q signals need to be viewed in an X versus Y fashion.

The reason that I and Q have not been displayed in this W and Y or vector diagram before is that traditional oscilloscopes have much lower X bandwidths than Y bandwidths. These limitations have been overcome with the introduction of the vector analyser.

There are three fundamental display modes of the vector analyser (see Fig. 13.27): it can display either of its two inputs or both simultaneously versus time ('eye diagram'), display both inputs in X versus Y fashion ('vector diagram'), or display both inputs in X versus Y fashion only at a specified time instant ('constellation diagram').

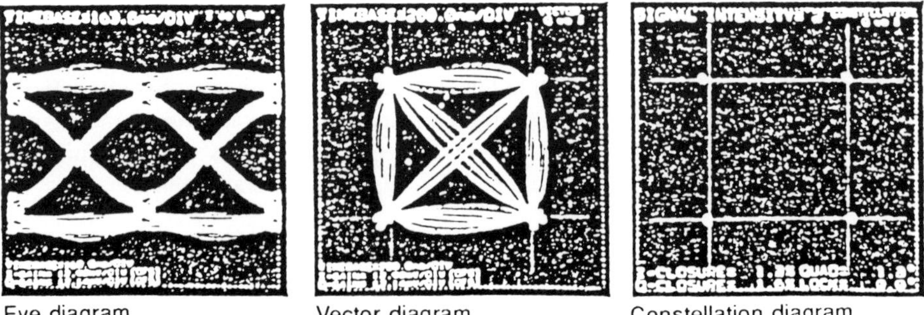

Fig. 13.27 Three fundamental display modes of the vector analyser.

13.6.4 Bit error rate measurement techniques

The overall performance of digital communication systems is often described by the system bit error rate or BER. The BER is simply the average number of bits transmitted.

The two commonly used degradations are gaussian noise and sinusoidal interference signals to provide carrier to noise (C/N) and carrier to interference (C/I) tests respectively.

There are basically two different methods of making BER versus C/N measurement: fade simulation using an attenuator at the receiver input and the additive noise method using a noise generator. Traditionally, the first technique has been used because of the lack of dedicated equipment for the second. With the introduction of noise and interference test sets, this is no longer a problem and the additive noise method is preferred.

One of the problems associated with BER tests is the separation of transmitter and receiver errors. Both C/N and C/I tests are usually performed on transmitter/receiver pairs whose IF or RF signals are connected 'back-to-back' making it impossible to separate the BER test results of the transmitter from the receiver. Testing the transmitter or receiver alone requires a corresponding 'ideal' receiver or transmitter. Isolation of receiver BER characteristics is made possible by using a vector signal generator.

Index